Pesticide Risk Assessment

Pesticide Risk Assessment

Edited by

Sumitra Arora

Principal Scientist, ICAR-National Research Centre for Integrated Pest Management, Pusa Campus, New Delhi, India

CABI

CABI is a trading name of CAB International

CABI
Nosworthy Way
Wallingford
Oxfordshire OX10 8DE
UK

Tel: +44 (0)1491 832111
Fax: +44 (0)1491 833508
E-mail: info@cabi.org
Website: www.cabi.org

CABI
745 Atlantic Avenue
8th Floor
Boston, MA 02111
USA

Tel: +1 (617)682-9015
E-mail: cabi-nao@cabi.org

A catalogue record for this book is available from the British Library, London, UK.

Library of Congress Cataloging-in-Publication Data

Names: Arora, Sumitra, author.
Title: Pesticide risk assessment / Sumitra Arora, Principal Scientist, ICAR-National Research Centre for Integrated Pest Management, Pusa Campus, New Delhi-110012, India.
Description: Boston, MA, USA : CAB International, 2019. | Includes bibliographical references and index.
Identifiers: LCCN 2018027229 (print) | LCCN 2018032515 (ebook) | ISBN 9781780646343 (ePDF) | ISBN 9781780646664 (ePub) | ISBN 9781780646336 (hbk: alk. paper)
Subjects: LCSH: Pesticides--Risk assessment. | Pesticides--Environmental aspects. | Pesticides--Risk mitigation. | Pesticides--Toxicology.
Classification: LCC QH545.P4 (ebook) | LCC QH545.P4 A75 2019 (print) | DDC 363.738/498--dc23

LC record available at https://lccn.loc.gov/2018027229

ISBN-13: 978 1 78064 633 6 (hbk)
978 1 78064 634 3 (PDF)
978 1 78064 666 4 (ePub)

Commissioning editor: Rachael Russell/Rebecca Stubbs
Editorial assistant: Emma McCann
Production editor: Tim Kapp

Typeset by AMA DataSet Ltd, Preston, UK
Printed and bound in the UK by CPI Group (UK) Ltd, Croydon, CR0 4YY

Dedicated to the Almighty and my family,
especially my daughter, Shreya,
and son, Nikunj.

Contents

Acknowledgements

I am extremely indebted from the bottom of my heart to my Shri Guru Maharaj-Ji, for showering blessings, and for the completion of any task assigned to me. No work or assignment is satisfactorily completed without moral support from your family. My strongest support has been my mother, Ms Ginni Devi; my daughter, Dr Shreya Arora; my son, Nikunj Arora; my son-in-law, Vaibhav Gupta; my sister and guru, Ms Premlata Baiji; sisters, Jaidevi and Draupati; my brother, Gurdas Saini; and other family members. I am also thankful to my research associate, Navin Mogha, for providing technical help in writing a few of the chapters of this book. My daughter, Shreya, has been my hero; without whose never-failing encouragement this book would not have been finished in time.

> Do not wait; the time will never be 'just right'. Start where you stand, and work with whatever tools you may have at your command, and better tools will be found as you go along.
>
> *George Herbert*

About the Author

Dr Sumitra Arora graduated from Ramjas College of Delhi University, India, in chemistry, and completed her postgraduate degree in organic chemistry at Delhi University in 1989. She joined the Indian Council of Agricultural Research (ICAR) through Agricultural Research Services (ARS) in 1993, as a scientist, and after the completion of her training at Hyderabad, was posted to the National Centre for Integrated Pest Management (NCIPM) in 1994. She finished her doctorate in agricultural chemicals at the Indian Agricultural Research Institute (IARI) in 2003, as an in-service candidate. She has also undergone postdoctoral studies in 'toxicity assessment of pesticides in mixtures, using in-vitro and in-vivo methods' at the Commonwealth Scientific and Industrial Research Organisation (CSIRO), in 2010. She has published a number of research articles in national and international refereed journals with high impact factors. She has published a number of books on pesticides, indigenous technologies in plant protection and field guides for farmers. Since 2011, she has been a focal point scientist for South Asian Association for Regional Cooperation (SAARC) member states, representing India, as part of the pesticide data sharing network. Her areas of work are varied and include pesticide residue analysis, indigenous technologies in plant protection (patented a technology), environmental impact assessment of pesticides, isolation of plant-based bio-active molecules and integrated pest management activities for various crops, and she is currently involved in research on storage fumigation of quarantine commodities. She has organized and conducted a number of training programmes and workshops for the farmers, pesticide dealers and researchers at national and international levels.

Preface

Pesticides are designed to kill pest populations by their deleterious or toxic effects. The agriculture sector is dependent on the use of pesticides to ensure food security in order to meet the demands of the growing population, and manage the evolution of pest control. However, injudicious use of these chemicals may lead to environmental contamination with residues detected in food, feed and fibre. They need special care during handling, storage, application, disposal and transportation. To reduce their toxicity, they are formulated for agricultural or commercial use. Domestic pesticides are also potentially hazardous to humans and other organisms if misused or released into the environment in significant amounts. There is a false perception that any pesticide registered by the United States Environmental Protection Agency (US EPA) and used according to the label must be environmentally safe. However, the use of integrated pest management (IPM), a strategy that uses a combination of methods to manage pests without solely relying on chemical pesticides, leads to safe and economic crop production. In agricultural programmes, pesticides were generally chosen based on their efficacy or cost rather than on their potential environmental impact. However, a formal method is needed to determine environmentally based pesticide recommendations, and to estimate the environmental impact of specific pesticides. Various models and approaches have been applied to assess pesticide toxicity.

Pesticide Risk Assessment is an effort to help readers understand the basic information on pesticides, including their impact on the environment, their mode of action, estimation methods, risk assessment, mixture toxicity, alternatives for risk reduction and their regulatory aspects. Therefore, it is hoped that this book will inculcate a kind of knowledge-based support and enthusiasm for adopting pest management practices which are embedded with a judicious use of pesticides in agriculture. This will enable the production of healthy crops with greater yields and minimal associated environmental implications. The aim is to provide all the essential and required information on the domain of pesticides and their risk assessment in a handy and encapsulated form. Suggestions and guidance from readers to improve the contents of the book are wholeheartedly welcomed and acknowledged. It is hoped that the information in this book will empower researchers, farmers and policy makers by enhancing their knowledge and strengthening their decision-making power at an international level.

Sumitra Arora

Introduction

We have been witnessing a silent chemical revolution over the past half-century. Innumerable chemicals have been manufactured that do not otherwise exist in nature but possess very important and useful properties for the life and work of human beings. Some of these useful substances are now, however, playing havoc with the lives of humans and with the environment because of their overuse.

Pesticides bring about widespread environmental contamination, with residues detected far from their site of application. Pesticides sprayed over paddy fields in Punjab can end up in planktonic life in Antarctica, or in seals and Arctic foxes in Greenland. Pesticidal residues have been found in ambient air and drinking water samples in several countries. In addition to persisting in the environment over long periods of time, they are also toxic to non-target insect species.

The suffix 'cide' means killing. 'Pesticides' is a general term for substances that are used to kill a pest or control, repel, prevent or reduce the damage they cause. There are several different kinds, which are grouped according to their function and the class of pests they kill; for example, disinfectants, fumigants, insecticides, herbicides, fungicides, rodenticides, wood preservatives and domestic pesticides. These pesticides are valuable to us in many ways. They help us control or reduce hundreds of pests in and around our homes, maintain hygiene and play an important role in improving world health.

Today, organic compounds and their formulations have become the leading pesticides world over. They are less hazardous, more universal and are required in much smaller quantities than inorganic chemicals to obtain the same effect. Only a millionth of a gram of dichlorodiphenyltrichloroethane (DDT) powder per square centimetre of ground area is enough to wipe out certain insects completely. Though DDT is banned in developed countries, it is still used for the control of mosquitoes and insects in some developing countries. As a result, it is detected in mothers' milk in concentrations much above the permissible limits according to the World Health Organization (WHO), and is transmitted to subsequent generations.

Hazardous Nature of Pesticides

To ensure food security to meet the demands of the world's growing population, and to manage the evolution of pest control, the agriculture sector is dependent on the use of pesticides, which has led to a sequential rise in the production and consumption of pesticides over the past three decades. The general practices of using pesticides against crop pests have led to them being treated as synonymous with plant protection chemicals, although they are used for non-agricultural purposes too. The prominent groups of pesticides are organochlorines, organophosphates, carbamates and synthetic pyrethroids. The organochlorines were observed as being highly persistent in the environment so have been replaced with less persistent organophosphate and carbamate pesticides. These pesticides are nerve poisons and toxic even at low dosages. The western part of the world has phased out use of almost all organophosphates because of their neurotoxicity, which is especially harmful to children. An increasing number of insects have developed high levels of resistance to pyrethroids. These pyrethroids are also highly toxic to aquatic organisms. We are well aware of the impacts of injudicious use of pesticides on the environment in the race towards high production.

Pesticides can be potential polluters of water, soil and air. They are hazardous to humans when they are handled, stored, applied and disposed of improperly. Domestic chemical pesticides are pest control products that are specifically formulated and registered only for domestic use in the home, kitchen or garden. Although they are less toxic than the formulations prepared for agriculture or commercial use, domestic pesticides are also potentially hazardous to humans and other organisms if misused or released into the environment in significant amounts. Pesticides are usually non-biodegradable and, on account of this fact, minuscule quantities are left either in the soil or they enter the food chain. Research is ongoing in the search for biodegradable formulations of pesticides or biopesticides.

For several years, increased attention has been focused on integrated pest management (IPM) programmes and alternative methods of pest control to reduce pesticide use in agricultural systems because of food safety issues, groundwater contamination and increased environmental awareness. By definition, IPM is a pest management strategy that uses a combination of methods (sampling, thresholds, forecasts, biological and cultural controls, etc.) to manage pests without solely relying on chemical pesticides to produce a safe, economic crop. If, however, no other control measure is effective in preventing pest damage, a chemical pesticide is recommended. In past IPM programmes, pesticides were generally chosen based on their efficacy or cost rather than on their potential environmental impact. Although some pest management practitioners did take into account the effect of the pesticides on the applicator or beneficial natural enemies, such as predatory mites, when making pesticide recommendations, no formal method was available to assist them in making environmentally based pesticide choices. Because there is no easy method to assess pesticide impacts, each individual had to rely primarily on their own judgement to make these decisions. Some organically approved growers felt that only natural pesticides should be used in agricultural production systems because they are naturally occurring and are perceived to be less harmful to the environment. Other growers felt that any pesticide registered by the United States Environmental Protection Agency (US EPA) and used according to the label must be environmentally safe. In addition, IPM programmes throughout the country use various methods (number of sprays, the amount of active ingredient or formulated product used per acre, dosage equivalents, etc.) to quantify pesticide use and environmental impact to compare different pest management strategies or programmes. None of these methods estimates the environmental impact of specific pesticides. Therefore, the purpose of this book is to organize the published environmental impact information on pesticides into a usable form to help growers and other IPM practitioners make more environmentally sound pesticide choices.

As per the US EPA Pesticide Industry Sales and Usage Report of 2006 and 2007, the world used approximately 5.2 billion lb (2.36 billion kg) of pesticides, with herbicides constituting the largest part of the world pesticide use at 40%, followed by insecticides (17%) and fungicides (10%). In 2006 and 2007 the US used approximately 1.1 billion lb (0.5 billion kg) of pesticides, accounting for 22% of the world total, including 857 million lb (388.7 billion kg) of conventional pesticides, which are used in the agricultural sector (80% of conventional pesticide use) as well as the industrial, commercial, governmental, and home and garden sectors. Pesticides are also found in the majority of US households, with 78 million out of the 105.5 million households indicating that they use some form of pesticide (Wikipedia, n.d.). As of 2007, there were more than 1055 active ingredients registered as pesticides (Goldman, 2007). According to the US Centers for Disease Control and Prevention (CDC), more than 17,000 individual pesticide products have been marketed in the US alone (CDC, n.d.).

The US used 2.2 lb/ha (0.998 kg/ha) of arable land compared with 10.3 lb/ha (4.67 kg/ha) in China, 3.0 lb/ha (1.36 kg/ha) in the UK, 0.2 lb/ha (0.09 kg/ha) in Cameroon, 13.1 lb/ha (5.94 kg/ha) in Japan and 5.6 lb/ha (2.54 kg/ha) in Italy. Insecticide use in the US has declined by more than half since 1980 (0.6%/year), mostly due to the near phase-out of organophosphates. In corn (maize) fields, the decline was even steeper, due to the switchover to transgenic Bt corn (*Science*, 2013). For the global market of crop protection products, market analysts forecast revenues of over US$52 billion in 2019 (Ceresana Research, 2012).

Although synthetic pesticides have become integral components of IPM, they are non-biodegradable and contaminate the environment in a number of ways. Chemical pesticides have a hazardous impact on human beings and non-target organisms including natural enemies, terrestrial and aquatic organisms, as well as affecting other ecological components of the environment. Although it is ensured before registration of any pesticide that it is unlikely to pose any unreasonable risk to humans or the environment or non-target organisms, reports of pesticides contaminating the environment and presenting risk to human health are increasing. This has prompted researchers to continuously re-evaluate the risk of using pesticides. Assessment is predominantly of individual pesticides, in spite of the fact that our environment is a complex system where contaminants exist in mixtures and it is often difficult to understand and predict their risks. Assessing the cumulative toxicity of pesticides in mixtures has therefore been an enduring challenge for environmental health research as well as eco-toxicology for the past few decades. The assessment of the effect of mixtures of contaminants is regarded as an issue of high priority by many international organizations, such as the US EPA and the European Commission.

The risk of environmental contamination and health problems depends not only on how toxic the pesticide ingredients are, but also on the amount of exposure to the product. Basically it is the dosage of pesticides and the level of exposure, which make them poisonous.

A number of diagnostic tools are available, such as dose/concentration addition and interaction studies, to assess the cumulative risks from exposure to multiple pesticides in mixtures, with the same and different modes of action. The pesticide risk to human health is dependent on various factors such as age, gender, individual sensitivity, etc. The pesticide estimation tools include conventional methods like multi-residue methods for estimation of pesticides using gas chromatography–mass spectrometry (GC-MS) or liquid chromatography–mass spectrometry (LC-MS).

There is, therefore, a need for the regulation of pesticides and associated risk assessment for the environment. Replacing high-risk pesticides with low-risk ones, with lower impact on health, lower toxicity to non-target organisms, lower potential for groundwater contamination and low-dosage molecules, and improving application technologies would be the possible ways to reduce pesticide risk.

Various pesticide-regulating authorities like the US EPA are responsible for regulating pesticides under the Federal Insecticide, Fungicide, and Rodenticide Act (FIFRA) and the Food Quality Protection Act (FQPA). The US EPA regulates pesticides to ensure that these products do not pose a risk of adverse effects on humans or the environment.

This book deals with all kinds of risks, such as risk to aquifers and aquatic organisms such as fish; risk to pollinators; risk to human health; risk to soil microflora and fauna; and risk to terrestrial organisms like arthropods and wildlife, associated with the use of individual pesticides and their mixtures, their method of estimation and risk assessment using various tools. Remediation and management details are duly elaborated. The book also includes regulatory perspectives and covers various aspects of remediation for reducing the risk associated with pesticides.

References

Centers for Disease Control and Prevention (CDC) (n.d.) National Institute for Occupational Safety and Health, Pesticide Illness and Injury Surveillance. Available at: https://www.cdc.gov/niosh/topics/pesticides/default.html (accessed 24 July 2018).

Ceresana Research (2012) Market study: crop protection (UC-2805). June 2012. 837. Available at: www.ceresana.com/en/market-studies/chemicals/crop-protection (accessed 24 July 2018).

Goldman, L.R. (2007) Managing pesticide chronic health risks: US policies. *Journal of Agromedicine* 12(1), 67–75.

Science (2013) Infographic: pesticide planet. *Science* 341(6147), 730–731.

Wikipedia (n.d.) Pesticide. Available at: https://en.wikipedia.org/wiki/Pesticide#cite_note-EPAPesticides-23 (accessed 24 July 2018).

Impact and Estimation of Pesticides

1 Pesticides

An enormous number of chemicals exist in nature, which are used as pesticides, for agricultural and non-agricultural purposes. Pesticides are introduced to manage populations of living organisms, called pests, which are deleterious to crops or the health of the human race. A literal meaning of pesticide is 'a substance used to kill undesired living organisms'. However the Food and Agriculture Organization (FAO, 2013) has defined pesticide as:

> any substance or mixture of substances intended for preventing, destroying, or controlling any pest, including vectors of human or animal disease, unwanted species of plants or animals, causing harm during or otherwise interfering with the production, processing, storage, transport, or marketing of food, agricultural commodities, wood and wood products or animal feedstuffs, or substances that may be administered to animals for the control of insects, arachnids, or other pests in or on their bodies. The term includes substances intended for use as a plant growth regulator, defoliant, desiccant, or agent for thinning fruit or preventing the premature fall of fruit. Also used as substances applied to crops either before or after harvest to protect the commodity from deterioration during storage and transport.

Pest management is also achieved by the use of other biological active agents, such as trichogramma. The management of pests actually disturbs the balance of nature, thereby affecting biodiversity on Earth. Pesticides can adversely affect non-target organisms as well as affecting other ecological components of the environment, which they are not intended to do.

1.1 History

Until the 1940s, during World War Two, pesticides were widely used for pest control to ensure food security by the introduction of inorganic chemicals such as calcium, lead arsenate and sulfur, and the use of common salt and sodium chlorate as herbicides. The organic molecules of this period were natural products such as nicotine (extracted from tobacco) and rotenone (extracted from derris roots). These were named *first-generation pesticides*. These pesticides, including heavy metals, were toxic to humans and agricultural plants. The growth in synthetic pesticides accelerated in the 1940s with the discovery of the effects of organochlorines, known as *second-generation pesticides*, such as dichlorodiphenyltrichloroethane (DDT), β-hexachlorocyclohexane (BHC), aldrin, dieldrin, endrin, chlordane, parathion, captan and 2,4-dichlorophenoxyacetic acid (2,4-D). The discovery of DDT by Dr Paul Muller, in 1939 was a turning point that revolutionized agricultural production. DDT was popular because of its broad-spectrum activity (Unsworth, 2008) and was widely used. It appeared to have low toxicity to mammals, and reduced the incidence of insect-borne diseases, such as malaria, yellow fever and typhus. Consequently, in 1949, Dr Paul Muller won the Nobel Prize in Medicine for discovering its insecticidal properties. However, in 1946, resistance to DDT in house flies was reported and, because of its widespread use, there were reports of harm to non-target plants and animals and problems with residues (Unsworth, 2008). These insecticides were observed to be toxic to humans and agricultural plants, due to their bioaccumulation and biomagnification properties.

Throughout the 1950s, consumers and most policy makers were not overly concerned about the potential health risks associated with using these pesticides, as these were considered safer than the forms of arsenic that were used and had killed people in the 1920s and 1930s (Wessel's Living History Farm, 1930). However, the indiscriminate use of these highly persistent organochlorines could generate a lot of problems as highlighted by Rachel Carson in her 1962 book *Silent Spring* (Carson, 2002 [1962]). The problems necessitated a search for safer and more environmentally friendly products, which ushered in the era of smarter pesticides. Organophosphates and carbamates replaced the highly persistent organochlorines (Table 1.1). In the 1930s, a German chemist Gerhard Schrader at IG Farben experimented with nerve poison insecticides, organophosphates, which were meant to be used as chemical warfare agents during World War II (López-Muñoz *et al.*, 2009). After World War II, the organophosphate pesticides were synthesized in large quantities. Parathion was among the first marketed, followed by malathion and azinphosmethyl. These insecticides became more popular after many of the organochlorine insecticides like DDT, dieldrin and heptachlor were banned in the 1970s. Although less persistent in the environment, they are more lethal even at low dosages.

During the 1970s and 1980s, a herbicide, 'glyphosate', currently the world's best-selling herbicide product, was introduced, followed by sulfonylurea and imidazolinone products. The insecticide industry marked the era of the third generation of pesticides, with the introduction of avermectins, benzoylureas and Bt (*Bacillus thuringiensis*). The triazole, morpholine, imidazole, pyrimidine and dicarboxamide families of fungicides were also introduced during this period.

The massive use of pesticides during the 1960s significantly increased agricultural productivity leading to the 'Green Revolution' and a safe and secure food supply, along with other secondary benefits. However, some of the first-generation pesticides had the drawback of persistence in soils and aquatic sediments, bioconcentration in the tissues of invertebrates, moving up trophic chains and affecting top predators. Many chemicals that have been identified as endocrine disruptors are pesticides. Of these, 46% are insecticides, 21% herbicides and 31% fungicides; some of them were withdrawn from general use many years ago but are still found in the environment (e.g. DDT and atrazine) (Mnif *et al.*, 2011). Endocrine disruptor compounds (EDCs) have been defined in 2002 by the World Health Organization (WHO): 'An endocrine disruptor is an exogenous substance or mixture that alters the function(s) of the endocrine system and consequently causes adverse health effects in an intact organism, or its progeny, or (sub)populations' (Combarnous, 2017).

Pesticides with common modes of action become more selective and generally lead to resistance problems. During the 1990s research continued to find new molecules with greater selectivity and better environmental and toxicological profiles. New families like strobilurins and azolone were introduced as fungicides, as well as fiproles and spinosyn as insecticides with lower dosages in grams instead of kilograms per hectare. The introduction of user-friendly and environmentally safer formulations has helped to combat the problem of resistance management. The integrated pest management (IPM) approach has led to a reduction in the load of synthetic pesticides because of

Table 1.1. History of pesticides according to their generation.

Category	Example	Period of marketing	Group of pesticides	Benefits	Drawbacks	Reference
First generation	Heavy metals such as As, Cu and Pb	World War I; prior to 1940	Inorganic and organic botanical compounds	Pest management	Toxic to humans and agricultural plants	Hummel, 1983
Second generation (synthetic)	DDT, methoxychlor, lindane	Post-World War I; 1940 onwards (widespread use, 1940–1960)	Organochlorines	Highly effective for malaria control	Highly persistent due to their bioaccumulation and biomagnification properties	Hummel, 1983
	Dimethoate, malathion; aldicarb, carbaryl	Early 1950s (widespread use, post-1970s)	Organophosphates and carbamates	Less persistent in environment	Acutely potent nerve toxins and more lethal even at low dosages	Hummel, 1983
	Permethrin, cypermethrin and deltamethrin	1970s	Synthetic pyrethroids	Knock-down activity	They are axonic excitotoxins; can be combined with the synergist piperonyl butoxide (PBO)	Hummel, 1983
Third generation	Insect growth regulator	By 1975	Juvenile hormones, chitin synthesis inhibitors, moulting hormones, precocenes, and compounds that mimic their effects	Disrupt and impede the life cycle of insects	Selective, less harmful to the environment, compatible with pest management systems	Hummel, 1983
	Microbials	Late 1970s Bt protein in 1990s	*Bacillus thuringiensis* (Bt)	Toxins produced following ingestion	No human health or environmental hazards	Gaalaas Mullaney, 2018
Fourth generation	Behaviour modifiers (Antifeedants, pheromones)	Late 1990s	Sex attractants, mating disruptors	Interfere with normal behaviour patterns, delay or prevent mating and producing offspring	Volatile, specific, potent, and non-toxic	Metcalf and Metcalf, 1994
Fifth generation	Promising leads (natural products, brain hormone antagonist)	Late 1990s	Nereistoxin (a neurotoxic substance isolated from the marine annelid *Lumbriconereis heteropoda*)	Model compound tor the synthesis of the insecticide cartap	Sensitive to sunlight, should be kept in dark (Fourouzan and Farrokh-Eslamlu, 2017)	Hummel, 1983
	Endocrine disruptors	Initiated from first generation until date	Man-made	Target a critical life cycle stage of insects	Alter the normal functioning of the endocrine system of both wildlife and humans	Mnif *et al.*, 2011

DDT, dichlorodiphenyltrichloroethane.

the other components for pest management that are involved. Resistant varieties, genetically engineered crops, cultural practices and biological control are a few of the components of the IPM strategy.

1.2 Global Status of Use of Pesticides

The first use of synthetic pesticides was in 1940. As per the EPA Pesticide Industry Sales and Usage Report of 2006 and 2007, the world used approximately 5.2 billion lb (2.4 billion kg) of pesticides, with herbicides constituting the largest part (40%) of world pesticide use, followed by insecticides (17%) and fungicides (10%). In 2006 and 2007 the US used approximately 1.1 billion lb (0.5 billion kg) of pesticides, accounting for 22% of the world total, including 857 million lb (388.7 million kg) of conventional pesticides, which are used in the agricultural sector (80% of conventional pesticide use) as well as the industrial, commercial, governmental, and home and garden sectors. Pesticides are also found in the majority of US households, with 78 million out of the 105.5 million households indicating that they use some form of pesticide (Wikipedia, December, 2014). In 1999 about 74% of households in the US were reported to use at least one pesticide in the home. The use of pesticides in the US doubled from 1960 to 1980, but total use has since remained stable or fallen. As of 2007, there were more than 1055 active ingredients registered as pesticides (Goldman, 2007), which yield over 20,000 pesticide products that are marketed in the US (CDC and P, 2015).

The consumption of pesticides has risen in developing countries, and the fastest growing markets are observed to be in Africa, Asia, South and Central America and the eastern Mediterranean region. There is high pesticide use on crops grown for export (Dr N. Besbelli, WHO, personal communication, July, 2008) (Laborde, 2008). Although consumption of pesticides in developing countries makes up only 25% of the pesticides produced worldwide, they account for 99% of the deaths. This is because the use of pesticides tends to be more intense and unsafe. The regulatory, health and education systems are weaker in developing countries, leading to a lack of awareness of the latest techniques in agriculture and safe handling of pesticides.

1.3 Classification of Pesticides

The pesticides could be broadly classified into two groups: agricultural and non-agricultural use, although some pesticides may be placed in both categories. In some countries such as the UK, the competent authority responsible for processing and approvals is different for each group. The approval processes for agricultural pesticides are managed by the Pesticides Safety Directorate at York, while non-agricultural pesticides are governed by the Health and Safety Executive (HSE) at Bootle. A similar kind of distinction is made between plant protection products and biocides by the European Union. Pesticides are classified based on their target pests (Table 1.2), chemical structure (Table 1.3), route of entry and mode of action (Table 1.4).

1.3.1 Classification according to their route of entry

- *Stomach poisoning* – the pesticide enters the body of pests via their mouthparts and digestive system and causes death by poisoning. For example, DDT, hexachlorocyclohexane (HCH), etc.

Table 1.2. Classification of pesticides based on their target organisms.

Group	Pesticide type	Target organism
1	Insecticide	Insects
2	Fungicides	Fungi
3	Herbicide	Weeds/plants
4	Molluscicide	Slugs, snails
5	Rodenticide	Rodents
6	Acaricide	Mites
7	Nematicides	Nematodes

Table 1.3. Classification of pesticides based on their chemical structure (with examples).

Group	Pesticides
Organochlorines	DDT, aldrin, dieldrin, heptachlor
Organophosphates	Phosphates, phosphonates, phosphinates, phosphorothioates, phosphorodithioates, phospharamides, phosphorothioamidates
Carbamates	n-methyl carbamates, dithiocarbamates, benzimidazoles
Pyrethrum and synthetic pyrethroids	Pyrethrins, cypermethrin, deltamethrin, fenvalerate
Phenols	Pentachlorophenols, dinoprop, DNOC
Morpholines	Fungicides such as amorolfine, fenpropimorph, dimethomorph, tridemorph
Chloroalkylthiols	Fungicides
Organometallics	Fungicides: mancozeb, maneb, zineb, metam sodium
Azoles	Fungicides: myclobutanil, tebuconazole, paclobutrazole, penconazole
Ureas, thioureas	Herbicides: sulfonylureas, linuron, benzthiazuron
Anilines	Pendimethalin, trifluralin, fluchloralin
Chloronitrile	Chlorothalonil, chloroxynil, dichlobenil

DDT, dichlorodiphenyltrichloroethane; DNOC, dinitro-ortho-cresol.

- *Contact poisoning* – the pesticide enters the body of pests via their epidermis through contact and generally controls a pest as a result of direct contact. Insects are killed when sprayed directly or when they come across surfaces treated with a residual contact insecticide. For example, pyrethrums, rotenone, toxaphene and chlordane.
- *Fumigation* – the pesticide in gaseous form enters the body of pests via their respiratory system and causes death by poisoning. For example, HCN, chloropicrin, methyl bromide, dichlorvos (DDVP), aluminium phosphide, etc. The soil may be fumigated to destroy grubs attacking roots or to control nematodes.
- *Systemic action* – these are the toxicants that are rapidly absorbed and translocated to various parts of the plants in amounts lethal to insects feeding on them. For example, phorate, aldicarb, carobfuran. Systemic herbicides move within the plant to untreated areas of the root, stem or leaves of plants. Translocation of these pesticides takes place mostly through the xylem vessels. Pesticides consumed by a host organism will stay in its body fluids. Pests feeding on the body fluids of the host organism will then be killed by poisoning.

1.3.2 Classification according to chemical structure (broad group-wise classification)

1. *Organochlorine pesticides*: these pesticides are synthetic organic compounds containing at least one covalently bonded atom of chlorine, which were the earliest discovered and used. Their characteristics are broad spectrum, highly persistent with relatively low toxicity. Prolonged use in large quantities can easily lead to environmental pollution and accumulation in fat tissues of mammals, resulting in cumulative poisoning or damage. Organochlorine pesticides were therefore banned under general circumstances and gradually replaced by other pesticides.

2. *Organophosphate pesticides*: the organophosphates or phosphate esters are the general names for esters of phosphoric acid. Organophosphate pesticides or acetylcholine esterase inhibitors are characterized by their multiple functions and the capacity for controlling a broad spectrum of pests. They are nerve poisons that can be used not only as stomach poison but also as contact poison and fumigants. These pesticides are also biodegradable, cause minimum environmental pollution and slow pest resistance. Temephos and fenitrothion are examples of organophosphate pesticides.

3. *Carbamate pesticides*: carbamate is an organic compound derived from carbamic acid (NH_2COOH). Carbamate esters (e.g. ethyl carbamate) and carbamic acids are functional groups that are interrelated structurally and often are interconverted chemically. These pesticides work on a similar principle to organophosphate pesticides by affecting the transmission of nerve signals resulting in the death of the pest by poisoning. They can be used as stomach and contact poisons as well as fumigant. Moreover, as their molecular structures are largely similar to those of natural organic substances, they can be degraded easily in a natural manner with minimum environmental pollution. Carbaryl and carbofuran are examples of carbamate pesticides.
4. *Synthetic pyrethroid*: synthetic pyrethroid pesticides are synthesized by imitating the structure of natural pyrethrins. They are comparatively more stable with longer residual effects than natural pyrethrins. Synthetic pyrethroid pesticides are highly toxic to insects but of only slight toxicity to mammals. Pyrethroids now constitute the majority of commercial household insecticides, being generally harmless to human beings in low doses, but they can harm sensitive individuals. They are usually broken apart by sunlight and the atmosphere in 1–2 days, and do not significantly affect groundwater quality. Allethrin and permethrin are examples of synthetic pyrethroid pesticides.
5. *Microbial insecticides*: microbial insecticides control pests by means of pathogenic micro-organisms including bacteria, fungi and viruses. *Bacillus thuringiensis israelensis* (B.t.i.) is an example of a microbial insecticide.
6. *Insect growth regulators*: insect growth regulators are compounds developed by copying insect juvenile hormone. The main functions are to interfere with the growth and hatching of larvae into adults, and to prevent the formation of an exoskeleton so as to prohibit the growth of the insect. As its ability to live as a living organism is curtailed, the insect as well as the whole insect population may die eventually. Methoprene is an example of an insect growth regulator.

1.3.3 Classification of pesticides according to their mode or site of action

Pesticides can also be classified according to their mode of action or the way a pesticide destroys or controls the target pest. This is also referred to as the primary site of action. For example, one insecticide may be an acetylcholinesterase inhibitor and affect an insect's nerves, while another may affect moulting. Similarly, a herbicide may mimic the plant's growth regulators and another may affect photosynthesis. A fungicide may affect cell division and another may slow the formation of an important compound in the fungus. The modes of action are limited in number but numbers of pesticides are many. Some pesticides have the same mode of action. The Insecticide Resistance Action Committee (IRAC) has classified pesticides according to their mode of action or the target site of action (Table 1.4).

1.4 Fungicides

The chemicals used to kill or inhibit the growth of fungi, the causal organisms of plant disease, are called fungicides. Although viruses, nematodes and bacteria also cause diseases in plants, fungi are the number one causes of crop loss worldwide. At times, the symptoms caused by pathogens resemble the symptoms caused by abiotic factors such as nutrient deficiency and air pollution.

Fungicides can be classified in a number of different ways, including (i) mobility in the plant, (ii) role in protection of plants, (iii) breadth of activity, (iv) mode of action, (v) chemical group and (vi) Fungicide Resistance Action Committee (FRAC) group (https://www.ncipmc.org/action/Fungicide%20Manual4.pdf).

1. *Mobility in the plant*. Contact fungicide: a fungicide that remains on the surface of the plant where it is applied but does not go deeper; these fungicides have no post-infection activity. Repeated applications are needed to protect new growth of the plant and to replace fungicide that has been washed off by rain or irrigation, or degraded by environmental factors such as sunlight.

Table 1.4. Classification of pesticides according to their mode of action.

Main group and primary site of action	Chemical subgroup or active ingredient	Active ingredients
Acetylcholinesterase (AChE) inhibitors Nerve action	1A Carbamates	Alanycarb, aldicarb, bendiocarb, benfuracarb, butocarboxim, butoxycarboxim, carbaryl, carbofuran, carbosulfan, ethiofencarb, fenobucarb, formetanate, furathiocarb, isoprocarb, methiocarb, methomyl, metolcarb, oxamyl, pirimicarb, propoxur, thiodicarb, thiofanox, triazamate, trimethacarb, XMC, xylylcarb
	1B Organophosphates	Acephate, azamethiphos, azinphos-ethyl, azinphos-methyl, cadusafos, chlorethoxyfos, chlorfenvinphos, chlormephos, chlorpyrifos, chlorpyrifos-methyl, coumaphos, cyanophos, demeton-s-methyl, diazinon, dichlorvos/DDVP, dicrotophos, dimethoate, dimethylvinphos, disulfoton, EPN, ethion, ethoprophos, famphur, fenamiphos, fenitrothion, fenthion, fosthiazate, heptenophos, imicyafos, isofenphos, isopropyl *O*- (methoxyaminothiophosphoryl) salicylate, isoxathion, malathion, mecarbam, methamidophos, methidathion, mevinphos, monocrotophos, naled, omethoate, oxydemeton-methyl, parathion, parathion-methyl, phenthoate, phorate, phosalone, phosmet, phosphamidon, phoxim, pirimiphos-methyl, profenofos, propetamphos, prothiofos, pyraclofos, pyridaphenthion, quinalphos, sulfotep, tebupirimfos, temephos, terbufos, tetrachlorvinphos, thiometon, triazophos, trichlorfon, vamidothion
GABA-gated chloride	2A Cyclodiene organochlorines	Chlordane, endosulfan
	2B Phenylpyrazoles (fiproles)	Ethiprole, fipronil
Sodium channel modulators Nerve action	3A Pyrethroids Pyrethrins	Acrinathrin, allethrin, d-*cis-trans* allethrin, d-*trans* allethrin, bifenthrin, bioallethrin, bioallethrin s-cyclopentenyl isomer, bioresmethrin, cycloprothrin, cyfluthrin, beta-cyfluthrin, cyhalothrin, lambda-cyhalothrin, gamma-cyhalothrin, cypermethrin, alpha-cypermethrin, beta-cypermethrin, theta-cypermethrin, zeta-cypermethrin, cyphenothrin [(1*R*)-*trans*- isomers], deltamethrin, empenthrin [(*EZ*)-(1*R*)- isomers], esfenvalerate, etofenprox, fenpropathrin, fenvalerate, flucythrinate, flumethrin, tau-fluvalinate, halfenprox, imiprothrin, kadethrin, permethrin, phenothrin [(1*R*)-*trans*- isomer], prallethrin, pyrethrins (pyrethrum), resmethrin, silafluofen, tefluthrin, tetramethrin, tetramethrin [(1*R*)-isomers], tralomethrin, transfluthrin
	3B DDT methoxychlor	DDT methoxychlor

continued

Table 1.4. Continued.

Main group and primary site of action	Chemical subgroup or active ingredient	Active ingredients
4 Nicotinic acetylcholine receptor (nAChR) agonists Nerve action	4A Neonicotinoids	Acetamiprid, clothianidin, dinotefuran, imidacloprid, nitenpyram, thiacloprid, thiamethoxam
	4B Nicotine	Nicotine
	4C Sulfoxaflor	
	4D Butenolides	Flupyradifurone
Nicotinic acetylcholine receptor (nAChR) allosteric activators Nerve action	Spinosyns	Spinetoram, spinosad
Chloride channel activators Nerve and muscle action	Avermectins, milbemycins	Abamectin, emamectin benzoate, lepimectin, milbemectin
7 Juvenile hormone mimics Growth regulation	7A Juvenile hormone analogues	Hydroprene, kinoprene, methoprene
	7B Fenoxycarb	Fenoxycarb
	7C Pyriproxyfen	Pyriproxyfen
8[a] Miscellaneous non-specific (multi-site) inhibitors	8A Alkyl halides	Methyl bromide and other alkyl halides
	8B Chloropicrin	Chloropicrin
	8C Sulfuryl fluoride	Sulfuryl fluoride
	8D Borates	Borax
	8E Tartar emetic	Tartar emetic
9 Modulators of chordotonal organs Nerve action	9B Pymetrozine	Pymetrozine
	9C Flonicamid	Flonicamid
10 Mite growth inhibitors Growth regulation	10A Clofentezine Hexythiazox Diflovidazin	Clofentezine, hexythiazox, diflovidazin
	10B Etoxazole	Etoxazole

11 Microbial disruptors of insect midgut membranes (includes transgenic crops expressing *Bacillus thuringiensis* toxins, however specific guidance for resistance management of transgenic crops is not based on rotation of modes of action)	11A *Bacillus thuringiensis* and the insecticidal proteins they produce	*Bacillus thuringiensis* sub sp. *israelensis*, *Bacillus thuringiensis* sub sp. *aizawai*, *Bacillus thuringiensis* sub sp. *kurstaki*, *Bacillus thuringiensis* sub sp. *tenebrionis* Bt crop proteins (midgut) Cry1Ab, Cry1Ac, Cry1Fa, Cry1A.105, Cry2Ab, Vip3A, mCry3A, Cry3Ab, Cry3Bb, Cry34Ab1/Cry35Ab1
	11B *Bacillus sphaericus*	*Bacillus sphaericus*
12 Inhibitors of mitochondrial ATP synthase Energy metabolism	12A Diafenthiuron	Diafenthiuron
	12B Organotin miticides	Azocyclotin, cyhexatin, fenbutatin oxide
	12C Propargite	Propargite
	12D Tetradifon	Tetradifon
13 Uncouplers of oxidative phosphorylation via disruption of the proton gradient Energy metabolism	Chlorfenapyr, DNOC sulfluramid	Chlorfenapyr, DNOC, sulfluramid
14 Nicotinic acetylcholine receptor (nAChR) channel blockers Nerve action	Nereistoxin analogues	Bensultap, cartap hydrochloride, thiocyclam, thiosultap-sodium
15 Inhibitors of chitin biosynthesis, type 0 Growth regulation	Benzoylureas	Bistrifluron, chlorfluazuron, diflubenzuron, flucycloxuron, flufenoxuron, hexaflumuron, lufenuron, novaluron, noviflumuron, teflubenzuron, triflumuron
16 Inhibitors of chitin biosynthesis, type 1 Growth regulation	Buprofezin	Buprofezin
17 Moulting disruptor, Dipteran Growth regulation	Cyromazine	Cyromazine
18 Ecdysone receptor agonists Growth regulation	Diacylhydrazines	Chromafenozide, halofenozide, methoxyfenozide, tebufenozide
19 Octopamine receptor agonists Nerve action	Amitraz	Amitraz
20 Mitochondrial complex III electron transport inhibitors Energy metabolism	20A Hydramethylnon	Hydramethylnon
	20B Acequinocyl	Acequinocyl
	20C Fluacrypyrim	Fluacrypyrim

continued

Table 1.4. Continued.

Main group and primary site of action	Chemical subgroup or active ingredient	Active ingredients
21 Mitochondrial complex I electron transport inhibitors Energy metabolism	21A	
	METI acaricides and insecticides	Fenazaquin, fenpyroximate, pyrimidifen, pyridaben, tebufenpyrad, tolfenpyrad
	21B Rotenone	Rotenone (derris)
22 Voltage-dependent sodium channel blockers Nerve action	22A Indoxacarb	Indoxacarb
	22B Metaflumizone	Metaflumizone
23 Inhibitors of acetyl CoA carboxylase Lipid synthesis, growth regulation	Tetronic and Tetramic acid derivatives	Spirodiclofen, spiromesifen, spirotetramat
24 Mitochondrial complex IV electron transport inhibitors	24A Phosphine	Aluminium phosphide, calcium phosphide, phosphine, zinc phosphide
	24B Cyanide	Cyanide
25 Mitochondrial complex II electron transport inhibitors Energy metabolism	Beta-ketonitrile derivatives	Cyenopyrafen, cyflumetofen
28 Ryanodine receptor modulators Nerve and muscle action	Diamides	Chlorantraniliprole, cyantraniliprole, flubendiamide
UN (compounds of unknown or uncertain mode of action)	Azadirachtin	Azadirachtin
	Benzoximate	Benzoximate
	Bifenazate	Bifenazate
	Bromopropylate	Bromopropylate
	Chinomethionat	Chinomethionat
	Cryolite	Cryolite
	Dicofol	Dicofol
	Pyridalyl	Pyridalyl
	Pyrifluquinazon	Pyrifluquinazon

DDT, dichlorodiphenyltrichloroethane; XMC, 3,5-xylyl methylcarbamate; DDVP, 2,2-dichlorovinyl dimethyl phosphate; EPN: ethyl p-nitrophenyl benzenethionophosphonate; GABA, gamma-aminobutyric acid; ATP, adenosine triphosphate; DNOC, dinitro-ortho-cresol; METI, mitochondrial electron transport inhibitors.

Chlorothalonil and captan are contact fungicides.

Systemic fungicide: a fungicide that is absorbed into plant tissue and may offer some post-infection activity. Very few fungicides are truly systemic (i.e. move freely throughout the plant); however, some are upwardly systemic (i.e. move only upwards in the plant through xylem tissue), and some are locally systemic (i.e. move into treated leaves and redistribute to some degree within the treated portion of the plant). Examples are tricyclazole and hexconazole.

2. *Role in protection* (some fungicides can fall into more than one of these categories). Preventative activity: in this activity, a fungicide is present on the plant as a protective barrier before the pathogen arrives or begins to develop, that is, the fungicide prevents infection from occurring (also referred to as a protective activity). For example, azoxystrobin, tebuconazole and pyraclostrobin.

Early-infection activity: occurs when the active ingredient of a fungicide can penetrate the plant and stop the pathogen in the plant tissues, usually most effective 24 to 72 h after infection occurs, depending on the fungicide. This type of activity is sometimes referred to as 'curative' or 'kickback' activity. Most fungicides that have early-infection activity also have preventative activity and are most effective when applied before infection occurs. Triazoles like tebuconazole and penconazole are applied as early-infection treatments.

Anti-sporulant activity: an ability to prevent spores from being produced. In this case, disease continues to develop (e.g. lesions continue to expand), but spores are not produced or released, so the amount of inoculum available to infect surrounding plants is reduced. For example, botanical fungicides.

3. *Breadth of metabolic activity.* Single-site fungicide: fungicide active against only one point or function in one of the metabolic pathways of a fungus or against a single critical enzyme or protein needed by the fungus. These fungicides tend to have systemic properties. Thiophanate-methyl and trifloxystrobin are examples.

Multi-site fungicide: fungicide that affects a number of different metabolic sites within the fungus. Examples are captan, chlorthalonil and Bordeaux.

4. *Mode of action.* How a fungicide kills or suppresses a target fungus, which is the specific biochemical process of the target fungus that is affected by a fungicide (Table 1.5). Examples are damaging cell membranes, inactivating critical enzymes or proteins, or interfering with key processes such as energy production or respiration.

5. *Chemical group or class.* The name given to a group of chemicals that share a common biochemical mode of action and may or may not have similar chemical structure. Fungicides approved for use on field crops fall into different chemical groups. The fungal growth inhibition by fungicides may be due to damage in their cell membranes, inactivation of critical enzymes or proteins, or interference with key processes such as energy production or respiration. Others impact specific metabolic pathways such as the production of sterols or chitin. For example, phenylamide fungicides bind to and inhibit the function of RNA polymerase in oomycetes, while the benzimidazole fungicides inhibit the formation of beta tubulin polymers used by cells during nuclear division.

6. *FRAC group.* The FRAC has classified fungicides as per their mode of action and resistance risk (Table 1.6). The intrinsic risk for resistance evolution to a given fungicide group is estimated to be low, medium or high according to the principles described in FRAC monographs.

1.5 Herbicide Classification (http://en.wikipedia.org/wiki/Herbicide)

A herbicide is a compound that is toxic to plants, and capable of either killing or severely injuring unwanted vegetation; and may thus be used for the elimination of plant growth or the killing of plant parts.

Herbicides can be grouped by: (i) activity; (ii) use; (iii) mechanism of action; (iv) chemical nature; (v) type of vegetation

Table 1.5. Classification of fungicides based on mode of action (Whitehead, 1995).

Chemical group	Compound	Different modes of action	Application	Other uses
Inorganic				
	Bordeaux mixture	Protectant	Spray	
	Copper oxychloride	Protectant		Also bactericide
	Mercurous chloride			
	Sulfur	Protectant		Also foliar feed
Anilides				
	Carboxim		Seed dressing	
Benzimidazole (mbc)				
	Thiabendazole	Protectant, curative, systemic	Spray	
Chlorophenyl				
	Chlorothalonil	Protectant	Spray	
	Quintozene	Protectant	Soil applied	
Conazole				
	Flusilazole	Protective, curative	Spray	
	Prochloraz	Protectant, eradicant	Spray	
	Propiconazole	Protectant, curative	Spray	
	Tebuconazole	Curative	Spray	
	Triadimefon	Protectant, curative, systemic	Spray	
Dicarboximides				
	Iprodione	Protectant, eradicant	Spray	
Dinitrophenyls				
	Dinocap	Protectant	Spray	
Dithicarbamate				
	Captan	Protectant	Spray	
	Mancozeb	Protectant	Spray	
	Maneb	Protectant	Spray	
	Thiram	Protectant		Also animal repellant
Morpholines				
	Fenpropidin	Protectant, curative, systemic	Spray	
	Fenpropimorph	Curative, systemic	Spray	
	Tridemorph	Protectant, eradicant, systemic	Spray	
Phenylamide				
	Metalaxyl	Protectant	Seed treatment	
Quinones				
	Dithianon	Protectant, eradicant	Spray	

controlled; and (vi) Herbicide Resistance Action Committee classification.

1.5.1 By activity

- *Contact herbicides* destroy only the plant tissue in contact with the chemical. Generally, these are the fastest acting herbicides. For broadleaf weeds it will kill the above-ground leafy part of the plants. It will not directly kill the belowground plant parts, such as roots, bulbs, tubers or rhizomes. They are less effective on perennial plants, which are able to regrow from roots or tubers. Glufosinate ammonium is a broad-spectrum contact herbicide and paraquat is a non-selective contact herbicide.
- *Systemic herbicides* are translocated through the plant, either from foliar application down to the roots, or from soil application up to the leaves.

Table 1.6. Commercial fungicides according to their mode of action and resistance risk.

Mode of action	Target site and code	Group name	Chemical group	Common name	Comments	FRAC code
A: Nucleic acids synthesis	**A1:** RNA polymerase I	PA – fungicides (phenyl amides)	Acylalanines	Benalaxyl Benalaxyl-M (=kiralaxyl) Furalaxyl Metalaxyl Metalaxyl-M (=mefenoxam)	Resistance and cross-resistance well known in various oomycetes but mechanism unknown. High-risk. See FRAC phenylamide guidelines for resistance management	4
			Oxazolidinones	Oxadixyl		
			Butyrolactones	Ofurace		
	A2: Adenosindeaminase	Hydroxy-(2-amino-) pyrimidines	Hydroxy-(2-amino-) pyrimidines	Bupirimate Dimethirimol Ethirimol	Medium-risk resistance and cross-resistance known in powdery mildews. Resistance management required	8
	A3: DNA/RNA synthesis (proposed)	Heteroaromatics	Isoxazoles Isothiazolones	Hymexazol Octhilinone	Resistance not known	32
	A4: DNA topoisomerase type II (gyrase)	Carboxylic acids	Carboxylic acids	Oxolinic acid	Bactericide. Resistance known Risk in fungi unknown Resistance management required	31
B: Mitosis and cell division	**B1:** β-tubuline assembly in mitosis	MBC –fungicides (methyl benzimidazole carbamates)	Benzimidazoles	Benomyl Carbendazim Fuberidazole Thiabendazole	Resistance common in many fungal species. Several target site mutations, mostly E198A/G/K, F200Y in β-tubulin gene Positive cross-resistance between the group members	1
			Thiophanates	Thiophanate thiophanate-methyl	Negative cross-resistance to NPhenylcarbamates High-risk	
	B2: β-tubulin assembly in mitosis	N-phenyl carbamates	N-phenyl carbamates	Diethofencarb	Resistance known. Target site mutation E198K. Negative cross-resistance to benzimidazoles High-risk. Resistance management required	10

continued

Table 1.6. Continued.

Mode of action	Target site and code	Group name	Chemical group	Common name	Comments	FRAC code
	B3: β-tubulin assembly in mitosis	Benzamides	Toluamides	Zoxamide	Low to medium risk	22
		Thiazole carboxamide	Ethylamino-thiazolecarboxamide	Ethaboxam	Resistance management required	
	B4: Cell division (proposed)	Phenylureas	Phenylureas	Pencycuron	Resistance not known	20
	B5: Delocalization of spectrin-like proteins	Benzamides	Pyridinylmethyl benzamides	Fluopicolide	Resistance not known	43
C: Respiration	**C1:** Complex I NADH oxido-reductase	Pyrimidinamines	Pyrimidinamines	Diflumetorim	Resistance not known	39
		Pyrazole-MET1	Pyrazole-5-carboxamides	Olfenpyrad		
	C2: Complex II: succinate dehydrogenase	SDHI (succinate dehydrogenase inhibitors)	Phenyl-benzamides	Benodanil Flutolanil Mepronil	Resistance known for several fungal species in field populations and lab mutants	7
			Phenyl-oxo-ethyl thiophene amide	Isofetamid	Target site mutations in *sdh* gene, e.g. H/Y (or H/L) at 257, 267, 272 or P225L, dependent on fungal species	
			Pyridinyl-ethyl-benzamides	Fluopyram		
			Furan-carboxamides	Fenfuram	Resistance management required	
			Oxathiin-carboxamides	Carboxinoxycarboxin	Medium to high risk	
			Thiazole-carboxamides	Thifluzamide	See FRAC SDHI guidelines for resistance management	

		Pyrazole-4-carboxamides	Benzovindiflupyr		
			Bixafen		
			Fluxapyroxad		
			Furametpyr		
			Isopyrazam		
			Penflufen		
			Penthiopyrad		
			Sedaxane		
		Pyridine-carboxamides	Boscalid		
C3: Complex III: cytochrome bc1 (ubiquinol oxidase) at Qo site (*cyt b* gene)	QoI fungicides (quinone outside inhibitors)	Methoxy-acrylates	Azoxystrobin	Resistance known in various fungal species. Target site mutations in *cyt b* gene (G143A, F129L) and additional mechanisms	11
			Coumoxystrobin		
			Enoxastrobin		
			Flufenoxystrobin		
			Picoxystrobin	Cross-resistance shown between all members of the QoI group	
			Pyraoxystrobin		
		Methoxy-acetamide	Mandestrobin		
		Methoxy-carbamates	Pyraclostrobin	High risk	
			Pyrametostrobin		
			Triclopyricarb	See FRAC QoI guidelines for resistance management	
		Oximino-acetates	Kresoxim-methyl		
			Trifloxystrobin		
		Oximino-acetamides	Dimoxystrobin		
			Fenaminstrobin		
			Metominostrobin		
			Orysastrobin		
		Oxazolidine-diones	Famoxadone		
		Dihydro-dioxazines	Fluoxastrobin		
		Imidazolinones	Fenamidone		
		Benzyl-carbamates	Pyribencarb		

continued

Table 1.6. Continued.

Mode of action	Target site and code	Group name	Chemical group	Common name	Comments	FRAC code
	C4: Complex III: cytochrome bc1(ubiquinone reductase) at Qi site	QiI fungicides (quinone inside inhibitors)	Cyano-imidazole	Cyazofamid	Resistance risk unknown but assumed to be medium to high (mutations at target site known in model organisms). Resistance management required	21
			Sulfamoyl-triazole	Amisulbrom		
	C5: Uncouplers of oxidative phosphorylation		Dinitrophenyl crotonates	Binapacryl Meptyldinocap Dinocap	Resistance not known. Also acaricidal activity	29
			2,6-Dinitroanilines	Fluazinam	Low risk. However, resistance claimed in *Botrytis* in Japan	
			(Pyr.-hydrazones)	(Ferimzone)	Reclassified to U 14 in 2012	
	C6: Inhibitors of oxidative phosphorylation, ATP synthase	Organo tin compounds	Tri-phenyl tin compounds	Fentin acetate Fentin chloride Fentin hydroxide	Some resistance cases known. Low to medium risk	30
	C7: ATP production (proposed)	Thiophene-carboxamides	Thiophene-carboxamides	Silthiofam	Resistance reported. Low-risk	38
	C8: Complex III: cytochrome bc1 (ubiquinone reductase) at Qo site, stigmatellin binding subsite	QoSI fungicides (quinone outside inhibitor, stigmatellin binding type)	Triazolo-pyrimidylamine	Ametoctradin	Not cross-resistant to QoI fungicides. Resistance risk assumed to be medium to high (single-site inhibitor). Resistance management required	45
D: Amino acids and protein synthesis	**D1:** Methionine biosynthesis (proposed) (*cgs* gene)	AP fungicides (anilino-pyrimidines)	Anilino-pyrimidines	Cyprodinil Mepanipyrim Pyrimethanil	Resistance known in *Botrytis* and *Venturia*, sporadically in *Oculimacula* Medium risk See FRAC anilinopyrimidine guidelines for resistance management	9
	D2: Protein synthesis	Enopyranuronic acid antibiotic	Enopyranuronic acid antibiotic	Blasticidin-S	Low to medium risk. Resistance management required	23

	D3: Protein synthesis	Hexopyranosyl antibiotic	Hexopyranosyl antibiotic	Kasugamycin	Resistance known in fungal and bacterial (*P. glumae*) pathogens. Medium risk Resistance management required	24
	D4: Protein synthesis	Glucopyranosyl antibiotic	Glucopyranosyl antibiotic	Streptomycin	Bactericide. Resistance known. High risk Resistance management required	25
	D5: Protein synthesis	Tetracycline antibiotic	Tetracycline antibiotic	Oxytetracycline	Bactericide. Resistance known. High risk Resistance management required	41
E: Signal transduction	**E1:** Signal transduction (mechanism unknown)	Aza-naphthalenes	Aryloxyquinoline Quinazolinone	Quinoxyfen Proquinazid	Resistance to quinoxyfen known. Medium risk. Resistance management required. Cross-resistance found in *Erysipho (Uncinula) necator* but not in *Blumeria graminis*	13
	E2: MAP/histidine-kinase in osmotic signal transduction (*os-2*, HOG1)	PP fungicides (phenylpyrroles)	Phenylpyrroles	Fenpiclonil Fludioxonil	Resistance found sporadically, mechanism speculative. Low to medium risk Resistance management required	12
	E3: MAP/histidine-kinase in osmotic signal transduction (*os-1*, *Daf1*)	Dicarboximides	Dicarboximides	Chlozolinate Iprodione Procymidone Vinclozolin	Resistance common in *Botrytis* and some other pathogens Several mutations in *OS-1*, mostly I365S Cross-resistance common between the group members Medium to high risk See FRAC dicarboximide guidelines for resistance management	2
F: Lipid synthesis and membrane integrity	**F1:**	Formerly dicarboximides				
	F2: Phospholipid biosynthesis, methyltransferase	Phosphoro-thiolates	Phosphorothiolates	Edifenphos Iprobenfos (IBP) Pyrazophos	Resistance known in specific fungi. Low to medium risk. Resistance management required if used for risky pathogens	6
		Dithiolanes	Dithiolanes	Isoprothiolane		

continued

Table 1.6. Continued.

Mode of action	Target site and code	Group name	Chemical group	Common name	Comments	FRAC code
	F3: Lipid peroxidation (proposed)	AH fungicides (aromatic hydrocarbons) (chlorophenyls, nitroanilines)	Aromatic hydrocarbons	Biphenyl Chloroneb Dicloran Quintozene (PCNB) Tecnazene (TCNB) Tolclofos-methyl	Resistance known in some fungi. Low to medium risk Cross-resistance patterns complex due to different activity spectra	14
		Heteroaromatics	1,2,4-Thiadiazoles	Etridiazole		
	F4: Cell membrane permeability, fatty acids (proposed)	Carbamates	Carbamates	Iodocarb Propamocarb Prothiocarb	Low to medium risk. Resistance management required	28
	F5:	Formerly CAA fungicides				
	F6: Microbial disrupters of pathogen cell membranes	Microbial (*Bacillus* sp.)	*Bacillus* sp. and the fungicidal lipopeptides produced	*Bacillus subtilis* syn. *B. amyloliquefaciens*[a] strain QST 713 *Bacillus amyloliquefaciens* strain FZB24 *Bacillus amyloliquefaciens* strain MBI600 *Bacillus amyloliquefaciens* strain D747	[a]Synonyms for *Bacillus amyloliquefaciens* are *Bacillus subtilis* and *B. subtilis* var. *amyloliquefaciens* (previous taxonomic classification) Resistance not known Induction of host plant defence described as additional mode of action for strain FZB24	44
	F7: Cell membrane disruption (proposed)	Plant extract	Terpene hydrocarbons and terpene alcohols	Extract from *Melaleuca alternifolia* (tea tree)	Resistance not known	46

Target	Group name	Chemical group	Common name	Comments	FRAC code
		Piperazines	Triforine	There are big differences in the activity spectra of DMI fungicides	
		Pyridines	Pyrifenox Pyrisoxazole		
		Pyrimidines	Fenarimol Nuarimol		
		Imidazoles	Imazalil Oxpoconazole Pefurazoate Prochloraz Triflumizole		
G: Sterol biosynthesis in membranes **G1:** C14-demethylase in sterol biosynthesis (*erg11/cyp51*)	DMI fungicides (demethylation inhibitors) (SBI: Class I)	Triazoles triazolinthiones	Azaconazole Bitertanol Bromuconazole Cyproconazole Difenoconazole Diniconazole Epoxiconazole Etaconazole Fenbuconazole Fluquinconazole Flusilazole Flutriafol Hexaconazole Imibenconazole Ipconazole Metconazole Myclobutanil Penconazole Propiconazole Prothioconazole Simeconazole Tebuconazole Tetraconazole Triadimefon Triadimenol Triticonazole	Resistance is known in various fungal species. Several resistance mechanisms are known incl. target site mutations in *cyp51* (erg 11) gene, e.g. V136A, Y137F, A379G, I381V; cyp51 promotor; ABC transporters and others Generally wise to accept that cross-resistance is present between DMI fungicides active against the same fungus DMI fungicides are sterol biosynthesis inhibitors (sbis), but show no cross-resistance to other SBI classes Medium risk See FRAC SBI guidelines for resistance management	3

continued

Table 1.6. Continued.

Mode of action	Target site and code	Group name	Chemical group	Common name	Comments	FRAC code
	G2: 14-reductase and 87-isomerase in sterol biosynthesis (*erg24*, *erg2*)	Amines ('morpholines') (SBI: Class II)	Morpholines	Aldimorph Dodemorph Fenpropimorph Tridemorph	Decreased sensitivity for powdery mildews Cross-resistance within the group generally found but not to other SBI classes Low to medium risk See FRAC SBI guidelines for resistance management	5
			Piperidines	Fenpropidin Piperalin		
			Spiroketal-amines	Spiroxamine		
	G3: 3-Keto reductase, C4-demethylation (*erg27*)	(SBI: Class III)	Hydroxyanilides	Fenhexamid	Low to medium risk. Resistance management required	17
			Amino-pyrazolinone	Fenpyrazamine		
	G4: Squalene-epoxidase in sterol biosynthesis (*erg1*)	(SBI: Class IV)	Thiocarbamates	Pyributicarb	Resistance not known, fungicidal and herbicidal activity	18
			Allylamines	Naftifine Terbinafine	Medical fungicides only	
H: Cell wall biosynthesis	**H3:** Trehalase and inositol biosynthesis	Alucopyranosyl antibiotic	Glucopyranosyl antibiotic	Validamycin	Resistance not known	26
	H4: Chitin synthase	Polyoxins	Peptidyl pyrimidine nucleoside	Polyoxin	Resistance known. Medium risk. Resistance management required	19
	H5: Cellulose synthase	CAA fungicides (carboxylic acid amides)	Cinnamic acid amides	Dimethomorph Flumorph Pyrimorph	Resistance known in *Plasmopara viticola* but not in *Phytophthora infestans*	40
			Valinamide carbamates	Benthiavalicarb Iprovalicarb Valifenalate	Cross-resistance between all members of the CAA group	
			Mandelic acid amides	Mandipropamid	Low to medium risk. See FRAC CAA guidelines for resistance management	

I: Melanin synthesis in cell wall	**I1:** Reductase in melanin biosynthesis	MBI-R (melanin biosynthesis inhibitors – reductase)	Isobenzo-furanone	Fthalide	Resistance not known	16.1
			Pyrrolo-quinolinone Triazolobenzo-thiazole	Pyroquilon Tricyclazole		
	I2: Dehydratase in melanin biosynthesis	MBI-D (melanin biosynthesis inhibitors – dehydratase)	Cyclopropane-carboxamide	Carpropamid	Resistance known Medium risk Resistance management required	16.2
			Carboxamide Propionamide	Diclocymet Fenoxanil		
P: Host plant defence induction	**P1:** Salicylic acid pathway	Benzo-thiadiazole (BTH)	Benzo-thiadiazole (BTH)	Acibenzolar-s-methyl	Resistance not known	P 1
	P2	Benzisothiazole	Benzisothiazole	Probenazole (also antibacterial and antifungal activity)	Resistance not known	P 2
	P3	Thiadiazole-carboxamide	Thiadiazole-carboxamide	Tiadinil Isotianil	Resistance not known	P 3
	P4	Natural compound	Polysaccharides	Laminarin	Resistance not known	P 4
	P5	Plant extract	Complex mixture, ethanol extract	Extract from *Reynoutria sachalinensis* (giant knotweed)	Resistance not known	P 5
Not classified	Unknown	Cyanoacetamide-oxime	Cyanoacetamide-oxime	Cymoxanil	Resistance claims described Low to medium risk. Resistance management required	27
	Unknown	Phosphonates	Ethyl phosphonates	Fosetyl-Al Phosphorous acid and salts	Few resistance cases reported in few pathogens Low risk	33

continued

Table 1.6. Continued.

Mode of action	Target site and code	Group name	Chemical group	Common name	Comments	FRAC code
	Unknown	Phthalamic acids	Phthalamic acids	Teclofthalam (bactericide)	Resistance not known	34
	Unknown	Benzotriazines	Benzotriazines	Triazoxide	Resistance not known	35
	Unknown	Benzene-sulfonamides	Benzene-sulfonamides	Flusulfamide	Resistance not known	36
	Unknown	Pyridazinones	Pyridazinones	Diclomezine	Resistance not known	37
	Unknown	Thiocarbamate	Thiocarbamate	Methasulfocarb	Resistance not known	42
	Unknown	Phenyl-acetamide	Phenyl-acetamide	Cyflufenamid	Resistance in *Sphaerotheca* Resistance management required	U 6
	Actin disruption (proposed)	Aryl-phenyl-ketone	Benzophenone	Metrafenone	Less sensitive isolates detected in wheat powdery mildew. Medium risk. Resistance management required	U 8
	Cell membrane disruption (proposed)	Guanidines	Guanidines	Dodine	Resistance known in *Venturia inaequalis*. Low to medium risk Resistance management recommended	U 12
	Unknown	Thiazolidine	Cyano-methylene-thiazolidine	Flutianil	Resistance not known	U 13
	Unknown	Pyrimidinone-hydrazones	Pyrimidinone-hydrazones	Ferimzone	Resistance not known Reclassified from C5 in 2012	U 14
	Oxysterol binding protein (OSBP) inhibition (proposed)	Piperidinyl-thiazole-isoxazolines	Piperidinyl-thiazole-isoxazolines	Oxathiapiprolin	Resistance risk assumed to be medium to high (single-site inhibitor). Resistance management required	U 15
	Complex III: cytochrome bc1, unknown binding site (proposed)	4-Quinolyl-acetate	4-Quinolyl-acetate	Tebufloquin	Not cross-resistant to QoI. Resistance risk unknown but assumed to be medium. Resistance management required	U 16

Not classified	Unknown	Diverse	Diverse	Mineral oils, organic oils, potassium bicarbonate, material of biological origin	Resistance not known	NC
Multi-site contact activity	Multi-site contact activity	Inorganic	Inorganic	Copper (different salts)	Generally considered as a low-risk group without any signs of resistance developing to the fungicides	M 1
		Inorganic	Inorganic	Sulfur		M 2
		Dithiocarbamates and relatives	Dithiocarbamates and relatives	Ferbam Mancozeb Maneb Metiram Propineb Thiram Zineb Ziram		M 3
		Phthalimides	Phthalimides	Captan Captafol Folpet		M 4
		Phloronitriles (phthalonitriles)	Chloronitriles (phthalonitriles)	Chlorothalonil		M 5
		Sulfamides	Sulfamides	Dichlofluanid Tolylfluanid		M6
		Guanidines	Guanidines	Guazatine Iminoctadine		M 7
		Triazines	Triazines	Anilazine		M 8
		Quinones (anthraquinones)	Quinones (anthraquinones)	Dithianon		M 9
		Quinoxalines	Quinoxalines	Chinomethionat/ quinomethionate		M 10
		Maleimide	Maleimide	Fluoroimide		M 11

They are capable of controlling perennial plants. They can destroy a greater amount of plant tissue than contact herbicides. Glyphosate is a systemic non-selective, while fluroxypyr is a systemic selective herbicide.

1.5.2 By use

Soil-applied herbicides are applied to the soil and are taken up by the roots of the target plant.

- *Pre-plant incorporated* herbicides are mechanically incorporated into the soil prior to planting. The objective for incorporation is to prevent dissipation through photodecomposition and/or volatility. Trifluralin and imazethapyr are pre-plant incorporated herbicides.
- *Pre-emergent herbicides* are applied to the soil before the crop emerges and prevent germination or early growth of weed seeds. Metolachlor and pendimethalin are pre-emergent herbicides, widely used for control of annual grasses.
- *Post-emergent herbicides* are applied after the crop has emerged. Imazamox, an imidazolinone, and carfentazone-ethyl are examples of post-emergent herbicides.

1.5.3 By mechanism of action

1. *ACCase inhibitors* are chemicals used to kill grasses, the monocots. Acetyl coenzyme A carboxylase (ACCase) is part of the first step of lipid synthesis. Thus, ACCase inhibitors affect cell membrane production in the meristems of the grass plant. The ACCases of grasses are sensitive to these herbicides, whereas the ACCases of dicot plants are not. Fluazifop-p-butyl, quizalofop-p-tefuryl and fenoxaprop-p-ethyl are some ACCase inhibitors.

2. *ALS inhibitors* are DNA synthesis inhibitors. The acetolactate synthase (ALS) enzyme (also known as acetohydroxyacid synthase, or AHAS) is the first step in the synthesis of the branched chain amino acids (valine, leucine and isoleucine). These herbicides slowly starve affected plants of these amino acids, which eventually leads to inhibition of DNA synthesis. They affect grasses and dicots alike. The ALS inhibitor family includes sulfonylureas, imidazolinones (IMIs), triazolopyrimidines (TPs), pyrimidinyl oxybenzoates (POBs) and sulfonylamino carbonyl triazolinones (SCTs).

3. *EPSPS inhibitors.* The enolpyruvylshikimate 3-phosphate synthase enzyme EPSPS is used in the synthesis of the amino acids tryptophan, phenylalanine and tyrosine. They affect grasses and dicots alike. Glyphosate (Roundup) is a systemic EPSPS inhibitor but is inactivated by soil contact.

4. *Synthetic auxin* is a synthetic analogue and mimics the plant hormone. Synthetic auxins were discovered in the 1940s after a long study of the plant growth regulator auxin. They have several points of action on the cell membrane, and are effective in the control of dicot plants. 2,4-D is a synthetic auxin herbicide.

5. *Photosystem II inhibitors.* These inhibit the electron transport system of photosynthesis. The transfer of electrons from Photosystem II to Photosystem I is essential for the production of photosynthetic energy. They reduce electron flow from water to NADPH2+ during the photochemical step. They bind to the Qb site on the D2 protein, and prevent quinone from binding to this site. Therefore, this group of compounds causes electrons to accumulate on chlorophyll molecules. As a consequence, oxidation reactions in excess of those normally tolerated by the cell occur, and the plant dies. The triazine herbicides (including atrazine) are PSII inhibitors.

6. *Photosystem I inhibitors* are considered to be contact herbicides and are often referred to as membrane disruptors. The destroyed cell membranes result in leakage of cell contents into the intercellular spaces. These are actually 'electron thieves' and divert electrons from the normal pathway through FeS to Fdx to NADP leading to direct discharge of electrons on oxygen. As a result, reactive oxygen species are produced and oxidation reactions in excess of those normally tolerated by the cell occur, leading to inhibitition of photosynthesis. Bipyridinium herbicides (such as diquat and paraquat) inhibit the Fe-S–Fdx step of that chain, while diphenyl ether herbicides (such as nitrofen, nitrofluorfen and acifluorfen) inhibit the Fdx–NADP step.

7. *HPPD inhibitors* inhibit 4-hydroxyphenylpyruvate dioxygenase, which is involved in tyrosine breakdown. The carotenoids, produced by tyrosine breakdown products, protect chlorophyll in plants from being destroyed by sunlight. Lack of carotenoids results in plants turning white due to complete loss of chlorophyll, and plant death. These inhibitors, discovered over 30 years ago, are the newest class of chemicals in the herbicide world. This chemistry was first introduced in Asia for weed control in rice, but has primarily been used in maize in the USA. Mesotrione and sulcotrione are herbicides in this class; a drug, nitisinone, was discovered in the course of developing this class of herbicides.

1.5.4 By chemical nature and composition

Compounds with chemical affinities are grouped together. This is useful in characterizing herbicides.

1. *Inorganic herbicides*: The first chemicals used for weed control before the introduction of the organic compounds and contain no carbon. Examples:

i) *Acids*: Arsenic acid, arsenious acid, arsenic trioxide, sulfuric acid.

ii) *Salts*: Borax, copper sulfate, ammonium sulfate, sodium chlorate, sodium arsenite, copper nitrate.

2. *Organic herbicides*: Oils and non-oils containing carbon and hydrogen atoms in their molecules. Examples:

i) *Oils*: Diesel oil, standard solvent, xylene-type, aromatic oils, polycyclic, aromatic oils etc.

ii) *Aliphatics*: Dalapon, TCA, Acrolein, glyphosate methyl bromide.

iii) *Amides*: Propanil, butachlor, alachlor, CDAA, Diphenamide, Naptalam, Propachlor.

iv) *Carbamates*: Propham, chloropham, barban.

I. *Thiocarbamates*: Butylate, dilate, triallate, s-ethyl dipropylthiocarbamate (EPTC), molinate, pebulate, vernolate, benthlocarb, aslum, cycolate.

II. *Dithiocarbamates*: CDEC, metham.

v) *Nitralin (benzonitrates)*: Dichlobenil, bromoxynil, ioxynil.

vi) *Ditroanilines (Toluidines)*: Benefin, nitralin, trifluralin, butralin, dinitramine, fluchlorine, oxyzalin, penoxalin.

vii) *Phenoxy*: 2,4-D; 2,4, 5-T; 4-(2-methyl-4-chlorophenoxy) butyric acid (MCPB); 2,4-DB; 2,4-DP; 2,4,5-TP (silvex)

viii) *Ureas*: Monuron, diuron, fenuron, neburon, flumeturon, methabenzthiazuron, buturon, chlorbromuron, chloroxuron, norea siduson, metoxuron.

ix) *Organic arsenicals*: Cacodlic acid, monosodium methyl arsenate (MSMA), disodium methyl arsenate (DSMA).

x) *Aromatic carboxylic acids*:

I. *Phenyl acetic acid*: Fenac or 2,3,6-trichlorophenyl acetic acid.

II. *Benzoic acid*: 2,3,6-trichlorobenzoic acid (2,3,6 TBA), amiben (2-amino-2,5-dichlorobenzoic acid).

III. *Phthalic acid*: Endothal, dimethyl ester of tetrachloroterephthalic acid (DCPA).

IV. NPA or N-1 napthyl pthalamic acid.

xi) *Aromatics*:

I. *Substituted phenols*: PCP or pentachlorophenol, PCP sodium salt, DNBP or dinitrophenols.

II. *Diphenyl ethers*: Nitrogen, flurodifen, oxyfluorfen.

xii) *Heterocyclic nitrogen derivatives*: Simazine, atrazine, maleic hydrazide, amitrol.

I. *Triazines*: Atrazine, simazine, ametryne, terbuteryne, cyprazinc, metribuzin, prometryn, propazine.

II. *Sulfonylureas*: Such as benzsulfuron, chlorimuron, chlorsulfuron, metsulfuron, sulfometuron, thiameturon.

III. *Imadazolinones*: Imazapyr, imazaquin, imazethapyr, imazapic.
IV. *Uracils*: Bromacil, terbacil.
V. *Bipyridyliums*: Paraquat, diquat.
xiii) *Others*: Bentazon, piclaram, pyrazon, pyrichlor, endothall, bensulphide, maleic hydrazide (MH), DCPA.

1.5.5 By type of vegetation controlled

It is important to know the life cycle of the vegetation or plant, that is, the length of time the plant lives, time of year it germinates and grows, and the type of reproductive capabilities the plant possesses. Based on their life cycles, plants or weeds are grouped mainly into two categories:

1. *Annuals* are plants that complete their life cycle in less than 1 year. Most annual weed seeds will remain viable in the soil for 1–7 years. The pre-emergent herbicides are used to manage these annual weeds.
2. *Perennials* are plants that live for more than 2 years.Usually perennials do not produce seed in the year of establishment and survive for many growing seasons. Most perennials reproduce by seed and many are able to spread by vegetative means. The most effective way to deal with perennial weeds is through the use of post-emergent herbicides.

The three categories that weeds fall into are broadleaf, grassy and sedges, as follows:

1. *Broadleaf weeds* have different shapes and sizes with various patterns to their leaves. They are generally wider than their length. The post-emergent herbicides such as 2,4-D, methylchlorophenoxypropionic acid (MCPP), 2-methyl-4-chlorophenoxyacetic acid (MCPA) and dicamba could be used to effectively control these weeds.
2. *Grassy weeds* tend to be narrower and longer, similar to grass. The leaf blades are longer than they are wide and have parallel veins. An example is crabgrass or dallisgrass. Proper pre-emergent herbicides like pendimethalin, trifluralin and dithiopyr need to be selected for their control. These are applied to the soil prior to weed seed germination. They provide good control of many annual grassy weeds and are the best weapon against crabgrass. As a general rule, most grassy weed herbicides will not control broadleaf weeds and vice versa.
3. *Sedge* is another unique weed, which is neither a broadleaf nor grassy. Sedges are difficult to control. They form tubers containing food reserves for plant survival. Herbicide treatments will provide control of the shoots and leaves of the plant, but often multiple herbicide applications need to be made as sedges can regenerate from the belowground nutlets. The new group of herbicides with low dosages, the sulfonylureas, are an important component of turfgrass weed management. The halosulfuron and trifloxysulfuron herbicides are used for sedge control in turfgrass.

1.5.6 By HRAC system

The Herbicide Resistance Action Committee (HRAC) system has been developed partly in cooperation with the Weed Science Society of America (WSSA) and new herbicides have been categorized jointly by HRAC and WSSA. For reference the numerical system of the WSSA is listed, too. The WSSA and HRAC systems differ only in minor ways. Herbicides in italics are listed in the HRAC classification system but are not listed in the WSSA classification (Table 1.7). (www.hrac-global.com/Education/ClassificationofHerbicideSiteofAction.aspx).

The aim of HRAC is to create a uniform classification of herbicide sites of action in as many countries as possible. Such a classification system could be useful for many instances but there are cases where weeds exhibit multiple resistance across many of the groups listed and in these cases the key may be of limited value. The system itself is *not* based on resistance risk assessment but can be used by the farmer or researcher as a tool to choose herbicides in different sites of action groups, so that mixtures or rotations of active ingredients can be planned.

Table 1.7. Classification of herbicide according to HRAC.

HRAC group	Site of action	Chemical family	Active ingredient	WSSA group
A	Inhibition of acetyl CoA carboxylase (ACCase)	Aryloxyphenoxy-propionate (informally termed as 'FOPs')	Clodinafop-propargyl, cyhalofop-butyl, diclofop-methyl, fenoxaprop-p-ethyl, fluazifop-p-butyl, haloxyfop-r-methyl, propaquizafop, quizalofop-p-ethyl	1
		Cyclohexanedione (informally termed as 'DIMs')	Alloxydim, butroxydim, clethodim, cycloxydim, *profoxydim*, sethoxydim, *tepraloxydin*, tralkoxydim	
		Phenylpyrazoline (informally termed as 'DEN')	Pinoxaden	
B	Inhibition of acetolactate synthase ALS (acetohydroxyacid synthase AHAS)	Sulfonylurea	Amidosulfuron, azimsulfuron, bensulfuron-methyl, chlorimuron-ethyl, chlorsulfuron, cinosulfuron, cyclosulfamuron, ethametsulfuron-methyl, ethoxysulfuron, flazasulfuron, flupyrsulfuron-methyl-Na, foramsulfuron, halosulfuron-methyl, *imazosulfuron*, iodosulfuron, mesosulfuron, metsulfuron-methyl, nicosulfuron, *oxasulfuron*, primisulfuron-methyl, prosulfuron, pyrazosulfuron-ethyl, rimsulfuron, sulfometuron-methyl, sulfosulfuron, thifensulfuron-methyl, triasulfuron, tribenuron-methyl, trifloxysulfuron, triflusulfuron-methyl, *tritosulfuron*	2
		Imidazolinone	Imazapic, imazamethabenz-methyl, imazamox, imazapyr, imazaquin, imazethapyr	
		Triazolopyrimidine	Cloransulam-methyl, diclosulam, florasulam, flumetsulam, *metosulam*, *penoxsulam*	
		Pyrimidinyl(thio) benzoate	Bispyribac-Na, pyribenzoxim, *pyriftalid*, pyrithiobac-Na, *pyriminobac-methyl*	
		Sulfonylaminocarbonyl-triazolinone	Flucarbazone-Na, propoxycarbazone-Na	
C1	Inhibition of photosynthesis at photosystem II	Triazine	Ametryne, atrazine, cyanazine, desmetryne, *dimethametryne*, prometon, prometryne, propazine, simazine, simetryne, terbumeton, terbuthylazine, *terbutryne*, trietazine	5
		Triazinone	Hexazinone, metamitron, metribuzin	
		Triazolinone	Amicarbazone	
		Uracil	Bromacil, *lenacil*, terbacil	

continued

Table 1.7. Continued.

HRAC group	Site of action	Chemical family	Active ingredient	WSSA group
		Pyridazinone	pyrazon = chloridazon	
		Phenyl-carbamate	Desmedipham, phenmedipham	
C2	Inhibition of photosynthesis at photosystem II	Urea	*Chlorobromuron*, chlorotoluron, *chloroxuron*, dimefuron, diuron, *ethidimuron*, *fenuron*, fluometuron (see F3), isoproturon, *isouron*, linuron, methabenzthiazuron, *metobromuron*, metoxuron, monolinuron, *neburon*, siduron, tebuthiuron	7
		Amide	Propanil, *pentanochlor*	
C3	Inhibition of photosynthesis at photosystem II	Nitrile	*Bromofenoxim*, bromoxynil, ioxynil	6
		Benzothiadiazinone	Bentazon	
		Phenyl-pyridazine	Pyridate, *pyridafol*	
D	Photosystem-I-electron diversion	Bipyridylium	Diquat, paraquat	22
E	Inhibition of protoporphyrinogen oxidase (PPO)	Diphenylether	Acifluorfen-Na, bifenox, *chlomethoxyfen*, *fluoroglycofen-ethyl*, fomesafen, *halosafen*, lactofen, oxyfluorfen	14
		Phenylpyrazole	*Fluazolate*, pyraflufen-ethyl	
		n-phenylphthalimide	Cinidon-ethyl, flumioxazin, flumiclorac-pentyl	
		Thiadiazole	Fluthiacet-methyl, *thidiazimin*	
		Oxadiazole	Oxadiazon, oxadiargyl	
		Triazolinone	Azafenidin, *carfentrazone-ethyl*, *sulfentrazone*	
		Oxazolidinedione	*Pentoxazone*	
		Pyrimidindione	*Benzfendizone*, butafenacil	
		Other	*Pyraclonil*, *profluazol*, flufenpyr-ethyl	
F1	Bleaching: inhibition of carotenoid biosynthesis at the phytoene desaturase step (PDS)	Pyridazinone	Norflurazon	12
		Pyridinecarboxamide	Diflufenican, picolinafen	
		Other	Beflubutamid, fluridone, flurochloridone, flurtamone	
F2	Bleaching: inhibition of 4-hydroxyphenyl-pyruvate-dioxygenase (4-HPPD)	Triketone	Mesotrione, sulcotrione	27
		Isoxazole	*Isoxachlortole*, isoxaflutole	
		Pyrazole	Benzofenap, pyrazolynate, pyrazoxyfen	
		Other	*Benzobicyclon*	

continued

Table 1.7. *Continued.*

HRAC group	Site of action	Chemical family	Active ingredient	WSSA group
F3	Bleaching: inhibition of carotenoid biosynthesis (unknown target)	Triazole	Amitrole (*in vivo* inhibition of lycopene cyclase)	11
		Isoxazolidinone	Clomazone	13
		Urea	Fluometuron (see C2)	
		Diphenylether	Aclonifen	
G	Inhibition of EPSP (5-enolpyruvylshikimate-3-phosphate) synthase	Glycine	Glyphosate, *sulfosate*	9
H	Inhibition of glutamine synthetase	Phosphinic acid	Glufosinate ammonium, *bialaphos = bilanaphos*	10
I	Inhibition of DHP (dihydropteroate) synthase	Carbamate	Asulam	18
K1	Microtubule assembly inhibition	Dinitroaniline	Benefin = benfluralin, *butralin*, *dinitramine*, ethalfluralin, oryzalin, pendimethalin, trifluralin	3
		Phosphoroamidate	*Amiprophos-methyl*, *butamiphos*	
		Pyridine	Dithiopyr, thiazopyr	
		Benzamide	Propyzamide = pronamide, *tebutam*	
		Benzoic acid	DCPA = chlorthal-dimethyl	3
K2	Inhibition of mitosis/ microtubule organization	Carbamate	*Chlorpropham*, *propham*, carbetamide	23
K3	Inhibition of VLCFAs (see table notes) (inhibition of cell division)	Chloroacetamide	Acetochlor, alachlor, butachlor	15
			Dimethachlor, dimethanamid, metazachlor, metolachlor, *pethoxamid*	
			Pretilachlor, propachlor, *propisochlor*, thenylchlor	
		Acetamide	*Diphenamid*, napropamide, *naproanilide*	
		Oxyacetamide	Flufenacet, mefenacet	
		Tetrazolinone	Fentrazamide	
		Other	Anilofos, *cafenstrole*, *piperophos*	
L	Inhibition of cell wall (cellulose) synthesis	Nitrile	Dichlobenil, *chlorthiamid*	20
		Benzamide	Isoxaben	21
		Triazolocarboxamide	*Flupoxam*	
		Quinoline carboxylic acid	quinclorac (for monocots) (also group O)	26
M	Uncoupling (membrane disruption)	Dinitrophenol	**DNOC*, *dinoseb*, dinoterb	24

continued

Table 1.7. Continued.

HRAC group	Site of action	Chemical family	Active ingredient	WSSA group
N	Inhibition of lipid synthesis – not ACCase inhibition	Thiocarbamate	Butylate, cycloate, *dimepiperate*, *EPTC, esprocarb, molinate, *orbencarb*, pebulate, prosulfocarb, thiobencarb = benthiocarb, *tiocarbazil*, triallate, vernolate	8
		Phosphorodithioate	bensulide	
		Benzofuran	*Benfuresate*, ethofumesate	
		Chloro-carbonic acid	**TCA*, *dalapon*, *flupropanate*	26
O	Action like indole acetic acid (synthetic auxins)	Phenoxy-carboxylic acid	Clomeprop, 2,4-D, 2,4-DB, dichlorprop = 2,4-DP, MCPA, MCPB, mecoprop = MCPP = CMPP	4
		Benzoic acid	Chloramben, dicamba, TBA	
		Pyridine carboxylic acid	Clopyralid, fluroxypyr, picloram, triclopyr	
		Quinoline carboxylic acid	Quinclorac, (also group L), quinmerac	
		Other	benazolin-ethyl	
P	Inhibition of auxin transport	Phthalamate Semicarbazone	Naptalam, diflufenzopyr-Na	19
R	. . .	. . .	. . .	
S	. . .	. . .	. . .	
	. . .	. . .	. . .	
Z	Unknown Note: While the site of action of herbicides in Group Z is unknown it is likely that they differ in site of action among themselves and from other groups.	Arylaminopropionic acid	Flamprop-m-methyl/-isopropyl	25
		Pyrazolium	Difenzoquat	26
		Organoarsenical	DSMA, MSMA	17
		Other	*Bromobutide*, *(chloro)-flurenol*	
			Cinmethylin	
			Cumyluron	
			Dazomet	
			Dymron = daimuron, *methyl-dimuron = methyl-dymron*, *etobenzanid*, fosamine, *indanofan*, metam, *oxaziclomefone*, *oleic acid*	
			Pelargonic acid, pyributicarb	

According to information and comments the following herbicides are classified in the January 2005 version in HRAC (WSSA) groups: B (2): cancelled: #9; #9; procarbazone (approved ISO name: propoxycarbazone); E (14): cancelled: pyrazogyl (approved name: pyraclonil).

DCPA, dimethyl tetrachloroterephthalate; VLCFA, very long chain fatty acid; DNOC, dinitro-ortho-cresol; EPTC, S-ethyl-N,N-dipropylthiocarbamate; TCA, 2,4,5-trichloroacetic acid; 2,4-D, 2,4-dinitrophenylhydrazine; 2,4-DB, 4-(2,4-dichlorophenoxy)butyric acid; 2,4-DP, 2-(2,4-dichlorophenoxy)propionic acid; MCPA, 2-methyl-4-chlorophenoxyacetic acid; MCPB, 4-(2-methyl-4-chlorophenoxy) butyric acid; MCPP/CMPP, methylchlorophenoxypropionic acid or mecoprop; DSMA, disodium methyl arsenate; MSMA, monosodium methyl arsenate.

Criteria for descriptors of the quality of mechanism of action information

Strong evidence that action at this protein (or protein complex) is responsible for insecticidal effects	Potent effects on the function of the target protein *and* either resistance due to mutation/overexpression/removal of this protein *or* correlation of potency between effects on the protein and biological activity for a set of analogues
Good evidence that action at this protein (or protein complex) is responsible for insecticidal effects	Highly potent effects on the function of the protein combined with clearly consistent physiological effects
Compounds affect the function of this protein, but it is not clear that this is what leads to biological activity	Compounds (or their metabolites) have moderate or low potency on the function of the protein, and there is little or no evidence associating this effect with biological activity. Compounds may be grouped because of similarity of structure and distinctive physiological effect
Target protein responsible for biological activity is unknown, or uncharacterized	Compounds may be grouped because of similarity of structure and distinctive physiological effect

Notes regarding subgroups

Subgroups represent distinct chemical classes that are believed to have the same mechanism of action but are different enough in chemical structure or mode of interaction with the target protein that the chance of selection for either metabolic or target site cross-resistance is reduced compared to close analogues. Subgroups may also distinguish compounds that are chemically similar but known to bind differently within the target or to have differential selectivity among multiple targets. Additional information regarding subgroups is provided in Table 1.8.

The cross-resistance potential between subgroups is higher than that between different groups, so rotation between subgroups should be avoided. In exceptional circumstances (i.e. when effective registered insecticides from other mode of action groups are unavailable), rotation may be considered following consultation with local expert advice and where cross-resistance does not exist. These exceptions should not be considered sustainable resistance management strategies, and alternative options should be sought to maintain pest susceptibility.

1.6 Classification of Pesticides According to Risk Involved

There is always a risk to human health when handling pesticides and their formulated products. Pesticide exposure can occur via ingestion, inhalation, dermal absorption or ocular contact. There is a stringent need to identify whether signs and symptoms of pesticide poisoning are due to the active ingredient (the pesticide itself), inactive ingredients, solvents or additives, which may vary by region, country, manufacturer or individual preference. In 1973, WHO suggested a tentative classification of pesticides to distinguish between the more and less hazardous forms of each pesticide (extract from WHO Chronicle, 1975), taking into account the views of members of the WHO Expert Advisory Panel on Insecticides and other expert advisory panels with special competence and interest in pesticide technology, as well as the comments of WHO Member States and of two international agencies. The WHO recommendations for classification of pesticides were based on their hazard levels. The proposal for recommendations was approved by the 28th World Health Assembly in 1975 and was published in the WHO Chronicle (1975) as an annex, but was not considered as part of the

Table 1.8. Additional information about particular subgroups.

Subgroups	Notes
3A and 3B	Because DDT is no longer used in agriculture, this is only applicable for the control of human disease vectors such as mosquitoes
4A, 4B,4C and 4D	Although these compounds are believed to have the same target site, current evidence indicates that the risk of metabolic cross-resistance between subgroups is low
10A	Hexythiazox is grouped with clofentezine because they exhibit cross-resistance, even though they are structurally distinct, and the target site for these compounds is unknown. Diflovidazin has been added to this group because it is a close analogue of clofentezine and is expected to have the same mode of action
11A	Different *Bacillus thuringiensis* products that target different insect orders may be used together without compromising their resistance management. Rotation between certain specific *Bacillus thuringiensis* microbial products may provide resistance management benefits for some pests. Consult product-specific recommendations
	Bt crop proteins: Where there are differences among the specific receptors within the midguts of target insects, transgenic crops containing certain combinations of the listed proteins provide resistance management benefits
22A and 22B	Although these compounds are believed to have the same target site, current evidence indicates that the risk of metabolic cross-resistance between subgroups is low

DDT, dichlorodiphenyltrichloroethane.

classification initially. However, the Member States and pesticide registration authorities suggested further guidance for classification of individual pesticides. Guidelines were first issued in 1978, and have since been revised and reissued every few years.

The pesticides are assigned to a Toxicity Class, based on their acute risk to human health (i.e. the risk of single or multiple exposures over a relatively short period of time). It takes into consideration the toxicity of the technical active substance and also describes methods for the classification of formulations (in particular, allowance is made for the lesser hazards from solids as compared with liquids). There is a list of common technical grade pesticides and recommended classifications together with a listing of active ingredients, believed to be obsolete or discontinued for use as pesticides, pesticides subject to the prior informed consent procedure (Rotterdam Convention), limitations to trade because of the Stockholm convention of persistant organic pollutants (POPs), and gaseous or volatile fumigants not classified under these recommendations. The classification is based primarily on the acute oral and dermal toxicity to the rat since these determinations are standard procedures in toxicology. Where the dermal LD value of a compound is such that it would place it in a more restrictive class than the oral LD value would indicate, the compound will always be classified in the more restrictive class.

WHO has categorized pesticides into five classes, according to the *risk* involved in their use (Table 1.9). The US EPA has a 'toxicity ranking' with four categories. The EPA warnings are often used on pesticide labels.

A colour code has also been introduced for pesticide containers. A red stripe on the label indicates that the chemical is extremely poisonous, yellow indicates highly poisonous, and blue and green indicate that they are moderately and slightly poisonous, respectively. These colours are used at the bottom of the pesticide label, along with pictures to show the type of protective clothing that should be used while handling them.

The WHO classification system has been modified and replaced with the latest classification, that is, the Globally Harmonized System of Classification and Labelling of Chemicals (GHS), which has been widely used for the classification and labelling of

Table 1.9. WHO classification of pesticides as per their hazard.

S. no.	WHO class	Medium LD by the oral route (acute toxicity) LD_{50} mg/kg body weight of test animals	Medium LD by the dermal route (dermal toxicity) LD_{50} mg/kg body weight of test animals	Colour of identification band on the label
		LD_{50} for the rat (mg/kg body weight)		
1	Ia Extremely hazardous	<5	<50	Bright red
2	Ib Highly hazardous	5–50	50–200	Bright yellow
3	II Moderately hazardous	50–2000	200–2000	Bright blue
4	III Slightly hazardous	>2000	>2000	Bright green
5	U Unlikely to present acute hazard	5000 or higher		

chemicals worldwide. For this revision of the classification, the WHO Hazard Classes have been aligned in an appropriate way with the GHS Acute Toxicity Hazard Categories for acute oral or dermal toxicity as the starting point for allocating pesticides to a WHO Hazard Class (IPCS, 2009). As has always been the case, the classification of some pesticides has been adjusted to take account of severe hazards to health other than acute toxicity. The GHS Acute Toxicity Hazard Category for each pesticide is now presented alongside the existing information. The WHO system could be further developed over time in consultation with countries, international agencies and regional bodies. The GHS meets this requirement as a classification system with global acceptance following extensive international consultation.

There are reports of acute pesticide poisoning (APP) cases, which account for significant morbidity and mortality worldwide, especially in developing countries (Jeyaratnam, 1990; Kishi and Ladou, 2001). The severity and likelihood of effects from APP can vary according to specific agent, dose, underlying physiologic reserve, comorbidities, route of exposure, organ system, age, poverty, education and other factors (Thundiyil *et al.*, 2008). It is advisable to avoid pesticide products, classified by the WHO as either extremely dangerous (WHO Class Ia) or highly hazardous (WHO Class Ib). However, it must be kept in mind that even WHO Class II products, which are considered less hazardous, should only be applied by trained and supervised operators.

1.7 Conclusion

After the pesticide era of inorganic molecules, during World War I, the organic molecules, based on natural products, such as nicotine (extracted from tobacco) and rotenone (extracted from derris roots), were introduced. These were known as first-generation pesticides. Second-generation pesticides included organochlorines. However, the indiscriminate use of these highly persistent organochlorines could generate a lot of problems as highlighted by Rachel Carson in 1962 in her book *Silent Spring*. The problems paved the way for safer and more environmentally friendly organophosphates including parathion, followed by malathion and azinphosmethyl. The herbicides glyphosate, sulfonylurea and imidazolinone came onto the market during 1970–1980. The introduction of pesticides, such as avermectins, benzoylureas and Bt, led to the era of third-generation pesticides. New families of pesticides with user-friendly and environmentally safer formulations, and lower dosages in grams instead of kilograms per hectare, were introduced; for example, strobilurins and azolone as fungicides; fiproles and spinosyn as insecticides, leading to the combatting of the problem of resistance management. The IPM approach has led to a reduction in the load of synthetic pesticides because of the other components for pest management involved. Resistance variety, genetically engineered crops, cultural practices and biological control are a few of the components of the IPM strategy.

The insecticides can be classified on the basis of their target organisms, chemical structure, route of entry and mode of site of action, while fungicides are classified according to their mobility in the plant, role in protection of plants, breadth of activity, mode of action, chemical group, and FRAC group. The herbicides can be grouped by their activity, use, mechanism of action, chemical nature, type of vegetation controlled and HRAC group. A risk to human health is always present when handling pesticides and their formulated products. Pesticide exposure can occur via ingestion, inhalation, dermal absorption or ocular contact. WHO has suggested a tentative classification of pesticides to distinguish between the more and less hazardous forms of each pesticide, based on their hazard level.

References

Carson, R. (2002 [1962]) *Silent Spring*, 40th Anniversary Edition. Houghton Mifflin Harcourt, Cambridge, Massachusetts.

CDC and P (2015) Pesticide illness and injury surveillance. The National Institute for Occupational Safety and Health, Centers for Disease Control and Prevention. Available at: https://www.cdc.gov/niosh/topics/pesticides/default.html (accessed 31 July 2018).

Combarnous, Y. (2017) Endocrine disruptor compounds (EDCs) and agriculture: The case of pesticides. *Comptes Rendus Biologies* 340 (9–10), 406–409.

FAO (2013) International Code of Conduct on the Distribution and Use of Pesticides (PDF). Food and Agriculture Organization of the United Nations, 2002. Archived from the original (PDF) on 4 April 2013.

Fourouzan, M. and Farrokh-Eslamlu Ali, M. (2017) Efficiency of some biorational insecticides on leafminer, *Liriomyza Sativae* Blanchard (Diptera: Agromyzidae), research and reviews. *Journal of Botanical Sciences* 6(3), 91–98.

Gaalaas Mullaney, E. (2018) *Bacillus thuringenesis* bacterium. Encyclopedia Britannica. Available at: https://www.britannica.com/science/Bacillus-thuringiensis (accessed 27 September 2018).

Goldman, L.R. (2007) Managing pesticide chronic health risks: US policies. *Journal of Agromedicine* 12(1), 67–75.

Hummel, H.E. (1983) Insecticides and their design. *Journal of Nematology* 15(4), 616–638.

IPCS (2009) International Programme on Chemical Society. The WHO recommended classification of pesticides by hazard, and guidelines to classification. WHO, Geneva, Switzerland.

Jeyaratnam, J. (1990) Acute pesticide poisoning: a major global health problem. *World Health Statistics Quarterly* 43, 139–144.

Kishi, M. and Ladou, J. (2001) International pesticide use. *International Journal of Occupational and Environmental Health* 7, 259–265.

Laborde, A. (2008) Children's Health and the Environment. WHO Training Package for the Health Sector, World Health Organization. Available at: www.who.int/ceh/capacity/Pesticides.pdf (accessed 27 September 2018)

López-Muñoz, F., García-García, P. and Alamo, C. (2009) The pharmaceutical industry and the German National Socialist Regime: I.G. Farben and pharmacological research. *Journal of Clinical Pharmacy and Therapeutics* 34(1), 67–77.

Metcalf, R.L. and Metcalf, R.A. (1994) Attractants, repellents, and genetic control in pest management. In: R.L. Metcalf and W.H. Luckmann (Eds) *Introduction to Insect Pest Management*, 3rd edition. John Wiley and Sons, New York.

Mnif, W., Ibn Hadj Hassine, A., Bouaziz, A., Bartegi, A., Thomas, O. and Roig, B. (2011) Effect of endocrine disruptor pesticides: A review. *International Journal of Environmental Research and Public Health* 8, 2265–2303. Available at: https://www.ncbi.nlm.nih.gov/pmc/articles/PMC3138025/pdf/ijerph-08-02265.pdf (accessed 27 September 2018).

Thundiyil, J.G., Stober, J., Besbelli, N. and Pronczuk, J. (2008) Acute pesticide poisoning: a proposed classification tool. *Bulletin of the World Health Organization* 86, 161–240.

Unsworth, J. (2008) History of pesticide use. Available at: http://agrochemicals.iupac.org/index.php?option=com_sobi2&sobi2Task=-sobi2Details&catid=3&sobi2Id=31 (accessed 31 July 2018).

Wessel's Living History Farm (1930) *Farming in the 1930s*. Wessel's Living History Farm, York, Nebraska.

Whitehead, R. (1995) *The UK Pesticide Guide*, CAB International and BCPC, Wallingford, UK.

WHO Chronicle (1975) *The WHO Recommended Classification of Pesticides by Hazard, and Guidelines to Classification, 2004*. IPCS, World Health Organization, Geneva, Switzerland.

Wikipedia (2014) Pesticide. Available at: https://en.wikipedia.org/wiki/Pesticide#cite_note-EPAPesticides-23 (accessed 31 July 2018).

2 Impact of Pesticides on the Environment

The use of pesticides in agriculture has become indispensable for increasing crop yield and quality of produce, controlling the pest population and improving the prospects for long-term sustainable food production. But indiscriminate use of pesticides has led to environmental contamination and various health-related issues. No segment of the population or compartment of the environment is completely protected against exposure to pesticides and their potentially serious health effects. A disproportionate burden is shouldered by the people of developing countries with high-risk groups (WHO, 1990). There has been an escalation in worldwide deaths and chronic illnesses due to pesticide poisoning. More recent World Health Organization (WHO) estimates showed that over 500,000 people died from self-harm using pesticides in South-east Asia and the Western Pacific during 2000 alone (WHO, 2001) (Fig. 2.1). Ideally, a pesticide should be lethal to the targeted pests only, but unfortunately many of them have lethal effects on non-target species too, including humans. Controversy over the use and abuse of pesticides has surfaced. The pandemic use of these chemicals, under the adage 'if a little is good, a lot more will be better' has played havoc with human and other life forms (Eleršek and Filipič, 2011). A variety of harmful effects (Table 2.1) are experienced by human beings, the aquatic ecosystem, animals, birds, wildlife, food, the environment and plants (cytogenetic hazards) because of indiscriminate application.

There has been overwhelming evidence for some of these pesticides posing potential risk to humans and other life forms as well as having unwanted side effects for the environment (Forget, 1993). The boon of pesticides is enhanced economic returns with increases in production of food, feed and fibre, and management of vector-borne diseases, while their bane is serious health implications for humans and adverse effects on the environment.

2.1 Direct Impact on Human Beings

Pesticide exposure is causing serious health issues in human beings. It is responsible for a range of neurological and possibly incurable health effects such as memory loss, loss of coordination, reduced speed of response to stimuli, reduced visual ability, etc. Young people are at higher risk from such exposure. At a developing stage, exposure to a toxin can permanently alter the way an individual's biological system operates. Toxic pesticide residues present in food can interact with essential minerals, thereby

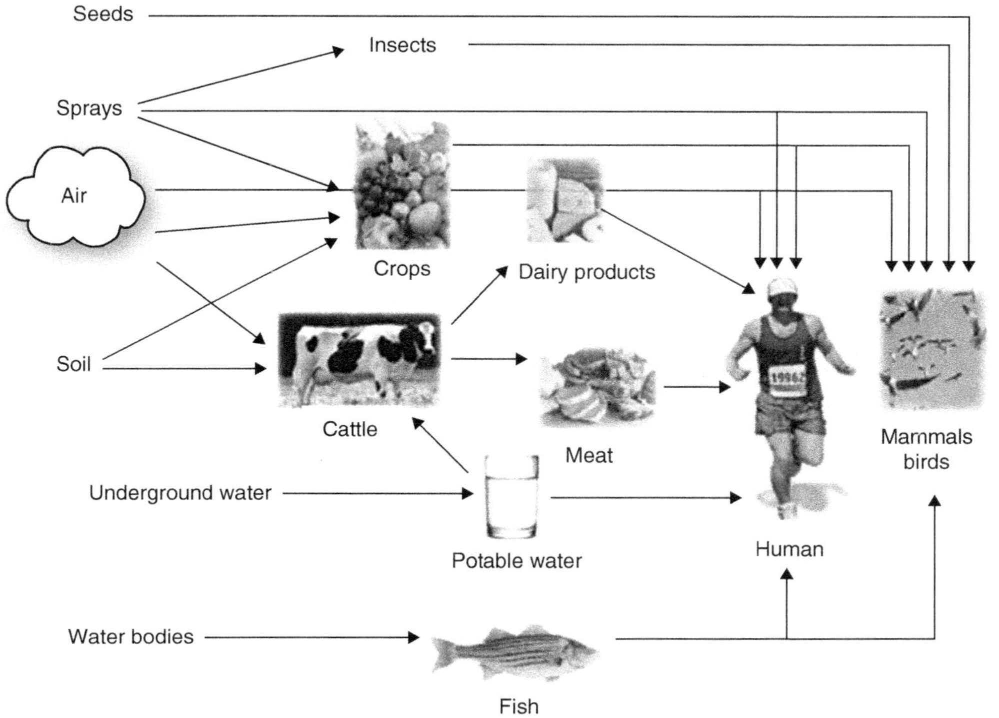

Fig. 2.1. Routes of exposure to OPs. (Reproduced from Eleršek and Filipič, 2011.)

Table 2.1. Common side effects of pesticides. (From Arora *et al.*, 2011.)

Environment component	Side effects
Abiotic	Residues in soil, water and air
Plants	Presence of residues; phytotoxicity; vegetation changes; (due to use of herbicides)
Animals, birds, insects, etc.	Residues in domestic animals and wildlife, physiological effects, mortality in certain wildlife species, mortality of beneficial predatory and parasitic insects, insect population changes (outbreak of secondary pests), genetic disorders
Humans	Biochemical changes, residues in tissues and organs, effect of occupational exposure, mortality and deformations
Food	Presence of residues
Aquatic	Gradual mortality of aquatic organisms
Ecosystem	Profound effect on water quality and aquatic ecosystems, contamination of water bodies and soil

affecting their bioavailability for the human body (Kaushik *et al.*, 2010). Chlorpyrifos is one of the world's most widely used organophosphate pesticides in agriculture, especially in developing countries like India, and has significantly reduced the bioavailability of copper (26%) and manganese (54%). It has been concluded that chlorpyrifos seems to exert its toxic effects by impairing essential micronutrient bioavailability. Use of this chemical poses a potential risk to humans and other life forms with unwanted side effects for the environment (Jeyaratnam, 1985; Igbedioh, 1991; Forget, 1993).

In developing countries, the high-risk groups exposed to pesticides include production workers, formulators, sprayers, mixers, loaders and agricultural farm workers. During manufacture and formulation, the possibility of hazards may be higher because the processes involved are not risk-free. Farm or industry workers may be at higher risk because of poor application technology and lack of awareness about use and impact of those chemicals on their health and environment. Worldwide deaths and incidence of chronic diseases due to pesticide poisoning number about 1 million per year (Environews Forum, 1999). A number of pesticide accidents have been observed in India, due to various reasons (Table 2.3).

A random sample of 213 farm workers in 24 communities and labour camps were examined by Coronado *et al.* (2003), in eastern Washington State, for association between occupational task and organophosphate pesticide residues in dust and organophosphate metabolite concentrations in urine samples of adult farm workers and their children. The farm tasks included harvesting or picking (79.2%), thinning (64.2%), loading plants or produce (42.2%), planting or transplanting (37.6%) and pruning (37.2%). Mixing, loading or applying pesticide formulations was reported to be part of their work by 20% of the collected sample. Thinning was associated with higher urinary pesticide metabolite concentrations in children (91.9% detectable, involved in thinning vs 81.3%, not involved in thinning; $P = 0.02$) but not in adults. Agricultural workers who are occupationally exposed to organophosphate pesticides have been observed to show adverse respiratory health and haematological alterations. Exposure alters lung function and leads to haematological abnormalities (Fareed *et al.*, 2013). Due to a lack of awareness of the hazardous health effects associated with application of pesticides in north India, the agricultural workers followed incorrect working practices without adopting any safety or precautionary measures for handling and spraying pesticides.

Mendola *et al.* (2004) conducted a series of pilot studies in an attempt to identify unrecognized pesticide-related illness. A higher level of urinary organophosphate metabolites was observed among pre-school children, aged 2–4 years, with flu-like symptoms, compared with levels for 6–11-year-old children in a US national sample from the National Health and Nutrition Examination Survey (NHANES). Although no clear high-risk profile for exposure or pesticide-related illnesses were observed among the children in the studies, pre-school children may have higher rates of exposure than children aged 6 and above in the United States, according to the findings. Pre-school children (129) of ages 20–66 months, from homes and day care centres in North Carolina, were observed to be exposed to 16 pesticides (including organochlorines, organophosphates, pyrethroids and acid herbicides) daily, through dietary and non-dietary ingestion (Morgan *et al.*, 2014). The inhalation routes were highest for 2,4-dichlorophenoxyacetic acid (2,4-D) and permethrin.

Six million children are at high risk of exposure to pesticides, used extensively in urban schools, homes and day care centres for control of roaches, rats and other vermin in America. The assays currently used for assessment of the toxicity of pesticides are insensitive and cannot accurately predict effects on children exposed *in utero* or in early postnatal life (Landrigan *et al.*, 1999). Prenatal exposure to chlorpyrifos has been observed to be associated with neurodevelopmental problems at 3 years of age (Rauh *et al.*, 2011). For each standard deviation increase in chlorpyrifos exposure (4.61 pg/g), full-scale IQ declined by 1.4% and working memory declined by 2.8%. Evidence of deficits in working memory index and full-scale IQ as a function of prenatal chlorpyrifos exposure at 7 years of age has been reported. Even low-level gestational exposure to organophosphate pesticides could impact neurobehaviour in early infancy. Yolton *et al.* (2013) reported that women who were white, married, had advanced education and reported more frequent consumption of fresh fruits and vegetables had higher concentrations of organophosphate metabolites than usual during pregnancy.

The levels of organophosphate metabolite among children living in a rice farming community in Thailand were observed to be strongly influenced by farming activity, household environments and child behaviours, suggesting that these are the primary pathways via which children are exposed to organophosphates (Rohitrattana *et al.*, 2014). A physically based probabilistic model (SHEDS-Pesticides, version 3) has been applied to estimate aggregate chlorpyrifos exposure and dosage to children in two age groups (0–4, 5–9 years) and over three time periods (<1 day, 1–7 days, 8–30 days), to address the aggregate exposure assessment needs of the Food Quality Protection Act (Zartarian *et al.*, 2000). Dermal contact with and non-dietary ingestion of residues from a recent pesticide application can be significant contributors to exposure and dose levels.

Organophosphate pesticides are commonly used in the US, leaving farm workers at risk of chronic exposure. Coronado *et al.* (2006) used a sample of 218 farmworkers in 24 communities and labour camps in eastern Washington State to examine the association between agricultural crop and organophosphate pesticide metabolite concentrations, in urine samples of adult farm workers and their children, and also organophosphate pesticide residues in house and vehicle dust samples. The crops were grouped into pome fruits (apples and pears) and non-pome fruit categories. Farm workers working on pome fruits had significantly higher concentrations of dimethyl pesticide metabolites in their urine and elevated azinphos-methyl concentrations in their homes and vehicles compared with workers who did not handle these crops. The children living with pome fruit workers were found to have higher concentrations of urinary dimethyl metabolites than did children of non-pome fruit workers. Significant correlations of adult urinary concentrations with both the vehicle and house dust azinphos-methyl concentrations were observed. Child urinary concentrations were correlated significantly with adult urinary concentrations and with the house dust azinphos-methyl concentration.

A study on workers ($n = 356$) in four units manufacturing hexachlorocyclohexane (HCH) in India revealed neurological symptoms (21%) that were related to the intensity of exposure (Nigam *et al.*, 1993). The magnitude of the toxicity risk involved in the spraying of methomyl, a carbamate insecticide, in field conditions was assessed by the National Institute of Occupational Health (NIOH) (Saiyed *et al.*, 1992). Significant changes were noticed in the electrocardiogram (ECG), the serum lactate dehydrogenase levels and cholinesterase (ChE) activities in the spraymen, indicating cardiotoxic effects of methomyl. Observations confined to health surveillance in male formulators engaged in the production of dust and liquid formulations of various pesticides (malathion, methyl parathion, dichlorodiphenyltrichloroethane (DDT) and lindane) in industrial settings revealed a high occurrence of generalized symptoms (headache, nausea, vomiting, fatigue, irritation of skin and eyes) as well as psychological, neurological, cardiorespiratory and gastrointestinal symptoms coupled with low plasma ChE activity (Gupta *et al.*, 1984).

The central nervous system (CNS) is adversely affected by acute organophosphate pesticide exposure, but little is known about the neurotoxicity of chronic exposures that do not result in acute poisoning. To examine this association, neurobehavioural (NB) tests were conducted by Starks *et al.* (2012) between 2006 and 2008 on licensed pesticide applicators enrolled in the Agricultural Health Study (AHS) in Iowa and North Carolina. A total of 701 male participants completed nine NB tests assessing their memory, motor speed and coordination, sustained attention, verbal learning, and visual scanning and processing. During the NB visit and before testing between 1993 and 2007, AHS interviews were also conducted to get data on ever-use and lifetime days of use of 16 organophosphate pesticides. The mean age of participants was 61 years (SD = 12). NB test performance was associated with lifetime days of use of some pesticides. All pesticides performed in different ways. For example, ethoprop was significantly associated with reduced performance on a test of motor speed and visual scanning. Malathion was significantly associated with poor performance on a test of visual scanning and processing. Conversely, chlorpyrifos, coumaphos, parathion, phorate and tetrachlorvinphos were associated with better verbal learning and memory; coumaphos was associated with better performance on a test of motor speed and visual scanning; and parathion was associated with better performance on a test of sustained attention. The results do not provide strong evidence for association of long-term organophosphate pesticide use with adverse CNS-associated NB test results.

Data on ever-use and lifetime days of use of 16 OP pesticides were obtained from AHS interviews conducted before testing between 1993 and 2007 and during the NB visit.

The organochlorines, now banned, are highly persistent, some of them with a half-life of more than 100 years. Most of them are fat soluble and deposit in the fat tissues of human beings, remaining there until their death. In fact residues of the parent compound and their metabolites are still reported in blood samples as well as the breast milk of mothers (Skaare *et al.*, 1988; Martínez *et al.*, 1993; Bergkvist *et al.*, 2012; Dewan *et al.*, 2013; Mannetje *et al.*, 2014). The US National Academy of Sciences stated that dichlorodiphenyldichloroethylene (DDE), a metabolite of DDT, causes eggshell thinning and that the bald eagle population in the United States declined primarily because of exposure to DDT and its metabolites (Liroff, 2000).

Organochlorine compounds could pollute the tissues of virtually every life form on Earth, the air, the lakes and the oceans, the fish that live in them and the birds that feed on the fish (Hurley *et al.*, 1998).

Wang *et al.* (2012) investigated levels of organophosphate exposure in pregnant women and the association between organophosphate exposure and perinatal outcomes in Shanghai, China, by enrolling 187 healthy pregnant women between September 2006 and January 2007. The mothers, after delivery, were assessed through a questionaire and analyses of maternal urinary non-specific metabolites of organophosphate pesticides. The results indicated that maternal urine pesticide levels in Shanghai were much higher than those reported in developed countries. An adverse association of prenatal dialkylphosphates (DAPs) with mental development and pervasive developmental problems at 24 months of age has been reported (Eskenazi *et al.*, 2007). The organophosphates are neurotoxic at high doses, but chronic exposure at lower levels could adversely affect children's cognitive development. Children in the highest quintile of maternal DAP concentrations had an average deficit of 7.0 IQ points compared with those in the lowest quintile (Bouchard *et al.*, 2011).

The levels of exposure to organophosphate pesticides were high in an Israeli study population compared to the general population in the US and Canada the intake of fruits is an important source of exposure (Berman *et al.*, 2013). The total molar dialkyl phosphate concentration increased with age and was higher in individuals with high income compared to individuals with the lowest income. Total diethyl metabolite concentrations were observed to be higher in females and in study participants whose fruit consumption was above the 75th percentile. The concentrations of organophosphate pesticide metabolites in adult urine could be due to demographic and dietary habits. Organophosphate pesticide metabolites were detected in a substantial proportion of urine samples of children. In particular, metabolites of parathion/methyl parathion, chlorpyrifos/chlorpyrifos-methyl and diazinon were detected by Arcury *et al.* (2007). The number of metabolites detected ranged from 0 to 7, with a mode of 4 detected (28.3%). Boys, children living in rented housing and children with mothers working part-time had more metabolites detected.

The chronic exposure of parents to organophosphate and organochlorine pesticides may be a potential risk factor for hypospadias, a congenital defect, in offspring (Michalakis *et al.*, 2014). Exposure to chlorpyrifos, an organophosphate, may lead to pseudocholinesterase deficiency, resulting in abdominal cramps, diarrhoea, sweating, excessive salivation, frequent urination and increased bronchial secretions (Lurati,

2013). Treatment with atropine, an antidote, could be the cure for this. Children of mothers residing near pyrethroid insecticide applications just before conception or during the third trimester may be at greater risk for both autism spectrum disorders (ASD) and developmental delay (DD), with odds ratios (OR) ranging from 1.7 to 2.3 (Shelton *et al.*, 2014). This study of ASD strengthens the evidence linking neurodevelopmental disorders with gestational pesticide exposure, particularly organophosphates. This provides novel results of ASD and DD associations with pyrethroids and carbamates, respectively.

Lacasaña *et al.* (2010b) demonstrated the ability of organophosphate pesticides to disturb thyroid gland function by experimental studies on animals. The study aimed to assess the association between exposure to organophosphate compounds and serum levels of thyroid hormones in floriculture workers. In this longitudinal study, 136 male subjects were selected from the State of Mexico and Morelos, Mexico, all of whom had been occupationally exposed to organophosphate pesticides, during application of both high (rainy season) and low (dry season) levels of agricultural pesticides. A questionnaire was also used to survey the general history profile of the workers. Urine and blood samples were taken the day after pesticide application. The results indicated an increase in both thyroid-stimulating hormone (TSH) and T4 hormones in serum associated with an increase in total dimethylphosphate levels (ΣDMP) in urine ($P < 0.001$) and a decrease in total T3 serum levels, with an increase of ΣDMP levels in the urine ($P = 0.053$). Exposure to organophosphate pesticides may be responsible for an increase in TSH and T4 serum hormone levels and a decrease in T3 serum hormone levels, therefore supporting the hypothesis that organophosphate pesticides act as endocrine disruptors in humans. The results are also supported by studies on floriculture workers by Lacasaña *et al.* (2010a). The paraoxonase-1 enzyme (PON1) plays an important role in the toxicity of some organophosphate pesticides; low PON1 activity being associated with higher pesticide sensitivity. When PON1 activity was increased, a decrease in the percentage of variation of TSH level was observed for each increment in one logarithmic unit of the ΣDAP levels. A stronger association has been observed between organophosphate pesticides and thyroid function in individuals with lower PON1 activity (Lacasaña *et al.*, 2010a).

Paraoxonase, a xenobiotic enzyme, plays a role in metabolizing some organophosphates. Carriers of the variant MM *PON1*-55 genotype exposed to organophosphates exhibited a greater than 2-fold increase in Parkinson's disease risk compared with persons who had the wildtype or heterozygous genotype and no exposure. The effect estimate for chlorpyrifos was more pronounced in younger-onset cases and controls. The risk is increased among PON1-55 variant carriers for specific organophosphates metabolized by PON1, which underscores the importance of considering susceptibility factors when studying environmental exposures in Parkinson's disease (Manthripragada *et al.*, 2010).

Certain environmental chemicals, including pesticides termed as endocrine disruptors, are known to elicit their adverse effects by mimicking or antagonizing natural hormones in the body, and it has been postulated that their long-term, low-dose exposure is increasingly linked to human health effects such as immune suppression, hormone disruption, diminished intelligence, reproductive abnormalities and cancer (Crisp *et al.*, 1998; Hurley *et al.*, 1998; Brouwer *et al.*, 1999)

Surajudeen *et al.* (2014) studied the role of oxidative stress in the pathogenesis of organophosphate pesticide toxicity in Nigeria. A group of 60 (30 pesticide applicators and 30 farmers) along with 30 healthy non-farmers were selected for this study. The team of researchers determined serum levels of malondialdehyde (MDA), catalase (CAT), glutathione peroxidase (GPx), reduced glutathione (GSH), myeloperoxidase (MPO), nitric oxide (NO) and serum activity of acetylcholinesterase (AChE) in farm workers exposed to organophosphate pesticides. The serum activities of AChE and

CAT were observed to be significantly lower, whereas MPO activity was significantly higher in pesticide applicators compared with controls. Similarly, farmers had significantly reduced serum AChE activity and significantly raised MPO activity compared with controls. However, serum activities of AChE, CAT and MPO were significantly lower, whereas the mean level of MDA was significantly higher in pesticide applicators compared with farmers. The study showed organophosphate pesticide applicators were more exposed to oxidative stress than farmers, thus indicating organophosphate pesticide applicators require more antioxidant supplements than farmers. Despite existing regulations, agricultural pesticide handlers continue to be exposed to organophosphate/carbamate (CB) insecticides at levels resulting in butyrylcholinesterase (BuChE) inhibition (Hofmann *et al.*, 2010). Reproductive toxicity was observed in males associated with the spraying of pesticides (organochlorines, organophosphates and carbamates) in cotton fields (Rupa *et al.*, 1991). In another study, the effects of spraying HCH in field conditions were evaluated by initiating short-term exposure to HCH in malaria spraymen (Gupta *et al.*, 1982).

A study on those affected by the Seveso disaster of 1976 in Italy during the production of 2,4,5-trichlorophenoxyacetic acid (2,4,5-T), a herbicide, concluded that chloracne (nearly 200 cases with a definite exposure dependence) was the only effect established with certainty as a result of dioxin formation (Pier *et al.*, 1998). Early health investigations including liver function, immune function, neurologic impairment and reproductive effects yielded inconclusive results. An excess mortality from cardiovascular and respiratory diseases was uncovered, possibly related to the psychosocial consequences of the accident in addition to the chemical contamination. An excess of diabetes cases was also found. Results of cancer incidence and mortality follow-up showed an increased occurrence of cancer of the gastrointestinal sites and of the lymphatic and haematopoietic tissue. Results cannot be viewed as conclusive, however, because of various limitations: limited individual exposure data, short latency period and small population size for certain cancer types. A similar study in 2001 observed no increase in all-cause and all-cancer mortality. However, the results support the notion that dioxin is carcinogenic to humans and corroborate the hypotheses of its association with cardiovascular- and endocrine-related effects (Pier *et al.*, 2001).

During the Vietnam War, US military forces sprayed nearly 19 million gallons (~72 million l) of herbicide on approximately 3.6 million acres (1.45 million ha) of Vietnamese and Laotian land to remove forest cover, destroy crops and clear vegetation from the perimeters of US bases. This effort, known as Operation Ranch Hand, lasted from 1962 to 1971. Various herbicide formulations were used, but most were mixtures of the phenoxy herbicides 2,4-D and 2,4,5-T. Approximately 3 million Americans served in the armed forces in Vietnam during the Vietnam War. Some of them (as well as some Vietnamese combatants and civilians, and members of the armed forces of other nations) were exposed to defoliant mixtures, including Agent Orange. There was evidence of a cancer risk to Vietnam veterans, workers occupationally exposed to herbicides or dioxins (since dioxins contaminated the herbicide mixtures used in Vietnam) and to the Vietnamese population (Frumkin, 2003).

2.2 Impact through Food Commodities

A programme called 'Monitoring of Pesticide Residues in Products of Plant Origin in the European Union' was initiated in 1996 to determine the extent of pesticide contamination in food stuffs. Initially, during 1996, seven pesticides (acephate, chlopyriphos, chlopyriphos-methyl, methamidophos, iprodione, procymidone and chlorothalonil) and two other groups of pesticides (benomyl group and maneb group, i.e. dithiocarbamates) were analysed in apples, tomatoes, lettuce, strawberries and grapes, constituting an average of about 9700 samples for each pesticide or pesticide group. A total of

5.2% of the samples were found to contain residues and 0.31% samples had residues higher than the respective maximum residue limit (MRL). Pesticide residues in lettuce crop exceeded MRL more frequently than any other crops investigated (Aktar *et al.*, 2009). The following year, 1997, some 6000 samples were analysed and 13 pesticides (acephate, carbendazin, chlorothalonil, chlopyriphos, DDT, diazinon, endosulfan, methamidophos, iprodione, metalaxyl, methidathion, thiabendazole, triazophos) were assessed in five commodities (mandarins, pears, bananas, beans and potatoes).

Hospital surveys have recorded most pesticide deaths in developing countries as a result of self-poisoning instead of food poisoning (Eddleston, 2000). The Global Burden of Disease Study estimated that 798,000 people died from deliberate self-harm in 1990, over 75% of whom were from developing countries (Murray and Lopez, 1996). More recent WHO estimates showed that over 500,000 people died from self-harm in Southeast Asia and the Western Pacific during 2000 alone (WHO, 2001). Suicide is the commonest cause of death in young Chinese women and Sri Lankan men and women (Sri Lankan Ministry of Health, 1995; Murray and Lopez, 1996; WHO, 2001).

The epidemiological studies demonstrate that pesticide residues in agricultural food increase the risk of cancer in the professional populations. The pesticide residues may also be at the origin of dysfunction of the reproductive systems, endocrinal and immune systems, and can also lead to ague attack, malformation of the vertebral column and deterioration of the lungs (Benzidane and Dahamna, 2013). During a study in Algeria, 61% of the analysed samples of apples, tomatoes, green and red pepper, and cucumber were observed to have stocks of residues of chlorpyrifos pesticide, broadly greater than the accepted maximum limit. The presence of chlorpyrifos and phosmet residues in fruits and vegetables was the major cause predicting urinary DAP metabolite concentrations in Chilean school children aged 6–12 years, as per analysis of samples collected from homes and schools (Muñoz-Quezada *et al.*, 2012). Increased levels of insecticides in the blood of vertebrates cause reproductive dysfunction. Food containing flesh, such as fish, chicken and goat, which are contaminated with pesticide residues beyond the permissible limit must be avoided by human beings (Singh *et al.*, 2008). Residues of organochlorine and organophosphate insecticides were investigated in the blood samples of fish, chicken, goat and man. The samples showed high levels of HCH and DDT and metabolites in chicken, goat and man. The order of bioaccumulation was goat > chicken > man > fish. DDT residues were very high in different tissues of fish (blood, liver, brain and ovary) compared with HCH residues.

The pesticide residues of organophosphates and synthetic pyrethroids were monitored in 810 samples of five different vegetables from the central, Volta, Greater Accra, Ashanti and eastern regions of Ghana, and a neighbouring country (Togo), collected from Agbogbloshie market in Accra, Ghana, during January 2009 through to December 2011 (Osei-Fosu *et al.*, 2014). Although 18% of the samples had no detectable residues and 62% were below the MRLs, 20% of samples exceeded the EU adopted MRL values. The lettuce contained the highest number of residues above their MRLs as well as cucumber and cabbage. Arora (2009) investigated residues of pesticides applied in non-integrated pest management (IPM) fields of okra and aubergine to compare with pesticides used under an IPM strategy. The residues of chlorpyrifos in soil were found to be 4.219 and 1.135 μg/g at harvest time in non-IPM and IPM fields of summer okra crops, from an initial value of 0.407 μg/g before sowing; while in the aubergine crop, it was not detected in soils of any trials. During the first year of study, the residues of chlorpyrifos and cypermethrin in okra were observed to be 5.75 and 0.625 μg/g, respectively, for non-IPM fields, and 0.104 μg/g of chlorpyrifos for IPM trials. The pesticide residues were found to be 0.77, 1.39, 0.4 and 0.32 μg/g for cypermethrin, chlorpyrifos, monocrotophos and dimethoate, respectively, for non-IPM okra in the second year. For aubergine, residues of cypermethrin and imidacloprid were not detected in IPM trials, while the levels were found to be 0.28 and 0.78 μg/g for cypermethrin and chlorpyrifos, respectively, in non-IPM trials.

In India the first report of poisoning due to pesticides was from Kerala in 1958, where over 100 people died after consuming wheat flour contaminated with parathion (Karunakaran, 1958). This prompted the Special Committee on Harmful Effects of Pesticides constituted by the Indian Council of Agricultural Research (ICAR) to focus attention on the problem (Report of the Special Committee of ICAR, 1972). In a multi-centric study to assess the pesticide residues in selected food commodities collected from different states of the country (Surveillance of Food Contaminants in India, 1993), DDT residues were found in about 82% of the 2205 samples of bovine milk collected from 12 states. About 37% of the samples contained DDT residues above the tolerance limit of 0.05 mg/kg (whole milk basis). The highest level of DDT residues found was 2.2 mg/kg. The proportion of the samples with residues above the tolerance limit was highest in Maharastra (74%), followed by Gujarat (70%), Andhra Pradesh (57%), Himachal Pradesh (56%) and Punjab (51%). In the remaining states, this proportion was less than 10%. Data on 186 samples of 20 commercial brands of infant formula showed the presence of residues of DDT and HCH isomers in about 70 and 94% of the samples, with their maximum levels of 4.3 and 5.7 mg/kg (fat basis), respectively. Measurement of chemicals in the total diet provides the best estimates of human exposure and of the potential risk. The risk of consumers may then be evaluated by comparison with toxicologically acceptable intake levels. The average total DDT and benzenehexachloride (BHC) consumed by an adult were found to be 19.24 mg/day and 77.15 mg/day, respectively, by Kashyap *et al.* (1994). Fatty food was the main source of these contaminants. In another study during 1992, the average daily intake of HCH and DDT by Indians was reported to be 115 and 48 mg/person, respectively, which was higher than levels observed in most developed countries (Kannan *et al.*, 1992).

2.3 Impact on Environment

The environmental impact of pesticides means the impact of pesticides on non-target organisms. Approximately 98% of sprayed insecticides and 95% of herbicides reach a destination other than their target species, because they are formulated to be sprayed or spread across entire agricultural fields (Tyler Miller, 2004). Pesticides can be carried into aquatic environments through runoff, and adjoining fields, grazing areas, human settlements and undeveloped areas through wind, potentially affecting other species. Other problems related to health issues emerge from poor production, transport and storage practices (Tashkent, 1998). Pest resistance and resurgence are other emerging issues resulting from overuse and repeated applications of pesticides.

In addition to killing insects or weeds, pesticides can be toxic to a host of other organisms including birds, fish, beneficial insects and non-target plants. Insecticides are generally the most acutely toxic class of pesticides, but herbicides can also pose risks to non-target organisms. Each pesticide is evaluated for a specific set of environmental concerns before being registered for use. The undesirable effects have led many pesticides to be banned, while regulations have limited and/or reduced the use of others. Over time, pesticides have generally become less persistent and more species-specific, reducing their environmental footprint. In addition, the introduction of low-dosage pesticides has reduced the amounts applied per hectare, in some cases by 99%. However, the global spread of pesticide use, including the use of older/obsolete pesticides that have been banned in some jurisdictions, has increased overall (Lamberth *et al.*, 2013).

2.4 Surface and Groundwater Contamination

Surface and groundwater pollution due to pesticides is a worldwide problem. In a study by the US Geological Survey (USGS) in the United States, pesticides were observed to pollute every stream and over 90% of wells sampled (Gilliom *et al.*, 2007). Approximately 143 pesticides and 21 metabolites have been detected in groundwater, including pesticides from every major chemical class, in more than 43 states, as per USGS studies (Singare and Dhabarde, 2014). Pesticide residues have also been found in rain and groundwater (Kellogg *et al.*, 2000). Pesticide concentrations have exceeded the permissible limits for drinking water in some samples of river and groundwater according to studies by the UK government (Bingham, 2007). Pesticide impacts on aquatic systems are often examined using a hydrology transport model to study movement and fate of chemicals in rivers and streams (States of Jersey, 2007). The four major routes through which pesticides reach the water are: drift outside of the intended area of spray; percolation or leaching through the soil; carry over to the water as runoff or spillage accidentally or through neglect; and through eroding soil (Papendick *et al.*, 1986). Factors that affect a pesticide's ability to contaminate water include its water solubility, the distance from application site to a body of water, weather, soil type, presence of a growing crop and the method used to apply the chemical (Pedersen, 1997).

During one survey in India, 58% of drinking water samples drawn from various hand pumps and wells around Bhopal were found to be contaminated with organochlorine pesticides above the US Environmental Protection Agency (EPA) standards (Kole and Bagchi, 1995). Once groundwater is polluted with pesticides, it may take years for dissipation of the contaminants or for it to be cleaned up. Clean up may also be very costly and complex, if not impossible (Gulati *et al.*, 2015; US EPA., 2001).

The hydrology, sediment and pesticide transport components were evaluated using Soil and Water Assessment Tool (SWAT) on the northern San Joaquin Valley watershed of California. The calibrated SWAT model was applied to simulate fate and transport processes of two organophosphate pesticides, diazinon and chlorpyrifos, at watershed scale. The study indicated that curve number was the major factor influencing the hydrology, while pesticide fate and transport were mainly affected by surface runoff and pesticide application (Luo *et al.*, 2008). The deleterious effect of pesticides on aquatic life and contamination of surface waters is the growing concern of the day. The potential impact of 16 selected pesticides belonging to the classes of triazines, phenylureas, organophosphates, chloroacetanilides and thiocarbamates on the aquatic organisms in surface waters from the Llobregat River (north-east Spain) and some of its tributaries (Anoia and Rubí) have been assessed (Köck-Schulmeyer *et al.*, 2012). Contamination increased downstream of the river and was clearly influenced by rainfall and hence river flow. The existing pesticide contamination posed a low to high ecotoxicological risk for aquatic organisms, algae and macro-invertebrates were at higher risk than fish; and the organophosphates diazinon and malathion and the phenylurea diuron were the major contributors to the overall toxicity and therefore the most problematic compounds.

Trends in concentrations of organophosphate insecticides (chlorpyrifos and diazinon) and herbicides (atrazine, S-ethyl dipropylthiocarbamate (EPTC), metolachlor, simazine and trifluralin) were determined for 15 streams in California, Oregon and Washington. Consistent upward trends in simazine at urban-influenced sites from 2000 to 2005, and downward trends in atrazine and EPTC at agricultural sites from the mid-1990s to 2005 were observed (Johnson *et al.*, 2011). The study briefly presents two examples showing that timing of peak concentrations for individual compounds varied greatly across this geographic gradient because of different application periods and the effects of local rain patterns, irrigation and soil drainage; and reconstructions of continuous diazinon concentrations at sites in California are used to evaluate compliance with total maximum daily load targets. To test the toxicity of sediments in California's Central Valley, the samples were analysed for nine hydrophobic pesticides. Abamectin, fenpropathrin and methyl parathion were found at concentrations of potential concern, but only in one or two samples. In three-quarters of over 100 samples from agriculture-affected waterways, pyrethroids exceeded the concentrations expected to cause toxicity. The pyrethroid bifenthrin in particular, as well as lambda-cyhalothrin, cypermethrin, esfenvalerate, permethrin and the organophosphate chlorpyrifos, were primarily responsible for the observed toxicity (Weston *et al.*, 2013).

The herbicides 2,4-D, diuron and prometon, and the insecticides chlorpyrifos and diazinon, all commonly used by urban homeowners and school districts, were among the 21 pesticides detected most often in surface and groundwater across the US in a study by USGS (1998). Trifluralin and 2,4-D were found in water samples collected in 19 out of the 20 river basins studied (Bevans *et al.*, 1998; Fenelon, 1998; Levings *et al.*, 1998; Wall *et al.*, 1998). The USGS also found that concentrations of insecticides in urban streams commonly exceeded guidelines for protection of aquatic life (USGS, 1999). According to USGS, 'in general more pesticides were detected in urban streams than in agricultural streams' (Bortleson and Davis, 1987–1995). The herbicide 2,4-D was the most commonly found pesticide, detected in 12 out of 13 streams. The insecticide diazinon, and the weedkillers dichlobenil, diuron, triclopyr and glyphosate were detected also in Puget Sound basin streams. Both diazinon and diuron were found at levels exceeding concentrations recommended by the National Academy of Sciences for the protection of aquatic life (Bortleson and Davis, 1987–1995).

Atrazine was found in groundwater and surface water in maize-producing areas of the Transvaal (South Africa) in 1991–1992 (Bottoni and Funari, 1992). Leaching of atrazine in Germany (Grandet *et al.*, 1989) and Denmark (Felding, 1992) was also reported. Contamination of well water by atrazine, alachlor and carbofuran has been found in Switzerland (Buser, 1990). Atrazine, simazine and cyanazine were found in wellwater in the US (Poinke and Glotfelty, 1989). Pesticides have also been found in wells in The Netherlands, Italy, Israel, Japan, Canada and Australia. Herbicides and nematicides are supposed to be the major groundwater pollutant pesticides as they are applied directly to the soil. There have been many reports published on groundwater contamination due to pesticides (Poinke and Glotfelty, 1989; Kolpin, *et al.*, 1997a, 1997b; Gomez *et al.*, 1998) all over the world. It was found that the values of the pesticides were higher than the permissible health values as given by US EPA (1995).

2.5 Soil Health

The use of pesticides decreases the general biodiversity in the soil. The use of non-chemical or biopesticides results in higher soil quality (Johnston, 1986), causing an increase in organic matter, which leads to higher water retention. This helps increase crop yields by 20–40% for farms in drought years, compared with their conventional counterparts according to Lotter *et al.* (2003). Many of the chemicals used in pesticides are persistent soil contaminants, whose impact may endure for decades and adversely affect soil conservation (US EPA, 2007). The persistence of pesticides in soil is influenced mainly by degradation and sorption. Depending on the chemical nature of the pesticide, these factors control the transportation from soil to water, and in turn to air and our food. The microorganisms present in the soil may also be responsible for breaking down and degradation of pesticides. Sorption affects bioaccumulation of pesticides, which are dependent on organic matter in the soil. The larger the organic matter content, the greater the adsorption of pesticides and metabolites The capacity of the soil to hold positively charged ions in an exchangeable form is important with paraquat and other pesticides that are positively charged. Soil pH is also of some importance. Adsorption increases with decreasing soil pH for ionizable pesticides (e.g. 2,4-D, 2,4,5-T, picloram and atrazine) (Andreu and Pico', 2004). Weak organic acids have been shown to be weakly sorbed by soil, because of pH and their mostly acidic structure. Sorbed chemicals have been shown to be less accessible to microorganisms. The longer pesticides reside in soil, the more resitstant they become to degradation and extraction, as their biological activity is lost gradually (Arias-Estévez *et al.*, 2008).

A large number of transformation products (TPs) have been documented from a wide range of pesticides (Barceló and Hennion, 1997; Roberts *et al.*, 1998; Roberts *et al.*, 1999). The persistence and movement of these pesticides and their TPs are determined by water solubility, soil sorption constant (Koc), the octanol/water partition coefficient (Kow) and half-life in soil (DT_{50}). Pesticides and TPs could be grouped into:

1. Non-polar hydrophobic, persistent and bioaccumulable pesticides, generally strongly bound to soil. The organochlorines like DDT, endosulfan, endrin, heptachlor, lindane and their TPs are in this category. Almost all of these are banned in agriculture but their residues are still present.
2. Polar hydrophilic and non-persistant pesticides including mainly herbicides but also including carbamates, fungicides and some organophosphate insecticide TPs. These pesticides could be responsible for contamination of drinking water due to their mobility from soil by runoff and leaching. The pesticides and their TPs are retained by soils to different degrees, depending on the interactions between soil and pesticide properties.

Earthworms have been called 'ecosystem engineers'. They play an important role in breaking down dead organic matter leading to its decomposition, which releases nutrients locked up in dead plants and animals making them available for use by living plants. In the course of this process they also allow bacteria and fungi to feed on it and release the nutrients. They are also responsible for incorporating organic matter into the soil, improving its fertility. Insecticides and fungicides are the most toxic pesticides impacting survival and reproduction, respectively, for earthworms. The pesticides disrupt enzymatic activities, increase individual mortality, decrease fecundity and growth, change individual behaviour such as feeding rate, and decrease the overall community biomass and density (Pelosi *et al.*, 2014).

Wang *et al.* (2012) conducted a study to investigate the comparative toxicity of 45 pesticides, including insecticides, acaricides, fungicides and herbicides, towards the epigeic earthworm *Eisenia fetida*. Results from a 48-h filter paper contact test indicated that clothianidin, fenpyroximate and pyridaben were supertoxic to *E. fetida*, with LC_{50} values ranging from 0.28 (0.24–0.35) to 0.72 (0.60–0.94) μg/cm^2, followed by carbaryl, pyridaphenthion, azoxystrobin, cyproconazole and picoxystrobin, with LC_{50} values ranging from 2.72 (2.22–0.3.19) to 8.48 (7.38–10.21) μg/cm^2. A decrease was observed in the weight of earthworms at all concentrations of the pesticide dichlorvos, whereas reproduction and avoidance behaviour, which are sensitive parameters, were found to be significantly affected by all three doses of dichlorovos (Farrukh and Ali, 2011).

Overuse of pesticides affects soil organisms, thereby affecting soil health in a similar way to human overuse of antibiotics. 'Indiscriminate use of chemicals might work for a few years, but after a while, there aren't enough beneficial soil organisms to hold onto the nutrients' (Singh and Iyer, 2017).

Common herbicides like triclopyr disrupt the process of transformation of atmospheric nitrogen into nitrates by soil microorganisms (Pell *et al.*, 1998); glyphosate reduces the growth and activity of free-living nitrogen-fixing bacteria in soil (Santos and Flores, 1995) and 2,4-D reduces nitrogen fixation by the bacteria that live on the roots of bean plants (Arias and Fabra, 1993; Fabra *et al.*, 1997), reduces the growth and activity of nitrogen-fixing blue-green algae (Singh and Singh, 1989; Tözüm-Çalgan and Sivaci-Güner, 1993) and inhibits the transformation of ammonia into nitrates by soil bacteria (Frankenberger and Tabatabai, 1991; Martens and Bremner, 1993). The mycorrhizal fungi, responsible for nutrient uptake of many plants, can be damaged by herbicides present in the soil. It has been observed that oryzalin and trifluralin both inhibited the growth of certain species of mycorrhizal fungi (Kelley and South, 1978). Glyphosate has been shown to be toxic to mycorrhizal fungi in *in vitro* studies, and some damaging effects were seen at concentrations lower than those found in soil following typical applications (Chakravarty and Sidhu, 1987; Estok *et al.*, 1989). Triclopyr has also been observed to be toxic to several species of mycorrhizal fungi (Chakravarty and Sidhu, 1987) and oxadiazon has been shown to reduce the number of mycorrhizal fungal spores (Moorman, 1989).

2.6 Atmosphere/Air Contamination

Pesticide sprays can directly hit non-target vegetation or can contribute to air pollution through drift by means of their suspended particles in air and are carried by wind to other areas, potentially contaminating them (Cornell University, 2007; US EPA, 2007). Wind velocity, temperature and relative humidity have a greater impact on spread of the pesticide in the air. During pesticide application, their vapours may be blown to nearby areas, potentially posing a threat to wildlife (National Park Service, 2006). Low relative humidity and high temperature result in more evaporation of sprayed pesticides. The level of contamination of the outdoor environment is therefore often dependent on the season. Ground spraying produces less pesticide drift than aerial spraying does, for obvious reasons (Palmer *et al.*, 2007). Some pesticide drift occurs even from ground equipment application (Glotfelty and Schomburg, 1989). Drift can account for a loss of 2–25% of the chemical being applied, which can spread over a distance of a few yards to several hundred miles. As much as 80–90% of an applied pesticide can be volatilized within a few days of application (Majewski and Capel, 1995). Pesticides that are used for fumigating the soil can give off volatile organic compounds, which can form tropospheric ozone, a pollutant, by reacting with other chemicals. Pesticide use accounts for about 6% of total tropospheric ozone levels (UC IPM Online, 2007).

Reports are available on consistency studies into the evidence of pesticide residues in air. According to the USGS, pesticides have been detected in the atmosphere in all sampled areas of the USA (Singh and Iyer, 2017). Nearly every pesticide investigated has been detected in rain, air, fog or snow across the nation at different times of the year (US Geological Survey, 1999).

The vapours of ester formulation herbicides, volatilizing off treated plants, have been shown to cause severe damage to other plants (Straathoff, 1986). The pesticides may also have sublethal effects on plants, as well as affecting non-target plants. The phenoxy herbicide 2,4-D can injure nearby trees and shrubs through drift or volatilization in the air (Dreistadt *et al.*, 1994). Glyphosate can severely reduce seed quality (Locke *et al.*, 1995). It can also increase the susceptibility of certain plants to disease (Brammall and Higgins, 1988). The US Fish and Wildlife Service has recognized 74 endangered plants that may be threatened by glyphosate alone (Cox, 1995). Clopyralid can reduce yields in potato plants (Lucas and Lobb, 1987). Volatilization of just 1% of applied clopyralid is enough to damage non-target plants.

Some of the pesticides, namely dacthal, chlorothalonil, chlorpyrifos, metolachlor, terbufos and trifluralin, have been detected in Arctic environmental samples (air, fog, water, snow) (Rice and Cherniak, 1997; Garbarino *et al.*, 2002). Some pesticides have been observed to have the ability to undergo short-range atmospheric transport (Muir *et al.*, 2004) to ecologically sensitive regions such as the Chesapeake Bay and the Sierra Nevada mountains (McConnell *et al.*, 1997; Le Noir *et al.*, 1999; Harman-Fetcho *et al.*, 2000; Thurman and Cromwell, 2000). A study also indicated presence of pesticides in the atmosphere of British Columbia dating from 1996. Belzer *et al.* (1998) investigated 57 chemicals at two sampling sites (Agassiz and Abbotsford) in the Fraser Valley, from February 1996 until March 1997. Chlorpyrifos and trace levels of endosulfan were the most frequently identified contaminants in seawater; chlorpyrifos and atrazine in marine ice; and chlorothalonil and trifluralin in surface microlayer samples. Chlorpyrifos, trifluralin, metolachlor, chlorothalonil, terbufos and endosulfan were detected in Arctic marine fog at concentrations several times higher than in adjacent waters or ice (Chernyak *et al.*, 1996).

Chlorpyrifos and eight pyrethroids were detected in the air samples of current-use pesticides (CUPs) from an urban community in Guangzhou, China (Li *et al.*, 2014a). Chlorpyrifos and cypermethrin were the most dominant CUPs detected in the atmosphere, accounting for 68% and 15% of the total CUPs, respectively. The peak concentrations were observed during summer. The chlorpyrifos was observed to have long-range transport. Evaluation of potential exposure from inhalation of atmospheric CUPs suggested that children, toddlers and infants had the highest exposure, but the risk quotients were low for all age groups when annual average concentrations were used as exposure metrics. Exposure risk was higher in summer due to higher atmospheric pesticide concentrations, longer exposure times and more pesticides being in the gaseous form. A study was also carried out to examine dynamic flux of organophosphate and pyrethroid pesticides for the air–(soil)–water–sediment transport system, where sediment was shown to act as a sink for pesticides (Li *et al.*, 2014b). A dynamic flux-based risk assessment indicated that inter-compartmental transport of chlorpyrifos decreased its atmospheric exposure, but had little influence on its aquatic toxicity. Instead, water-to-sediment transport of pyrethroids increased their sediment toxicity. Among CUPs at Alberta, triallate and trifluralin herbicides were detected in air concentrations at four sampling sites (Kumar, 2001) during 1999. Concentrations of CUPs in water, zooplankton, precipitation and air samples as well as stereoisomer fractions of metolachlor were determined in water samples collected from 10 remote inland lakes in Ontario, Canada, between 2003 and 2005 (Kurt-Karakus *et al.*, 2011). High detection frequency of CUPs in lake water and precipitation was attributed to high usage amounts, but some CUPs like ametryn and disulfoton, which were not used in Ontario, were also detected.

Malathion, chlorpyrifos, diazinon and endosulfan insecticides were detected intermittently with concentrations in the range 20–780 pg/m^3, south of Regina, Saskatchewan, in 1989 and 1990. Concentrations of 2,4-D also increased at the end of June (Waite *et al.*, 2002). Triallate, dicamba and bromoxynil concentrations were also higher in 1989 (peak concentration of 4.2 ng/m^3 in mid-June) compared with 1990 (600–700 pg/m^3 in mid-June). Waite *et al.* (2005) observed spatial variations of selected herbicides on two agricultural sites – Bratt's Lake, located 35 km south-west of Regina and Hafford to the north – and a background site at Waskesiu. A total of 16 pesticides (including six fungicides, seven insecticides and three herbicides) were detected in particulate matter air in a rural site of the Valencia region, during July to September in 2012 and from May to July in 2013. The total concentrations in the particulate phase (total suspended particulate; TSP) ranged from 3.5 to 383.1 pg/m^3. Carbendazim, tebuconazole, chlorpyrifos-ethyl and chlorpyrifos-methyl were found accumulated in the ultrafine–fine (<1 μm) and coarse (2.5–10 μm) particle size fractions. Others like omethoate, dimethoate and malathion presented only in the ultrafine–fine size fraction (<1 μm).

Some acid herbicides were also investigated in South Tobacco Creek, Manitoba during 1993–1996. Maximum concentrations were observed during periods of local use (Rawn *et al.*, 1999). Atrazine, a neutral herbicide, was also investigated in 1995 (Rawn *et al.*, 1998). It was first detected in mid-April, peaking in mid-June at about 300 pg/m^3, until the end of October. The insecticide dacthal was identified throughout the sampling periods in 1994, 1995 and 1996 (Rawn and Muir, 1999) in spite of the fact that it was not used in this area (<20–300 pg/m^3).

Zabik and Seiber (1990) assessed atmospheric transport of organophosphate pesticides from California's Central Valley to the Sierra Nevada mountains. Their findings indicate occurrence of atmospheric transport of pesticides applied in the valley to the Sierra Nevada mountains, but the levels decrease as distance and elevation increase from the valley floor. The oxon analogues of organophosphate pesticides, with increased potency, have been identified in environmental samples from air and surfaces following agricultural spray applications in California and Washington State (Armstrong *et al.*, 1994). These findings were of major concern because of potential health risks associated with these exposures. The pesticide drift from neighbouring vineyards was found to be affecting children attending school on a farm adjacent to an urban suburb of Cape Town, Western Cape in South Africa (Dalvie *et al.*, 2014). Eleven different pesticides were detected in samples of air, dust and grass cuttings at the pre-school and primary school located on the farms during September–December in 2010. Endosulfan, dimethomorph, penconazole, cyprodinil, boscalid and bromopropylate were on the spraying schedules of the two farms neighbouring the schools. The timing and location of detection were generally consistent with farm application. Chlorpyrifos, cypermethrin and permethrin, which were detected in the samples, are commonly used in household applications and one pesticide (pyriproxifen) is used in pet treatment agents. Kresoxim-methyl, the other pesticide detected, is likely to come from drift from other farms in the area.

2.7 Impact on Animals

The pesticide residues on food and feed may poison animals including humans, for example when wild animals enter sprayed fields or nearby areas shortly after spraying (Palmer *et al.*, 2007). Pesticides, especially the herbicides, affect biodiversity, thus eliminating some animals, the essential food sources, causing the animals to relocate, change their diet or starve. Birds are also affected when they eat insects or worms treated or contaminated with pesticides (Cornell University, 2007). Earthworms act as bioindicators of soil activity, as they digest organic matter and increase nutrient content in the top layer of soil. Pesticides have been shown to have harmful effects on growth and reproduction of earthworms (Yasmin and d'Souza, 2010). Some pesticides are bioaccumulated up to toxic levels once they enter in the bodies of organisms (Cornell University, 2007). The natural biodegradable pyrethrins and synthetic pyrethroid insecticides have been replacing the neurotoxic organophosphates/carbamates as animal/premise insect control agents. Horse stables often have mist sprayers that emit permethrin at set intervals. Dairy barns and swine confinement operations often control fly populations with these agents. The pyrethrins and persistent pyrethroids are supposed to have low mammalian and avian acute toxicity (an exception is permethrin, which is highly toxic to cats). However, both natural and synthetic forms are considered toxic to aquatic organisms. Snakes, amphibians and bees are also extremely vulnerable to pyrethrins and pyrethroids. Contamination of surface water with runoff of pyrethrin or pyrethroids poses hazards to aquatic life.

2.7.1 Birds

> It was a spring without voices. On the mornings that had once throbbed with the dawn chorus of robins, catbirds, doves, jays, wrens, and scores of other bird voices there was now no sound; only silence over the fields and woods and marsh.
>
> ('A Fable for Tomorrow' from Rachel Carson's *Silent Spring*)

Although birds are included in preregistration toxicity tests, they are still threatened. Birds or fish have greater visibility than the invertebrates, whose heavier losses from pesticides go largely unrecognized. The organophosphate and carbamate insecticides pose the greatest direct threat, particularly to birds. The complex mixtures of pesticides used in agriculture lead to lethal effects, as seen by the reports of deaths of an estimated 20,000 Swainson's hawks in Argentina in 1995, resulting from the use of monocrotophos and other organophosphates, sprayed to control grasshoppers (NRA Review, 2000) on sunflowers. Over the next 2 years, thousands more hawk deaths occurred, setting the stage for scientists as well as environmentalists and multinational corporations from Argentina, the US and Canada to join hands, resulting in withdrawl of monocrotophos from the markets of Argentina and Australia in 2001. As per US EPA guidelines, the database for avian species shows that monocrotophos can be rated as very highly toxic to birds by both the acute oral, based on reports for 13 species (LD_{50} of 0.19–6.49 mg/kg), and dietary routes of exposure, in three species (LC_{50} range 2.4–32 ppm).

Birds continue to be harmed by pesticide use. The US Fish and Wildlife Service estimate that 72 million birds are killed by pesticides in the United States each year (Fimrite, 2011). Bald eagles are common examples of non-target organisms that are impacted by pesticide use. Rachel Carson's book *Silent Spring* dealt with damage to bird species due to pesticide bioaccumulation. Another effect following repeated applications of chemical pesticides is degradation of the land to such an extent that no bird populations can be supported in the surrounding areas. In the farmland of the UK, populations of 10 different bird species declined by 10 million breeding individuals between 1979 and 1999, allegedly from loss of plant and invertebrate species on which the birds feed. Throughout Europe, 116 species of birds were threatened as of 1999. Reductions in bird populations have been found to be associated with times and areas in which pesticides are used (Kerbs *et al.*, 1999). The European and North American bird populations were affected by egg shell thinning caused by DDE, a metabolite of DDT (Vos *et al.*, 2000). Some fungicides used in groundnut farming, which are slightly toxic to birds and mammals, may kill earthworms, which can in turn reduce the populations of the birds and mammals that feed on them (Palmer *et al.*, 2007).

Feeding on seeds treated with chemical pesticides or sprayed insects also leads to high bird mortalities. However, there is no evidence from the field studies that dermal exposure is a significant route of mortality. Pesticide can kill birds directly, poison them without killing directly or reduce their food or habitat resources. Abnormal behaviour altering the ability of animals to survive or reproduce may be the symptoms of pesticide poisoning instead of instant death. In birds, for example, exposure can impact the bird's ability to sing and therefore decrease its chances of successfully attracting a mate or establishing a territory. Or it could impact the bird's care for its young, causing the nestlings to starve to death. Sufficient quantities of pesticide residues remain in the stomach of birds and mammals poisoned with organophosphates or carbamates to kill predators and scavengers such as eagles and crows.

Dr Pimentel has estimated the direct losses of birds from pesticides, which has risen from 67 million in 1998 to 72 million in 2001. But it remains unnoticed by the general public that pesticides may be involved in the death of birds. Unless there is a reliable method carried out by knowledgeable professionals for investigating and compiling such reports the public could become frustrated and cynical about the value of vigilance. Herbicides intended for plants can indirectly affect insects. Insecticides, by killing off pollinating insects, can lead to reductions in plants. Both herbicides and insecticides have indirectly reduced bird populations.

The Migratory Bird Treaty Act (MBTA), a law protecting migratory birds, applies to 'almost all native North American birds, with the exception of upland gamebirds. Unless permitted by regulation, it is unlawful to pursue, hunt, take, capture, kill, possess, sell, barter, purchase, ship, export, or import any migratory birds . . . or any part, nests, eggs, or product thereof' (US Fish and Wildlife Service, 2010). Under this law the killing of migratory birds with pesticides can be considered a violation even if the intent was not to do so and the pesticide was being used according to the label. Migratory birds are closer to the public eye (aided by binoculars) and dearer to the public heart than most other wildlife. Some chemicals may be banned in a country due to their highly toxic nature but may be freely available in other countries, threatening migratory birds journeying thousands of miles from the Arctic to the southern hemisphere. As described in Rachel Carson Council's (RCC) brochure 'A Nature Lover's Alert' an Oak Brook, Illinois, lawn care company was fined for being responsible for the deaths of 47 mallard ducks after they legally applied diazinon to the turf of a condominium in Indianapolis. They were fined US$4700 and found guilty of a misdemeanour under the Migratory Bird Treaty Act (Mayo *et al.*, 2004).

In general, herbicides are non-toxic to birds. However, birds exposed multiple times to trifluralin, considered 'practically non-toxic to birds' in acute toxicity studies, experienced diminished reproductive success in the form of cracked eggs (US EPA, 1996). The hatching of chicken eggs was reduced by exposing them to 2,4-D (Duffard *et al.*, 1981) and caused feminization or sterility in pheasant chicks (Lutz and Lutz-Ostertag, 1972).

Effects of some organochlorines on fish-eating water birds and marine mammals have been documented in North America and Europe (Cooke, 1979; Kubiak *et al.*, 1989; Barron *et al.*, 1995). Despite the continuing usage, little is known about the impacts of organochlorines on bird populations in developing countries. Among the developing countries, India has been one of the major producers and consumers of organochlorines in recent years. As a consequence, wild birds in India are exposed to great amounts of organochlorine pesticides (Tanabe *et al.*, 1998). In addition to resident birds, migratory birds are also exposed to these organochlorines when they visit tropical regions in winter. The Indian subcontinent has been a host to a multitude of birds from western Asia, Europe and Arctic Russia in winter (Woodcock, 1980). Hundreds of species of waterfowl, including wading birds such as plovers, terns and sandpipers, migrate each winter to India covering long distances (Grewal, 1990). However, concentrations of organochlorines in prey items and in eggs of Indian birds have not been reported. The concentrations of organochlorine pesticides in whole-body homogenates of birds have been reported elsewhere (Tanabe *et al.*, 1998).

A few exposure studies on organochlorines have reported a decline in the populations of bats in various parts of the world (Clark and Lamont, 1976; Geluso *et al.*, 1976; Jefferies, 1976; Altenbach *et al.*, 1979; Clark, 1981; Clark and Krynitsky, 1983; Thies and McBee, 1994). The world population of bats declined to approximately 200,000 in 1973 from an estimated population of 8.7 million in 1936 (Geluso *et al.*, 1976). It has recovered slightly to an estimated number of 700,000 in 1991. High tissue concentrations of p,p'-dichlorodiphenyldichloroethene (p,p'-DDE) found in bats in Carlsbad Caverns in Mexico and in New Mexico in the USA have been found to be the reason for the population decline (Geluso *et al.*, 1976; Thies and McBee, 1994). Occurrence of stillbirths in little brown bats exposed to high concentrations of PCBs, p,p'-DDE and/or oxychlordane has been documented (Clark and Lamont, 1976; Jefferies, 1976). These observations indicate that bats can accumulate high concentrations of organochlorines and may be affected by their potential toxic effects.

The flying fox or the new world fruit bat, short-nosed fruit bat and Indian pipistrelle bat are resident species and are very common in south India. Their habitat is mainly agricultural areas, rock caves and abandoned houses in domesticated areas (Aktar *et al.*, 2009). Insects serve as an important diet for many bats, leading to transfer of organochlorines into their body (McBee and Bickham, 1992). There are reports of organochlorine pesticides and PCBs found in the livers and eggs of birds in developed countries (Becker, 1989; Cade *et al.*, 1989; Bernardz *et al.*, 1990; Castillo *et al.*, 1994; Mora, 1996; 1997). Similarly, several studies reported organochlorines in a variety of biota including humans and wildlife from India (Senthilkumar *et al.*, 2000). However, no studies analysing whole-body homogenates of birds are available, which is important to evaluate the biomagnification features and body burdens of organochlorines (McBee and Bickham, 1992). Earlier studies used specific body tissues to estimate the biomagnification of organochlorines.

However, theoretically, estimation of biomagnification factors requires whole-body concentrations rather than specific tissue concentrations.

2.7.2 Aquatic, marine organisms and aquifers

Tiny, obscure but essential aquatic invertebrates are highly vulnerable to commonly occurring low levels of pesticides in streams, lakes and rivers. Even marine mammals with few natural enemies are at risk from the persistent pesticides that gradually accumulate in their tissues. The organochlorine pesticides and their metabolites along with other contaminants are routinely found in the tissues of marine mammals such as bottlenose dolphins, Baikal seals, harbour seals, striped dolphins and Beluga whales. Certain contaminants can be passed to the offspring from the mother's body through milk. The fact that male dolphins do not lactate contributes to higher contaminant levels in adult males than in adult females. These organochlorines also reduce immune function and increase cancer rates.

Acclimation to hypersaline conditions enhances the acute toxicity of certain thioether organophosphate and carbamate pesticides in some species of euryhaline fish (Maryoung *et al.*, 2014). Phillips *et al.* (2012) reported both agriculture and urban land uses are contributing toxic concentrations of organophosphate (water and sediment) and pyrethroid pesticides (sediment) to adjacent watersheds, the portions of the Santa Maria River and Oso Flaco Creek watersheds in central California, USA. Approximately 67% of the nearly 300 native freshwater mussel species (family *Unionidae*), the most imperilled faunal group in North America, are listed as endangered, threatened (Bringolf *et al.*, 2007). The atrazine formulation (Aatrex) was more toxic than technical-grade atrazine in chronic tests with juvenile *Lampsilis siliquoidea*. For chlorpyrifos and permethrin pesticides, acute and chronic toxicities of technical-grade pesticides were similar to the toxicity of pesticide formulations. High pesticide ecotoxicological risk was noticed in many Lebanese surface waters; this risk was driven mainly by diazinon, chlorpyrifos, fenpropathrin and bifenthrin insecticides (Kouzayha *et al.*, 2013). Among CUPs, three fungicides (azoxystrobin, boscalid and pyraclostrobin), one herbicide (propyzamide) and two organophosphate insecticides (chlorpyrifos and diazinon) were detected frequently in an agriculturally dominated central California coastal estuary, in water, sediment and tissue of resident aquatic organisms (Smalling *et al.*, 2013). Fungicides and insecticides were detected frequently in fish and invertebrates collected near the mouth of the estuary and the contaminant profiles differed from the sediment and water collected. Domagalski and Kuivila (1993) analysed suspended sediment and water samples from San Francisco Bay in 1991 during low river discharge and after spring rains, for organophosphate, carbamate and organochlorine pesticides, petroleum hydrocarbons, biomarkers and polynuclear aromatic hydrocarbons. Most of the measured diazinon, at least 98% for all samples, was in the dissolved phase. The results suggested that concentrations of diazinon sorbed to suspended sediments, therefore, were independent of the patterns of sediment movement.

2.7.3 Wildlife

Wildlife is an essential element of a healthy ecosystem, including mammals, birds and fish, on which we all depend. Therefore, any harm or damage to them should be of concern to all humanity, even without considering their worth or lack of it. Wildlife serve as sentinels for our society and environment. Wildlife are most vulnerable to pesticide toxicity and most exposed to chemical pesticides on a daily basis, yet they are the least sheltered from these hazards. Wildlife will continue to be affected by pesticide toxicity if use of these chemicals continues at the present rate. Wildlife may eat the pesticides in granular form, mistaking them for grains of food. A few granules of a pesticide may be enough to kill a small bird (Palmer *et al.*, 2007). Currently, pesticides are being applied at nearly twice the rate they were 35 years ago when Rachel Carson wrote *Silent Spring*. This 'chemically based paradigm for pest control' (M.L. Winston in *Nature Wars*, 1997, p. 448; Horrigan *et al.*, 2002) is especially threatening to wildlife. Lacking airtight shelters, protective clothing and unable to read warning signs, birds, fish and insects neither have man-made safeguards nor for the most part their own defences against the chemical barrage.

Approximately 50% of pests are controlled in agricultural and natural ecosystems by natural processes (Crowder *et al.*, 2010). For example, birds help in maintaining the balance of the ecosystem by providing valuable service to growers and to the public through controlling insects; still there is a correlation between bird deaths by pesticides and pest populations. Injudicious use of pesticides, therefore, results in economic losses as well as affecting ecosystems and wildlife. We know that many vital services from nature are threatened by chemical pesticides, and that biological alternatives to the use of chemicals for control of unwanted species are underutilized.

The injudicious use of pesticides in India has brought down the vulture population from 40 million to 60,000 due to poisoning (Prakash *et al.*, 2007). Carrion-consuming birds are now on the critically endangered list of BirdLife International, a global network of conservation groups based in Britain (www.glo-wild.com). The vultures actually act as scavengers and their disappearance has resulted in an abundance of carcasses left untouched, bringing in a whole host of new scavengers, such as rats, stray dogs, etc., in ever-increasing numbers. These new scavengers are not as effective as vultures in clearing up the carcasses or removing disease from them. When these new scavengers, like dogs, scavenge from cows, flesh is left on the bones, resulting in an increased risk of diseases such as tuberculosis, anthrax and foot and mouth disease. These dogs are reported to be 'hunting' humans as well as wildlife and bring with them the threat of rabies. Due to an increase in the number of feral dogs in India, the human infection rate with rabies has reached a world record level, over 20,000 deaths per year (Markandya *et al.*, 2008). This was controlled by sterilization of about 8 million dogs over a decade. The contamination of water and the water table, through half-eaten and rotten flesh, has potential dire consequences in a country where water is a scarce and valued resource.

There's an interesting story about the Parsis of India, who practise Zoroastrianism, and rely on vultures to dispose of their dead bodies, as their faith and fears over hygiene prevents them from burying, burning or submerging. The dead bodies were left in the open for the vultures to consume and this could usually be done (to the bone) in around an hour by a large group of vultures. This was practised for centuries. A sharp fall in vulture populations left half-consumed bodies lying for days leading to further risk of infection in other communities. As well as this, the ever-increasing feral dog population has developed a taste for human flesh. Around 1000 roaming dogs in Bikaner in western Rajasthan state have reportly attacked humans (www.glo-wild.com). The indiscriminate use of DDT, both as an agricultural pesticide and in malaria control, had been observed to be responsible for plummeting vulture populations, reported a few years ago in Keoladeo National Park in western Rajasthan state. DDT, used to control an outbreak of malaria, facilitated its entry into the food chain. Tests carried out by the Indian Council of Medical Research (ICMR) in Delhi showed unacceptably high amounts of DDT in the flesh of cows, which is left largely to vultures to feed on because of Hindu religious taboos concerning the cow.

2.7.4 Amphibians

Pesticides negatively affect many amphibian species by causing sublethal effects, such as reduced growth and increased susceptibility to disease, leading sometimes to mortality. It has been observed that pesticides and fertilizers had a negative effect on survival of −0.9027 and growth of −0.0737 across all reported amphibian species. However, the effect of pesticides and fertilizers varies among amphibian species. The inorganic fertilizers, organophosphates, chloropyridinyl, phosphonoglycines, carbamates and triazines negatively affect amphibian survival, while organophosphates and phosphonoglycines negatively affect amphibian growth. Baker *et al.* (2013) suggested that pesticides and fertilizers are an important stressor for amphibians in agriculturally dominated systems.

2.7.5 Endangered species

The species designated as endangered, threatened or listed are to be given the highest level of protection from pesticides of any wildlife, according to the Endangered Species Act. Instead of populations, the endangered wildlife is to be protected as individuals, along with their habitat. The insecticide carbofuran has killed a number of bald eagles when applied according to the label and it is still registered for use. A recent lawsuit by environmentalists seeks better US EPA protection of salmon, a listed species, from levels of the insecticide diazinon too low to be lethal but able to damage the salmons' ability to avoid predation and to find the streams of their birth through their legendary homing ability (Center for Biological Diversity, 2013).

There is always a need to observe the early signs of pesticide contamination in endangered wildlife in our environment. For example, before anyone knew how poisonous DDT was to humans, scientists discovered the chilling consequences the chemical had for bald eagles and other wildlife species. The Endangered Species Act requires that the government reduce the impacts of pesticides on imperilled wildlife like salmon, frogs and salamanders, which are especially sensitive to pesticides. The more we can do to reduce pesticides that harm wildlife, the more we do to protect human exposure to those same pesticides.

In 2011 the Center for Biological Diversity and Pesticide Action Network North America filed the most comprehensive legal action ever taken to protect endangered species from harmful pesticides. The lawsuit seeks protection for 214 threatened or endangered species in 49 states and two US territories. The Center continued to advance the lawsuit in 2013 to focus their legal strategy on some of the worst pesticides harming endangered species.

River dolphins are among the world's most seriously endangered species, resulting in their dwindling populations and facing the threat of extinction. The Yangtze river dolphin (*Lipotes vexillifer*) in China and the Indus river dolphin (*Platanista minor*) in Pakistan are already close to extinction (Perrin *et al.*, 1989; Renjun, 1990; Reeves *et al.*, 1991; Reeves and Chaudhry, 1998). As well as habitat degradation (such as construction of dams) (Reeves and Leatherwood, 1994), boat traffic, fishing, incidental and intentional killings, and chemical pollution have all been threats to the health of river dolphins (Kannan *et al.*, 1993a, b; 1994; 1997; Senthilkumar *et al.*, 1999).

2.8 Non-target Organisms

Through soil, air, water and non-target organisms in urban landscapes which are contaminated with pesticides, the plants and animals harmed include beneficial soil microorganisms and insects, non-target plants, fish, birds and other wildlife. Glyphosate or glyphosate-containing products can cause sublethal effects such as erratic swimming and laboured breathing, which increase the fish's chance of being eaten (Liong *et al.*, 1988). 2,4-D herbicides cause physiological stress responses in sockeye salmon (McBride *et al.*, 1981) and reduce the food-gathering abilities of rainbow trout (Little, 1990). Several cases of pesticide poisoning of dolphins have been reported worldwide.

In addition to fish, other marine or freshwater animals are endangered by pesticide contamination. The reproductive and immunological functions of captive or wild aquatic mammals are adversely affected by exposure to persistent, bioaccumulative and toxic contaminants such as DDT and PCBs. (Helle *et al.*, 1976; Reijnders, 1986; Martineau *et al.*, 1987; Kannan *et al.*, 1993b; Ross *et al.*, 1995; Colborn and Smolen, 1996). Aquatic mammals inhabiting freshwater systems, such as otters and mink, have been reported to be sensitive to chemical contamination (Leonards *et al.*, 1995; 1997). Products containing 2,4-D have been shown to be harmful to shellfish (Cheney *et al.*, 1997) and other aquatic species (Sanders, 1969; US EPA, 2005). The common symptoms along with the wildlife systems affected by pesticide poisoning are listed in Table 2.2.

Algae are the staple organism in the food chain of aquatic ecosystems. Atrazine and alachlor have been found to damage cells, block photosynthesis, and stunt growth of algae and diatoms even at fairly low levels (Kumar, 2011; Spawn *et al.*, 1997). Dr Tyrone Hayes of UC Berkeley discovered that 10% of male frogs raised in atrazine-contaminated water developed into females. Male frogs that were genetically males phenotypically developed ovaries within their testes (Environmental Impacts, n.d.). Ambrosi *et al.* (1978) reported that oxadiazon severely reduced algae growth. Pesticides also affect beneficial insects such as parasitoids and predators. The pollinators, honeybees, are also affected. Pesticide residues are reported even in honey. These friendly insects are very helpful in reducing the population of various insect pests. By using pesticides, the population of these beneficial insects is also reduced. Chlorpyriphos is toxic to a wide variety of beneficial arthropods including bees, ladybirds and parasitic wasps.

2.8.1 Honeybees

Honeybees are responsible for the pollination of agricultural crops, but are greatly affected by insecticides when foraging in the field. The bees are highly sensitive to pesticide contamination and immediately die, unable to return to their hive. In such cases, the queen, brood and nurse bees are not contaminated and the colony survives. The second more deadly way honeybees are affected is when the bee comes into contact with insecticide and transports it back to the colony, either as contaminated pollen or nectar or on its body. This leads to a large number of dead bees in front of the hives, the main symptom of contamination. Another symptom is sudden loss of the colony's field force. The colony may also suffer from blood disease and chilled brood (Henry *et al.*, 2012).

An imidacloprid-containing product has been associated with 'mad bee' disease in France. Beekeepers have reported that it has killed millions of bees and destroyed the insects' sense of direction. Imidacloprid can travel throughout the plant and accumulate in the nectar and the pollen. This insecticide has been reported as being highly toxic to honeybees, when sprayed onto a flowering crop or used as seed treatment. The insecticide spinosads, produced by microorganisms, has been observed to be highly toxic to honeybees and very highly toxic to the

Table 2.2. Symptoms in wild mammals caused by long exposure to pesticides. (From Hudson *et al.*, 1984.)

Category	System affected	Common symptoms
Respiratory	Nose, trachea, lungs	Irritation, coughing, choking, tight chest
Gastrointestinal	Stomach, intestine	Nausea, diarrhoea, vomiting
Renal	Kidney	Back pain, urination (more or less than usual)
Dermatological	Skin, eye	Rashes, itching, redness, swelling
Reproductive	Ovaries, testes	Infertility, miscarriage
Neurological	Brain, spinal cord	Behaviour confusion, depression, dizziness, convulsion, coma

eastern oyster. The herbicide oxadiazon has also been found to be toxic to bees (Washington State Department of Transportation, 1993). Herbicides may hurt insects or spiders indirectly when they destroy the foliage that these animals need for food and shelter. For example, spider and carabid beetle populations declined when 2,4-D applications destroyed their natural habitat according to Asteraki *et al.* (1992). The dangers of insect growth regulators, especially fenoxycarb to honeybee (*Apis mellifera*) broods have also been reported (Anonymous, 1990).

Scientists placed blue fish in test tanks that had received small doses of malathion that equate to the amount that has been found to occur in the environment after routine spraying. After 24 h mild degenerative changes were seen in the gills of the fish. After 48 h, the damage was more pronounced and after 96 h there was severe damage resulting in a condition known as 'bulbing'. Martin *et al.* (1993) also reported heart defects, blood clotting disorders, oscillating blood in the heart and pericardial oedema in the fish due to the effects of low levels of malathion. Damage to the liver, kidneys and intestines of lizards exposed to malathion at very low levels (1 mg/kg body weight) was reported by Miller and Clark (1997). Hill and Hoofman (1997) found that birds that had died as a result of DDT toxicity showed petechial and ecchymotic haemorrhages in the heart and the myocardium appeared whitish in colour on necropsy. The lungs appeared congested and darkened in colour and in some cases there were blood-tinged exudates in the bronchioles. The liver may be affected with fatty degeneration and focal necrosis was reported.

The organochlorines (DDT, dieldrin and lindane) were found to be toxic to vultures. Diagnostic testing of 42 samples of recently deceased vultures included histopathology, toxicology and electron microscopy of the 14 vultures, which showed enteritis, arteriolitis, tracheitis and visceral gout. Organochlorines were detected in fat samples of some vultures, which indicated exposure to organochlorines (Edge and Ketchun, 1992).

The clinical signs of DDT toxicity in birds include ataxia, wing droop, jerkiness in gait, and when symptoms are prolonged for a few days, acute toxic tubular nephrosis, enteritis and dehydration were reported (Hill and Fleming, 1982; LaBonde, 1992). The symptoms of monocrotophos toxicity in avian species include depression, zigzag movement, ataxia, frequent diarrhoea, hypersalivation, lacrymation and paralysis of leg according to Brar and Sandhu (1991).

Table 2.3. Salient pesticide accidents in India.

Pesticide	Place	Year	Causes
Parathion	Kerala	1958	Food contaminated due to leakage
	Bombay	1962	Inhalation in manufacturing plant
HCH	Bombay	1963	Rice contaminated
Endrin	Bombay	1964	Food contaminated
DDT	Punjab	1965	Chutney contaminated
Diazinon	Pune	1968	Food contaminated
HCH	UP (Lakhimpur)	1976	Mixed with wheat
Endrin	Karnataka	1977	Contaminated crabs in rice field
Aluminium phosphide	Rajasthan	1983	Contaminated food grain
Methyl isocyanate	Bhopal	1984	Storage tank leakage
Cartap hydrochloride	Tamil Nadu	1988	Factory workers
Endosulfan	Kerala	1997	Aerial spray of endosulfan
Phorate	Kerala	2001	Spray drifts from banana field
Endosulfan	Jabalpur (MP)	2002	Contaminated wheat flour

The clinical signs of carbamate and organophosphate toxicity in waterfowl have been reported as hyperexcitability and restlessness initially, followed by depression, excessive lachrymation and salivation, diarrhoea and urination, dyspnoea with open-mouthed breathing, cyanosis, lethargy, tremors and convulsions (Hill and Fleming, 1982; LaBonde, 1992).

2.9 Adverse Effects of Pesticides on Physical and Biological Environment

The indiscriminate use of broad-spectrum chemicals has resulted in reduction of natural enemies, risk of contamination of food and fodder, damage to ecosystems and hazards to the environment including human beings through the food chain and groundwater. The majority of pesticide poisonings and deaths occur in the developing world, although far greater quantities of pesticides are used in the developed world. The reasons are as follows:

- The developing countries have a higher proportion of the population involved in agriculture.
- Poorer application of technology and pesticide handling practices.
- Use of improper equipment (e.g. leaky backpack sprayers).
- Generally employ more toxic pesticides than those used in developed countries.

According to studies, around three new synthetic chemicals are introduced each day in the agricultural sector (Anonymous, 2000). A variety of harmful effects are experienced by human beings, aquatic ecosystems, animals, birds, food, the environment and plants (cytogenetic hazards) because of indiscriminate application. Moreover, the consignments sent by at least 40 Indian companies were rejected due to contamination of these commodities with pesticides like BHC, heptachlor and cypermethrin (Dureja, 1999).

2.9.1 Storage

Chlorpyriphos, a broad-spectrum organophosphate insecticide, affects the nervous system by inhibiting an enzyme that is important in the transmission of nerve impulses. It is widely used for controlling termites, cockroaches and white ants in urban areas. It is added in large quantities to the foundations of buildings. It has been voluntarily withdrawn from the market by the manufacturer, DowElanco in the US. The US EPA tried to phase it out by 2005, but in India demand is increasing The people involved in storage of vegetable produce, from the farmers' fields, intentionally apply or even dip the produce in pesticidal solutions to maintain the colour and quality of produce. In addition to the market problem, the problem of the existing stocks of unused, unusable and unwanted products, often stored in damaged containers under highly inadequate conditions, is still there.

Reformulation and triangulation (re-export to another user) of obsolete stocks can only partially solve the problem. The only solution for this problem is to detoxify the products by any of the chemical, enzymatic or photochemical methods and proper disposal of them in a safe place.

2.9.2 Cytogenetic hazards

Fuberidazole, a fungicide, affects meiosis and seed set when applied to the grain. Thiabendazole and oxycarboxim at 3 moles/l significantly reduce seed germination.

Chiasma frequencies were significantly reduced in barley (*Hordeum vulgare*) by 16 pesticides, of which the fungicides ditalimphos and chlorbromuron caused the most severe reductions in chiasma frequency (Behera and Sharma, 1990; Panneerselvam and Sharma, 1990). Plants are also damaged by exposure to chlorpyriphos. Delayed seedling emergence, fruit deformities and abnormal cell division all result from chlorpyriphos exposure (Gilani *et al.*, 2016).

2.9.3 Other environment impacts of pesticides

Development of insect resistance

The intensive use of pesticides in agriculture is a cause of serious concern. The problem is especially serious because of the development of resistance to pesticides in important pests and the presence of pesticide residue in agricultural and dairy products. Pesticide resistance in agriculture was first noticed in India in 1963 when a number of serious pests were reported to have become resistant to DDT and HCH (two of the most commonly used pesticides during the 1960s and 1970s). Since then the number of pests with pesticide resistance has increased. The most serious resistance problem was witnessed in cotton, for which American bollworm was a serious pest. The bollworm developed resistance to almost all pesticides in a number of regions, and was particularly serious in parts of Punjab, Haryana, Andhra Pradesh, Karnataka and Maharashtra. Other important pests of cotton, white fly and jassid, have also developed pesticide resistance in some places. Malaria vectors in India are resistant to DDT alone or double resistant to DDT and HCH or triple resistant to DDT, HCH and malathion, and quadruple resistant to DDT, HCH, malathion and deltamethrin. Growing pesticide resistance has meant that a large proportion of agricultural production is lost to pests. According to some estimates, these losses amount to between 20 and 30% of total production (Mehrotra, 1989). The losses are particularly serious in cotton. For example, cotton production in Punjab declined by about 50% during 1997 and 1998 causing a number of cotton farmers to commit suicide in the affected areas. Diamondback moth (DBM) has become resistant to cypermethrin, fenvalerate and deltamethrin in Punjab. For the past 10 years, persistent DBM resistance to quinalphos in various regions of Punjab was calculated after determining the baseline toxicity. Pesticide resistance has mainly been caused by excessive and indiscriminate use of pesticides (Jayaraj, 1989). Pesticides of spurious quality, which are commonly sold in small towns and villages, have also contributed to resistance in many areas. For example, in Bidar (Karnataka) where the problem of pest resistance has become extremely serious, more than 50 brands of pesticides were found to be substandard in 1998–1999. In another incident, the licences of 115 pesticide producers were cancelled in Punjab for selling substandard pesticide. Substandard pesticides contribute to resistance as the pests are repeatedly exposed to a low concentration of pesticides. This contributes to the build up of resistance, without destroying the pests.

Resurgence

Resurgence is an abnormal increase in the pest population resulting from the insecticide application, often far exceeding the economic injury level. The first report of pest resurgence as induced by insecticides was by the red spider mite on cotton with usage of DDT in 1970. The most extreme reports of resurgence of insect pests are in planthoppers, especially brown planthopper (BPH) of rice. BPH was a minor pest during the 1950s but later became a major pest of rice due to extensive use of insecticides. The factors implicated in BPH resurgence include reduction of nymphal stage, longer oviposition period, shortened lifecycle, enhanced reproductive rate, higher feeding rate and destruction of natural enemies, especially *Cyrtorhinus lividipennis*.

Conclusion

The agriculture sector is dependent on the use of pesticides to meet the demands of the growing population, leading to a sequential rise in production and consumption of pesticides over the past three decades. However, injudicious use of chemical pesticides leads to contamination of the environment, leaving residues in food, feed and fibre. The indiscriminate use of pesticides has led to environmental contamination, polluting water, air and soil, as well as causing various health-related issues. Exposure to pesticides and the potentially serious health effects have not spared any segment of the population. The toxic pesticide residues present in food can interact with essential minerals thereby affecting their bioavailability for the human body. These can adversely affect the peripheral and CNS, reproductive systems, endocrinal and immune systems and cause deterioration of the lungs. The pesticide residues in agricultural food have been demonstrated to increase the risk for cancer in the professional populations. They can contaminate aquatic environments including groundwater, contaminating other organisms including birds, fish, beneficial insects and non-target plants. The use of certain pesticides adversely affects soil conservation and decreases its general biodiversity. Wildlife may eat the pesticides in granular form, mistaking them for grains of food. A few granules of a pesticide may be enough to kill a small bird. Wildlife including endangered species will continue to be affected by pesticide toxicity if use of these chemicals continues at the present rate. The other non-target organisms, natural enemies, pollinators, predators and soil microorganisms need to be saved from the toxic effect of these chemicals. In addition to residues of these pesticides, resurgence and resistance are major issues. Hospital surveys, however, have recorded most pesticide deaths in developing countries as a result of self-poisoning rather than food poisoning.

References

Aktar, M.W., Sengupta, D. and Chowdhury, A. (2009) Impact of pesticides use in agriculture: Their benefits and hazards. *Interdisciplinary Toxicology* 2(1), 1–12.

Altenbach, J.S., Geluso, K.N. and Wilson, D.E. (1979) Population size of Tadarida brasiliensis at Carlsbad Caverns in 1973. In: Genoways, H.H. and Baker, R.J. (eds) *Biological Investigations in Guadelupe Mountains National Park, Texas.* Proceedings of a symposium held at Texas Tech University, Lubbock, Texas, 4–5 April 1975. Proceedings and Transactions Series No. 4, pp. 341–348. National Park Service, Washington DC.

Ambrosi, D., Isensee, A. and Macchia, J. (1978) Distribution of oxadiazon and phoslone in an aquatic model ecosystem. *American Chemical Society* 26(1), 50–53.

Andreu, V. and Picó, Y. (2004) Determination of pesticides and their degradation products in soil: Critical review and comparison of methods. *Trends in Analytical Chemistry* 23(10–11), 772–789.

Anonymous (1990) Report on 'Fourth International Symposium on the harmonization of methods for testing the toxicity to bees' held in Czechoslovakia on 15–18 May 1990.

Anonymous (2000) Deadly increase. *Down to Earth*, June 15, 4.

Arcury, T.A., Grzywacz, J.G., Barr, D.B., Tapia, J., Chen, H. and Quandt, S.A. (2007) Pesticide urinary metabolite levels of children in eastern North Carolina farmworker households. *Environmental Health Perspective* 115(8), 1254–1260.

Arias, R.N. and Fabra de Peretti, A. (1993) Effects of 2,4-dichlorophenoxyacetic acid on *Rhizobium* sp. growth and characterization of its transport. *Toxicology Letters* 68, 267–273.

Arias-Estévez, M., López-Periago, E., Martínez-Carballo, E., Simal-Gándara, J., Mejuto, J.C. and García-Río, L. (2008) The mobility and degradation of pesticides in soils and the pollution of groundwater resources. *Agriculture, Ecosystems and Environment* 123, 247–260.

Armstrong, J.L., Fenske, R.A., Yost, M.G., Galvin, K., Tchong-French, M. and Yu, J. (1994) Presence of organophosphorus pesticide oxygen analogs in air samples. *Atmospheric Environment* 2013 February 1, 66, 145–150.

Arora, S. (2009) Analysis of insecticides in okra and brinjal from IPM and non-IPM fields. *Environmental Monitoring and Assessment* 151, 311–315.

Arora, S., Kanojia, K. and Bambawale, O.M. (2011) *Pesticides: Status, Latest WHO ranking, Regulation and Label Claims in India.* National Centre for Integrated Pest Management, New Delhi, India, p. 292.

Asteraki, E.J., Hanks, C.B. and Clements, R.O. (1992) The impact of the chemical removal of the hedge-based flora on the community structure of carabid beetles (Col. Carabidae) and spiders (Araneae) of the field and hedge bottom. *Journal of Applied Entomology* 113, 398–406.

Baker, N.J., Bancroft, B.A. and Garcia, T.S. (2013) A meta-analysis of the effects of pesticides and fertilizers on survival and growth of amphibians. *Science of the Total Environment* 449, 150–156.

Barceló, D. and Hennion, M.C. (1997) *Trace Determination of Pesticides and Their Degradation Products in Water.* Elsevier, Amsterdam, The Netherlands.

Barron, M.G., Galbraith, H. and Beltman, D. (1995) Comparative reproduction and developmental toxicology of birds. *Comparative Biochemistry and Physiology* 112c, 1–14.

Becker, P.H. (1989) Seabirds as monitor organisms of contaminants along the German North Sea coast. *Hel Meerr* 43, 395–403.

Behera, B.N. and Sharma, C.B.S.R. (1990) Cytogenetic hazards from agricultural chemicals. *Biologisches zentralblatt* 109(4), 307–311.

Belzer, W., Evans, C. and Poon, A. (1998) Atmospheric concentrations of agricultural chemicals in the Lower Fraser Valley. FRAP study report, 1998, Aquatic and Atmospheric Science Division, Environment Canada, Vancouver, BC.

Benzidane, C. and Dahamna, S. (2013) Chlorpyrifos residues in food plant in the region of Setif-Algeria. *Communications in Agricultural and Applied Biological Sciences* 78(2), 157–160.

Bergkvist, C., Aune, A., Nilsson, I., Sandanger, T.M., Hamadani, J.D., Tofail, F., Oyvind-Odland, J., Kabir, I. and Vahter, M. (2012) Occurrence and levels of organochlorine compounds in human breast milk in Bangladesh. *Chemosphere* 88, 784–790.

Berman, T., Goldsmith, R., Göen, T., Spungen, J., Novack, L., Levine, H., Amitai, Y., Shohat, T. and Grotto, I. (2013) Urinary concentrations of organophosphate pesticide metabolites in adults in Israel: Demographic and dietary predictors. *Environment International* 60, 183–189.

Bernardz, J.C., Klem, D., Goodrich, L.J. and Senner, S.E. (1990) Migration counts of raptorat Hawk Mountain, Pennsylvania, as indicators of population trends, 1934–1986. *Auk* 107, 96–109.

Bevans, H.E., Lico, M.S. and Lawrence, S.J. (1998) Water quality in the Las Vegas Valley area and the Carson and Truckee Riverbasins, Nevada and California, 1992–1996. *US Geological Survey Circular* 1170, Reston, Virginia.

Bingham, S. (2007) Pesticides in rivers and groundwater. Environment Agency, UK. Available at: https://wikividly.com/wiki/Environmental_impact_of_pesticides (accessed 28 September 2018).

Bortleson, G. and Davis, D. (1987–1995) *Pesticides in Selected Small Streams in the Puget Sound Basin.* US Geological Survey and Washington State Department of Ecology, Lacey, Washington, pp. 1–4.

Bottoni, P. and Funari, E. (1992) Criteria for evaluating the impact of pesticides on groundwater quality. *Science of the Total Environment* 123–124, 581–590.

Bouchard, M.F., Chevrier, J., Harley, K.G., Kogut, K., Vedar, M., Calderon, N., Trujillo, C., Johnson, C., Bradman, A., Barr, D.B. and Eskenazi, B. (2011) Prenatal exposure to organophosphate pesticides and IQ in 7-year-old children. *Environmental Health Perspectives* 119(8), 1189–1195.

Brammall, R.A. and Higgins, V.J. (1988) The effect of glyphosate on resistance of tomato to *Fusarium* crown and root rot disease and on the formation of host structural defensive barriers. *Canadian Journal of Botany* 66, 1547–1555.

Brar, R.S. and Sandhu, H.S. (1991) Acute monocrotophos toxicity in domestic fowl. *Indian Journal of Animal Science* 61, 297–298.

Bringolf, R.B., Cope, W.G., Barnhart, M.C., Mosher, S., Lazaro, P.R. and Shea, D. (2007) Acute and chronic toxicity of pesticide formulations (atrazine, chlorpyrifos, and permethrin) to glochidia and juveniles of *Lampsilis siliquoidea. Environmental Toxicology and Chemistry* 26(10), 2101–2107.

Brouwer, A., Longnecker, M.P., Birnbaum, L.S., Cogliano, J., Kostyniak, P., Moore, J., Schantz, S. and Winneke, G. (1999) Characterization of potential endocrine related health effects at low-dose levels of exposure to PCBs. *Environmental Health Perspectives* 107, 639.

Buser, H.R. (1990) Atrazine and other triazine herbicides in lakes, rains and groundwater in Switzerland. *Environmental Science and Technology* 24, 1049.

Cade, T.J., Lincer, J.L., White, C.M., Rosenau, D.G. and Swartz, L.G. (1989) DDE residues and eggshell changes in Alaskan falcons and hawks. *Science* 172, 955–957.

Castillo, L., Thybaud, E., Caquet, T. and Ramade, F. (1994) Organochlorine contaminants in common tern (*Sterna hirundo*) eggs and young from the Rhine River area (France). *Bulletin of Environmental Contamination and Toxicology* 53, 759–764.

Center for Biological Diversity (2013) Pesticides and US Endangered Species. Available at: www.biologicaldiversity.org/campaigns/pesticides_reduction/national_map (accessed 28 September 2018).

Chakravarty, P. and Sidhu, S.S. (1987) Effects of glyphosate, hexazinone and triclopyr on in vitro growth of five species of ectomycorrhizal fungi. *European Journal of Forest Pathology* 17, 204–210.

Cheney, M.A., Fiorillo, R. and Criddle, R.S. (1997) Herbicide and estrogen effects on the metabolic activity of Elliptiocomplanata measured by calorespirometry. *Comparative Biochemistry and Physiology* 118C, 159–164.

Chernyak, S.M., Rice, C.P. and McConnell, L.L. (1996) Evidence of currently-used pesticides in air, ice, fog, seawater and surface microlayer in the Bering and Chukchi seas. *Marine Pollution Bulletin* 32(5), 410–419.

Clark, D.R. (1981) Death of bats from DDE, DDT or dieldrin diagnosis via residues in carcass fat. *Bulletin of Environmental Contamination and Toxicology* 26, 367–371.

Clark, D.R. and Krynitsky, A.J. (1983) DDT: Recent contamination in New Mexico and Arizona. *Environment* 25, 27–31.

Clark, D.R. and Lamont, T.G. (1976) Organochlorine residues in females and nursing young of the big brown bats. *Bulletin of Environmental Contamination and Toxicology* 15, 1–8.

Colborn, T. and Smolen, M.J. (1996). Epidemiological analysis of persistent organochlorine contaminants in cetaceans. *Reviews of Environmental Contamination and Toxicology* 146, 91–172.

Cooke, A.S. (1979) Egg shell characteristic of gannets *Sula bassana*, shags *Phalacrocorax aristotelis* and great black-backed gulls *Larus marinus* exposed to DDE and other environmental pollutants. *Environmental Pollution* 19, 47–65.

Cornell University (2007) Pesticides in the environment. Pesticide fact sheets and tutorial, module 6. Pesticide Safety Education Program. Retrieved on 11 October 2007. Available at: http://psep.cce.cornell.edu/Tutorials/core-tutorial/module06/index.aspx (accessed 28 September 2018)

Coronado, G.D., Thompson, B., Strong, L., Griffith, W.C. and Islas, I. (2003) Agricultural task and exposure to organophosphate pesticides among farmworkers. *Environmental Health Perspectives* 112(2), 142–147.

Coronado, G.D., Vigoren, E.M.,Thompson, B., Griffith, W.C. and Faustman, E.M. (2006) Organophosphate pesticide exposure and work in pome fruit: Evidence for the take-home pesticide pathway. *Environmental Health Perspectives* 114(7), 999–1006.

Cox, C. (1995) Glyphosate, Part 2: human exposure and ecological effects. *Journal of Pesticide Reform* 15(4). Available at: www.sequoiaforestkeeper.org/Exhibit_B_Glyphosate_Ecological_Effects_2.pdf (accessed 28 September 2018).

Crisp, T.M., Clegg, E.D., Cooper, R.L., Wood, W.P., Anderson, D.G., Baeteke, K.P., Hoffmann, J.L., Morrow, M.S., Rodier, D.J., Schaeffer, J.E., Touart, L.W., Zeeman, M.G. and Patel, Y.M. (1998) Environmental endocrine disruption: An effects assessment and analysis. *Environmental Health Perspectives* 106, 11.

Crowder, D.W., Northfield, T.D., Strand, M.R. and Snyder, W.E. (2010) Organic agriculture promotes evenness and natural pest control. *Nature* 466, 109–112.

Dalvie, M.A., Sosan, M.B., Africa, A., Cairncross, E. and London, L. (2014) Environmental monitoring of pesticide residues from farms at a neighbouring primary and pre-school in the Western Cape in South Africa. *Science of the Total Environment* 466–467, 1078–1084.

Dewan, P., Jain, V., Gupta, P. and Banerjee, B.D. (2013) Organochlorine pesticide residues in maternal blood, cord blood, placenta, and breastmilk and their relation to birth size. *Chemosphere* 90(5), 1704–1710.

Domagalski, J.L. and Kuivila. K.M. (1993) Distributions of pesticides and organic contaminants between water and suspended sediment, San Francisco Bay, California. *Estuaries* 16(3), 416–426.

Dreistadt, S.H., Clark, J.K. and Flint, M.L. (1994) Pests of landscape trees and shrubs. *An Integrated Pest Management Guide*. University of California Division of Agriculture and Natural Resources. Publication 3359.

Duffard, R., Traini, L. and Evangelista de Duffard, A. (1981) Embryotoxic and teratogenic effects of phenoxy herbicides. *Acta Physiol Latinoam* 31, 39–42.

Dureja P. (1999) US find pesticides in Indian food. PRJ News. *Pesticide Research Journal* 11(2), 231–232.

Eddleston, M. (2000). Patterns and problems of deliberate self-poisoning in the developing world. *The Quarterly Journal of Medicine* 93, 715–731.

Edge, W. and Ketchun, L. (1992) Endangered species and pesticide. Oregon State University Cooperative Extension Service. VTP006.

Eleršek, T. and Filipič, M. (2011) Organophosphorus pesticides – mechanisms of their toxicity. In: Stoytcheva, M. (ed.) *Pesticides – the Impacts of Pesticide Exposure*. InTech, London, pp. 243–260.

Environews Forum (1999) Killer environment. *Environmental Health Perspectives* 107, A62.

Environmental Impacts (n.d.) In: Pesticide Action Network. Available at: www.panna.org/issues/persistent-poisons/environmental-impacts (accessed 20 December 2014).

Eskenazi, B., Marks, A.R., Bradman, A., Harley, K., Barr, D.B., Johnson, C., Morga, N. and Jewell,

N.P. (2007) Organophosphate pesticide exposure and neurodevelopment in young Mexican-American children. *Environmental Health Perspectives* 115(5), 792–798.

Estok, D., Freedman, B. and Boyle, D. (1989) Effects of the herbicides 2,4-D, glyphosate, hexazinone, and triclopyr on the growth of three species of ectomycorrhizal fungi. *Bulletin of Environmental Contamination and Toxicology* 42, 835–839.

Fabra, A., Duffard, R. and Evangelista de Duffard, A. (1997) Toxicity of 2,4-dichlorophenoxyacetic acid in pure culture. *Bulletin of Environmental Contamination and Toxicology* 59, 645–652.

Fareed, M., Pathak, M.K., Bihari, V., Kamal, R., Srivastava, A.K. and Kesavachandran, C.N. (2013) Adverse respiratory health and hematological alterations among agricultural workers occupationally exposed to organophosphate pesticides: A cross-sectional study in North India. *PLoS ONE* 8(7), e69755.

Farrukh, S. and Ali, A.S. (2011) Effects of dichlorovos organophosphate on growth, reproduction, and avoidance behavior of earthworm *Eisenia foetida*. *Iranian Journal of Toxicology* 5(14), 495–501.

Felding, G. (1992) Leaching of atrazine into groundwater. *Pesticide Science* 35, 39.

Fenelon, J.M. (1998) Water quality in the White River Basin, Indiana, 1992–96. USGS. US Geological Survey Circular 1150, Reston, Virginia.

Fimrite, P. (2011) Suit says EPA fails to shield species from poisons. *The San Francisco Chronicle*, June 27, 2011.

Forget, G. (1993) Balancing the need for pesticides with the risk to human health. In: Forget, G., Goodman, T. and de Villiers, A. (eds) *Impact of Pesticide Use on Health in Developing Countries*, IDRC, Ottawa, p. 2.

Frankenberger, W.T. Jr and Tabatabai, M.A. (1991) Factors affecting L-asparaginase activity in soils. *Biology and Fertility of Soils* 11, 1–5.

Frumkin, H. (2003) Agent orange and cancer: An overview for clinicians. *CA: Cancer Journal for Clinicians* 53, 245.

Garbarino, J.R., Snyder-Conn, E., Leiker, T.J. and Hoffman, G.L. (2002) Contaminants in Arctic snow collected over northwest Alaskan sea ice. *Water, Air and Soil Pollution* 139, 183–214.

Geluso, K.N., Altenbach, J.S. and Wilson, D.E. (1976) Bat mortality: Pesticide poisoning and migratory stress. *Science* 194, 184–186.

Gilani, R.A., Rafique, M., Rehman, A., Munis, M.F., Rehman, S.U. and Chaudhary, H.J. (2016) Biodegradation of chlorpyrifos by bacterial genus *Pseudomonas*. *Journal of Basic Microbiology* 56(2), 105–119.

Gilliom, R.J., Barbash, J.E., Crawford, G.G., Hamilton, P.A., Martin, J.D., Nakagaki, N., Nowell, L.H., Scott, J.C., Stackelberg, P.E., Thelin, G.P. and Wolock, D.M. (2007) *The quality of our nation's waters: Pesticides in the nation's streams and ground water, 1992–2001*. US Geological Survey. Retrieved on 13 September 2007.

Glotfelty, D.E. and Schomburg, C.J. (1989) Volatilization of pesticides from soil. In: Sawhney, B.L. and Brown, K. (eds) *Reactions and Movements of Organic Chemicals in Soil.* Soil Science Society of America Special Pub, Madison, Wisconsin, 22, pp. 181–209.

Gomez, D.E., Barreda, D.J.R., Vila, M.G., Rueda, E.L., Olmo, A.S., Gomez, D.E., Barred, A.D., Garcia De La Cuadra, J., Ten, A. and Peris, C. (1998) Dissipation of some citrus selective residual herbicides in an irrigation well. *Journal of Chromatography* 795, 125–131.

Grandet, M., Quentin, K.E. and Weil, L.Z. (1989) Studies of groundwater pollution by atrazine. *Wasser Abwasser Forsch* 22, 231.

Grewal, B. (1990) *Birds of India*. Guide Book Co. Ltd, Hong Kong, in conjunction with Gulnohr Press Pvt Ltd., New Delhi, p. 193.

Gulati, K., Thakur, S. and Jindal, T. (2015) Chlorpyrifos toxicology and persistence in environment: An Indian perspective. *International Journal of Multidisciplinary Research and Development* 2(7), 1–6. Available at: www.allsubjectjournal.com/vol2/issue7/PartA/pdf/85.1.pdf (accessed 28 September 2018).

Gupta, S.K., Parikh, J.R., Shah, M.P., Chatterjee, S.K. and Kashyap, S.K. (1982) Changes in serum hexachlorocyclohexane (HCH) residues in malaria spraymen after short term occupational exposure. *Archives of Environmental Health* 37, 41.

Gupta, S.K., Jani, J.P., Saiyed, H.N. and Kashyap, S.K. (1984) Health hazards in pesticide formulators exposed to a combination of pesticides. *Indian Journal of Medicinal Research* 79, 666.

Harman-Fetcho, J.A., McConnell, L.L., Rice, C.P. and Baker, J.E. (2000) Wet deposition and air–water gas exchange of currently used pesticides to a subestuary of the Chesapeake Bay. *Environmental Science and Technology* 34, 1462–1468.

Helle, E., Olsson, M. and Jensen, S. (1976) DDT and PCB levels and reproduction in ringed seal from the Bothnian Bay. *Ambio* 5, 188–189.

Henry, M., Béguin, M., Requier, F., Rollin, O., Odoux, JF, Aupinel, P., Aptel, J., Tchamitchian, S. and Decourtye, A. (2012) Response to comment on 'A Common Pesticide Decreases

Foraging Success and Survival in Honey Bees'. *Science* 337, 1453.

Hill, E.F. and Fleming, W.J. (1982) Anticholinesterase poisoning of birds: Field monitoring and diagnosis of acute poisoning. *Environmental Toxicology and Chemistry* 1, 27–38.

Hill, E. and Hoofman, D. (1997) Avian models for toxicity testing. *Journal of the American College of Toxicology* 3, 357–376.

Hofmann, J.N., Keifer, M.C., De Roos, A.J., Fenske. R.A., Furlong, C.E., van Belle, G. and Checkoway, H. (2010) Occupational determinants of serum cholinesterase inhibition among organophosphate-exposed agricultural pesticide handlers in Washington State. *Occupational and Environmental Medicine* 67(6), 375–386.

Horrigan, L., Lawrence, R.S. and Walker, P. (2002) How sustainable agriculture can address the environmental and human health harms of industrial agriculture. *Environmental Health Perspectives* 110(5), 445–456.

Hudson, R., Tucker, R. and Haegele, M. (1984) *Handbook of Toxicity of Pesticides to Wildlife.* Resource Publication 153. United States Department of the Interior Fish and Wildlife Service, Washington, DC.

Hurley, P.M., Hill, R.N. and Whiting, R.J. (1998) Mode of carcinogenic action of pesticides inducing thyroid follicular cell tumours in rodents. *Environmental Health Perspectives* 106, 437.

Igbedioh, S.O. (1991) Effects of agricultural pesticides on humans, animals and higher plants in developing countries. *Archives of Environmental Health* 46, 218.

Jayaraj, S. (1989) Advances in biological means of pest control. In: *The Hindu Survey of Indian Agriculture.* Hindu Newspaper Group, India.

Jefferies, D.J. (1976) Organochlorine insecticide residue in British bats and their significance. *Journal of Zoology* 166(1), 245–251.

Jeyaratnam, J. (1985) Health problems of pesticide usage in the third world. *British Journal of Industrial Medicine* 42(8), 505.

Johnson, H.M., Domagalski, J.L. and Saleh, D.K. (2011) Trends in pesticide concentrations in streams of the Western United States, 1993–2005. *Journal of American Water Resources Association* 47(2), 265–286.

Johnston, A.E. (1986) Soil organic-matter, effects on soils and crops. *Soil Use Management* 2(3), 97–105.

Kannan, K., Tanabe, S., Ramesh, A., Subramanian, A. and Tatsukawa, R. (1992) Persistent organochlorine residues in food stuffs from India and their implications on human dietary exposure. *Journal of Agricultural and Food Chemistry* 40, 518.

Kannan, K., Sinha, R.K., Tanabe, S., Ichihashi, H. and Tatsukawa, R. (1993a) Heavy metals and organochlorine residues in Ganges river dolphins from India. *Marine Pollution Bulletin* 26, 159–162.

Kannan, K., Tanabe, S., Borrell, A., Aguilar, A., Focardi, S. and Tatsukawa, R. (1993b) Isomer-specific analysis and toxic evaluation of polychlorinated biphenyls in striped dolphins affected by an epizootic in the western Mediterranean Sea. *Archives of Environmental Contamination and Toxicology* 25, 227–233.

Kannan, K., Tanabe, S., Tatsukawa, R. and Sinha, R.K. (1994) Biodegradation capacity and residue pattern of organochlorines in Ganges river dolphins from India. *Toxicology and Environmental Chemistry* 42, 249–261.

Kannan, K., Senthilkumar, K. and Sinha, R.K. (1997) Sources and accumulation of butyltin compounds in Ganges river dolphin, *Platanista gangetica. Applied Organometallic Chemistry* 11, 223–230.

Karunakaran, C.O. (1958) The Kerala food poisoning. *Journal of Indian Medical Association* 31, 204.

Kashyap, R., Iyer, L.R. and Singh, M.M. (1994) Evaluation of daily dietary intake of dichlorodiphenyltrichloroethene (DDT) and benzenehexachloride (BHC) in India. *Archives of Environmental Health* 49, 63.

Kaushik, G., Satya, S., Naik, S.N. and B. Tripathi (2010) Impact of chlorpyrifos residues on micronutrient bioavailability in chickpea. *Food Digestion* 1(1–2), 57–61.

Kelley, W.D. and South, D.B. (1978) In vitro effects of selected herbicides on growth and mycorrhizal fungi. In: *Weed Science Society of America Meeting.* Auburn University, Auburn, Alabama, p. 38.

Kellogg, R.L., Nehring, R., Grube A., Goss, D.W. and Plotkin, S. (February 2000) Environmental indicators of pesticide leaching and runoff from farm fields. *United States Department of Agriculture Natural Resources Conservation Service.* Retrieved on 3 October 2007.

Kerbs, J.R., Wilson, J.D., Bradbury, R.B. and Siriwardena, G.M. (1999) The second silent spring. Commentary in *Nature* 400, 611–612.

Köck-Schulmeyer, M., Ginebreda, A., González, S., Cortina, J.L., de Alda, M.L. and Barceló, D. (2012) Analysis of the occurrence and risk assessment of polar pesticides in the Llobregat River Basin (NE Spain). *Chemosphere* 86(1), 8–16.

Kole, R.K. and Bagchi, M.M. (1995) Pesticide residues in the aquatic environment and their possible ecological hazards. *Journal of Inland Fisheries Society of India* 27(2), 79–89.

Kolpin, D.W., Kalkhoff, S.J., Goolsby, D.A., Sneck-Fahrer, D.A. and Thurman, E.M. (1997a) Occurence of selected herbicides and herbicide degradation products in Iowa's groundwater, 1995. *Ground Water* 35, 679.

Kolpin, D.W., Squillace, P.J., Zogrski, J.S. and Barbash, J.F. (1997b) Pesticides and volatile organic compounds in shallow urban groundwater of USA. Groundwater Urban Environment Proceedings, IAH Congress, 27th, 1, 469.

Kouzayha, A., Al Ashi, A., Al Akoum, R., Al Iskandarani, M., Budzinski, H. and Jaber, F. (2013) Occurrence of pesticide residues in Lebanon's water resources. *Bulletin of Environmental Contamination and Toxicology* 91(5), 503–509.

Kubiak, T.J., Harris, H.J., Smith, L.M., Schwartz, T.R., Stalling, P.L., Trick, L., Sielo, D.E., Pocherty, P.D. and Erdman, T.C. (1989) Microcontaminants and reproductive impairment of the Forster's tern in Green Bay, Lake Michigan. *Archives of Environmental Contamination and Toxicology* 18, 706–727.

Kumar, S. (2011) Aquatic weeds problems and management in India. *Indian Journal of Weed Science* 43(3/4), 118–138.

Kumar, Y. (2001) Pesticides in ambient air in Alberta. Report prepared for the Air Research Users Group, Alberta Environment, Edmonton, Alta.

Kurt-Karakus, P.B., Teixeira, C., Small, J., Muir, D. and Bidleman, T.F. (2011) Current-use pesticides in inland lake waters, precipitation, and air from Ontario, Canada. *Environmental Toxicology and Chemistry* 30(7), 1539–1548.

LaBonde, J.J. (1992) Two clinical cases of exposure to household use of organophosphate and carbamate insecticides. Association of Avian Veterinarians Conference, New Orleans, 113–118.

Lacasaña, M., López-Flores, I., Rodríguez-Barranco, M., Aguilar-Garduño, C., Blanco-Muñoz, J., Pérez-Méndez, O., Gamboa, R., Gonzalez-Alzaga, B., Bassol, S. and Cebrian, M.E. (2010a) Interaction between organophosphate pesticide exposure and PON1 activity on thyroid function. *Toxicology and Applied Pharmacology* 249(1), 16–24.

Lacasaña, M., López-Flores, I., Rodríguez-Barranco, M., Aguilar-Garduño, C., Blanco-Muñoz, J., Pérez-Méndez, O., Gamboa, R., Bassol, S. and Cebrian, M.E. (2010b) Association between organophosphate pesticides exposure and thyroid hormones in floriculture workers. *Toxicology and Applied Pharmacology* 243(1), 19–26.

Lamberth, C., Jeanmart, S., Luksch, T. and Plant, A. (2013) Current challenges and trends in the discovery of agrochemicals. *Science* 341(6147), 742–746.

Landrigan, P.J., Claudio, L., Markowitz, S.B., Berkowitz, G.S., Brenner, B.L., Romero, H, Wetmur, J.G, Matte, T.D, Gore, A.C., Godbold, J.H. and Wolff, M.S. (1999) Pesticides and inner-city children: Exposures, risks, and prevention. *Environmental Health Perspectives* 107(3), 431–437.

Le Noir, J.S., McConnell, L.L., Fellers, G.M., Cahill, T.M. and Seiber, J.N. (1999) Summertime transport of current use pesticides from California's central valley to the Sierra Nevada mountain range, USA. *Environmental Toxicology and Chemistry* 18, 2715–2722.

Leonards, P.E.G., de Vries, T.H., Minnaard, W., Stuijfzand, S., de Voogt, P., Cofino, W.P., van Straalen, N.M. and van Hattum, B. (1995). Assessment of experimental data on PCB-induced reproduction in mink, based on an isomer- and congener-specific approach using 2,3,7,8-tetrachlorodibenzo-p-dioxin toxic equivalency. *Environmental Toxicology and Chemistry* 14, 639–652.

Leonards, P.E.G., Zierikzee, Y., Brinkman, U.A.T., Cofino, W.P., van Straalen, N.M. and van Hattum, B. (1997) The selective dietary accumulation of planar polychlorinated biphenyls in the otter (*Lutra lutra*). *Environmental Toxicology and Chemistry* 16, 1807–1815.

Levings, G.W., Healy, D.F., Richey, S.F. and Carter, L.F. (1998) Water quality in the Rio Grande Valley, Colorado, New Mexico, and Texas, 1992–1995. US Geological Survey Circular 1162, Reston, Virginia.

Li, H., Ma, H., Lydy, M.J. and You, J. (2014a) Occurrence, seasonal variation and inhalation exposure of atmospheric organophosphate and pyrethroid pesticides in an urban community in South China. *Chemosphere* 95, 363–369.

Li, H., Wei, Y., Lydy, M.J. and You, J. (2014b) Intercompartmental transport of organophosphate and pyrethroid pesticides in South China: Implications for a regional risk assessment. *Environmental Pollution* 190, 19–26.

Little, E.E. (1990) Behavioral indicators of sublethal toxicity of rainbow trout. *Archives of Environmental Contamination and Toxicology* 19, 380–385.

Liong, P.C., Hamzah, W.P. and Murugan, V. (1988) Toxicity of some pesticides towards freshwater fishes. *Malaysian Agriculture Journal* 54(3), 147–156.

Liroff, R.A. (2000) Balancing risks of DDT and malaria in the global POPs treaty. *Pesticide Safety News* 4, 3.

Locke, D., Landivar, J.A. and Moseley, D. (1995) The effects of rate and timing of glyphosate applications of defoliation effciency, regrowth inhibition, lint yield, fiber quality and seed quality. *Proceedings of Beltwide Cotton Conference, National Cotton Council of America*, 1088–1090.

Lotter, D.W., Seidel. R. and Liebhardt, W. (2003) The performance of organic and conventional cropping systems in an extreme climate year. *American Journal of Alternative Agriculture* 18(3), 146–154.

Lucas, W.J. and Lobb, P.G. (1987) Response of potatoes, tomatoes and kumaras to foliar applications of MCPA, MCPB, 2,4-D, clopyralid, and amitrole. *Proceedings 40th N.Z. Weed and Pest Control Conference*, 59–63.

Luo, Y., Zhang, X., Liu, X., Ficklin, D. and Zhang, M. (2008) Dynamic modeling of organophosphate pesticide load in surface water in the northern San Joaquin Valley watershed of California. *Environmental Pollution* 156, 1171–1181.

Lurati, A.R. (2013) Organophosphate exposure with pseudocholinesterase deficiency. *Workplace Health and Safety* 61(6), 243–245.

Lutz, H. and Lutz-Ostertag, Y. (1972) The action of different pesticides on the development of bird embryos. *Advances in Experimental Medicine and Biology* 27, 127–150.

Majewski, M. and Capel, P. (1995) *Pesticides in the Atmosphere: Distribution Trends, and Governing Factor.* Ann Arbor Press Inc., Chelsea, Michigan.

Mannetje, A., Coakley, J., Bridgen, P., Smith, A.H., Read, D., Pearce, N. and Douwes, J. (2014) Estimated infant intake of persistent organic pollutants through breast milk. *New Zealand Medical Journal* 127(1401), 56–68.

Manthripragada, A.D., Costello, S., Cockburn, M.G., Bronstein, J.M. and Ritz, B. (2010) Paraoxonase 1 (PON1), agricultural organophosphate exposure, and Parkinson disease. *Epidemiology* 21(1), 87–94.

Markandya, A., Taylor, T., Long, A., Murty, M.N., Murty, S. and Dhavala, K. (2008) Counting the cost of vulture decline – an appraisal of the human health and other benefits of vultures in India. *Ecological Economics* 67, 194–204.

Martens, D.A. and Bremner, J.M. (1993) Influence of herbicides on transformations of urea nitrogen in soil. *Journal of Environmental Science and Health, B* 28, 377–395.

Martin, A., Whitford, F., Becovitz, J. and Rew, C. (1993) Pesticides and applicator certification. Purdue University Cooperative Extension Service.

Martineau, D., Béland, P., Desjardins, C. and Lagacé, A. (1987). Levels of organochlorine chemicals in tissues of beluga whales (*Delphinapterus leucas*) from the St. Lawrence Estuary, Québec, Canada. *Archives of Environmental Contamination and Toxicology* 16, 137–147.

Martínez Montero, El., Romanos Lezcano, A., Praena Crespo, M., Repetto Jiménez, M. and Martínez Ruiz, D. (1993) Organochlorine compounds: blood levels in mothers, newborns, and infants, in maternal milk and formula. Study in the province of Huelva. *Anales Espanoles De Pediatria* 39(1), 46–52.

Maryoung, L., Lavado, R. and Schlenk, D. (2014) Impacts of hypersaline acclimation on the acute toxicity of the organophosphate chlorpyrifos to salmonids. *Aquatic Toxicology* 152, 284–290.

Mayo, L., Johnson, B., Olson, E., Bartlett, B., Olcott, M., Garlow, C., Skulnik, G., Stennett, E. and Breakey, B. (2004) Not in my back yard: Having a lawn that's green, and green. *Maryland Chapter of the Sierra Club*, Chesapeake, Early Summer 2004, p. 14. Available at: https://www.sierraclub.org/sites/www.sierraclub.org/files/sce/maryland-chapter/Chesapeake/2004%20Early%20Summer.pdf (accessed 28 September 2018).

McBee, K. and Bickham, J. (1992) Mammals as bioindicators of environmental toxicity. In: Genoways, H.H. (ed.) *Current Mammalogy*, Vol. 2, Plenum Publishing Corporation, New York, p. 37.

McBride, J.R., Dye, H.M. and Donaldson, E.M. (1981) Stress response of juvenile sockeye salmon (*Oncorhynnchus nerka*) to the butoxyethanol ester of 2,4-dichlorophenoxyacetic acid. *Bulletin of Environmental Contamination and Toxicology* 27, 877–884.

McConnell, L.L., Nelson, E., Rice, C.P., Baker, J.E., Johnson, W.E., Harman, J.A. and Bialek, K. (1997). Chlorpyrifos in the air and surface water of Chesapeake Bay: Predictions of atmospheric deposition fluxes. *Environmental Science and Technology* 31, 1390–1398.

Mehrotra, K.N. (1989) Pesticide resistance in insect pests: Indian scenario. *Pesticide Research Journal* 1(2), 95–103.

Mendola, P., Barr, D., Walsh, D., Hern, S.N., Rhoney, S., Needham, L., Hilborn, E.D., Gonzales, M., Robertson, G., Carty, C. and Creason, J.P. (2004) Organophosphate pesticide exposures – where are the high risk children? *Presented at International Society for Environmental Epidemiology*, New York, 1–4 August.

Michalakis, M., Tzatzarakis, M.N., Kovatsi, L, Alegakis, A.K., Tsakalof, A.K., Heretis, I. and Tsatsakis, A. (2014) Hypospadias in offspring is associated with chronic exposure of parents to organophosphate and organochlorine pesticides. *Toxicology Letters* 230(2), 139–145.

Miller, B.K. and Clark, D.M. (1997) New CRP provisions provide good return on marginal acres while managing for wildlife. *Hoosier Farmland Wildlife Notes: Fostering Wildlife in Agriculture* 3(1), FNR-157. Forestry and Natural Resources Cooperative Extension Service, Purdue University, West Lafayette, Indiana. Available at: https://www.extension.purdue.edu/extmedia/FNR/FNR-157.pdf (accessed 28 September 2018).

Moorman, T.B. (1989) A review of pesticide effects on microorganisms and microbial processes related to soil fertility. *Journal of Production Agriculture* 2(1), 14–23.

Mora, M.A. (1996) Conger-specific polychlorinated biphenyl patterns in eggs of aquatic birds from the Lower Laguna, Madre, Texas. *Environmental Toxicology and Chemistry* 15, 1003–1010.

Mora, M.A. (1997) Transboundary pollution: Persistent organochlorine pesticides in migrant birds of the southeastern United States and Mexico. *Environmental Toxicology and Chemistry* 16, 3–11.

Morgan, M.K., Wilson, N.K. and Chuang, J.C. (2014) Exposures of 129 preschool children to organochlorines, organophosphates, pyrethroids, and acid herbicides at their homes and daycares in North Carolina. *International Journal of Environmental Research and Public Health* 11(4), 3743–3764.

Muir, D.C.G., Teixeira, C. and Wania, F. (2004) Empirical and modeling evidence of regional atmospheric transport of current-use pesticides. *Environmental Toxicology and Chemistry* 23, 2421–2432.

Muñoz-Quezada, M.T., Iglesias, V., Lucero, B., Steenland, K., Barr, D.B., Levy, K., Ryan, P.B., Alvarado, S. and Concha, C. (2012) Predictors of exposure to organophosphate pesticides in school children in the Province of Talca, Chile. *Environment International* 15, 47, 28–36.

Murray, C.J.L. and Lopez, A.D. (1996) The global burden of disease: A comprehensive assessment of mortality and disability from diseases, injuries and risk factors in 1990 and projected to 2020 [Volume 1 of 10 in the Global Burden of Disease and Injury Series]. Harvard School of Public Health, Cambridge, Massachusetts.

National Park Service (2006) US Department of the Interior (1 August 2006), Sequoia & Kings Canyon National Park: Air quality – airborne synthetic chemicals. Nps.gov (accessed 19 September 2007).

Nigam, S.K., Karnik, A.B., Chattopadhyay, P., Lakkad, B.C., Venkaiah, K. and Kashyap, S.K. (1993) Clinical and biochemical investigations to evolve early diagnosis in workers involved in the manufacture of hexachlorocyclohexane. *International Archives of Occupational and Environmental Health* 65, S193.

NRA Review (2000) The NRA Review of monocrotophos. January 2000. National Registration Authority for Agricultural and Veterinary Chemicals, NRA Review Series 00.1, Canberra, Australia.

Osei-Fosu, P., Donkor, A.K., Nyarko, S., Nazzah, N.K., Asante, I.K., Kingsford-Adabo, R. and Arkorful, N.A. (2014) Monitoring of pesticide residues of five notable vegetables at Agbogbloshie market in Accra, Ghana. *Environmental Monitoring and Assessment* 186(11), 7157–7163.

Palmer, W.E., Bromley, P.T. and Brandenburg, R.L. (2007) Wildlife and pesticides – peanuts. *North Carolina Cooperative Extension Service* (accessed 11 October 2007).

Panneerselvam, N. and Sharma, C.B.S.R. (1990) Cytogenetic hazards of agricultural chemicals. 8. Chiasma frequencies are reduced in barley (*Hordeum vulgare* L.) by sixteen pesticides. *Biologisches Zentralblatt* 109(4), 303–306.

Papendick, R.I., Elliott, L.F. and Dahlgren, R.B. (1986) Environmental consequences of modern production agriculture: How can alternative agriculture address these issues and concerns? *American Journal of Alternative Agriculture* 1(1), 3–10.

Pedersen, T.L. (June 1997) Pesticide residues in drinking water. Available at: http://extoxnet.orst.edu (accessed 15 September 2007).

Pell, M., Stenberg, B. and Torstensson, L. (1998) Potential denitrification and nitrification tests for evaluation of pesticide effects in soil. *Ambio* 27, 24–28.

Pelosi, C., Barot, S., Capowiez, Y., Hedde, M. and Vandenbulcke, F. (2014) Pesticides and earthworms: A review. *Agronomy for Sustainable Development* 34, 199–228.

Perrin, W.F., Brownell, R.L. Jr, Kaiya, Z. and Jiankang, L. (1989) Biology and conservation of the river dolphins. In: Perrin, W.F., Brownell, R.L. Jr, Kaiya, Z. and Jiankang, L. (eds) *Proceedings of the Workshop on Biology and Conservation of the Platanistoid Dolphins*, Wuhan, China, 28–30 October, 1986.

Phillips, B.M., Anderson, B.S., Hunt, J.W., Siegler, K., Voorhees, J.P., Tjeerdema, R.S. and McNeill, K. (2012) Pyrethroid and organophosphate pesticide-associated toxicity in two coastal watersheds (California, USA). *Environmental Toxicology and Chemistry* 31(7), 1595–1603.

Pier, A.B., Bernucci, I., Brambilla, G., Consonni, D. and Pesatori, C.A. (1998) The Seveso studies on early and long-term effects of dioxin exposure: A review. *Environmental Health Perspective* Suppl 106(S2), 5–20.

Pier, A.B., Consonni, D., Bachetti, S., Rubagotti, M., Baccarelli, A., Zocchetti, C. and Pesatori, A.C. (2001) Health effects of dioxin exposure: A 20-year mortality study. *American Journal of Epidemology* 153(11), 1031–1044.

Poinke, H.B. and Glotfelty, D.E. (1989) Nature and extent of groundware contamination by pesticides in an agricultural watershed. *Water Research* 23, 1031.

Prakash, V., Green, R.E., Pain, D.E., Ranade, S.P., Saravanan, S., Prakash, N., Venkitachalam, R., Cuthbert, R., Rahmani, A.R. and Cunningham, A.A. (2007) Recent changes in populations of resident Gyps vultures in India. *Journal of Bombay Natural History Society* 104, 129–135.

Rauh, V., Arunajadai, S., Horton, M., Perera, F., Hoepner, L., Barr, D.B. and Whyatt, R. (2011) Seven-year neurodevelopmental scores and prenatal exposure to chlorpyrifos, a common agricultural pesticide. *Environmental Health Perspectives* 119(8), 1196–1201.

Rawn, D. and Muir, D. (1999) Sources of chlorpyrifos and dacthal to a small Canadian prairie watershed. *Environmental Science and Technology* 33, 3317–3323.

Rawn, D.F.K., Halldorson, T.H.J. and Muir, D.C.G. (1998) Triazines herbicides risk assessment. In: Ballantine, L.G., McFarland, J.E. and Hackett, D.S. (eds) *ACS Symposium Series* 683, 158–176. Americal Chemical Society, Washington DC.

Rawn, D.F.K., Halldorson, T.H.J., Lawson, B.D. and Muir, D.C.G. (1999) A multi year study of four herbicides in air and precipitation from a small prairie watershed. *Journal of Environmental Quality* 28, 898–906.

Reeves, R.R. and Chaudhry, A.A. (1998) Status of the Indus river dolphin *Platanista minor*. *Oryx* 32, 35–44.

Reeves, R.R. and Leatherwood, S. (1994) Dams and river dolphins: Can they coexist? *Ambio* 23, 172–175.

Reeves, R.R., Chaudhry, A.A. and Khalid, U. (1991) Competing for water on the Indus plain: Is there a future for Pakistan's river dolphins? *Environment Conservation* 18341–18350.

Reijnders, P.J.H. (1986) Reproductive failure in common seals feeding on fish from polluted coastal waters. *Nature* 324, 456–457.

Renjun, L. (1990) New advances on population status and protective measures for *Lipotes vexillifer* and *Neophocaena phocaenoides* in the Changjiang River. *Aquatic Mammals* 17, 181–183.

Report of the Special Committee of ICAR (1972) Harmful effects of pesticides. In: Wadhwani, A.M. and Lall, I.J., *Report of the Special Committee of ICAR*. Indian Council of Agricultural Research, New Delhi, India, p. 44.

Rice, C.P. and Cherniak, S.M. (1997) Marine arctic fog: An accumulator of currently used pesticide. *Chemosphere* 35, 867–878.

Roberts, T.R., Hutson, D., Lee, P.W., Nicholls, P.H., Plimmer, J.R. *et al.* (1998) *Metabolic Pathway of Agrochemicals, Part I: Herbicides and Plant Growth Regulators.* The Royal Society of Chemistry, Cambridge, UK.

Roberts, T.R., Hutson, D., Lee, P.W., Nicholls, P.H., Plimmer, J.R. *et al.* (1999) *Metabolic Pathway of Agrochemicals, Part II: Insecticides and Fungicides.* The Royal Society of Chemistry, Cambridge, UK.

Rohitrattana, J., Siriwong, W., Tunsaringkarn, T., Panuwet, P., Ryan, P.B., Barr, D.B., Robson, M.G. and Fiedler, N. (2014) Organophosphate pesticide exposure in school-aged children living in rice and aquacultural farming regions of Thailand. *Journal of Agromedicine* 19(4), 406–416.

Ross, P.S., de Swart, R.L., Reijnders, P.J.H., Loveren, H.V., Vos, J.G. and Osterhaus, A.D.M.E. (1995) Contaminant-related suppression of delayed-type hypersensitivity and antibody responses in harbor seals fed herring from the Baltic Sea. *Environmental Health Perspectives* 103, 162–167.

Rupa, D.S., Reddy, P.P. and Reddi, O.S. (1991) Reproductive performance in population exposed to pesticides in cotton fields in India. *Environmental Research* 55, 123–128.

Saiyed, H.N., Sadhu, H.G., Bhatnagar, V.K., Dewan, A., Venkaiah, K. and Kashyap, S.K. (1992) Cardiac toxicity following short term exposure to methomyl in spraymen and rabbits. *Human and Experimental Toxicology* 11, 93.

Sanders, H.O. (1969) Toxicity of pesticides to the crustacean *Gammarus lacustris*. Technical Papers of the Bureau of Sport Fisheries and Wildlife No. 25. US Department of Interior Fish and Wildlife Service, Washington DC (January).

Santos, A. and Flores, M. (1995) Effects of glyphosate on nitrogen fixation of free-living heterotrophic bacteria. *Letters in Applied Microbiology* 20, 349–352.

Senthilkumar, K., Kannan, K., Sinha, R.K., Tanabe, S. and Giesy, J.P. (1999) Bioaccumulation profiles of polychlorinated biphenyl congeners and organochlorine pesticides in Ganges River dolphins. *Environmental Toxicology and Chemistry* 18, 1511–1520.

Senthilkumar, K., Kannan, K., Subramanian, A. and Tanabe, S. (2000) Accumulation of organochlorine pesticides and polychlorinated biphenylsin sediments, aquatic organisms, birds, bird eggs and bat collected from South India. *Environmental Science and Pollution Research* 7, 1–13.

Shelton, J.F., Geraghty, E.M., Tancredi, D.J., Delwiche, L.D., Schmidt, R.J., Ritz, B., Hansen, R.L. and Hertz-Picciotto, I. (2014) Neurodevelopmental disorders and prenatal residential proximity to agricultural pesticides: The CHARGE study. *Environmental Health Perspectives* 122, 1103–1109.

Singare, P.U. and Dhabarde, S.S. (2014) Pollution scenario due to discharge of effluent from agrochemicals and pesticides manufacturing industries of Dombivali industrial belt of Mumbai, India. *International Letters of Chemistry, Physics and Astronomy* 22, 8–15.

Singh, J.B. and Singh, S. (1989) Effect of 2,4-dichlorophenoxyacetic acid and maleic hydrazide on growth of bluegreen algae (cyanobacteria) *Anabaena doliolum* and *Anacystis nidulans*. *Scientific Culture* 55, 459–460.

Singh, P.B., Singh, V. and Nayak, P.K. (2008) Pesticide residues and reproductive dysfunction in different vertebrates from north India. *Food and Chemical Toxicology* 46(7), 2533–2539.

Singh, S. and Iyer, P.R. (2017) Isolation and characterization of microorganisms from industrial effluents that degrade dimethoate. *International Journal of Research Studies in Microbiology and Biotechnology* 3(3), 6–11.

Skaare, J.U., Tuveng, J.M. and Sande, H.A. (1988) Organochlorine pesticides and polychlorinated biphenyls in maternal adipose tissue, blood, milk, and cord blood from mothers and their infants living in Norway. *Archives of Environmental Contamination and Toxicology* 17(1), 55–63.

Smalling, K.L., Kuivila, K.M., Orlando, J.L., Phillips, B.M., Anderson, B.S., Siegler, K., Hunt, J.W. and Hamilton, M. (2013) Environmental fate of fungicides and other current-use pesticides in a central California estuary. *Marine Pollution Bulletin* 73(1), 144–153.

Sri Lankan Ministry of Health (1995) Annual health bulletin, Sri Lanka Ministry of Health, Columbo.

Spawn, R. L. Hoagland, K. D and Siegfried, B.D. (1997) Effects of alachlor on an algal community from a Midwestern agricultural stream. *Environmental Toxicology and Chemistry* 16(4), 785–793.

Starks, S.E., Gerr, F., Kamel, F., Lynch, C.F., Jones, M.P., Alavanja, M.C., Sandler, D.P. and Hoppin, J.A. (2012) Central nervous system function and organophosphate insecticide use among pesticide applicators in the Agricultural Health Study. *Neurotoxicology and Teratology* 34(1), 168–176.

States of Jersey (2007) Environmental protection and pesticide use. In: Deepa, T.V., Lakshmi, G., Lakshmi, P.S. and Sreekanth, S.K. (2011) Ecological effects of pesticides, pesticides in the modern world – pesticides use and management. In: Stoytcheva, M. (ed.) *Pesticides – the Impacts of Pesticide Exposure*. InTech, London. Available at: https://www.intechopen.com/books/pesticides-in-the-modern-world-pesticides-use-and-management/ecological-effects-of-pesticide-s (accessed 28 September 2018).

Straathoff, H. (1986) Investigations on the phytotoxic relevance of volatilization of herbicides. *Mededelingen van de Faculteit Landbouwwetenschappen, Rijksuniversiteit Gent* 51(2A), 433–438.

Surajudeen, Y.A., Sheu, R.K., Ayokulehin, K.M. and Olatunbosun, A.G. (2014) Oxidative stress indices in Nigerian pesticide applicators and farmers occupationally exposed to organophosphate pesticides. *International Journal of Applied and Basic Medical Research* 4(S1), S37–S40.

Surveillance of Food Contaminants in India (1993) Report of an ICMR Task ForceStudy (Part 1). In: Toteja, G.S., Dasgupta, J., Saxena, B.N. and Kalra, R.L., *Surveillance of Food Contaminants in India*. Indian Council of Medical Research, New Delhi.

Tanabe, S., Senthilkumar, K., Kannan, K. and Subramanian, A.N. (1998) Accumulation features of polychlorinated biphenyls and organochlorine pesticides in resident and migratory birds from south India. *Archives of Environmental Contamination and Toxicology* 34, 387–397.

Tashkent (1998) Part 1. Conditions and provisions for developing a national strategy for biodiversity conservation. *Biodiversity Conservation National Strategy and Action Plan of Republic of Uzbekistan*, prepared by the National Biodiversity Strategy Project Steering Committee with the Financial Assistance of The Global Environmental Facility (GEF) and Technical Assistance of United Nations Development Programme (UNDP) (accessed 17 September 2007).

Thies, M.L. and McBee, K. (1994) Cross-placental transfer of organochlorine pesticides in Mexican free-tailed bats from Oklahoma and New Mexico. *Archives of Environmental Contamination and Toxicology* 27, 239–242.

Thurman, E.M. and Cromwell, A.E. (2000) Atmospheric transport, deposition and fate of triazine herbicides and their metabolites in pristine areas at Isle Royale National Park. *Environmental Science and Technology* 34, 3079–3085.

Tözüm-Çalgan, S.R.D. and Sivaci-Güner, S. (1993) Effects of 2,4-D and methylparathion on growth and nitrogen fixation in cyanobacterium, Gloeocapsa. *International Journal of Environmental Studies* 23, 307–311.

Tyler Miller, G. (2004) *Sustaining the Earth: An Integrated Approach.* January 2004. Thomson/Brooks/Cole, Pacific Grove, California, pp. 211–216.

UC IPM Online (2007) (11 August 2006), What's up, doc? Maybe less air pollution. Statewide IPM Program, Agriculture and Natural Resources, University of California, Available at: http://ipm.ucdavis.edu (accessed 15 October 2007).

US EPA (1996) Reregistration eligibility decision (RED): Trifluralin. Office of Prevention, Pesticides, and Toxic Substances, Washington, DC. National Center for Environmental Publications and Information (EPA/NCEPI), Cincinnati, Ohio. Available at: https://archive.epa.gov/pesticides/reregistration/web/pdf/0179fact.pdf (accessed 28 September 2018).

US EPA (2001) Managing small-scale application of pesticides to prevent contamination of drinking water. *Water Protection Practices Bulletin.* Office of Water, Washington, DC (July). EPA 816-F-01-031.

US EPA (2005) Reregistration eligibility decision (RED) decision for 2,4-D. List A Case 0073. The Office of Pesticide Programs. National Technical Information Service (NTIS) Springfield, Virginia. Available at: https://archive.epa.gov/pesticides/reregistration/web/pdf/24d_red.pdf (accessed 28 September 2018).

US EPA (2007) Sources of common contaminants and their health effects. Emergency Response Program. United States Environmental Protection Agency, Washington, DC.

US Fish and Wildlife Service (2010) Students and educators. US Fish and Wildlife Service. Forensics Laboratory. Last updated 2 February 2010. Sources of common contaminants and their health effects. Available at: https://www.fws.gov/lab/wildlife_laws.php (accessed 10 October 2007).

US Geological Survey (1998) Pesticides in surface and ground water of the United States: Results of the National Water Quality Assessment Program (NAWQA). Pesticide National Synthesis Project. National Water Quality Assessment US GS. Available at: https://ne.water.usgs.gov/Nawqa/pubs/ContaminantsFS/allsum.html.newx (accessed 16 January 2019).

US Geological Survey (1999) The quality of our nation's waters – nutrients and pesticides: US Geological Survey Circular 1225, 82 p. Reston, Virginia. Available at: https://pubs.usgs.gov/circ/circ1225/pdf/front.pdf (accessed 28 September 2018).

Vos, J.G., Dybing, E., Greim, H.A., Ladefoged, O., Lambré, C., Tarazona, J.V., Brandt, I. and Vethaak, A.D. (2000). Health effects of endocrine-disrupting chemicals on wildlife, with special reference to the European situation. *Critical Reviews in Toxicology* 30(1), 71–133.

Waite, D.T., Cessna, A.J., Grover, R., Kerr, L.A. and Snihura, A.D. (2002) Environmental concentrations of agricultural herbicides: 2,4-D and triallate. *Journal of Environmental Quality* 31, 129–144.

Waite, D.T., Bailey, P., Sproull, J.F., Quiring, D.V., Chau, D.F., Bailey, J. and Cessna, A.J. (2005) Atmospheric concentrations and dry and wet deposits of some herbicides currently used on the Canadian prairies. *Chemosphere* 58, 693–703.

Wall, G.R., Riva-Murray, K. and Phillips, P.J. (1998) Water quality in the Hudson River Basin, New York and adjacent states, 1992–1995. US Geological Survey Circular 1165 USGS, Reston, Virginia.

Wang, Y., Wu, S., Chen, L., Wu, C., Yu, R., Wang, Q. and Zhao, X. (2012) Toxicity assessment of 45 pesticides to the epigeic earthworm *Eisenia fetida. Chemosphere* 88(4), 484–491.

Washington State Department of Transportation (1993) Draft roadside vegetation management environmental impact statement, appendix B, B2–10.

Weston, D.P., Ding, Y., Zhang, M. and Lydy, M.J. (2013) Identifying the cause of sediment toxicity in agricultural sediments: The role of pyrethroids and nine seldom-measured hydrophobic pesticides. *Chemosphere* 90, 958–964.

Winston, M.L. (1997) *Nature Wars: People vs. Pests.* Harvard University Press, Harvard.

World Health Organization (WHO) (1990) *Impact of Pesticides Used in Agriculture.* Geneva Public Health. WHO, Geneva.

WHO (2001) *Mental Health: New Understanding, New Hope. The World Health Report.* WHO, Geneva.

Woodcock, M.W. (1980) *Birds of India, Nepal, Pakistan, Bangladesh and Sri Lanka.* Harper Collins, London, p. 176.

Yasmin, S. and d'Souza, D. (2010) Effects of pesticides on the growth and reproduction of

earthworm: A review. *Applied and Environmental Soil Science*, 1. Article ID 678360, 9pp.
Yolton, K., Xu, Y., Sucharew, H., Succop, P., Altaye, M., Popelar, A., Montesano, M.A., Calafat, A.M. and Khoury, J.C. (2013) Impact of low-level gestational exposure to organophosphate pesticides on neurobehavior in early infancy: A prospective study. *Environmental Health* 12, 79.
Zabik, J.M. and Seiber, J.N. (1990) Atmospheric transport of organophosphate pesticides from California's Central Valley to the Sierra Nevada Mountains. *Journal of Environmental Quality* 22(1), 80–90.
Zartarian, V., Ozkaynak, A.H., Xue, J., Glen, W.G. and Furtaw, Jr, E.J. (2000) Modeling aggregate chlorpyrifos exposure and dose to children. *Presented at ISEA Exposure Analysis in the 21st Century: Integrating Science, Policy and Quality of Life*, Monterey Peninsula, California, 24–27 October 2000.

II Pesticide Risk and Assessment

3 Toxicity of Pesticides Based on their Mode of Action

Pesticides are widely used in agriculture for increasing crop yields, as well as in homes and gardens. Farmers spend approximately US$4.1 billion on pesticides annually, with the justification of direct dollar return of $3–5 for every $1 spent on pesticides as per US EPA, 1991 (Pimentel, 1991). The annual application of synthetic pesticides to food crops in the European Union alone exceeds 140,000 tonnes (t) (Eurostat Statistical Books, 1992–2003), an amount that corresponds to 280 g/EU citizen/year. The data indicate no decrease in consumption pattern. Pesticides have been reported as contaminants of food products, fruits, vegetables and cereals, processed food, baby food and drinking water; they can be neurotoxic, immunotoxic or endocrine disruptors.

3.1 Toxicity of Neurotoxic Pesticides

Many pesticides target the nervous system of insect pests. These pesticides are also likely to be neurotoxic to humans, because of the similarity in neurochemical processes. Children with their developing brains are much more vulnerable to injury caused by neurotoxic pesticides than adults (Dobbing, 1968).

The nervous system is the control and communication system of the body. It controls all our thoughts and movements, with the important job of sending and receiving messages throughout the body. It is, therefore, the basis for learning and thinking; sensory perception and movement; behaviour and emotion; and regulation of many of the important functions of the cardiovascular system and other internal organs. Pesticides inducing nervous system damage (*neurotoxic pesticides*) can produce a variety of effects, collectively referred to as *neurologic effects*, and can encompass any of the above functions and behaviours. The nervous system is further subdivided anatomically into the *central nervous system* (CNS), which includes the brain and spinal cord, and the *peripheral nervous system* (PNS), which includes nerves connecting organs and tissues with the spinal cord and brain. The latter include *sensory neurons* and *motor neurons*.

Neurotoxic pesticides have an adverse effect on the CNS or PNS. The insecticides are most often responsible for neurotoxic effects in mammals including humans, as they target the insect's nervous system. The neurotoxicity could be due to interference with chemical neurotransmission or ion channels (Costa *et al.*, 2008). Some herbicides and fungicides have also been shown to possess neurotoxic properties. Pesticides can lead to acute neurotoxicity or chronic neurodegenerative disorders, most notably Parkinson's disease. Besides pesticides, a large number of chemicals have also been reported to cause neurotoxicity including metals (e.g. lead), industrial chemicals (e.g. acrylamide), solvents (e.g. toluene), natural toxins (e.g. domoic acid), pharmaceutical drugs (e.g. doxorubicin) and recreational drugs (e.g. 'ecstasy') (Spencer *et al.*, 2000). The neurotoxicants have been divided into four groups (Anthony *et al.*, 2001):

1. Causing neuronopathies; loss of neurons either by necrosis or by apoptosis.
2. Targeting the axon causing axonopathies. The axon along with the myelin sheath surrounding it degenerates, keeping the cell body intact. Most often it results in a clinical condition called peripheral neuropathy, in which sensations and motor strength are first inpaired in the feet and hands (Anthony *et al.*, 2001).
3. Inducing myelinopathies; the neurons are structurally unaffected, but their functions are altered.
4. Affecting neurotransmission; by inhibiting release of neurotransmitters (e.g. acetylcholine release) (Costa, 1988a).

The target sites for insecticides in insects are also available in mammals, hence insecticides are not species-selective with respect to targets of toxicity, and mammals, including humans, are highly sensitive to their toxicity. The most widely used insecticides are cholinesterase inhibitors (organophosphates and carbamates), followed by pyrethroids and other recently developed compounds, such as the neonicotinoids. Most of the organochlorine compounds (e.g. dichlorodiphenyltrichloroethane, DDT) were banned in the mid-1970s; however, some are still used in certain countries, and they have high environmental persistence, which results in exposure to the general population. Various groups of pesticides responsible for neurotoxicity are discussed here.

3.1.1 Organochlorine compounds

The era of synthetic pesticides started in the 1940s with the introduction of DDT, developed by P. Muller in 1939. The organochlorine insecticides include DDT and its analogues; cyclodienes such as chlordane, aldrin or dieldrin; and the hexachlorocyclohexanes (HCHs), such as lindane. Prior to the mid-1980s, these insecticides were widely used in agriculture, malaria control programmes and forestry. They are cheap to manufacture and effective against target pests. These insecticides are acute neurotoxic chemicals. They are responsible for the disruption of the pumping of calcium, chloride, sodium and potassium ions across the nerve cells. They have moderate acute toxicity but chronic exposure may be associated with adverse health effects, particularly in the liver and the reproductive system (Costa *et al.*, 2008). The poisoning symptoms include headache, dizziness, nausea, vomiting, incoordination, tremor and mental confusion. More severe poisoning causes myoclonic jerking movements, then generalized tonic–clonic convulsions.

The organochlorines may be divided into two subclasses on the basis of distinctly different sites of toxic action, mechanisms and specific symptomology: DDT-type insecticides and chlorinated cyclodienes. DDT and its analogues act primarily on the PNS, and alter membrane permeability for potassium and sodium, and the calcium-mediated processes (Heinzow and Andersen, 2006). Holan (1969) investigated the mechanism of action for DDT-type insecticides. DDT exerts its toxic action at the sodium gates of the axon by preventing the deactivation or closing of that gate after activation and membrane depolarization. The DDT molecules bind to sodium channels in neurons, increasing permeability to sodium. This increased permeability facilitates uncoordinated discharge of neurons. This uncoordinated discharge can lead to CNS alterations including myoclonus and seizures. The mechanism results in leakage of Na^+ ions through the nerve membrane, creating a destabilizing negative after potential. The hyperexcitability of the nerve results in trains of repetitive discharges in the neuron after a single stimulus and/or occurring spontaneously (Narahashi, 1976). Cyclodiene organochlorine pesticides inhibit gamma aminobutyric acid (GABA)ergic neurotransmission by blocking the chloride channel of GABA(A) receptors. They have a CNS stimulating mode of action. They permit only partial repolarization of the neuron and, thus, uncoordinated nervous excitation. This inhibition results in the accumulation of intracellular free calcium. Cyclodienes (dieldrin, heptaclor mirex, aldrin) may induce seizures for days after overexposure.

Target organisms

According to US Environmental Protection Agency (US EPA) reports, fat and fat solvents enhance gastrointestinal, and probably dermal, absorption of organochlorines. Once the organochlorine enters the body, a significant portion of the parent compound is stored in fat tissue. Nearly all organochlorines get dechlorinated, oxidized and conjugated to some extent, yielding measurable urinary metabolites. Biliary is the chief route of excretion and the chemical may be reabsorbed (enterohepatic circulation), slowing excretion. However, much of the unmetabolized organochlorine is efficiently reabsorbed by the intestine, substantially retarding faecal excretion. Storable lipophilic compounds are likely to be excreted in maternal milk (Solomon *et al.*, 1977; Stevens *et al.*, 1993; Rogan, 1996).

The organochlorines are partitioned and stored mainly in adipose tissues because of their lipophilicity, where they remain biologically inactive. Experimental studies have indicated that neuroactive-contact insecticides must be highly lipophilic to penetrate the insect cuticle as well as the nerve sheath. The use of hydrophobicity parameters in quantitative structure-activity relationship (QSAR) analyses has attempted to confirm and describe that importance (Coats, 1990).

Coats (1990) reported that the acute toxic effects of organochlorine insecticides in animals are principally due to hyperexcitation in the nervous system, resulting in the disruption of nervous system function followed by respiratory failure leading to death. Neurophysiological observations have revealed hyperexcitation and repetitive discharges in insect nerves.

Non-target organisms

The organochlorine pesticides were deliberately synthesized to persist in the environment with very slow degradation and solubility in fat tissues for easy intake by insects, with the expectation that they would be harmless to mammals. But the properties of high persistence and fat solubility were actually highly undesirable, leading to their entry into the food chain, where they bioaccumulated in the fat tissues of large animals and humans and were passed on to their offspring. This resulted in contamination of not only avians but also wildlife around the globe, including the Arctic and Antarctic regions, where these insecticides are rarely used. Unexpectedly, DDT caused a thinning of birds' egg shells and resulted in the death of their developing young (Lah, 2011).

Lindane, aldrin, dieldrin, endrin and toxaphene often produce convulsions as the first sign of injury. Among organochlorines, the cyclodienes, such as endrin, dieldrin and isobenzan, are observed as the most toxic to humans and perhaps >10-fold more acutely toxic than DDT (Joy, 1982). These organochlorines have been banned in most countries owing to ecological considerations, However, some, such as DDT, are being reintroduced in parts of the world for malaria control. The organochlorines are reported to impair foetal growth and affect birth size. In humans, there is no placental barrier against the transfer of organochlorine pesticides across the maternal–foetal axis. The transfer of organochlorine pesticides across the placenta could impair the anthropometric development of the foetus, reducing the birthweight, length, head circumference, chest circumference and ponderal index, which is associated with reduced birth size of the offspring. HCH was most consistently associated with decreased birth size of the neonate (Dewan *et al.*, 2013). Some of the symptoms of organochlorine poisoning are dizziness, tremor of the extremities, confusion and vomiting, while convulsions occur only in severe poisoning. DDT interferes with the sodium channels in the axonal membrane via a mechanism similar to that of pyrethroids, both in insects and mammals (Vijveberg *et al.*, 1982). Treatment for DDT poisoning focuses on the nervous system. In animals, DDT-induced tremors and mortality have been found to be reduced by phenytoin and calcium gluconate, respectively. In humans, diazepam or phenobarbital, given orally, may be beneficial to control seizures and myoclonic movements that sometimes persist for several days following acute poisoning by the more slowly excreted organochlorines.

Overexposure to organochlorine pesticides has led to a range of symptoms associated with health, such as cancer. Exposure to DDT has been linked to pancreatic cancer and non-Hodgkin's lymphoma (Cantor *et al.*, 1992; Garabrant *et al.*, 1992). Increased risk of breast cancer is also associated with exposure to DDT early in life (Cohn *et al.*, 2007). Women from an agricultural region in India with breast cancer were shown to have significantly higher concentrations of organochlorines in their blood, irrespective of their age, diet, and geographic distribution (Mathur *et al.*, 2002). Prenatal exposure to the organochlorine pesticide chlordane has been linked to the reduction of testosterone in adult female rats and behavioural changes in both sexes (Cassidy *et al.*, 1994). Organochlorine pesticide exposure is associated with neurodevelopmental health effects in humans too. Exposure to organochlorine pesticides has been associated with decreased memory, attention and verbal skills in children (Jurewicz and Hanke, 2008; Korrick and Sagiv, 2008). Organochlorine exposure has also been found to be associated with the development of autism (Roberts *et al.*, 2007). The concentration of dichlorodiphenyldichloroethylene (DDE, a metabolite of DDT) in a mother's blood was observed to be associated with preterm birth and baby size. The higher the concentration, the more likely she was to have a preterm birth and the baby was more likely to be small for its gestation age (Longnecker *et al.*, 2001). These pesticides have been found to alter levels of maternal thyroid hormones during pregnancy (Chevrier *et al.*, 2008). Many organochlorine chemicals, including DDT, endosulfan and lindane, are known to produce anti-thyroid effects (Román, 2007). Thyroid deficiencies during pregnancy and postpartum are known to cause altered development, retardation, decreased intellectual capacity, psychomotor delays and deafness. These thyroid-disrupting pesticides may play a role in the development of autism (Román, 2007). Neurodegenerative diseases such as Parkinson's disease and Alzheimer's disease are more common in people with general pesticide exposures, including organochlorine pesticide exposure (Kamel and Hoppin, 2004).

3.1.2 Organophosphate and carbamate compounds: mode of action and toxicity

Pesticides of both these groups are acetylcholinesterase enzyme (AChE) inhibitors, leading to poisoning in humans, animals and wildlife; however, they exhibit different toxicological features. The organophosphates were introduced in the early 1940s, followed by carbamates in the 1950s, as insecticides (Costa, 1988b), and gained wide popularity due to their lower persistence in the environment and mammalian systems, and also due to lower development of resistance in insects, as compared with organochlorines (Marrs, 2012). The organophosphate insecticides had been a cost-effective approach for managing crop-destroying, disease-carrying and residential insect pests. The organophosphates do not persist in the environment due to their rapid degradation via hydrolysis after exposure to sunlight, air and soil, and hence they were thought to be a safer and more attractive alternative to highly persistent organochlorines; however, their anticholinesterase activity categorized them as toxic to beneficial and non-target organisms (Cohen, 2014).

Target organisms

AChE is responsible for hydrolysing acetylcholine, a major neurotransmitter in the central and peripheral (autonomic and motor-somatic) nervous systems, producing choline and acetic acid. Inhibition of AChE by organophosphates causes accumulation of acetylcholine at cholinergic synapses, with overstimulation of cholinergic receptors of the muscarinic and nicotinic type. This results in 'cholinergic syndrome', symptoms of which include increased sweating and salivation, profound bronchial secretion, bronchoconstriction, miosis, increased gastrointestinal motility, diarrhoea, tremors, muscular twitching and various CNS effects.

Most commonly used organophosphate insecticides contain a P = S (thione) analogue, which needs to be metabolically bioactivated to oxon P = O moiety, as only oxon analogues are potential inhibitors of AChE. This bioactivation involves an oxidative desulfuration mediated by enzymes of the cytochrome P450 family (CyPs), which leads to the formation of an oxygen analogue (oxon) of the parent insecticide. Biotransformations other than this are detoxication reactions, as they lead to metabolites of lesser or no toxicity (Costa, 1988b; Lotti, 2000). The organophosphate insecticides, although less persistent than organochlorines, have high acute toxicity, with oral LD_{50} values in rat often below 50 mg/kg. The value is still lower for some organophosphates like malathion. The P = O moiety in organophosphates phosphorylates a hydroxyl group on serine in the active (esteratic) site of the enzyme, thus restricting its action on the physiological substrate.

Non-target organisms

Monocrotophos, one of the organophosphate insecticides used in agriculture, is one of the most bird-toxic agricultural pesticides in use. Avian mortalities have been documented to have reached over 100,000, including mass-poisoning of nearly 6000 Swainson's hawk (*Buteo swainsoni*) in Argentina during 1995/96, since the introduction of monocrotophos, in 1964 (Goldstein *et al.*, 1999). Despite this, it remains one of the most heavily used pesticides in the world. In another incident in Keoladeo National Park, northern India, 15 Sarus crane (*Grus antigone*) were found dead in a field near their wintering grounds in November 2000 (Pain *et al.*, 2004). The deaths occurred when cranes had eaten recently sown monocrotophos-treated wheat (Meine and Archibald, 1996). In another poisoning incident in Mongolia in 2002, about 3500 km^2 of steppe were treated with the rodenticide bromadiolone, following a population explosion of voles, resulting in more than 340 deaths of birds at several localities, including 145 Demoiselle crane (*Grus virgo*) and large raptors such as steppe eagle (*Aquila nipalensis*) and Saker falcon (*Falco cherrug*) (Natsagdorj and Batbayar, 2002). The actual scale of the mortality was presumably much larger than was observed. Unfortunately, intentional poisoning of birds with pesticides is a common practice in some parts of the world, for example, South Africa, where farmers are attempting to prevent crop damage or to control perceived predators of small livestock. This type of poisoning is one of the greatest threats to the vulnerable blue crane (*Grus paradisea*) (Barnes, 2000), and in Europe numerous raptor species, including a number of globally threatened birds, are deliberately poisoned in this way (Mineau *et al.*, 1999).

The organophosphates are responsible for respiratory failure, caused by inhibition of respiratory centres in the brainstem, bronchoconstriction and increased bronchial secretion, and flaccid paralysis of respiratory muscles, leading ultimately to death (Lotti, 2000; 2001).

The other non-target species affected by organophosphates are mostly invertebrates, fish, birds (including migratory), mammals, fungi and plants. Beneficial insects (except for honeybees) and beneficial fungi are also affected by organophosphates. These organophosphates have caused a loss of 67 million birds yearly, according to US reports (Tesfahunegny and Mulualem, 2016) .

Chemicals like oximes can enhance the slow hydrolysis of phosphorylated AChE by water, and hence, they are a common treatment for organophosphate poisoning in human medical literature (Sungur and Güven, 2001; Marrs and Vale, 2006; Jokanovi, 2009; Verma and Ahmad, 2009). The oximes become ineffective in reactivating phosphorylated AChE once the enzyme-inhibitor complex has 'aged', due to loss of one of the two alkyl groups. The enzyme inhibition reaction becomes irreversible, once phosphorylated AChE has aged (Sun *et al.*, 1979; Carletti *et al.*, 2008, 2010), and the only means of replacing its activity is through synthesis of new enzyme, a process that may take days. Atropine, a cholinergic muscarinic antagonist and the main antidote for organophosphate poisoning, blocks muscarinic receptors and prevents the accumulation of acetylcholine on these receptors (Lotti, 2001).

Carbamates were developed in the 1950s and are still used today (Wheeler, 2002). Carbamate insecticides have varying degrees of acute oral toxicity, ranging from moderate to low toxicity (carbaryl), to extremely high toxicity (e.g. aldicarb; $LD_{50} = 0.8$ mg/kg). The signs and symptoms of carbamate poisoning are the same as observed with organophosphate intoxication, based on excessive cholinergic stimulation (Marrs, 2012). The main differences in the modes of action of carbamates are: (i) some carbamates are potent inhibitors of aliesterase (miscellaneous aliphatic esterases whose exact functions are not known), and their selectivity is sometimes more pronounced against the ChE of different species; and (ii) ChE inhibition by carbamates is reversible (Ware and Whitacre, 2004).

When ChE is inhibited by a carbamate, it is said to be *carbamylated*, as when an organophosphate results in the enzyme being *phosphorylated*. In insects, the effects of organophosphates and carbamates are primarily those of poisoning of the CNS, since the insect neuromuscular junction is not cholinergic, as in mammals. The only cholinergic synapses known in insects are in the CNS. (The chemical neuromuscular junction transmitter in insects is thought to be glutamic acid.)

The structural differences between organophosphates and carbamates affect the efficiency and degree to which the acetylcholinesterase is blocked. Muscle weakness, dizziness, sweating and slight body discomfort are commonly reported early symptoms. Headache, salivation, nausea, vomiting, abdominal pain and diarrhoea are often prominent at higher levels of exposure. Contraction of the pupils with blurred vision, incoordination, muscle twitching and slurred speech have also been reported (Fishel, 2013). However, acute intoxication by carbamates is generally resolved within a few hours. The reason being, unlike organophosphates, carbamate poisonings tend to be of shorter duration because the inhibition of AChE is reversible due to reactivation of carbamylated AChE. The carbamylated AChE does not undergo the ageing reaction that occurs with organophosphates, resulting in more rapid metabolization. Carbamates are, therefore, an alternative to irreversible phosphorylation of AChE, physostigmine (eserine, natural alkaloid) by organophosphates (Gupta, 2006). The carbamates penetrate the cuticle of plant tissue providing some resistance to wash off during rains in comparison to organophosphates, which have limited penetrative potential (Wise, 2014). Carbamates generally have low vapour pressure and low water solubility, making them slow to evaporate and less soluble in water.

These insecticides are rapidly detoxified and excreted in warm-blooded animals and, in general, they are selective against targeted insect pests (Sogorb and Vilanova, 2002). However, carbamates are toxic to pollinators, such as honeybees. They have low persistence in soil, plants and the environment, requiring several applications over a growing season (Wheeler, 2002). There is no evidence of carbamates causing delayed neurotoxicity (Wheeler, 2002); also they are not regarded as mutagenic, carcinogenic or teratogenic substances. However, long-term exposure could be associated with neuropsychological function impairment, and there is evidence of a chronic effect of cumulative high exposure to these compounds (Hernández *et al.*, 2004; Roldán-Tapia *et al.*, 2005).

Atropine, a muscarinic antagonist, is the antidote for carbamates, as well as organophosphates. Oximes have been shown to aggravate the toxicity of carbaryl, but may have beneficial effects in the case of other carbamates, such as aldicarb (Ecobichon, 2001).

3.1.3 Pyrethroids: mode of action and their toxicity

Pyrethroids, developed in the 1970s, are synthetic versions of pyrethrins, extracted from the chrysanthemum plant, and chemically engineered to be more toxic with longer breakdown times. Pyrethroids are widely used as agricultural and household insecticides. Due to their low mammalian and avian toxicity, the World Health Organization (WHO) has recommended treating bednets with pyrethroids, and they play a crucial role in vector-based malaria intervention strategies (WHOPES, 2005; Raghavendra *et al.*, 2011). The pyrethroids have more stability in light, heat and moisture, and more persistance than their parent compounds, pyrethrins, in spite of their chemical and toxicological similarity (Gosselin *et al.*, 1984). Pyrethroids react with voltage-gated sodium channels in nerves, prolonging the time during which the channels are open. Incoordination, convulsions and paralysis are caused by exposure to toxic doses of these compounds (Soderlund and Bloomquist, 1989). Based on observed symptoms and mode of action on sodium channels, pyrethroids are generally divided into two classes, type I and II compounds. A key structural difference between type I and type II pyrethroids is the presence of a cyano group in the latter. Exposure to each type produces different symptoms. However, certain pyrethroids produce a combination of the two syndromes (Vijveberg *et al.*, 1982; Ray and Fry, 2006). Symptoms include continuous tingling or pricking or, when more severe, burning sensations in the skin. Paraesthesia is presumably due to abnormal pyrethroid-induced repetitive activity in skin nerve terminals (Ray and Fry, 2006). Unlike organophosphate pesticides, pyrethroids exhibit low toxicity to mammals and birds, however they have strong selectivity for insects and invertebrates (Ecobichon, 1996; Fishel, 2005). They have been reported to have low potential to bioaccumulate and proficient detoxicification in mammalian receptors. According to a WHO report, the synthetic pyrethroids are neuropoisons acting on the axons in the PNS and CNS by interacting with sodium channels in mammals and/or insects. The metabolism of pyrethroids includes breakage of the ester bond by esterase action and oxidation on various parts of the molecule (Reigart and Roberts, 1999). The mechanism of action of pyrethroids is the same in insects and in mammals.

Target organisms

The ketoalcoholic esters present in pyrethrins and pyrethroids are responsible for their insecticidal properties. Being strongly lipophilic, these acids rapidly penetrate many insects and paralyse their nervous system (Reigart and Roberts, 1999). Because of their residual activity, pyrethroids provide excellent control of pest invertebrates, resulting in strong repellency, and high toxicity (Hammond, 1996). These pesticides are often formulated with synergists to increase their potency and persistence in the environment, retaining low mammalian toxicity and compromising the human body's ability to detoxify the pesticide. The processes critical to physiological growth of insect pests and their ability to feed are disrupted by pyrethroid exposure (Juarez, 1995). Dietary exposure to synthetic pyrethroids including cypermethrin, deltamethrin, fenvalerate and permethrin decreased feeding rates in the red flour beetle (*Tribolium castaneum*), a stored grain pest, resulting in reduced larval weight, delayed pupation and delayed adult emergence (Ishaaya *et al.*, 1983).

The pyrethroids have also been reported to disrupt chemical communication among invertebrates, besides interfering with growth processes. Adult tobacco hornworm

(*Manduca sexta*) females exposed to permethrin were less likely to attract males than were non-exposed females (Haynes, 1988). Similarly, permethrin was found to interfere with the chemical communication of male and female pink bollworm moth (*Pectinophora gossypiella*) (Haynes and Baker, 1985). The sensitivity of sex pheromone in both males and females is greater than the disruption of flight abilities. Pyrethroid exposure can also reduce the number and viability of eggs produced by pest invertebrates. Doses as low as 0.05 μg cypermethrin per tick have been shown to reduce average ovary weight, average oocyte length and ovary vitellogenin (Friesen and Kaufman, 2003). Likewise, deltamethrin reduced the production of both ootheca and viable offspring in German cockroaches (*Blattella germanica*) (Lee *et al*., 1998). Deltamethrin significantly reduces fecundity by reducing longevity of exposed male and female roaches.

Non-target organisms

Both pyrethroids and pyrethrins are moderately or slightly toxic to birds but extremely toxic to aquatic organisms, such as the bluegill and lake trout fish, at the recommended levels of application. Lobster, shrimp, mayfly nymphs and zooplankton are the most susceptible non-target aquatic organisms (Mueller-Beilschmidt, 1990; Fishel, 2005), with exposure leading to damage of gills and behavioural changes. The predators and beneficial insects may become susceptible to lower dosages of pyrethroids. High toxicity of pyrethroids to aquatic organisms could be attributed to their accumulation in sediments. Pyrethroids have been detected in sediments of water bodies adjacent to residential/urban areas, in a recent water quality monitoring study in California (US EPA, 2009). The accumulation of pyrethroids in sediments could be due to their movement into non-target areas through runoff or spray drift, which may occur during application (US EPA, 2014). The low vapour pressures and high octanol–water coefficients of pyrethroids indicate a low tendency to volatilize and a high affinity for organic matter, soils and clay (Oros and Werner, 2005). The availability of these insecticides on application sites is retained; however, their bioavailability to non-target organisms is reduced due to this sorption property.

Similar to pyrethrins, pyrethroids may act as dermal and respiratory allergens. Several signs and symptoms of poisoning by pyrethroids have been reported. Exposure to them can cause contact dermatitis and asthma-like reactions. Persons, especially children, with a history of allergies or asthma are particularly sensitive, and a strong cross-reactivity with ragweed pollen has been recognized (Beyond pesticides, Daily News Blog, Washington DC). Other symptoms of acute toxicity due to inhalation include sneezing, nasal stuffiness, headache, nausea, incoordination, tremors, convulsions, facial flushing and swelling, and burning and itching sensations. The most severe cases of poisoning have been reported in infants, who are not able to efficiently break down pyrethroids (ETN, Pyrethroids, 1994). Death in these cases is due to respiratory failure. Symptoms of acute exposure last about 2 days. They are also reported to disrupt the endocrine system, affecting reproduction and sexual development, and the immune system, increasing the risk of breast cancer (Garey and Wolff, 1998).

The pyrethroids are highly toxic to honeybees. Exposure to them could alter honeybee behaviour associated with maintenance, feeding and communication. The intoxicated bees spent less time walking, giving food and antennae touching, and had altered flight patterns and homing abilities as compared to non-exposed bees, according to Vandame *et al*. (1995). It was observed during the study that approximately 90% of control bees returned to the hive within 30 s of flight, however, only 9.0% of treated bees were able to return within this time period. This delay is most likely related to the affected bees' inability to correctly orient themselves using the sun after being exposed to deltamethrin. Dietary exposures also cause behavioural aberrations, and decrease fecundity. Daily egg production was reduced by consumption of both bifenthrin and deltamethrin, and time in the egg stage was extended. Deltamethrin exposure also reduced capping rate and extended the duration of the immature stage (Dai *et al.*, 2010). Other pyrethroid effects occurred sporadically, as effects on rate of development, number of feeding larvae (unsealed brood stage), hatch rate, and number of post-feeding larvae and pupae (sealed brood stage), were not significant.

The parasitoid wasps, a crop beneficial organism, use agricultural pests as their nests to complete their life cycle. They lay their eggs inside aphids, larval moths and beetles, and consume the host from the inside, eventually causing death. The species-specific nature of the parasitoid–host relationship increases their economic benefit. However, exposure to sublethal concentrations of pyrethroids has been observed to interfere with parasitoid host-finding behaviours, reaction to mating pheromones and reproduction. Olfactory orientation towards prey populations was reduced in aphid parasitoids following exposure to sublethal concentrations of lambda-cyhalothrin (Desneux *et al.*, 2007). Similarly, addition of deltamethrin to honeydew residues caused significant parasitoid repellency (Longley and Jepson, 1996a; 1996b). Deltamethrin exposure at 10% of the LD_{50} also altered sex pheromone production by the female moth egg parasitoid *Trichogramma brassicae*; males localized less precisely to pheromone extracts from exposed females than from unexposed females (Delpuech *et al.*, 1999). Although an inability to locate hosts or mates can cause a reduction in parasitoid fecundity, there is evidence that pyrethroid exposure can also alter other aspects of parasitoid reproduction. When exposed to lambda-cyhalothrin prior to oogenesis, the trichogramma wasp (*Trichogramma pretiosum*) experienced a reduction in fecundity (Desneux *et al.*, 2007). The sublethal deltamethrin and lambda-cyhalothrin exposures altered male:female ratios in exposed wheat aphid parasitoids (*Aphidius uzbekistanicus*) and resulted in a decreased number of female offspring. It was observed in hymenopteran species that females are produced by fertilized eggs, while unfertilized eggs produce males, and that the insecticide exposure might have impacted the behaviours associated with egg fertilization (Desneux *et al.*, 2007). However, data from field studies indicate that sublethal effects of pyrethroid exposure may not have large-scale impacts on overall parasitoid population viability (Longley, 1999).

The piperonyl butoxide (PBO) added to formulate pyrethroids has also been shown to be toxic to non-target organisms. It inhibits important liver enzymes responsible for the breakdown of some toxins, drugs and other chemicals including the active ingredients of pesticides. Heavy exposure to pyrethroid synergist may make the person vulnerable to many toxic insults. The symptoms of PBO poisoning include anorexia, vomiting, diarrhoea, intestinal inflammation, pulmonary haemorrhage and perhaps mild CNS depression. Exposure has also caused slight skin irritation, and increased liver weights. Takahashi *et al.* (1994) reported its toxicity as carcinogenic in animals, which is not considered a concern by US EPA.

3.1.4 Other insecticides

Several other classes of insecticides are also reported as neurotoxic to non-target species. *In vivo* and *in vitro* studies have shown formamidines to be selective agonists at alpha 2-adrenergic receptors (Costa *et al.*, 1988). Several cases of acute poisoning with amitraz have been reported in children in Turkey (Caksen *et al.*, 2003). Similar signs and symptoms of poisoning to those caused by alpha 2-adrenergic receptor agonists were recorded, such as nausea, hypotension, hyperglycaemia, bradycardia and miosis. The alpha 2-adrenoceptor antagonist, yohimbine, has proved useful as an antidote in animals (Andrade and Sakate, 2003), but has not been evaluated in humans.

Plant based

Rotenone, present in the roots of derris plants, is used as an agricultural insecticide/acaricide, particularly in organic farming (Isman, 2006). It has the ability to inhibit the mitochondrial respiratory chain, by blocking electron transport in nicotinamide adenine dinucleotide (NADH)-ubiquinone reductase, the energy conserving enzyme complex. Nicotine, an alkaloid extracted from the leaves of tobacco plants, is a systemic insecticide effective for a wide range of insects. Nicotine activates nicotinic acetylcholine receptors in target and non-target organisms, leading to stimulation followed by protracted depolarization, which results in receptor paralysis. The LD_{50} of nicotine is usually below 50 mg/kg (Ujvary, 2001). The main systems affected by nicotine are nervous, cardiovascular, respiratory, endocrine and gastrointestinal tract. Signs and symptoms of poisoning include nausea, vomiting, muscle weakness, respiratory effects, headache, lethargy and tachycardia. Nicotine can be absorbed by all routes of exposure. The neonicotinoids, a new class of insecticides, have been developed by chemical modifications of nicotine (Matsuda *et al.*, 2001). The neonicotinoids have similar mammalian toxicity to that of nicotine. Contrarily, they are very selective and specific towards insect nicotinic receptors (Tomizawa and Casida, 2005). Some of the neonicotinoids may act as pro-insecticides, for example, derivatives of imidacloprid are converted to imidacloprid.

Herbicides

Herbicides are the chemicals used to kill or severely injure undesired plants or vegetation. Although herbicides have lower acute toxicity than insecticides, various classes of herbicides have been associated with adverse effects ranging from carcinogenicity, target organ toxicity, to endocrine disruption. Examples are 2,4-dinitrophenylhydrazine (2,4-D) and paraquat herbicides. The chlorophenoxy herbicide 2,4-D is widely used for the control of broadleaved weeds. It has a simple aliphatic carboxylic acid moiety attached to a chlorine- (or methyl-) substituted aromatic ring by an ether bond. *In vivo*, the salts and esters are rapidly dissociated or hydrolysed, and therefore, the toxicity of each chlorophenoxy compound depends principally on the acid form of the pesticide (Charles *et al.*, 1996).

Extensive investigations have suggested that paraquat, after absorption, accumulates mainly in the lung and then in the kidney, irrespective of the route of exposure (Bus and Gibson, 1984; Smith, 1987). The reduced form of paraquat, a free radical, is reoxidized to cationic form with a concomitant production of superoxide anion (O^{2-}), a process known as 'redox cycling'. This superoxide anion is responsible for cytotoxicity that initiates lipid peroxidation, ultimately leading to cell death (Bus and Gibson, 1984). Several reports are cited in the literature of acute poisoning with paraquat in humans leading to high mortality (Wesseling *et al.*, 2001). Chronic paraquat exposure has been suggested as a possible aetiological factor in the development of Parkinson's disease. Paraquat can also cause CNS effects in animals (McCormack *et al.*, 2002). Glyphosate exposure is also associated with oxidative damage and neurotoxicity. Cattani *et al.* (2014) investigated whether acute exposure to Roundup® (glyphosate) increases $^{45}Ca^{2+}$ influx by activating N-methyl-D-aspartate (NMDA) receptors and voltage-dependent Ca^{2+} channels, leading to oxidative stress and neural cell death. It was observed that both acute and chronic exposure to glyphosate decreased 3H-glutamate uptake and metabolism, while inducing $^{45}Ca^{2+}$ uptake and ^{14}C-methylaminoisobutyric acid (^{14}C-MeAIB) accumulation in immature rat hippocampus. The glyphosate-induced neurotoxicity also involved the activation of calmodulin-dependent protein kinase II (CaMKII) and extracellular signal-regulated kinase (ERK).

Fungicides

Fungicides are the chemicals extensively used in crop protection against fungi and moulds, and generally comprise a large number of compounds with diverse chemical structures. Several fungicides are carcinogenic (e.g. captan), while others may have skin or eye irritant properties (e.g. chlorothalonil). Dithiocarbamates and some organic metal compounds, no longer used now, are considered as neurotoxic (Clarkson, 2001). The organic mercury compounds, such as methylmercury, were used extensively as fungicides in the past for the prevention of seed-borne diseases in grains and cereals. However, their use is banned due to their high neurotoxicity and risk of human poisoning (Bakir *et al.*, 1973). The symptoms include parasthesia, ataxia, fatigue, hearing and vision loss, plasticity and tremors (Clarkson and Magos, 2006).

Dithiocarbamates, such as maneb (Mn), ziram and zineb (Zn) and mancozeb (Mn and Zn), have low acute toxicity; however, chronic exposure is associated with adverse effects that may be due to the dithiocarbamic acid or the metal moiety present in the structure. The toxicity and teratogenicity of the dithiocarbamates is due to the action of the metabolite ethylenthiourea (ETU) on the thyroid. The impact on thyroid function leads to potential developmental neurotoxicity, as thyroid hormones have an important role in brain development (Chan and Kilby, 2000). The dithiocarbamates may also be neurotoxic via a mechanism that doesn't involve ETU. One of the symptoms of a high dose may be hindlimb paralysis, which is possibly related to the release of the carbon disulfide moiety (Johnson *et al.*, 1998). Chronic exposure to the manganese moiety in maneb, rather than the dithiocarbamate, has been observed to be associated with Parkinsonism (Ferraz *et al.*, 1988; Meco *et al.*, 1994). The structure of dithiocarbamate fungicides resembles that of disulfiram, a compound used therapeutically to produce intolerance to alcohol, by virtue of its ability to inhibit aldehyde dehydrogenase. Interactions of dithiocarbamates with alcohol, leading to elevation in acetaldehyde levels, have been reported.

3.2 Toxicity of Immunotoxic Pesticides

The *immune system* consists of biological structures and processes constituting a set of first defence agents including the mononuclear phagocytic cells (macrophages in tissues, monocytes in circulation) and the natural killer cells, as well as specific lymphoid tissues dispersed throughout the body. To function properly, an immune system must detect a wide variety of agents, known as pathogens, from viruses to parasitic worms, or cells that have been rendered abnormal from ageing, injury or disease, including certain abnormalities associated with cancerous cells, and distinguish them from the organism's own healthy tissue. Abnormal cells, once identified, are destroyed and replaced. This process underlies the body's recognition and healing of wounds. Every human being consists of a highly organized collection of cells in an environment shared with cells of other living organisms. The immune system is able to distinguish cells of the individual (*self*) from cells of other organisms (*foreign*), and destroy foreign cells (*immune response*) that enter the body inadvertently, which might otherwise produce lethal disease.

Immunotoxic pesticides are chemicals that disrupt the function of the immune system. These immunotoxicants can either impair or stimulate the immune responses, depending on the mechanism of action (Durkin and Diamond, 2002). The dysregulation of the immune system is represented by dysregulation of cytokine gene expression, making them valuable markers to study the alteration of the immune system. The analysis of cytokine gene expression, therefore, has been widely employed for immunotoxicity testing, that is, evaluating the toxic effects of pesticides on the immune system (Oliveira *et al.*, 1995; Corsini *et al.*, 1998; Dastych *et al.*, 1999; Vandebriel *et al.*, 1999; Voccia *et al.*, 1999; House and Descotes, 2007). The cytokines are an important component of the immune system connecting its dispersed elements. Insecticides belonging to the organochlorine, organophosphate, carbamate and pyrethroid groups, and some herbicides and fungicides, are reported as immunotoxic pesticides.

3.2.1 Organochlorines

A diet with 100–400 ppm of DDT for 5 weeks decreased the humoral immune response to bovine serum albumin according to Glick (1974). Similarly 200 ppm of DDT or its metabolite suppressed the generation of both humoral and cellular immune responses to ovalbumin in rats (Banerjee *et al.*, 1996). Lindane administered at a dosage of 3 mg/kg/day for 5 weeks decreased the humoral immune response to bacterial antigen in the rabbit (Desi *et al.*, 1978). The organochlorines alter the membrane fluidity; and modulation of membrane-mediated changes in cells is due to an immunotoxic mechanism of organochlorines (Rodgers, 2001). A study of a chemical mixture of polychlorinated biphenyls (PCB), organochlorine pesticides and methylmercury (MeHg) in children revealed that early exposure to these toxicants manifests itself as aberrant development of children's Th1/Th2 immune response. Similarly, amino acid derivatives induce up-regulation of interleukin-5 (IL-5) expression in lymphocytes *in vivo* (Yamaoka *et al.*, 1994) and chemokines in keratinocytes, respectively (Yamaoka *et al.*, 1994; Hayley *et al.*, 2011). Histopathological analysis revealed decreased cellularity of the periarteriolar lymphoid sheath (PALS; T-cell area) and decreased germinal centre (GC) development in all three strains of mice treated with 300 mg/kg of methoxychlor (MXC), suggesting MXC has an immune-suppressive effect in mice (Hayashi *et al.*, 2013). Similarly, endosulfan depressed humoral and cell-mediated immunity (CMI) in rodents at doses that do not show any other overt signs of toxicity (Banerjee and Hussain, 1986; 1987). The *in vitro* exposure of endosulfan in peripheral sheep leukocytes indicated genotoxic and immunotoxic effects (Pistl *et al.*, 2001).

3.2.2 Organophosphates

The main toxicity of organophosphates is neurotoxicity, through inhibition of acetylcholinesterase, but they also affect the immune response. It has been found through experimental and observational studies that organophosphate pesticides can be immunotoxic to humans and animals. The toxic effects of organophosphates on the immune system can be direct or indirect and are reflected in different immune organ pathologies by lowering humoral and/or cell immunity. Direct immunotoxic effects of organophosphates can be due to:

- Inhibition of serine hydrolases (complement system) or esterases (lymphocyte and monocyte membranes) in the immune system.
- Oxidative damage of immune system organs.
- Changes in signal transduction pathways that control proliferation and immune cell differentiation.

Indirect immunotoxicity of organophosphates is expressed as changes in the nervous system or chronic effects of altered metabolism on the immune system. Selecting appropriate biomarkers and biological methods is very difficult since the physiological diversity of immune systems among organisms is very high (Galloway and Handy, 2003). Persons exposed to chlorpyriphos for a period of 1–5 years were more likely to face allergies and sensitivities to antibiotics along with copiousness of certain lymphocytes (a decrease in T-cells and an increase in cluster of differentiation 26 (CD26) cells) (Thrasher *et al.*, 1993).

Li (2007) reported immunotoxicity of organophosphates, observing their effects on antibody production, IL-2 production, T-cell proliferation, decrease in CD5 cells, and increases in CD26 cells and autoantibodies, Th1/Th2 cytokine profiles, the inhibition of natural killer (NK) cells, lymphokine-activated killer (LAK) cells and cytotoxic T lymphocyte (CTL) activities. It has been reported that NK cells, LAK cells and CTLs induce cell death in tumours or virus-infected target cells. The organophosphates impair the granule exocytosis pathway of NK cells, LAK cells and CTLs by inhibiting the activity of granzymes, and by decreasing the intracellular levels of perforin, granzyme A and granulysin, which were mediated by inducing degranulation of NK cells and by inhibiting the transcription of the mRNAs of perforin, granzyme A and granulysin. The organophosphates impair the

FasL/Fas pathway of NK cells, LAK cells and CTLs, as investigated by using perforin-knockout mice, in which the granule exocytosis pathway of NK cells does not function and only the FasL/Fas pathway remains functional. Organophosphates induce apoptosis of immune cells.

Organophosphate pesticide exposure reduces host resistance to both viral and bacterial infections, and may contribute to an increased risk of cancer development (Galloway and Handy, 2003; Dallaire *et al.*, 2004). The toxic effects of organophosphate pesticides such as parathion, chlorpyrifos, malathion and diazinon on immune functions of invertebrates, fish and higher vertebrate wildlife leads to pathology of immune organs, and decreased humoral and/or CMI. Organophosphate pesticide exposure is also associated with decreased natural killer cell activity (NKCA) and increased cytokine production (Corsini *et al.*, 2008; Udoji *et al.*, 2010). Organophosphates and organochlorines increase IL-2, IL-4 and tumour necrosis factor alpha (TNF-α), with decrease in interferon gamma (IFN-γ) in serious cases of human poisoning (Seth *et al.*, 2005; 2008). The serum level of cytokines can also be altered by occupational exposure to these pesticides (Corsini *et al.*, 2008). Organophosphates decrease immune function and probably act through inhibition of serine esterases (Rodgers, 2001).

3.2.3 Carbamates

Carbamates are known for their low persistence in the environment. Carbaryl selectively increased serum IgG1 and IgG2b levels in mice following oral exposure according to Andre *et al.* (1983). Although humoral immune response was not affected by oral or dermal exposure of carbaryl, it could suppress the inhalation exposure in rats, with doses up to 335 mg/kg. Carbofuran could decrease the humoral immune response to neutral and pathogenic antigens but did not affect the generation of a delayed-type hypersensitivity (DTH) *in vivo* response (Fournier *et al.*, 1988).

The herbicide propanil has been found to increase plasma IgG1 levels and regulate IL-6, IL-10 and IFN release. This was also associated with a reduced IFN:IL-4 ratio (Corsini *et al.*, 2007). The adverse health effects in children due to environmental toxicants were found to be closely associated with altered Th1/Th2 cytokine profile (Duramad *et al.*, 2006; Duramad *et al.*, 2007).

3.2.4 Pyrethroids

Cypermethrin, a pyrethroid and constituent of many household insecticides, caused significant suppression of phorbol myristate acetate (PMA)/ionomycin-induced green fluorescent protein (GFP) expression within IL-2, IFN-γ, IL-4 and IL-10 reporter cell lines at IC-10 and IC-10/10 concentration (Thakur *et al.*, 2013). The study is evidence for the immunosuppressive action of this compound. Similarly, a significant decline in antibody titre was observed in cypermethrin-treated rats (Sankar *et al.*, 2010). The insecticide caused significant suppression of CMI in mice and goats (Stelzer and Gordon, 1984; Tamang *et al.*, 1988) and suppression in DTH reaction in broiler chickens (Khurana *et al.*, 1999). Varshneya *et al.* (1992) observed a significant decrease in spleen weights and increase in adrenal weights in rats receiving the highest test level of cypermethrin. However, low doses (5 and 10 mg/kg) did not have any adverse effect on the immunocompetence of rats. The metabolites of synthetic pyrethroids are more immunotoxic to target organs, and hence need to be considered along with their parent compound for comprehensive health risk assessment for immunotoxicity (Zhang *et al.*, 2010). Exposure to synthetic pyrethroids and their metabolites also led to changes in the secretion levels of TNF-α and ILs, and again the metabolites showed stronger effects than the parent compounds.

In malaria-prone countries, small-meshed bednets, treated with pyrethroids (insecticide treated bednets, ITNs), are being used to repel or kill mosquitoes to control malaria (Watts, 2012). WHO has recommended alpha-cypermethrin, cyfluthrin, deltamethrin, etofenprox, lambda-cyhalothrin and permethrin as insecticides. Recent studies of children have reported immunotoxic effects following exposure to pyrethroids in ITNs, with increased incidence of anti-nuclear antibodies associated with autoimmune disease (US EPA, 1996). Piperonyl butoxide (PBO), not a pyrethroid but a synergist, added for formulating pyrethroids has also been shown to be immunotoxic to non-target organisms. The major target organs for pyrethroids are thyroid, liver, kidney; they increase kidney and liver weight, and also oxidative stress.

Owing to its high hydrophobic nature, deltamethrin may accumulate in body systems and increase the production of reactive oxygen species leading to oxidative stress-induced toxicity (Sharma *et al.*, 2013). Deltamethrin and α-cypermethrin were observed to increase the number of spleen cells producing antibody to sheep red blood cells (SRBCs) and in NK cell activity were reported at oral doses of 5 and 10 mg/kg body weight per day for 28 days in male F344 rats (Madsen *et al.*, 1996). These effects were accompanied at the highest dose by reduced body weight and increased adrenal weight, suggesting that effects at the highest dose were accompanied by generalized toxicity. Emara and Draz (2007) reported that repeated non-continuous inhalation of a commonly utilized insecticide that contains imiprothrin and deltamethrin in rats can cause a variety of immunotoxic effects in sites distal to the lungs.

3.2.5 Herbicides

Hassuneh *et al.* (2012) assessed acute immunotoxicity of paraquat (PQ), a herbicide, in BALB/c mice to examine its effect on cytokine gene expression profile. The single sub-toxic oral doses of 2, 4 and 20 mg/kg of PQ significantly inhibited the *in vitro* mitogen-induced proliferation in both T- and B-cells. These acute doses also reduced IgM plaque-forming cell counts significantly. The mRNA expression of several pro-inflammatory cytokines were elevated at 2.0 and 0.006 μg/ml of PQ, as per the RT^2-PCR array for common cytokine expression. At concentrations equivalent to ADI, PQ was observed to cause a marked up-regulation of the IL-17 cytokine family including IL-17B, C, E and F. mRNA expression of the T(H)-17-promoting cytokines IL-9 and -21 was also up-regulated. The PQ was observed to deplete immune functions at higher concentrations, but at lower concentrations it can alter immune responses towards the pro-inflammatory profiles as that of T(H)-17. The authors emphasized the importance of applying strict regulation on the use of PQ considering its MRLs set by FAO and WHO for certain food and animal feed crops in the range between 0.005 and 10.0 mg/kg.

Ren *et al.* (2013) studied the toxic effects of simazine on mice spleen immune cells and the underlying mechanisms. Mice were given simazine at 0, 90, 200 or 400 mg/kg by gastric gavage for 3 weeks. The authors then measured immune cell proliferation and the expressions of apoptosis-related proteins (Bcl-2, Bax, Fas and Caspase-3), spleen cell intracellular (Ca^{2+} concentration), cellular oxidative stress level and immune functions. After 3 weeks, mice exposed to simazine had reduced proliferation of both spleen T- and B-cells. The number of spleen CD4$^+$ T-lymphocytes decreased with simazine exposure, while CD8$^+$ T-cells remained unchanged. Exposure to simazine resulted in reduced immune function, higher intracellular Ca^{2+} concentration and oxidative stress. Finally, simazine induced spleen immune cells apoptosis by reducing Bcl-2, while increasing Fas and Caspase-3 level. Overall, the immunotoxicity of simazine may involve the induction of immune cell apoptosis and alterations in the immune and physiological functions of spleen cells.

3.2.6 Fungicides

Captan, a non-systemic phthalimide fungicide, caused significant suppression of PMA/ionomycin-induced GFP expression from IL-2, IFN-γ, IL-4 and IL-10 reporter cell lines at IC-10 concentration, confirming its immunosuppressive action, studied in rats and mice (Lafarge-Frayssinet and Decloître, 1982).

The US EPA (Holland *et al.*, 2004–2008) has suggested that cytokine dose–response studies should be carried out to determine whether pesticide and allergen exposures interact synergistically, additively or antagonistically. This would generate new information about the effects of pesticide exposures on the immune system. The study would also lead to risk evaluation of low-dose pesticide exposures to children and also identify key allergens that initiate immune responses that potentially lead to asthma. Finally, neuropathy target esterase (NTE) studies may have important clinical uses in assessing adverse health outcomes in cases of pesticide poisonings and chemical warfare, and in understanding the mechanism of neurobehavioural effects of pesticide exposure in children.

Exposure to pesticides may result in decreased immunocompetence leading to more severe and prolonged infections, and development of cancer, or immunostimulation, which can lead to immune-mediated diseases, hypersensitivity reactions, inflammatory responses, asthma, allergies and autoimmune diseases (Corsini *et al.*, 2008; Hertz-Picciotto *et al.*, 2008). It has also been reported that disruption to critical events in the development of the immune system may result in neurobehavioural and psychiatric disorders (Hertz-Picciotto *et al.*, 2008), suggesting a strong connection between development of the immune system and that of the CNS. Foetal exposure to immunotoxic pesticides crossing the placenta may permanently damage the development of the immune system, leading to deficient immunity and consequent conditions such as asthma, type 1 diabetes, recurrent otitis media, paediatric coeliac disease, decreased resistance to infectious disease, reduced ability to fight viruses, bacteria and parasites, tumour cells, autoimmune disease, hypersensitivity reactions and chronic diseases later in life (Peden 2000; Miller *et al.*, 2002; Dietert 2011; Winans *et al.*, 2011).

People with weak immune systems that cannot adequately detoxify or metabolize pesticides such as organophosphates are more susceptible to, for example, breast cancer, or at greater risk than average persons. However, children are at the greatest risk because of their relatively greater exposure and greater vulnerability to the effects of pesticides (WHO, 2006).

3.3 Toxicity of Endocrine Disruptor Pesticides

The endocrine system consists of a set of glands (thyroid, gonads, adrenal and pituitary) and the hormones they produce (thyroxine, oestrogen, testosterone and adrenaline), which help guide the development, growth, reproduction and behaviour of animals, including humans. Hormones are the signalling chemicals, travelling through the bloodstream and eliciting responses in other parts of the body. Hormones can be proteins, polypeptides, amino acids or steroids. The most well known hormones are the sex steroids oestrogen, produced in the ovaries, and testosterone, produced in the testes. Oestrogen and testosterone are also produced in the adrenal glands of both sexes. Other hormones include thyroxin, produced in the thyroid, and insulin, produced in the pancreas. The pituitary and hypothalamus in the brain release a variety of hormones that affect other organs, including the sex glands (Extoxnet, 1998).

Endocrine-disrupting chemicals (EDCs) are compounds that alter the normal functioning of the endocrine system of both wildlife and humans (Mnif *et al.*, 2011). A huge number of chemicals, including several pesticides, have been identified as endocrine disruptors. The EDCs can interfere with the body's hormones or chemical messengers, leading to an increase in birth defects, sexual abnormalities and reproductive failure (Campion *et al.*, 2012). According to US EPA's working definition, endocrine disruptors 'interfere with the synthesis, secretion, transport, binding, action, or elimination of natural hormones in the body that are responsible for the maintenance of homeostasis (normal cell metabolism), reproduction, development, and/or behavior' (Li *et al.*, 2017). Many endocrine disruptors are thought to mimic hormones.

The organochlorines such as DDT, endosulfan, dieldrin and by-products like PCBs, dioxins and phenols have been reported to have unintentional hormone-like activity. Humans and animals may be exposed to a wide range of endocrine active substances through diet and environmental exposure (water, soil, air). Recent studies have shown that exposure to high doses can result in malformed reproductive organs, consistent with sex hormone imbalance at a critical stage of foetal development. Adverse effects on sexual development have been observed on direct feeding very high doses of DDT to rats. EDCs have been reported in semen, the ovarian follicle, the womb and in breast milk at elevated concentrations. Endocrine disruptors have been linked to attention deficit hyperactivity disorder (ADHD), Parkinson's disease, Alzheimer's disease, diabetes, cardiovascular disease, obesity, early puberty, infertility and other reproductive disorders, and childhood and adult cancers (EDC Daily Blog, n.d.). Endocrine-mediated adverse effects and mechanisms of action may be found at exposure levels lower than those that cause other adverse effects. The findings of Sobolewski *et al.* (2014) show the potential for enhanced behavioural effects of EDC mixtures in males and underscore the need for animal studies to fully investigate mixtures, including chemicals that converge on common physiological substrates to examine potential mechanisms of toxicity with full dose–effect curves to assist in interpretations of relevant mechanisms.

3.4 Sex Hormone Disruptors

DDT and its metabolites, methoxychlor, chlordecone, dieldrin, endosulfan and toxaphene have been found to bind to the oestrogen (female hormone) receptor, while dicarboximide fungicides, vinclozolin and procymidone; p'p'-DDE; certain pyrethroids; and the herbicide linuron have been found to bind to the androgen (male hormone) receptor.

- *Pesticides affecting steroid synthesis and metabolism:* various imidazole pesticides (such as propiconazole, epoziconazole and ketoconazole), fenarimol and several organochlorine pesticides have been reported to affect hormone synthesis and/or metabolism.
- *Thyroid hormone disruptors*: pesticides such as amitrole, ioxynil and the dithiocarbamates (such as maneb, mancozeb and zineb) appear to interfere with thyroid hormone synthesis and can cause cancer of the thyroid.
- *Effects on brain:* both organophosphate and carbamate pesticides are acetylcholine esterase enzyme inhibitors, and hence block nerve impulses. Aldicarb, an extremely toxic systemic carbamate insecticide, is also suspected of being an endocrine disruptor. It has also been suggested that there may be a link between low-level exposure and effects on the immune system.
- *Assessing mode of action*: the nervous, endocrine and immune systems are interconnected. Apart from the sex and thyroid hormones, many other hormones are involved, including retinoids, progestins and corticosteroids. Furthermore, apart from hormone messengers, there are many other signalling processes involved.

A list of endocrine disruptors and route of exposure for humans are provided in Table 3.1 and 3.2.

3.5 Conclusion

Pesticides have been reported as contaminants of food products, including processed and baby food, fruits, vegetables, cereals and drinking water. They can be neurotoxic, immunotoxic or endocrine disruptors.

Table 3.1. Endocrine-disrupting effects of various pesticides. (From: Lyons, 1999.)

S. no.	Pesticides	Group of pesticides	Human exposure routes
1.	Atrazine	Herbicide	Has been found in drinking water on occasion
2.	Beta-HCH	Organochlorine	In the past, high levels have been found in rabbit imported from China, and it has also been found in other meat and fish and butter oil. Also found in breast milk, and still found in 1996/97 UK samples, although levels have decreased
3.	Chlordane	Organochlorine	On occasion has been found in breast milk, as has oxychlordane, the stable metabolite. Due to atmospheric transport, Inuit women tend to have a diet highly contaminated with chlordane
4.	DDT	Organochlorine	Residues found in beef (from Brazil and Zimbabwe). In the past, found in rabbit (from China), cow's milk and butter. Recently found in fish (tuna) and imported lamb's liver. Found in breast milk and metabolites still found in UK samples, although declined since banned
5.	Dicofol	Organochlorine	May be found in apples and lemon products. Has been found as a contaminant in human body fat in US surveys in the 1980s
6.	Dieldrin	Organochlorine	A 1992 survey in the UK found very high levels in eels. Also in 1993 it was found in samples of UK cow's milk and butter oil. Dieldrin has been found in breast milk, and is still found in some UK samples, although levels have declined since usage was banned
7.	Endosulfan	Organochlorine	Residues have been found in 1998 UK samples of peas, mange tout, tomatoes and soft fruit, such as plums and blackcurrants. In the early 1990s, endosulfan was found in baby foods in the US
8.	HCB (hexachlorobenzene)	Organochlorine	Found in rabbit from China. Has also been found in earlier surveys of eels, meat, cow's milk and cheese. HCB has also been found in breast milk, and still found in some UK samples although levels have declined
9.	Lindane (gamma-HCH)	Organochlorine	A 1998 survey found lindane in 75% of samples of chocolate with a high cocoa butter content. Also found in 1998 samples of cereal grains, mushrooms and in earlier surveys in fish, meat, butter, cheese, flour, bread and some vegetables, such as onions. Has also been found in cow's milk in the UK, although not detected in the latest survey in 1998. Still detected in some UK breast milk samples
10.	Pyrethroids	SP (synthetic pyrethroid)	When used for timber treatment, exposure may occur via inhalation. Some have been found in rivers
11.	Triazines – atrazine, simazine, cyanazine, amitrole and prometryn are triazine herbicides	Herbicides	Like atrazine, simazine has occasionally been found in drinking water
12.	Vinclozolin – a dicarboximide fungicide used on rape, beans, peas, turf and apple blossom. In UK, some illegal usage on winter lettuce. P+B		Residues found in UK lettuce in 1997 samples due to illegal usage. Also found in 1998 samples of kiwi fruit, peaches/nectarines and tomatoes, and earlier samples of peas, peppers, orange, beans, cress, garlic, grape juice, salmon, and sultanas. A related anti-androgenic fungicide, procymidone, has similarly been found in UK lettuce, and also in aubergines, peas, pears, wine and tomatoes

P, persistent; B, bioaccumulative. These symbols are used as found in the Oslo and Paris Commissions (OSPAR) document.

Table 3.2. Pesticides in environment with reproductive and/or endocrine-disrupting effects. (From: Fleşeriu, 2010.)

S. no.	Pesticides	Examples
1	Herbicides	2,4-D, 2,4,5-T, acetochlor, alachlor, amitrole, atrazine, bromacil, bromoxynil, cyanazine, DCPA (dacthal), ethiozin, glufosinate-ammonium, ioxynil, linuron, metribuzin, molinate, nitrofen, oryzalin, oxyacetmide/fluthamide (FOE 5043), paraquat, pendimethalin, picloram, prodiamine, pronamide, simazine, terbutryn, thiazopyr, triclorobenzene, trifluralin
2	Fungicides	Benomyl, etridiazole, fenarimol, fenbuconazole, hexachlorobenzene, mancozeb, maneb, metiram, nabam, pentachloronitrobenzene, pentachlorophenol, triadimefon, tributyltin, vinclozolin, zineb, ziram
3	Insecticides	Aldicarb, aldrin, bifenthrin, carbaryl, carbofuran, chlordane, chlordecone, chlorfentezine, 8-cyhalothrin, DDT and metabolites DDE, DDD, deltamethrin, dicofol, dieldrin, dimethoate, dinitrophenol, endosulfan (a and b), endrin, ethofenprox, fenitrothion, fenvalerate, fipronil, α-HCH, heptachlor and H-epoxide, lindane (g-HCH), malathion, methomyl, methoxychlor, mirex, oxychlordane, parathion (methylparathion), photomirex, pyrethrins, synthetic pyrethroids, ronnel (fenchlorfos), toxaphene, transnonachlor
4	Nematicide	DBCP
5	Rodenticide	N-2-fluorenylacetamide

2,4-D, 2,4-Dichlorophenoxyacetic acid; 2,4,5-T, 2,4,5-trichlorophenoxyacetic acid; DDT, dichlorodiphenyltrichloroethane; DDE, dichlorodiphenyldichloroethylene; DDD, dichlorodiphenyldichloroethane; α-HCH, hexachlorocyclohexane; DBCP, 1,2-dibromo-3-chloropropane.

The highly persistant neurotoxic organochlorines are responsible for disruption of pumping of calcium, chloride, sodium and potassium ions across the nerve cells. DDT and its analogues act primarily on the PNS; cyclodiene organochlorine pesticides inhibit GABA-ergic neurotransmission by blocking the chloride channel of GABA(A) receptors. Many of the unmetabolized organochlorines are efficiently reabsorbed by the intestine, substantially retarding faecal excretion. The organophosphates are degraded rapidly via hydrolysis after exposure to sunlight, air and soil, and are hence a safer and more attractive alternative to highly persistent organochlorines. However, their anticholinesterase activity categorizes them as toxic to beneficial and non-target organisms. The oxon analogues of organophosphates and carbamates are potential inhibitors of AChE. However, ChE inhibition by carbamates is reversible in contrast to that of organophosphates. The oximes can enhance the slow hydrolysis of phosphorylated AChE by water, and hence, they are a common treatment for organophosphate poisoning in human medical literature. Pyrethroids, developed in the 1970s, are known for their knockdown effect with longer breakdown times, and low mammalian and avian toxicity. They have strong selectivity for insects and invertebrates, and are reported to have low potential to bioaccumulate and proficient detoxicification in mammalian receptors. However, the synthetic pyrethroids are neuropoisons acting on the axons in the PNS and CNS by interacting with sodium channels in mammals and/or insects. The neonicotinoids, a new group of insecticides, have similar mammalian toxicity to that of nicotine, but they are very selective and specific towards insect nicotinic receptors. However, they have been reported unsafe for pollinators. Although herbicides have lower acute toxicity than insecticides, various classes of herbicides have been associated with adverse effects ranging from carcinogenicity, target organ toxicity, to endocrine disruption. Fungicides generally comprise a large number of compounds, with diverse chemical structures. Several fungicides are carcinogenic (e.g. captan), while others may have skin or eye irritant properties (e.g. chlorothalonil).

Dithiocarbamates and some organic metal compounds, no longer used now, are considered as neurotoxic. The organochlorine, organophosphate, carbamate, and pyrethroid group insecticides, and some herbicides and fungicides are reported as immunotoxic pesticides. Humans and animals may be exposed to a wide range of endocrine active substances through diet and environmental exposure (water, soil, air). The organochlorines and their byproducts have been reported to have unintentional hormone-like activity. Exposure to high doses can result in malformed reproductive organs, consistent with sex hormone imbalance at a critical stage of foetal development. EDCs have been reported in semen, the ovarian follicle, the womb and in breast milk at elevated concentrations. These organochlorines and their metabolites have been found to bind to the oestrogen (female hormone) receptor, while dicarboximide fungicides, vinclozolin and procymidone; p′p′-DDE, certain pyrethroids and the herbicide linuron have been found to bind to the androgen (male hormone) receptor.

References

Andrade, S.F. and Sakate, M. (2003) The comparative efficacy of yohimbine and atipamezole to treat amitraz intoxication in dogs. *Veterinary and Human Toxicology* 45, 124–127.

Andre, F., Gillon, F., Andre, C., Lafont, S. and Jourdan, G. (1983) Pesticide containing diet augments anti-sheep red blood cells non reaginic antibody responses in mice but may prolong murine infection with *Giardia muris*. *Environmental Research* 32, 145–150.

Anthony, D.C., Montine, T.J., Valentine, W.M. and Graham, D.G. (2001) Toxic responses of the nervous system. In: Klaassen, C.D. (ed.), *Casarett and Doull's Toxicology. The Basic Science of Poisons*. McGraw-Hill, New York, pp. 535–563.

Bakir, F., Damluji, S.F., Amin-Zaki, L., Murthada, M., Khalidi, A., al-Rawi, N.Y., Tikriti, S., Dahahir, H.I., Clarkson, T.W., Smith, J.C. and Doherty, R.A. (1973) Methylmercury poisoning in Iraq. *Science* 181, 230–241.

Banerjee, B.D. and Hussain, Q.Z. (1986) Effect of sub-chronic endosulfan exposure on humoral and cell-mediated immune responses in albino rats. *Archives of Toxicology* 59(4), 279–284.

Banerjee, B.D. and Hussain, Q.Z. (1987) Effects of endosulfan on humoral and cell-mediated immune responses in rats. *Bulletin of Environmental Contamination and Toxicology* 38(3), 435–441.

Banerjee, B.D., Ray, A. and Pasha, S.T. (1996) A comparative evaluation of immunotoxicity of DDT and its metabolites in rats. *Indian Journal of Experimental Biology* 34(6), 517–522.

Barnes, K.N. (2000) *The Eskom Red Data Book of Birds of South Africa, Lesotho and Swaziland*. BirdLife South Africa, Johannesburg.

Bus, J.S. and Gibson, J.E. (1984) Paraquat: Model for oxidant-initiated toxicity. *Environmental Health Perspectives* 55, 37–46.

Caksen, H., Odabas, D., Arslan, S., Akgun, C., Atas, B., Akbayram, S. and Tuncer, O. (2003) Report of eight children with amitraz intoxication. *Human and Experimental Toxicology* 22, 95–97.

Campion, S., Catlin, N., Heger, N., McDonnell, E.V., Pacheco, S.E., Saffarini, C., Sandrof, M.A. and Boekelheide, K. (2012) Male reprotoxicity and endocrine disruption. *NIH Public Access* 101, 315–360.

Cantor, K.P., Blair, A., Everett, G., Gibson, R. and Burmeister, L.F. (1992) Pesticides and other agricultural risk factors for non-Hodgkin's lymphoma among men in Iowa and Minnesota. *Cancer Research* 52, 2447–2455.

Carletti, E., Li, H., Li, B., Ekström, F., Nicolet, Y., Loiodice, M., Gillon, E., Froment, M.T., Lockridge, O., Schopfer, L.M., Masson, P. and Nachon, F. (2008) Aging of cholinesterases phosphylated by tabun proceeds through O-dealkylation. *Journal of American Chemical Society* 130(47), 16011–16020.

Carletti, E., Colletier, J.P., Dupeux, F., Trovaslet, M., Masson, P. and Nachon, F. (2010) Structural evidence that human acetylcholinesterase inhibited by tabun ages through O-dealkylation. *Journal of Medicinal Chemistry* 53(10), 4002–4008.

Cassidy, R.A., Vorhees, C.V., Minnema, D.J. and Hastings, L. (1994) The effects of chlordane exposure during pre- and postnatal periods at environmentally relevant levels on sex steroid-mediated behaviors and functions in the rat. *Toxicology and Applied Pharmacology* 126, 326–337.

Cattani, D., de Liz Oliveira Cavalli, V.L., Heinz Rieg, C.E., Domingues, J.T., Dal-Cim, T., Tasca, C.I., Mena Barreto Silva, F.R. and Zamoner, A. (2014) Mechanisms underlying the neurotoxicity induced by glyphosate-based herbicide in immature rat hippocampus: Involvement of glutamate excitotoxicity. *Toxicology* 320, 34–45.

Chan, S. and Kilby, M.D. (2000) Thyroid hormone and fetal nervous system development. *Journal of Endocrinology* 165, 1–8.

Charles, J.M., Cunny, H.C., Wilson, R.D. and Bus, J.S. (1996) Comparative subchronic studies on 2,4-dichlorophenoxyacetic acid, amine, and ester in rats. *Fundamental and Applied Toxicology* 33, 161–165.

Chevrier, J., Eskenazi, B., Holland, N., Bradman, A. and Barr, D.B. (2008) Effects of exposure to polychlorinated biphenyls and organochlorine pesticides on thyroid function during pregnancy. *American Journal of Epidemiology* 168(3), 298–310.

Clarkson, T.W. (2001) Inorganic and organometal pesticides. In: Krieger, R. (ed.) *Handbook of Pesticide Toxicology*, Academic Press, San Diego, pp. 1357–1428.

Clarkson, T.W. and Magos, L. (2006) The toxicology of mercury and its chemical compounds. *Critical Reviews in Toxicology* 36, 609–662.

Coats, J.R. (1990) Mechanisms of toxic action and structure activity relationships for organochlorine and synthetic pyrethroid insecticides. *Environmental Health Perspectives* 87, 255–262.

Cohen, E. (2014) *Advances in Insect Physiology: Target Receptors in the Control of Insect Pests: Part II*, vol. 46. Academic Press, Elsevier, Oxford, UK.

Cohn, B., Wolff, M., Cirillo, P. and Sholtz, R. (2007) DDT and breast cancer in young women: New data on the significance of age at exposure. *Environmental Health Perspectives* 115(10), 1410–1414.

Corsini, E., Primavera, A., Marinovich, M. and Galli, C.L. (1998) Selective induction of cell-associated interleukin-1alpha in murine keratinocytes by chemical allergens. *Toxicology* 129, 193–200.

Corsini, E., Codecà, I., Mangiaratti, S., Birindelli, S., Minoia, C. *et al.* (2007) Immunomodulatory effects of the herbicide propanil on cytokine production in humans: *in vivo* and *in vitro* exposure. *Toxicology and Applied Pharmacology* 222, 202–210.

Corsini, E., Liesivuori, J., Vergieva, T., Van Loveren, H. and Colosio, C. (2008) Effects of pesticides exposure on the human immune system. *Human and Experimental Toxicology* 27(9), 671–680.

Costa, L.G. (1988a) Interactions of neurotoxicants with neurotransmitter systems. *Toxicology* 49, 359–366.

Costa, L.G. (1988b) Organophosphorus compounds. In: Galli, C.L., Manzo, L., Spencer, P.S. (eds) *Recent Advances in Nervous System Toxicology*. Plenum Publishing Corp., New York, pp. 203–246.

Costa, L.G., Olibet, G. and Murphy, S.D. (1988) Alpha2-adrenoceptors as a target for formamidine pesticides: *in vitro* and *in vivo* studies in mice. *Toxicology and Applied Pharmacology* 93, 319–328.

Costa, L.G., Giordano, G., Guizzetti, M. and Vitalone, A. (2008) Neurotoxicity of pesticides: A brief review. *Frontiers in Bioscience* 13, 1240–1249.

Dai, P.-L., Wang, Q., Sun, J.-H., Liu, F., Wang, X., Wu, Y.-Y. and Zhou, T. (2010) Effects of sublethal concentrations of bifenthrin and deltamethrin on fecundity, growth, and development of the honeybee *Apis mellifera ligustica*. *Environmental Toxicology and Chemistry* 29, 644–649.

Dallaire, F., Dewailly, E., Muckle G., Vezina, C., Jacobson, S.W., Jacobson, J.L. and Ayotte, P. (2004) Acute infections and environmental exposure to organochlorines in Inuit infants from Nunavik. *Environmental Health Perspectives* 112, 1359–1365.

Dastych, J., Walczak-Drzewiecka, A., Wyczolkowska, J. and Metcalfe, D.D. (1999) Murine mast cells exposed to mercuric chloride release granule-associated N-acetyl-beta-D-hexosaminidase and secrete IL-4 and TNF-alpha. *Journal of Allergy and Clinical Immunology* 103, 1108–1114.

Delpuech, J.M., Legallet, B., Terrier, O. and Fouillet, P. (1999) Modifications of the sex pheromonal communication of *Trichogramma brassicae* by a sublethal dose of deltamethrin. *Chemosphere* 38, 729–739.

Desi, I., Varga, L. and Karkas, I. (1978) Studies of the immunosuppresive effect of organochlorine and organophosphoric pesticides in subacute experiments. *Journal of Hygiene, Epidemiology, Microbiology and Immunology* 22, 115–122.

Desneux, N., Decourtye, A. and Delpuech, J.M. (2007) The sublethal effects of pesticides on beneficial arthropods. *Annual Review of Entomology* 52(1), 81–106.

Dewan, P., Jain, V., Gupta, P. and Banerjee, B.D. (2013) Organochlorine pesticide residues in maternal blood, cord blood, placenta, and breastmilk and their relation to birth size. *Chemosphere* 90(5), 1704–1710.

Dietert, R.R. (2011) Role of developmental immunotoxicity and immune dysfunction in chronic disease and cancer. *Reproductive Toxicology* 31(3), 319–326.

Dobbing, J. (1968) Vulnerable periods in developing brain. In: Davison, A.N. and Dobbing J. (eds) *Applied Neurochemistry*. Davis, Philadelphia, Pennsylvania, pp. 287–316.

Duramad, P., Harley, K., Lipsett, M., Bradman, A., Eskenazi, B. *et al.* (2006) Early environmental exposures and intracellular Th1/Th2 cytokine profiles in 24-month-old children living in an agricultural area. *Environmental Health Perspectives* 114, 1916–1922.

Duramad, P., Tager, I.B. and Holland, N.T. (2007) Cytokines and other immunological biomarkers in children's environmental health studies. *Toxicology Letters* 172, 48–59.

Durkin, P. and Diamond, G. (2002) Neurotoxicity, immunotoxicity, and endocrine disruption with specific commentary on glyphosate, triclopyr, and hexazinone: Final report. Submitted to Forest Service, Leslie Rubin, COTR, Animal and Plant Health Inspection Service (APHIS), Biotechnology, Biologics and Environmental Protection, Environmental Analysis and Documentation, United States Department of Agriculture.

Ecobichon, D.J. (1996) Toxic effects of pesticides. In: Klaassen, C.D. (eds) *Casarett and Doull's Toxicology: The Basic Science of Poisons* (6th edn). McGraw-Hill, New York, pp. 666–669.

Ecobichon, D.J. (2001) Carbamate insecticides. In: Krieger, R. (ed.) *Handbook of Pesticide Toxicology.* Academic Press, San Diego, California, pp. 1087–1106.

EDC Daily Blog (n.d.) Pesticide-induced diseases: Endocrine disruption. What is the endocrine system? Beyond Pesticides. Available at: www.beyondpesticides.org/resources/pesticide-induced-diseases-database/body-burden (accessed 31 July 2018).

Emara, A.M. and Draz, E.I. (2007) Immunotoxicological study of one of the most common over-the-counter pyrethroid insecticide products in Egypt. *Inhalation Toxicology* 19(12), 997–1009.

ETN (Extension Toxicology Network) (1994) Pyrethroids. Pesticide Information Profiles. Available at: http://extoxnet.orst.edu/pips/pyrethri.htm (accessed 28 September 2018).

Eurostat Statistical Books (1992–2003) The use of plant protection products in the European Union. Data 1992–2003. Available at: https://ec.europa.eu/eurostat/documents/3217494/5611788/KS-76-06-669-EN.PDF (accessed 28 September 2018).

Extoxnet (1998) Questions about endocrine disruptors. Available at: http://extoxnet.orst.edu/faqs/pesticide/endocrine.htm (accessed 31 July 2018).

Ferraz, H.B., Bertolucci, P.H., Pereira, J.S., Lima, J.G. and Andrade, L.A. (1988) Chronic exposure to the fungicide maneb may produce symptoms and signs of CNS manganese intoxication. *Neurology* 38, 550–553.

Fishel, F.M. (2005) Pesticide toxicity profile: Synthetic pyrethroid pesticides. Agronomy Department, Florida Cooperative Extension Service, Institute of Food and Agricultural Sciences, University of Florida. Report #PI-54.

Fleşeriu, A. (2010) Endocrine disrupting pesticides and their impact on wildlife and human health. *HVM Bioflux* 2(1), 1–4.

Fishel, F.M. (2013) Pesticide toxicity profile: Carbamate pesticides. USDA, UF/IFAS Extension Service, University of Florida.

Fournier, M., Chevalier, G., Nadeau, D., Trotter, B. and Krzystyniak, K. (1988) Virus-pesticide interactions with murine cellular immunity after sublethal exposure dieldrin and aminocarb. *Journal of Toxicology and Environmental Health* 1, 103–118.

Friesen, K.J. and Kaufman, W.R. (2003) Cypermethrin inhibits egg development in the ixodid tick, *Amblyomma hebraeum. Pesticide Biochemistry and Physiology* 76, 25–35.

Galloway, T. and Handy, R. (2003) Immunotoxicity of organophosphate pesticides. *Ecotoxicology* 12, 345.

Garabrant, D.H., Held, J., Langholz, B., Peters, J.M. and Mack, T.M. (1992) DDT and related compounds and risk of pancreatic cancer. *Journal of the National Cancer Institute* 84, 764–771.

Garey, J. and Wolff, M. (1998) Estrogenic and antiprogestagenic activities of pyrethroid insecticides. *Biochemical and Biophysical Research Communications* 251(3), 855–859.

Glick, B. (1974) Antibody-mediated immunity in the presence of mirex and DDT. *Poultry Science* 53, 1476–1485.

Goldstein, M.I., Lacher, T.E., Woodbridge, B., Bechard, M.J., Canavelli, S.B., Zaccagnini, M.E., Cobb, G.P., Scollon, E.J., Tribolet, R. and Hooper, M.J. (1999) Monocrotophos – induced mass mortality of Swainson's hawks in Argentina, 1995–96. *Ecotoxicology* 8, 201–214.

Gosselin, R.E., Smith, R.P. and Hodge, H.C. (1984) *Clinical Toxicology of Commercial Products* (5th edn). Williams and Wilkins, Baltimore, Maryland.

Gupta, R.C. (ed.) (2006) *Toxicology of Organophosphate and Carbamate Compounds.* Academic Press, Abingdon, UK.

Hammond, R.B. (1996) Residual activity of lambda-cyhalothrin against bean leaf beetle (Coleoptera: Chrysomelidae) in soybeans. *Journal of Agricultural Entomology* 13, 365–373.

Hassuneh, M.R., Albini, M.A. and Talib, W.H. (2012) Immunotoxicity induced by acute subtoxic doses of paraquat herbicide: Implication of shifting cytokine gene expression toward T-helper

(T(H))-17 phenotype. *Chemical Research in Toxicology* 25(10), 2112–2126.

Hayashi, K., Fukuyama, T., Ohnuma, A., Tajima, Y., Kashimoto, Y., Yoshida, T. and Kosaka, T. (2013) Immunotoxicity of the organochlorine pesticide methoxychlor in female ICR, BALB/c, and C3H/He mice. *Journal of Immunotoxicology* 10(2), 119–124.

Hayley, S., Mangano, E., Crowe, G., Li, N. and Bowers, W.J. (2011) An *in vivo* animal study assessing long-term changes in hypothalamic cytokines following perinatal exposure to a chemical mixture based on Arctic maternal body burden. *Environmental Health* 10, 65.

Haynes, K.F. (1988) Sublethal effects of neurotoxic insecticides on insect behavior. *Annual Review of Entomology* 33, 149–168.

Haynes, K.F. and Baker, T.C. (1985) Sublethal effects of permethrin on the chemical communication system of the pink bollworm moth, *Pectinophora gossypiella*. *Archives of Insect Biochemistry and Physiology* 2, 283–293.

Heinzow, B. and Andersen, H.R. (2006) Biocides and pesticides. In: Duffus, J.H. and Worth, H.G.J. (eds) *Fundamental Toxicology*. RSC Publishing, Cambridge, UK.

Hernández, A., Gómez, M.A., Pena, G., Gil, F., Rodrigo, L., Villanueva, E. and Pla, A. (2004) Effect of long-term exposure to pesticides on plasma esterases from plastic greenhouse workers. *Journal of Toxicology and Environmental Health*, Part A 67, 1095–1108.

Hertz-Picciotto, I., Park, H.Y., Dostal, M., Kocan, A., Trnovec, T. and Sram, R. (2008) Prenatal exposure to persistent and non-persistent organic compounds and effects on immune system development. *Basic and Clinical Pharmacology and Toxicology* 102(2), 146–154.

Holan, G. (1969) New halocyclopropane insecticides and the mode of action of DDT. *Nature* 221, 1025–1029.

Holland, N.T., Casida, J., Lipsett, M. and Tager, I. (2004–2008) Center for Children's Environmental Health Research – mechanisms of pesticide neuro- and immunotoxicity. US EPA, Washington, DC.

House, V.R. and Descotes, J. (2007) Cytokines in human health: Immunotoxicology, pathology, and therapeutic applications. Part of the *Methods in Pharmacology and Toxicology* book series (MIPT). Humana Press, Springer Nature Switzerland AG, Cham, Switzerland.

Ishaaya, I., Elsner, A., Ascher, K.R.S. and Casida, J.E. (1983) Synthetic pyrethroids: Toxicity and synergism on dietary exposure of *Tribolium castaneum* (Herbst) larvae. *Pesticide Science* 14, 367–372.

Isman, M.B. (2006) Botanical insecticides, deterrents and repellents in modern agriculture and an increasingly regulated world. *Annual Review of Entomology* 51, 45–66.

Johnson, D.J., Graham, D.G., Amarnath, V., Amarnath, K. and Valentine, W.M. (1998) Release of carbon disulfide s a contributing mechanism in the axonopathy produced by N,N-diethyldithiocarbamate. *Toxicology and Applied Pharmacology* 148, 288–296.

Jokanovi, M. (2009) Medical treatment of acute poisoning with organophosphorus and carbamate pesticides. *Toxicology Letters* 190, 107–115.

Joy, R.M. (1982) Mode of action of lindane, dieldrin and related insecticides in the central nervous system. *Neurobehavioral Toxicology and Teratology* 4, 813–823.

Juarez, M.P. (1995) The effect of sublethal doses of insecticides on *Triatoma infestans* lipid synthesis. *Pesticide Biochemistry and Physiology* 52, 81–89.

Jurewicz, J. and Hanke, W. (2008) Prenatal and childhood exposure to pesticides and neurobehavioral development: Review of epidemiological studies. *International Journal of Occupational Medicine and Environmental Health* 21(2), 121–132.

Kamel, F. and Hoppin, J. (2004) Association of pesticide exposure with neurologic dysfunction and disease. *Environmental Health Perspectives* 112(9), 950–958.

Khurana, S., Chauhan, R.S. and Mahipal, S.K. (1999) Immunotoxic effects of cypermethrin on delayed type hypersensitivity reaction in chicken. *Indian Veterinary Journal* 76, 1055–1057.

Korrick, S. and Sagiv, S. (2008) Polychlorinated biphenyls, organochlorine pesticides and neurodevelopment. *Current Opinion in Pediatrics* 20(2), 198–204.

Lafarge-Frayssinet, C. and Decloître, F. (1982) Modulatory effect on the pesticide captan on the immune response in rats and mice. *Journal of Immunopharmacology* 4, 43–52.

Lah, K. (2011) Biological properties of insecticides. *Toxipedia*, updated 27 April 2011.

Lee, C.Y., Yap, H.H. and Chong, N.L. (1998). Sublethal effects of deltamethrin and propoxur on longevity and reproduction of German cockroaches, *Blattella germanica*. *Entomologia Experimentalis et Applicata* 89, 137–145.

Li, Q. (2007) New mechanism of organophosphorus pesticide-induced immunotoxicity. *Journal of Nippon Medical School* 74(2), 92–105.

Li, X., Gao, Y., Wang, J., Ji, G., Lu, Y., Yang, D., Shen, H., Dong, D., Pan, L., Xiao, H. and Zhu, B. (2017) Exposure to environmental endocrine disruptors and human health. *Journal of Public Health and Emergency* 1(8), 1–11.

Longley, M. (1999) A review of pesticide effects upon immature aphid parasitoids within mummified hosts. *International Journal of Pest Management* 45, 139–145.

Longley, M. and Jepson, P.C. (1996a) Effects of honeydew and insecticide residues on the distribution of foraging aphid parasitoids under glasshouse and field conditions. *Entomologia Experimentalis et Applicata* 81, 189–198.

Longley, M. and Jepson, P.C. (1996b) The influence of insecticide residues on primary parasitoid and hyperparasitoid foraging behaviour in the laboratory. *Entomologia Experimentalis et Applicata* 81, 259–269.

Longnecker, M.P., Klebanoff, M.A., Zhou, H. and Brock, J.W. (2001) Association between maternal serum concentration of the DDT metabolite DDE and preterm and small-for-gestational-age babies at birth. *The Lancet* 358, 110–114.

Lotti, M. (2000) Organophosphorus compounds. In: Spencer, P.S., Schaumburg, H.H. and Ludolph, A.C. (eds) *Experimental and Clinical Neurotoxicology*. Oxford University Press, Oxford, UK, pp. 898–925.

Lotti, M. (2001) Clinical toxicology of anticholinesterases in humans. In: R. Krieger (ed.) *Handbook of Pesticide Toxicology*. Academic Press, San Diego, California, pp. 1043–1085.

Lyons, G. (1999) Endocrine disrupting pesticides. *Pesticides News* No. 46, December 1999, pp. 16–19. Reproduced on the Pesticide Action Network UK website. Available at: www.pan-uk.org/pestnews/Actives/endocrin.htm (accessed 6 February 2011).

Madsen, C., Claesson, M.H. and Röpke, C. (1996) Immunotoxicity of the pyrethroid insecticides 31 deltamethrin and alpha-cypermethrin. *Toxicology* 107, 219–227.

Marrs, T.C. (2012) Mammalian toxicology of insecticides. In: Marrs, C.T. (ed.) *Royal Society of Chemistry, 2012 – Science*. Royal Society of Chemistry, London, p. 490.

Marrs, T.C. and Vale, J.A. (2006) Management of organophosphorus pesticide poisoning. In: Gupta, R.C. (ed.) *Toxicology of Organophosphorus and Carbamate Compounds*. Elsevier Academic Press, Amsterdam, pp. 715–733.

Mathur, V., Bhatnagar, P., Sharma, R.G., Acharya, V. and Sexana, R. (2002) Breast cancer incidence and exposure to pesticides among young women originating from Jaipur. *Environment International* 28, 331–336.

Matsuda, K., Buckingam, S.D., Kleier, D., Rauh, J.J., Grauso, M. and Sattelle, D.B. (2001) Neonicotinoids: Insecticides acting on insect acetylcholine receptors. *Trends in Pharmacological Sciences* 22, 573–580.

McCormack, A.L, Thiruchelvam, M., Manning-Bog, A.M., Thiffault, C., Langston, J.W., Cory-Slechta, D.A. and Di Monte, D.A. (2002) Environmental risk factors and Parkinson's disease: Selective degeneration of nigral striatal dopaminergic neurons caused by the herbicide paraquat. *Neurobiology of Disease* 10, 119–127.

Meco, G., Bonifati, V., Vanacore, N. and Fabrizio, E. (1994) Parkinsonism after chronic exposure to the fungicide maneb (manganese ethylene-bis-dithiocarbamate). *Scandinavian Journal of Work, Environment and Health* 20, 3011–3305.

Meine, C.D. and Archibald, G.W. (eds) (1996) The cranes: Status survey and conservation action plan. IUCN, Gland, Switzerland/Cambridge, UK.

Miller, M.D., Marty, M.A., Arcus, A., Brown, J., Morry, D. and Sandy, M. (2002) Differences between children and adults: Implications for risk assessment at California EPA. *International Journal of Toxicology* 21, 403–418.

Mineau, P., Fletcher, M.R., Glaser, L.C., Thomas, N.J., Brassard, C., Wilson, L.K., Elliott, J.E., Lyons, L.A., Henny, C.J. and Bollinger, T. (1999) Poisoning of raptors with organophosphorus and carbamate pesticides with emphasis on Canada, U.S. and U.K. *Journal of Raptor Research* 33, 1–37.

Mnif, W., Hassine, A.I.H., Bouaziz, A., Bartegi, A., Thomas, O. and Roig, B. (2011) Effect of endocrine disruptor pesticides: A review. *International Journal of Environmental Research and Public Health* 8, 2265–2303.

Mueller-Beilschmidt, D. (1990) Toxicology and environmental fate of synthetic pyrethroids. *Journal of Pesticide Reform* 10(3), 32–37.

Narahashi, T. (1976) Nerve membrane as a target for pyrethroids. *Pesticide Science* 7, 267–272.

Natsagdorj, T. and Batbayar, N. (2002) The impact of rodenticide used to control rodents on demoiselle crane (*Anthropoides virgo*) and other animals in Mongolia. Unpublished report. Cited in *BirdLife International* (2004) Pesticides continue to poison birds on a large scale. Presented as part of the BirdLife State of the World's birds website. Available at: www.birdlife.org/datazone/sowb/casestudy/158 (accessed 28 September 2018).

Oliveira, D.B., Gillespie, K., Wolfreys, K., Mathieson, P.W., Qasim, F. *et al.* (1995) Compounds that induce autoimmunity in the brown Norway rat sensitize mast cells for mediator release and interleukin-4 expression. *European Journal of Immunology* 25, 2259–2264.

Oros, D.R. and Werner, I. (2005). Pyrethroid insecticides: An analysis of use patterns, distributions, potential toxicity and fate in the Sacramento-San Joaquin delta and central valley. In: White Paper for the Interagency Ecological Program. SFEI Contribution 415. San Francisco Estuary Institute, Oakland, California.

Pain, D.J., Gargi, R., Cunningham, A.A., Jones, A. and Prakash, V. (2004) Mortality of globally threatened Sarus cranes *Grus antigon* from monocrotophos poisoning in India. *Science and Total Environment* 326, 55–61.

Peden, D.B. (2000) Development of atopy and asthma: Candidate environmental influences and important periods of exposure. *Environmental Health Perspectives* 108(Suppl. 3), 475–482.

Pimentel, D. (1991) *CRC Handbook of Pest Management in Agriculture*, 2nd edn. CRC Press, Boca Raton, Florida.

Pistl, J., Kovalkovicova, N., Kacmar, P., Kusova, I., Mikula, I. and Sutiakova, I. (2001) Effect of endosulfan on peripheral sheep leukocytes *in vitro*. *Veterinary and Human Toxicology* 43(2), 78–82.

Raghavendra, K., Barik, T.K., Sharma, P., Bhatt, R.M., Srivastava, H.C., Sreehari, U. and Dash, A.P. (2011) Chlorfenapyr: A new insecticide with novel mode of action can control pyrethroid resistant malaria vectors. *Malaria Journal* 10(1), 16.

Ray, D.E. and Fry, J.R. (2006) A reassessment of the neurotoxicity of pyrethroid insecticides. *Pharmacology and Therapeutics* 111, 174–193.

Reigart, J.R. and Roberts J.R. (1999) *Recognition and Management of Pesticide Poisonings* (5th edn). US Environmental Protection Agency, EPA 735-R-98-003, 1999. Washington, DC. Available at: www.epa.gov/pesticides/safety/healthcare (accessed 28 September 2018).

Ren, R., Sun, D.J., Yan, H., Wu, Y.P. and Zhang, Y. (2013) Oral exposure to the herbicide simazine induces mouse spleen immunotoxicity and immune cell apoptosis. *Toxicologic Pathology* 41(1), 63–72.

Roberts, E.M., English, P.B., Grether, J.K., Windham, G.C., Somberg, L. and Wolff, C. (2007) Maternal residence near agricultural pesticide applications and autism spectrum disorders among children in the California Central Valley. *Environmental Health Perspective* 115, 1482–1489.

Rodgers, K.E. (2001) Immunotoxicity of pesticides. In: Krieger, R. (ed.) *Handbook of Pesticide Toxicology: Principles and Agents*. Academic Press, San Diego, California, pp. 769–782.

Rogan, W.J. (1996) Pollutants in breast milk. *Archives of Pediatrics and Adolescent Medicine* 150, 981–990.

Roldán-Tapia, L., Parrón, T. and Sánchez-Santed, F. (2005) Neuropsychological effects of long-term exposure to organophosphate pesticides. *Neurotoxicology and Teratology* 27(2), 259–266.

Román, G.C. (2007) Autism: Transient *in utero* hypothyroxinemia related to maternal flavonoid ingestion during pregnancy and to other environmental antithyroid agents. *Journal of the Neurological Sciences* 262, 15–26.

Sankar, P., Telang, A.G., Manimaran, A. and Malik, J.K. (2010) Immunoprotective effect of curcumin on cypermethrin-induced toxicity in rats. *Toxicological and Environmental Chemistry* 92, 1909–1917.

Seth, V., Ahmad, R.S., Suke, S.G., Pasha, S.T., Bhattacharya, A. and Banerjee, B.D. (2005) Lindane induced immunological alterations in human poisoning cases. *Clinical Biochemistry* 38, 678–680.

Seth, V., Banerjee, B.D., Ahmad, R.S., Bhattacharya, A. and Pasha, S.T. (2008) Alterations in immunoglobulins and cytokine levels in blood of malathion poisoning cases. *Indian Journal of Biochemistry and Biophysics* 45, 209–211.

Sharma, P., Jan, M. and Singh, R. (2013) Deltamethrin toxicity: A review. *Indian Journal of Biological Studies and Research* 2(2), 91–107.

Smith, L.L. (1987) Mechanism of paraquat toxicity in the lung and its relevance to treatment. *Human Toxicology* 6, 31–36.

Sobolewski, M., Conrad, K., Allen, J.L., Weston, H., Martin, K., Lawrence, B.P. and Cory-Slechta, D.A. (2014) Sex-specific enhanced behavioral toxicity induced by maternal exposure to a mixture of low dose endocrine-disrupting chemicals. *Neurotoxicology* 45, 121–130.

Soderlund, D.M. and Bloomquist, J.R. (1989) Neurotoxic actions of pyrethroid insecticides. *Annual Review of Entomology* 34, 77–96.

Sogorb, M.A. and Vilanova, E. (2002) Enzymes involved in the detoxification of organophosphorus, carbamate and pyrethroid insecticides through hydrolysis. *Toxicology Letters* 128(1–3), 215–228.

Solomon, L.M., Fahrner, L. and West, D.P. (1977) Gamma benzene hexachloride toxicity. *Archives of Dermatology* 113, 353–357.

Spencer, P.S., Schaumburg, H.H. and Ludolph, A.C. (eds) (2000) *Experimental and Clinical Neurotoxicology*. Oxford University Press, Oxford.

Stelzer, K.J. and Gordon, M.A. (1984) Effects of pyrethroids on lymphocyte mitogenic responsiveness. *Research Communications in Chemical Pathology and Pharmacology* 46, 137–150.

Stevens, M.F., Ebell, G.F. and Psaila-Savona, P. (1993) Organochlorine pesticides in Western

Australian nursing mothers. *Medical Journal of Australia* 158(4), 238–241.

Sun, M., Chang, Z., Shau, M., Huang, R. and Chou, T. (1979) The mechanism of ageing of phosphorylated acetylcholinesterase. *European Journal of Biochemistry* 100, 527–530.

Sungur, M. and Güven, M. (2001) Intensive care management of organophosphate insecticide poisoning. *Critical Care* 5, 211–215.

Takahashi, O., Oishi, S., Fujitani, T., Tanaka, T. and Yoneyama, M. (1994) Chronic toxicity studies of piperonyl butoxide in F344 rats: Induction of hepatocellular carcinoma. *Fundamental and Applied Toxicology* 22, 293–303.

Tamang, R.K., Jha, G.J., Gupta, M.K., Chauhan, H.V. and Tiwary, B.K. (1988) *In vivo* immunosuppression by synthetic pyrethroid (cypermethrin) pesticide in mice and goats. *Veterinary Immunology and Immunopathology* 19, 299–305.

Tesfahunegny, W. and Mulualem, G. (2016) Impact of pesticides on birds from DDT to current fatality: A literature review. *The Journal of Zoology Studies* 3(2), 45–56. Available at: https://pdfs.semanticscholar.org/eee3/b25eef04dab9455f1d82243c0b0bdf1a3e29.pdf (accessed 28 September 2018).

Thakur, D., Singh, P., Tripathi, C., Bhadauria, S., Jain, S.K. and Tripathi, R.K. (2013) *In vitro* immunotoxicity testing of pesticides using human cytokine promoter based reporter cell lines. *Clinical and Experimental Pharmacology* S4: 001. doi:10.4172/2161-1459.S4-001.

Thrasher, J.D., Madison, R. and Broughton, A. (1993) Immunologic abnormalities in humans exposed to chlorpyrifos: Preliminary observations. *Archives of Environmental Health* 48, 89–93.

Tomizawa, M. and Casida, J.E. (2005) Neonicotinoid insecticide toxicology: Mechanisms of selective action. *Annual Review of Pharmacology and Toxicology* 45, 247–268.

Udoji, F., Martin, T., Etherton, R. and Whalen, M.M. (2010) Immunosuppressive effects of triclosan, nonylphenol, and DDT on human natural killer cells *in vitro*. *Journal of Immunotoxicology* 7, 205–212.

Ujvary, I. (2001) Pest control agents from natural products. In: Krieger, R. (ed.) *Handbook of Pesticide Toxicology*. Academic Press, San Diego, California, pp. 109–179.

US Environmental Protection Agency (EPA) (1996) Memorandum. Etofenprox (ethofenprox): Decision regarding reclassification consideration by the HED Carcinogenicity Peer Review Committee. From Williams-Foy SY, Health Effects Division to Heyward A/ Larocca G Registration Division. United States Environmental Protection Agency, Washington, DC. Available at: www.epa.gov/pesticides/chem_search/cleared_reviews/csr_PC-128965_29-Aug-96_009.pdf (accessed 28 September 2018).

US EPA (2009) Lable changes to reduce risk: Ecological risk mitigation. Pyrethrins and pyrethroids reregistration and labelling. Minutes of meeting relevant to the total release fogger labeling improvements effort. Available at: https://www.epa.gov/ingredients-used-pesticide-products/pyrethrins-and-pyrethroids-reregistration-and-labeling (accessed 28 September 2018).

US EPA (2014) Pyrethroids and pyrethrins. Pesticides: Regulating Pesticides. Available at: http://www.epa.gov/oppsrrd1/reevaluation/pyrethroids-pyrethrins.html (accessed 16 January 2019).

Vandame, R., Meled, M., Colin, M.E. and Belzunces, L.P. (1995) Alteration of the homing flight in the honey bee *Apis mellifera L.* exposed to sublethal dose of deltamethrin. *Environmental Toxicology and Chemistry* 14, 855–860.

Vandebriel, R.J., Spiekstra, S.W., Hudspith, B.N., Meredith, C. and Van Loveren, H. (1999) *In vitro* exposure effects of cyclosporin A and bis(tri-n-butyltin)oxide on lymphocyte proliferation, cytokine (receptor) mRNA expression, and cell surface marker expression in rat thymocytes and splenocytes. *Toxicology* 135, 49–66.

Varshneya, C., Singh, T., Sharma, L.D., Bahga, H.S. and Garg, S.K. (1992) Immunotoxic responses of cypermethrin, a synthetic pyrethroid insecticide in rats. *Indian Journal of Physiology and Pharmacology* 36(2), 123–126.

Verma, S.K. and Ahmad, S. (2009) High dose pralidoxime in organophosphorus poisoning: A critical appraisal. *API Medicine Update* 19, 448–452.

Vijveberg, H.P.M., van der Zalm, J.M. and Van der Bercken, J. (1982) Similar mode of action of pyrethroids and DDT on sodium channel gating in myelinated nerves. *Nature* 295, 601–603.

Voccia, I., Blakley, B., Brousseau, P. and Fournier, M. (1999) Immunotoxicity of pesticides: A review. *Toxicology and Industrial Health* 15, 119–132.

Ware, G.W. and Whitacre, D.M. (2004) An introduction to insecticides (4th edn). In: Radcliffe, E.B. and Hutchison, W.D. (eds) *IPM World Textbook*. University of Minnesota, Saint Paul, Minnesota. Extracted from *The Pesticide Book* (6th edn) (2004), published by MeisterPro Information Resources, a division of Meister Media Worldwide, Willoughby, Ohio. Available at: https://ipmworld.umn.edu/ware-intro-insecticides (accessed 28 September 2018).

Watts, M. (2012) Pyrethroids in bed nets: deltamethrin, etofenprox, lambda-cyhalothrin permethrin, piperonyl butoxide. In: *Highly Hazardous Pesticides*, A PAN AP Factsheet series. Pesticide Action Network Asia and the Pacific, Penang, Malaysia. Available at: www.pananz.net/wp-content/uploads/2014/09/Pyrethroids.pdf (accessed 28 September 2018).

Wesseling, C., van Wendel De Joode, B., Ruepert, C., Leon, C., Monge, P., Hermosillo, H. and Partanen, T.J. (2001) Paraquat in developing countries. *International Journal of Occupational and Environmental Health* 7, 275–286.

Wheeler, W.B. (2002) Role of research and regulation in 50 years of pest management in agriculture. *Journal of Agricultural and Food Chemistry* 50(15), 4151–4155.

World Health Organization (WHO) (2006) Principles for Evaluating Health Risks in Children Associated with Exposure to Chemicals. Environmental Health Criteria 237. World Health Organization, Geneva. Published under the joint sponsorship of UNEP, ILO and WHO. Available at: www.inchem.org/documents/ehc/ehc/ehc237.pdf (accessed 28 September 2018).

World Health Organization Pesticide Evaluation Scheme (WHOPES) (2005) Safety of pyrethroids for public health use. WHO Pesticide Evaluation Scheme, Geneva.

Winans, B., Humble, M.C. and Lawrence, B.P. (2011) Environmental toxicants and the developing immune system: A missing link in the global battle against infectious disease? *Reproductive Toxicology* 31(3), 327–336.

Wise, J. (2014) Rainfast characteristics of insecticides on fruit. *Michigan State University Extension*, 10 June 2014.

Yamaoka, K.A., Miyasaka, N., Inuo, G., Saito, I., Kolb, J.P., *et al.* (1994) 1,1'-Ethylidenebis(tryptophan) (Peak E) induces functional activation of human eosinophils and interleukin 5 production from T lymphocytes: Association of eosinophilia-myalgia syndrome with a L-tryptophan contaminant. *Journal of Clinical Immunology* 14, 50–60.

Zhang, Y., Zhao, M., Jin, M., Xu, C., Wang, C. and Liu, W. (2010) Immunotoxicity of pyrethroid metabolites in an *in vitro* model. *Environmental Toxicology and Chemistry* 29(11), 2505–2510.

4 Diagnostic Tools for Pesticide Detection

4.1 Introduction

Pesticides are used worldwide during cultivation and postharvest storage to improve the quantities and quality of food crops (Fenik *et al.*, 2011). Pesticides are used for managing agricultural pests or unwanted plants to reduce destruction of edible crops, and to ensure the production of adequate food supplies for the increasing world population (Cserhati *et al.*, 2004; Hiemstra and de Kok, 2007; Guan *et al.*, 2010). Although the use of pesticides helps to reduce crop losses both before and after harvest (Sassolas *et al.*, 2012), it has led to severe risk of adverse consequences for human health and environmental contamination (Chowdhury *et al.*, 2014). They can easily enter the food chain, contaminating fruits, vegetables and other food crops, leaving higher levels of pesticide residues, posing hazards to human beings (Osman *et al.*, 2010). Residues above the prescribed maximum residue level (MRL) may pose health risks to unwary consumers (Dogheim *et al.*, 1996; Mukherjee and Gopal, 1996; Fillion *et al.*, 2000; Sinha *et al.*, 2012). This has raised serious concerns and makes it more necessary than ever to detect the minutest levels of contamination. A number of methods are available to diagnose, detect and estimate the presence of pesticides in food and environmental samples in order to take precautionary measures.

Among the various diagnostic tools, detection and estimation of pesticide residues in environmental samples using analytical methods is very common and is categorized under monitoring studies.

4.2 Monitoring of Pesticide Residue Analysis

Various methods are available for monitoring pesticide residues in food and feed. One should opt for a method providing results that are cost-effective, timely, reliable and verifiable (Herdman *et al.*, 1988a).

In the US the federal jurisdiction over pesticide residues is divided among four bodies, namely, the US Environmental Protection Agency (US EPA), the Food and Drug Administration (FDA) of the US Department of Health and Human Services, the Food Safety and Inspection Services (FSIS) and the Agricultural Marketing Services (AMS) of the US Department of Agriculture (USDA). Their authority for this work comes from five laws – FIFRA, Federal Food, Drug, and Cosmetic Act (FFDCA), FMIA, the Poultry Products Inspection Act and the Egg

Products Inspection Act (Herdman *et al.*, 1988b). The documents from various bodies responsible for jurisdiction over pesticide residues describe the method validation and analytical quality control requirements to support the validity of data used for checking compliance with MRLs, enforcement actions or assessment of consumer exposure to pesticides. The method must also be able to detect pesticides at, and often below, tolerance levels.

As per US EPA guidelines, the method should give reliable results within 24 h, require readily available chemicals and equipment, identify each pesticide in a mixture and detect residues at or below their tolerances. As per FSIS guidelines, a method is suitable for regulatory use if it does not take more than 2–4 h of analytical time per sample, the minimum proficiency level is at or below the tolerances, it has been developed with a quality assurance plan and it is successfully validated at an interlaboratory level (FSIS, 1987). FDA does not have a formal listing of guidelines for its methods, but it uses many of the same criteria in evaluating them (P. Lombardo, Center for Food Safety and Applied Nutrition, FDA, Washington, DC, personal communication, 1 August 1988; Yu, 2002). The key objectives are:

1. To provide a harmonized cost-effective quality assurance system in the European Union (EU).
2. To ensure the quality and comparability of analytical results.
3. To ensure that acceptable accuracy is achieved.
4. To ensure that false positives or false negatives are not reported.
5. To support compliance with ISO/IEC 17025 (accreditation standard).

A variety of analytical methods are available for detection of pesticide residues, however, all of them contain certain basic steps for processing laboratory samples in the procedure of pesticide residue analysis. These are:

1. *Sample preparation*: this involves chopping, grinding, macerating or separation of plant parts.
2. *Extraction*: this is basically the removal of pesticides from the sample to be extracted.
3. *Separation*: the sample co-extractives are separated.
4. *Clean-up*: this involves the removal of impurities, like plant pigments, that may interfere with the analysis of pesticide residues. It also includes partitioning and purification.
5. *Estimation*: the cleaned and purified sample is analysed using gas or liquid chromatography, for the detection and quantification of each pesticide.

Regardless of the selectivity and sensitivity in the determinative step, sample preparation (extraction and clean-up) often remains the main limiting step in the analysis of pesticide residues (Cunha *et al.*, 2010). Thus, several factors need to be taken into consideration when designing a sample preparation method, such as: (a) physicochemical properties of the pesticide residue(s); (b) detection and quantification limits required; (c) analytical instrumentation available for use; and (d) composition of the sample matrices (Lehotay, 2000).

4.2.1 QuEChERS, multi-residue method

QuEChERS (pronounced 'catchers'), an acronym for Quick, Easy, Cheap, Effective, Rugged and Safe, covers a variety of sample preparation and clean-up techniques for the analysis of multiple pesticide residues in agricultural matrices. It was originally designed for multi-residue analysis of fruits and vegetables, but now includes a wide range of agricultural products. Since its development and publication by scientists at the USDA in 2003 (Anastassiades *et al.*, 2003), QuEChERS has gained significant popularity as the method of choice. It combines several sample preparation steps and extends the range of analytes recovered compared with older, tedious extraction methods. A driving force in the growth of QuEChERS is the emerging need to determine trace amounts of analytes in a high-throughput environment. Due to the large number of QuEChERS methods now published, QuEChERS is considered an 'approach' rather than a 'method'. A key advantage is that QuEChERS has a wide scope of analysis and proven quality of results.

It is a very flexible approach that has been adapted for the analysis of many pesticide residues in food and animal tissues (Lehotay *et al.*, 2005a; Lehotay, 2007; Nguyen *et al.*, 2007; Paya *et al.*, 2007; Cajka *et al.*, 2008) as well as in other applications (Posyniak *et al.*, 2005; Mastovska and Lehotay, 2006; Plossl *et al.*, 2006). The main concept entails the use of an organic solvent, typically acetonitrile but alternatively ethyl acetate or acetone, for extraction of the pre-homogenized sample, which already contains water, or water is added, followed by addition of anhydrous $MgSO_4$ and NaCl to induce partitioning of the extract (upper phase) from the water (lower phase). The extraction is typically conducted by shaking, but blending is also an option. The use of buffering with either acetic acid/acetate (Lehotay *et al.*, 2005b) or citric acid/citrate salts (Paya *et al.*, 2007) during extraction is also an option depending on the analytes and matrices. After centrifugation, an aliquot of the extract is mixed with sorbent(s) in a centrifuge tube in an approach known as dispersive solid-phase extraction (SPE) clean-up. The final extract is analysed for pesticides by chromatographic methods, with detection ideally using mass spectrometry (MS) techniques.

The expansion of the QuEChERS methodology indicates not only its power for sample extraction and clean-up but also addresses the concern about detecting a vast array of pesticides, herbicides, fungicides, antibiotics and other compounds throughout the entire food supply. QuEChERS in its basic form involves three steps:

1. Liquid micro-extraction.
2. Solid-phase clean-up.
3. Liquid chromatography/tandem mass spectrometry (LC/MS/MS) or gas chromatography/mass spectrometry (GC/MS) analysis.

The extraction of targeted pesticide(s) from specific matrices followed by appropriate pre-concentration (if necessary, depending on the sensitivity of the method) and determination of extracted pesticides are two important protocols for residue analysis. Among various advanced techniques, supercritical fluid extraction (SFE) and microwave-assisted extraction (MAE) have been reported as advanced extraction techniques.

4.2.2 Supercritical fluids

For every substance there is a temperature above which it can no longer exist as a liquid, no matter how much pressure is applied. Similarly, there is a pressure above which the substance can no longer exist as a gas no matter how high the temperature is raised. These points are called the supercritical temperature and supercritical pressure, respectively, and are the defining boundaries on a phase diagram (Fig. 4.1) for a pure substance, beyond which the substance has properties intermediate between a liquid and a gas and is called a supercritical fluid (SCF). In this region the fluid has good solvating power and high diffusivity, which make it a good choice as a mobile phase in chromatography. SCFs are produced by heating a gas above its critical temperature or compressing a liquid above its critical pressure. A substance such as CO_2 can exist in solid, liquid and gaseous phases under various combinations of temperature and pressure, and there is a point, the critical temperature, where the liquid and vapour have the same density. In this physical state,

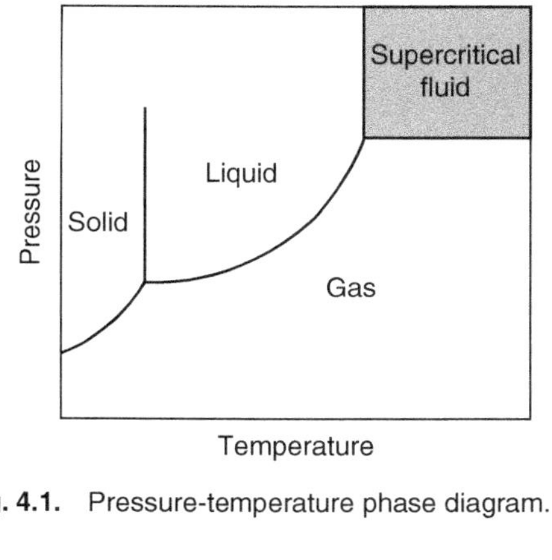

Fig. 4.1. Pressure-temperature phase diagram.

CO_2 is a good solvent for many organic substances. It also has valuable properties as a mobile phase in chromatography as its viscosity remains similar to that of a gas, permitting high flow rates. The benefits of using SCFs are their liquid-like densities offering higher solubility and increased column loading.

4.2.3 Supercritical fluid extraction

Although analytical extraction approaches have been improved, most of them rely on time-consuming procedures such as Soxhlet extraction, which require large volumes of expensive and toxic solvents (Antunes *et al.*, 2003; Luque de Castro and Herrera, 2003). SFE is rather a new technique for the extraction of analytes from sample matrices. The main advantages of using SCFs for extractions are that they are inexpensive, contaminant free and less costly to dispose of safely than organic solvents. Because of the distinct properties of SCFs, such as low viscosity, high diffusivity and adjustable density, SFE has gained increased attention as a better alternative to conventional extraction methods (Hawthorne, 1990). Carbon dioxide is the commonly used extraction solvent due to its suitable critical temperature (31.2°C) and pressure (72.8 atm), and its non-toxic and non-flammable properties (Luque de Castro and Herrera, 2003; Hawthorne, 1990). Moreover, CO_2 can easily be removed by reducing the pressure. SCFs can have solvating powers similar to those of organic solvents, but with higher diffusivities, lower viscosity and lower surface tension. The solvating power can be adjusted by changing the pressure or temperature, or adding a modifier to the SCF. A common modifier is methanol (typically 1–10%), which increases the polarity of supercritical CO_2. Water must be removed before performing SFE, because a highly water-soluble analyte will prefer to partition into the aqueous phase, resulting in a poor SFE recovery. Proper choice of the extraction fluid will also allow the analyst to conduct the extraction at low temperatures, a key feature to extend the application of SFE for the extraction of thermally liable substances (McHugh and Krukonis, 1986; Stahle *et al.*, 1988).

The instrumentation of SFE consists of a source of fluid in a pressurized cylinder, a fluid delivery module, an extracting vessel, a backpressure regulating device, modifier pumps and a collector for procuring the extract after SFE. Analytes are trapped by letting the solute-containing SCF decompress into an empty vial, through a solvent or onto a solid sorbent material. Extractions are done in dynamic, static or combination modes. In dynamic extraction, the SCF continuously flows through the sample in the extraction vessel and out of the restrictor to the trapping vessel. In static mode, the SCF circulates in a loop containing the extraction vessel for a period of time before being released through the restrictor to the trapping vessel. In the combination mode, a static extraction is performed for a period of time followed by a dynamic extraction. Co-extraction of unwanted solutes along with the target analyte may occur in SFE. These interferences can be removed either by conventional sample clean-up methods or by using a sorbent column downstream from the SFE device.

Rissato *et al.* (2005) used SFE for the extraction of various pesticides, including organochlorines, organophosphates, organonitrogens and pyrethroids, from soil samples before their estimation by GC. SFE has been used for the extraction of 17 organochlorine and organophosphate pesticides from a table-ready food containing crude vegetables, white bread and vegetable oil (Aguilera *et al.*, 2003). El-Saeid (2003) has applied SFE for the extraction of various pesticides from canned food, fruits and vegetables. Aguilera *et al.* (2005) have extracted 22 pesticides from rice samples using SFE, with recoveries ranging from 74–98%. They used 15 ml CO_2 volume, 50°C temperature, 200 atm pressure and methanol as the modifier (Aguilera *et al.*, 2005). SFE has been applied for the extraction of different pesticides from honey; the optimal conditions were reported to be 400 bar pressure, 90°C temperature, with acetonitrile as the modifier (Rissato *et al.*, 2004). Zhao *et al.* (2002) used SFE for the extraction of 12 organochlorine pesticides from *Angelicae sinensis*. The optimized extraction conditions were found to be pure CO_2 (without modifier), 15 MPa pressure, 60°C temperature and an extraction time of 20 min. Several organophosphate pesticides, including fenitrothion, chlorpyrifos, diazinon, methamidophos, edifenphos, mevinphos, fenthion and acephate, have been extracted from agro wastewater. This procedure was performed at 90°C and 325 atm for 20 min for static extraction followed by 40 min of dynamic extraction (Yu, 2002). King and Zhang (2002) have reported a rapid SFE methodology for the extraction of carbamate pesticides, including carbaryl, 3-hydroxycarbofuran, carbofuran, aldicarb and methiocarb. Sun and Lee (2003) have extracted four carbamates (propoxur, propham, chlorpropham and methiocarb) from soils using supercritical CO_2 at 300 kg/cm^3 modified with 10% methanol. The extraction was performed at 60°C and lasted for 30 min, resulting in up to 92% recoveries.

4.2.4 Microwave-assisted extraction

The extraction of organic compounds by microwave irradiation is a recent technique, first used in 1986 by Ganzler *et al.* (1986). In MAE the sample is suspended in a suitable extraction solvent and irradiated in a microwave oven (2450 MHz) for an appropriate time, without allowing the suspension to boil. This frequency of microwaves corresponds to a wavelength of 12.2 cm and energy of 0.23 cal/mol. This frequency can cause molecular rotations without altering the molecular structure. The electrical component of the wave changes 4.9×10^9 times per second, which generates a disorganized movement of polar molecules, causing rapid heating (Letellier and Budzinski, 1999). Non-polar solvents such as hexane and toluene are not affected by microwave energy and therefore require polar additives for use with MAE (Barnabus *et al.*, 1995). The main advantages of MAE are reductions in extraction time and solvent quantity; hence, the technique is environment friendly.

MAE has earlier been reported to be superior to ultrasonic as well as Soxhlet methods for the extraction of chlorinated pesticides from animal feed (Gfrerer *et al.*, 2004), whereas Barriada-Pereira *et al.* (2003) have found comparable recoveries of organochlorine pesticides from vegetable samples using MAE (81.5–108.4%) and Soxhlet (75.5–132.7%). Diagne *et al.* (2002) also observed comparative recoveries of fenitrothion residues in beans using MAE (89.8%) and Soxhlet (88.4%). A comparative study of five extraction techniques revealed the ability of MAE to retrieve a high concentration of pesticides from dietary composites (Rosenblum *et al.*, 2002). Dichlorvos has been extracted from vegetables using 10% aqueous ethylene glycol as extractant and medium microwave power for 10 min (Chen *et al.*, 2002). Singh *et al.* (2004) used MAE for simultaneous extraction of thiamethoxam, imidacloprid and carbendazim in fresh and cooked vegetable samples, with recoveries ranging from 68–106%. There was no breakdown of pesticides in cooked vegetables, and the parent compounds were extracted intact (Singh *et al.*, 2004). Bouaid *et al.* (2000) have reported a simple and rapid (9 min) MAE method for extraction of organophosphate pesticides from 1.5–2.5 g of orange peel using 10 ml of hexane/acetone mixture at 90°C and microwave power set at 50% (475 W). The recovery of pesticides ranged between 93 and 101%. Vryzas *et al.* (2002) applied MAE for extraction of carbamates from tobacco and peaches, with a recovery range of 80–100%. Sun and Lee (2003) have achieved 85–105% recoveries of carbamates in soil using 30 ml methanol as extractant and 6 min microwave heating at 80°C. Although more than 85% recoveries were obtained by both MAE and SFE, slightly higher recoveries of propoxur, propham and methiocarb were achieved using MAE, whereas SFE showed slightly higher recovery for chlorpropham (Sun and Lee, 2003). Various classes of pesticides, including parathion, methyl parathion, dichlorodiphenyldichloroethylene (DDE), hexachlorobenzene

(HCB), simazine and paraquat, have been extracted from soil samples using MAE, with recoveries of more than 80% (de Andrea *et al.*, 2001). MAE has been used for the extraction of organic contaminants, including pesticides, polychlorinated biphenyls and polycyclic hydrocarbons, from marine sediments and tissues (Jayaraman *et al.*, 2001; Numata *et al.*, 2004). The herbicide methabenzthiazuron has been extracted from soil using MAE; the recoveries of herbicides are dependent on soil type (Baez *et al.*, 2003). Triazine herbicides have been extracted from soil samples using MAE with water containing 1% methanol as extractant (Shen and Lee, 2003). MAE was performed at 105°C for 3 min using 80% output of maximum power (1200 W), resulting in 76.1–87.2% recoveries for triazine (Shen and Lee, 2003). MAE has been used for the extraction of multi-residue phenoxyalkanoic acid herbicides and their phenolic conversion products from soil using aqueous methanol as extraction medium; the recoveries were found to be above 80% (Patsias *et al.*, 2002). El-Saeid *et al.* (2005) have performed a comparative evaluation of SFE and MAE for extraction of atrazine from food samples.

4.2.5 SCF chromatography

In recent years, SCF chromatography (SFC) has emerged as a powerful green technology in industries such as pharmaceutical, agricultural, food and environmental sciences. The main difference between SFC and other chromatographic techniques (GC and high-performance liquid chromatography; HPLC) is the use of a SCF as the mobile phase in SFC. The SCF demonstrates unique characteristics that can greatly enhance the efficiency of the process. SFC has several advantages over other conventional chromatographic techniques compared with HPLC. SFC provides rapid separations without the use of organic solvents. With the desire for environmentally conscious technology, the use of organic chemicals as in HPLC could be reduced with the use of SFC. Because SFC generally uses CO_2, which is collected as a by-product of other chemical reactions or is collected directly from the atmosphere, it contributes no new chemicals to the environment. In addition, SFC separations can be done faster than HPLC separations, because the diffusion of solutes in SCFs is about 10 times greater than that of liquids and about three times less than that of gases. SFC has been shown to be superior to conventional HPLC in terms of high resolution, lower analysis time and better separation potential for drugs and their metabolites (Toribio *et al.*, 2005; Wang *et al.*, 2006). SFC has gained wide recognition as 'the future of HPLC'. The higher efficiency, economic impact, safety benefits and ease of use have made SFC a highly promising technology.

Part of the theory of separation in SFC is based on the density of the SCF, which corresponds to the solvating power. As the pressure in the system is increased, the SCF density increases and correspondingly its solvating power increases. Therefore, as the density of the SCF mobile phase is increased, components retained in the column can be made to elute. This is similar to temperature programming in GC or using a solvent gradient in HPLC. In SFC, the mobile phase is initially pumped as a liquid and is brought into the supercritical region by heating it above its supercritical temperature before it enters the analytical column. It passes through an injection valve, where the sample is introduced into the supercritical stream and then into the analytical column. It is maintained as supercritical as it passes through the column and into the detector via a pressure restrictor placed either after the detector or at the end of the column. The restrictor is a vital component as it keeps the mobile phase supercritical throughout the separation and often must be heated to prevent clogging. Although there are a number of possible fluids that may be used in SFC as the mobile phase, CO_2 is widely used due to its low cost, low interference with chromatographic detectors and good physical properties (non-toxic, non-flammable, suitable critical temperature). However, the main disadvantage of CO_2 is its inability to elute very polar or ionic compounds. This drawback can be overcome by adding a small portion of a second fluid called a modifier fluid. The modifier is generally an organic fluid that is completely miscible with CO_2. The addition of the modifier fluid improves the solvating ability of the SCF and sometimes enhances the selectivity of the separation. It can also improve the separation efficiency by blocking some of the highly active sites on the stationary phase.

Toribio *et al.* (2004) have performed chiral separation of six triazole pesticides, including cyproconazole, propiconazole, diniconazole, hexaconazole, tebuconazole and tetraconazole, using SFC equipped with a Chiralpak AD column. The same column type was also used for the separation of triadimefon and triadimenol enantiomers and diastereoisomers using SFC (del Nozal *et al.*, 2003). Dost *et al.* (2000) have developed a sensitive SFC-MS method for the separation and quantification of three classes of pesticides, including triazines, carbamates and sulfonylureas, in soil samples. A multiresidue method based on SFC and solid-phase extraction has been reported for the analysis of 35 common contaminants including pesticides in water; the detection limit ranged from 0.4–2.6 μg/l (Toribio *et al.*, 1998). SFC coupled with SFE has been used for the analysis of sulfonylureas, their precursors and metabolites in complex matrices such as soil, vegetation and cell culture medium (McNally and Wheeler, 1988). Wheeler and McNalley (1987) carried out a comparative evaluation of packed-column versus capillary column SFC for the analysis of herbicides and insecticides.

Since conventional extraction techniques such as Soxhlet extraction, sonication and mechanical shaking are laborious, time consuming and require large volumes of toxic organic solvents, much attention is being paid to developing more efficient and environmentally friendly techniques for rapid extraction of pesticides from complex matrices. Both SFE and MAE have emerged as efficient and rapid techniques for extraction of pesticides from a wide range of matrices, for analysis of various groups of pesticides from food samples. The chromatographic separation and determination of pesticides using SFC offer the advantages of high sensitivity, greater resolution, minimal use of organic solvents and versatility for analysing different groups of pesticides in food samples.

4.2.6 Single residue methods

Samples can be successfully analysed using single residue methods (SRMs) or multi-residue methods (MRMs). SRMs are designed to analyse a single pesticide at a time, and in many instances, its metabolites or degradation products (Herdman *et al.*, 1988a). A number of techniques are involved for SRMs, which vary widely in terms of reliability, efficiency, throughput (samples per day), degree of validation and practicality for regulatory use. These are basically required by the private sector for submission to US EPA as part of the tolerance-setting process. So, a method exists for every pesticide along with its tolerance, although methods for some pesticides (primarily the older ones) may not be effective. The SRMs, which are published in a peer-reviewed journal of high quality (similar to the ones approved by EPA but adapted for other commodities) and have undergone EPA laboratory evaluation, are followed (Kovacs *et al.*, 1987).

But for routine monitoring of pesticides, we generally do not come across single pesticides, rather a mixture of unknown pesticides, for which it is not feasible to use SRMs. Moreover, there is variation in the methodologies for SRMs for pesticides of the same class, and a wide array of laboratory devices and chromatographic detectors are required. Because of this the performance of some SRMs is not satisfactory (US Congress, 1988). These are, therefore, not considered adequate for routine monitoring by the regulatory agencies. We need an MRM that can measure many pesticides using the same resources. Some chemists prefer to develop their own methods or adapt existing methods developed for pesticides of similar structure. However, SRMs are not as capable of identifying unidentified analytical responses (UARs) as MRMs.

4.2.7 Multi-residue methods

The MRMs, as the name indicates, are designed to detect and quantify multiple pesticides and their toxicological metabolites in a range of environmental samples, under monitoring studies. These methods are quite sensitive, precise and accurate enough to be useful for regulatory purposes, and also economical. If an MRM is not able to quantify a particular pesticide but is able to detect its presence, the pesticide could be quantified using SRM. MRMs record the presence of unidentified chemicals, known as a UAR. These UARs can be identified by matching them with the results of known chemicals with similar chromatographic results or by MS. MRMs, therefore, help in identifying the presence of possible hazardous chemicals, which otherwise would have been overlooked in environmental samples. For example, during analysis of the organochlorine pesticides in meat and animal feed samples, the UARs on the chromatograms were first identified as polychlorinated biphenyls (PCBs), which led to their discovery.

The processing of samples in MRMs includes preparation, extraction, clean-up, chromatographic separation and detection, as mentioned above. GC or HPLC is used for the determination of pesticides for all MRMs. Of the 10 MRMs used under FDA and USDA, eight rely on GC as the determinative step (Table 4.1). The MRMs of FDA are considered of high quality and capable of providing data that will withstand challenge during court litigation (Sawyer, 1988). However, FDA's MRMs have the drawback of not detecting every pesticide.

For example, of the 316 pesticides with tolerances, only 163 of them could be analysed with FDA's five routinely used MRMs. A second weakness is that some MRMs require a great deal of time to perform, thereby reducing the number of samples analysed and the speed of analysis. For example, certain foods, such as those with high concentrations of fats and oils, are difficult to analyse in a timely manner.

In an attempt to reduce the need to use SRMs, US EPA now requires that all pesticides requiring a new tolerance be evaluated to see if they can be detected by FDA and USDA MRMs. Only FDA has developed the testing protocols to support such testing. FDA has also devised a 'decision tree' showing the order in which the FDA MRMs should be used to test the new pesticide to minimize research time. The results confirm whether or not that particular pesticide can be recovered through one of the MRMs. Earlier it was not clear whether the US EPA will require development and submission of an SRM if the pesticide can be analysed by an MRM (C. Trichilo, Residue Chemistry Branch, Environmental Protection Agency, Washington, DC, personal communication, 25 April 1988). However, SRMs are sometimes needed to analyse pesticides that are not adequately recovered using standard MRMs or to target specific pesticide/commodity combinations (PRMP, 2014).

Various regulatory agencies, such as FDA under the United States FFDCA has responsibility for enforcing tolerances established by US EPA in food (except meat and poultry) and animal feed moving in interstate commerce and for enforcing prohibition of a pesticide residue in food or feed for which no tolerance has been set or exemption given (FFDCA: 21 U.S.C. Sections 331–337, 1982 and Supp. IV 1986).

4.3 Other Diagnostic Tools for Detection of Pesticides

The conventional methods are labour intensive, time consuming and incur high costs for set-up; as well as requiring skilled labour and extensive sample preparation steps. Therefore, more analytical and simpler techniques could be considered desirable for pesticide detection. New, relatively inexpensive analytical test kits have been developed for the rapid testing of pesticides in foods and water (Villani, 1995). Many manufacturers

Table 4.1. Multi-residue methods routinely used by FDA, USDA and CDFA. (Adapted from US Congress, Office of Technology Assessment, 1988.)

Agency	Method	Food type analysed	Pesticide groups detected
FDA	GC-multiple detectors (Luke method)	Non-fatty	Organochlorines Organophosphates, organonitrogens
	GC-multiple detectors (Mills method)	Fatty	Organochlorines Organophosphates
	GC-multiple detectors (Mills *et al.*, 1963, method)	Non-fatty	Organochlorines Organophosphates
	GC-multiple detectors (Storherr method)	Non-fatty	Organophosphates
	HPLC-fluorescence (Krause method)	Non-fatty	N-methyl carbamates
USDA-FSIS	GC-ECD	Fat	Chlorinated hydrocarbons
	GC-ECD	Liver and fat	Chlorinated organophosphates
	(western method)	Liver	Organophosphates
	GC-NPD		
	(eastern method)	Liver	Carbamates
	HPLC-fluorescence		
USDA-AMS	GC-ECD	Fatty (raw egg products)	Chlorinated hydrocarbons
CDFA	GC-ECD	Non-fatty	Organochlorines
	GC-NPD or FPD	Non-fatty	Organophosphates
	HPLC-fluorescence	Non-fatty	Carbamates

Methods are identified by the combination of the instruments used for chromatographic separation and detection. FDA, Food and Drug Administration; GC, gas chromatography; HPLC, high-performance liquid chromatography; ECD, electron capture detector; NPD, nitrogen–phosphorus detector; FPD, flame photometric detector; USDA-FSIS, United States Department of Agriculture Food Safety and Inspection Service; USDA-AMS, United States Department of Agriculture Agricultural Marketing Service; CDFA, California Department of Food and Agriculture.

of these kits boast that users can get results faster and at lower cost than ever before, and that the kits can detect pesticide levels as low as or lower than the conventional solvent extraction/GC-based MRMs.

4.3.1 Immunoassays

The application of immunoassays to pesticide residue analysis dates back to 1971, when Ercegovich (1971) discussed his early work on immunochemical means for detection of pesticides and its importance for routine analyses. According to published reports, more and more researchers paid attention to pesticide immunoassay following Ercegovich's work. The Association of Official Analytical Chemists (AOAC) held its first workshop on immunoassay in 1990, and offered an immunoassay workshop at its national meeting again in 1991 (Herzog, 1997), which is evidence for the usefulness of this work.

Immunoassay of a substance relies on an antibody, or mixture of antibodies, as the analytical reagent, which binds to a molecule, called an antigen. Antibodies are actually proteins with the unique ability to bind with high specificity to one or a very limited group of molecules. Besides their binding specificity, immunoassays produce a measurable signal in response to a specific binding (Fig. 4.2). Most immunoassays today depend on the use of an analytical reagent that is associated with a detectable label, such as radioactive elements, enzymes and so on. Immunoassay has been a widely accepted technique, particularly in the clinical area (Parmar and Kocher, 2012). The phenomenon of competition between labelled and unlabelled insulin in the diabetic patient was first described by using

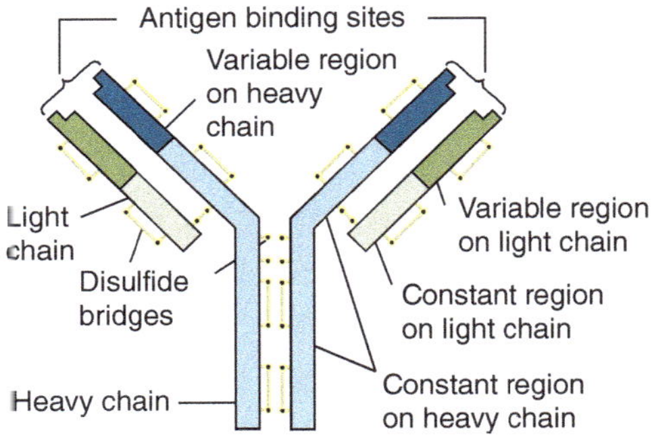

Fig. 4.2. Basic structure of antigen–antibody immunoglobulin molecule. (From Parmar and Kocher, 2012.)

radioimmunoassay and this work was considered as the beginning of modern immunoassay (Herzog, 1997).

4.3.2 Enzyme-linked immunosorbent assay

Approximately 90% of the developed immunoassays for the pesticide residue analysis use the enzyme-linked immunosorbent assay (ELISA) technique. It uses a solid-phase enzyme immunoassay (EIA) to detect the presence of a pesticide and focuses primarily on hapten preparation and conjugation strategies. It involves generation of antigen-specific antibodies followed by capturing the antigen present in a biological sample or vice versa. Availability of antibodies with the desired affinity and specificity is a prerequisite step for the development of immunoassays, and preparation of an immunogen with molecular weight appropriate enough to target the animal's immune response is the foremost step in antibody production.

ELISA may be run in a qualitative or quantitative mode. The qualitative results provide positive or negative results (yes or no) in sample, while in quantitative ELISA, the optical density (OD) of the sample is compared to a standard curve, which is typically a serial dilution of a known-concentration solution of the target molecule. For example, if a test sample returns an OD of 1.0, the point on the standard curve that gave OD = 1.0 must be of the same analyte concentration as the sample.

There is 'direct ELISA' and 'indirect ELISA' for analysing samples. For analysis of an antigen, 'direct ELISA' uses a labelled primary antibody, while 'indirect ELISA' detects a labelled secondary antibody. In the latter case a sandwich ELISA is clearly distinct from an indirect ELISA. When the 'primary' antibody is of interest, for example, in the case of immunization analyses, this antibody is directly detected by the secondary antibody and the term direct ELISA applies to a setting with two antibodies.

The cross-reactivities of the antibody with haptens similar to the analyte may be observed frequently depending on the conjugate and the class of chemicals under investigation. These cross-reactivities can produce false positive values for single-target immunoassay, which could be resolved by sophisticated sample preparation technology and/or antibody purification technology. For example, an antibody for alachlor was found to react very strongly to the sulfonic acid metabolite using an alachlor screening kit, but Aga *et al.* (1994) solved this problem by using solid-phase extraction prior to immunoassy and sequential elution of the two compounds with different organic solvents (Figs 4.3 and 4.4).

Various commercial kits are available for accurate and rapid determination of many *in situ* samples. These are tailored to target different analytes, thus eliminating in some cases the need for clean-up steps (Gabaldón *et al.*, 1999).

Sandwich ELISA

A 'sandwich' ELISA, is used to detect a sample antigen. The sample with an unknown amount of antigen is immobilized on a solid support (usually a polystyrene microtitre plate) either non-specifically (via adsorption to the surface) or specifically by capturing another antibody specific to the same antigen. After the antigen is immobilized, the detection antibody is added, forming a complex with the antigen. The detection antibody can be covalently linked to an enzyme,

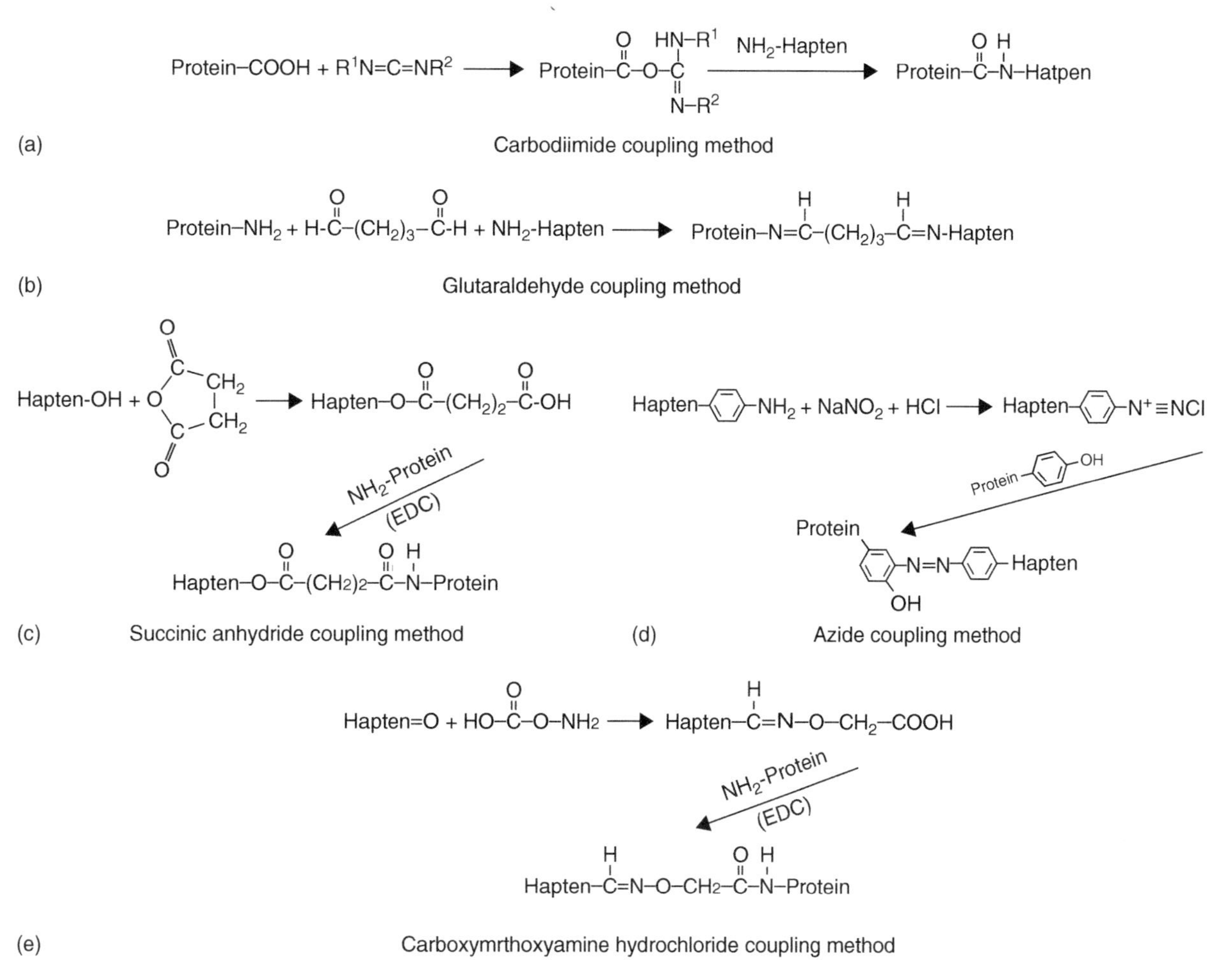

Fig. 4.3. General principles of some frequently used immunogen synthetic methods. (Modified after Tong *et al.*, 2007.)

or can itself be detected by a secondary antibody that is linked to an enzyme through bioconjugation. A mild detergent solution is used to wash the plate, between each step, to remove any proteins or antibodies that are left non-specifically bound. After the final step of washings, the plate is finally developed by adding an enzymatic substrate to produce a visible signal, indicating the quantity of antigen in the sample (Fig. 4.5).

Dot ELISA

Dot ELISA is a micro ELISA, utilizing antigen 'dotted' onto nitrocellulose filter discs – a technique that has been used for more than 25 years. It is a qualitative ELISA test, which can be performed quickly without the need for equipment or technical expertise. Because of its relative speed and simplicity, the dot ELISA is an attractive alternative to standard ELISA (Parmar and Kocher, 2012). This technique can even detect at the nanogram scale among targeted compounds *in situ*.

ELISA could also be used through competitive binding, where unlabelled antibody is incubated in the presence of its antigen. The complex is then added to an antibody-coated well. After washing the plate, a

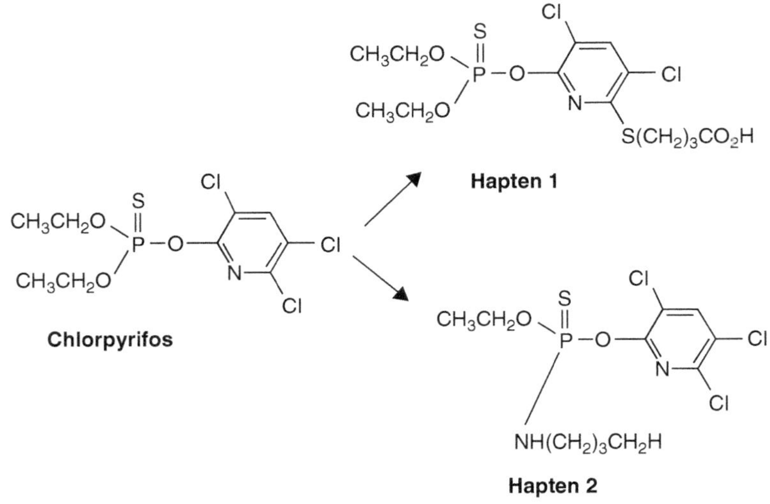

Fig. 4.4. Synthesis of chlorpyrifos hapten (introduction of carboxylic group). (From Goel, 2013.)

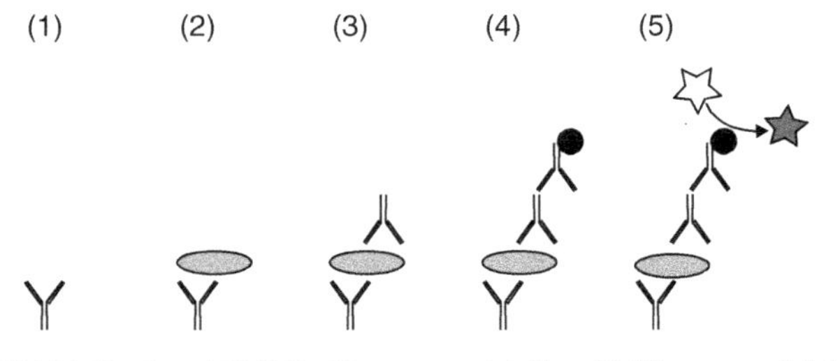

Fig. 4.5. Sandwich ELISA ('indirect' ELISA). (Courtesy of Jeffrey M. Vincour, published under a CC-BY SA 2.5 licence via Wikimedia Commons.) (1) Plate is coated with a capture antibody; (2) sample is added, and any antigen present binds to capture antibody; (3) detecting antibody is added, and binds to antigen; (4) enzyme-linked secondary antibody is added, and binds to detecting antibody; (5) substrate is added, and is converted by enzyme to detectable form.

secondary antibody, coupled to an enzyme specific to the primary antibody is added. A chromogenic or fluorescent signal is observed after the addition of the substrate. Some competitive ELISA kits include enzyme-linked antigen rather than enzyme-linked antibody. The less antigen in the sample, the more labelled antigen is retained in the well and the stronger the signal (Priyadharshini, 2012).

Multiple and portable ELISA is a new technique (EP 1 499 894 B1 in *EPO Bulletin* 25.02.209 N. 2009/09; USPTO 7510687 in *USPTO Bulletin* 31 March 2009; ZL 03810029.0 in *SIPO PRC Bulletin* 08.04.2009) that uses a solid phase made up of an immunosorbent polystyrene rod with 8–12 protruding ogives. The entire device is immersed in a test tube containing the collected sample and the following steps (washing, incubation in conjugate and incubation in chromogens) are carried out by dipping the ogives in microwells of standard microplates filled with reagents.

The major steps in the development of an immunoassay include:

1. Preparation of hapten and immunogen.
2. Characterization of the hapten.
3. Conjugation of hapten to macromolecular carrier.
4. Purification of conjugate.

The production of antibodies involves monoclonal and polyclonal antibody production. Monoclonal antibodies are produced through a fusion of mouse or rat B lymphocyte spleen cells with myeloma tumour cells to produce hybridoma cells, a small percentage of which will produce the desired antibody.

Monoclonal antibodies, however, are not necessarily the better choice for a specific immunoassay. Polyclonal-based immunoassay may be adequate for an immunoassay and, in some cases, they are the more sensitive of the two. But for other pesticides, monoclonal may be necessary (Hammock *et al.*, 1988). Production of monoclonal antibodies requires more time, labour, equipment and training than polyclonal antibodies and can add (Zhenchao *et al.*, 2012) 25% to development costs (K. Hunter, Westinghouse Bio-Analytic Systems Company, Rockville, MD, personal communication, 15 April, 1988) (Herdman *et al.*, 1988c). Polyclonal antibody production is conveniently carried out in warm blooded animals, for example, rabbits, sheep, mice or chickens.

Comparison of dot ELISA with standard ELISA

In the case of dot ELISA, an unknown amount of antigen is affixed to a surface while a specific antibody linked with an enzyme is washed over the surface allowing it to bind to the surface antigen. A substance is added that triggers a detectible signal from the enzyme. In contrast, standard ELISA utilizes a microtitre plate format and requires a relatively high amount of reagent and sample, heating, repetitive reagent dispensing, agitation and spectrophotometers for enzyme detection. Dot ELISA is known for its qualitativeness, and surface treatments with antigens of various viscosities. The incubation times vary depending on the positive and negative controls, and it reduces the signal to noise ratio (Parmar and Kocher, 2012). The dot ELISA can detect as little as 0.055 ng/ml microfilarial antigen added to normal human sera, whereas the lower limit with the sandwich ELISA is 10 ng/ml parasite antigen. Additionally, dot ELISA does not require radioactivity or sophisticated equipment and, therefore, can be performed in virtually all filariasis-endemic areas.

Application of immunoassay in pesticide detection

A variety of pesticides have been determined in fruit, vegetable and grain food samples by immunoassay. For example, residues of carbaryl (Abad and Montoya, 1995), azinphos-methyl (Mercader and Montoya, 1997), carbofuran (Abad *et al.*, 1997) and imidacloprid (Watanabe *et al.*, 2007) in fruit juice, and procymidone (Fernandez-Alba *et al.*, 1995), fenitrothion (Cho *et al.*, 2004), isofenphos (Lee *et al.*, 2006) and iprodione (Watanabe and Miyake, 2007) in vegetables have been detected by immunochemical analysis. The pesticides in liquid food sample can be detected directly, however, for solid food sample, the target pesticides must be extracted with an organic solvent, such as acetone, ether, petroleum ether, methanol, acetonitrile or hexane. Pigments, lipids, proteins or polyphenols or co-extractives in the matrix of food samples may interfere during sample preparation (Fan and He, 2011). The same could be avoided by further diluting the sample with buffer prior to the immunoassay for eliminating the matrix effect. For example, in the work of Abad and Montoya (1995) a monoclonal antibody-based immunoassay is applied to direct determination of carbaryl in apple and grape juices. The most accurate and precise results are obtained with a dilution of 1:100. In some cases a further clean-up step is needed, using solid-phase extraction columns, to separate analyte from the matrix.

The residues of procymidone are estimated at ppb level, in pepper samples using an ELISA commercial kit (Fernández-Alba *et al.*, 1995). In this work, ethyl acetate-sodium sulfate has been used for extracting the samples and clean-up is accomplished through a silica gel solid phase disposable cartridge. The eluted fraction is evaporated and dissolved with sonication in 2 ml of water before immunoassay analysis.

Another application of immunoassay is pesticide residue analysis in soil and water. Johnson *et al.* (2001) coupled SFE with a commercial semi-quantitative ELISA for detection of PCBs in an oily soil matrix. For water samples, sample preparation is not required as per the liquid food sample.

A very important application of pesticide immunoassay is biomonitoring, involving measurement of a parent chemical and/or metabolites or a product of its reaction with cellular components in selected tissues, body fluids such as blood, milk, urine or sweat, or expired breath of an exposed individual (Fan and He, 2011). Sample preparation is generally not required for most of the analytes except for dilution with water or buffer. Chuang *et al.* (2005) diluted the samples of human urine with phosphate-buffered saline containing 0.05% Tween and 0.02% sodium azide, for quantitative measurement of 2,4-dichlorophenoxyacetic acid (2,4-D) by immunoassay. Recovery rates of 2,4-D in spiked urine samples are over 70%. Validation of the results obtained by immunoassay can be done using different established method like HPLC, GC or GC/MS. Many groups have used this approach and have usually obtained correlation coefficients of >0.9. Often a slight overestimation of the immunoassay in comparison with HPLC or GC is observed owing to cross-reactivities of the antibody or matrix effects (Dankwardt, 2001).

Dipstick immunoassay is a convenient format with the advantage of being easy to use without sample preparation, instead involving dipping the strips into the water. The dipstick techniques involve competitive immuno-reaction followed by colour development. The primary antibody is coated on the test strip followed by incubation with the specific antibody. Once the strip is prepared, the immunoassay dipstick is dipped into the solution and the enzyme tracer, and then incubated for a certain time. The enzyme substrate is added and incubated if necessary. The dipstick is then removed from the solution and quantitative measurements can be obtained photometrically (Hennion and Barcelo, 1998).

Gold nanoparticles (GNPs) have also been used, on a dip strip, for the determination of an organochlorine insecticide, dichlorodiphenyltrichloroethane (DDT), by conjugating GNP to anti-DDT antibodies (IgY), which served as the detecting reagent. Depending on the concentration of free DDT in the sample, the binding of GNP conjugated anti-DDT antibodies to the immobilized dimethyl dioctadecyl ammonium bromide on the antigen bovine serum albumin (DDA-BSA) varied and was detected by the development of a red colour (due to gold nanoparticles) in the detection zone of nitrocellulose membrane-containing strips. The intensity of colour development is inversely proportional to the DDT concentration, with maximum intensity at zero DDT concentration. The lowest detection limit for DDT is determined to be 27 ng/ml in optimized conditions (Lisa *et al.*, 2009).

4.4 Biosensor-based Detection

The conventional methods used for detection of pesticides such as GC-MS and HPLC consume a lot of time, energy and labour in laboratory processing followed by analysis using expensive instruments. Immunoassays like ELISA have the same issues. These methods can detect only single compounds, require a long time for analysis (1–2 h) and extensive sample handling (large number of washing steps), making them unsuitable for process control monitoring assays where fast and repetitive analyses are required for the samples contaminated with pesticides (Mulchandani *et al.*, 2001a).

Over the past decade, the development of biosensors for the detection of pesticides has been considered as a promising alternative. A biosensor is a self-contained device that integrates an immobilized biological element (e.g. enzyme, DNA probe, antibodies or microorganisms) that recognizes the analyte (e.g. enzyme substrate, complementary DNA, antigen), and a transduction element used to convert the (bio) chemical signal resulting from the interaction of the analyte with the bioreceptor into an electronic one (Sassolas *et al.*, 2012). Thus, a continuous consumption of substrate(s) is achieved by the immobilized biocatalyst incorporated into the sensor: transient or steady-state responses are monitored by the integrated detector. The transducer part of the sensor serves to transfer the signal from the output domain of the recognition system to, mostly, the electrical domain (Mostafa, 2010). These are a low-cost, fast solution with high portability, and they do not require trained technicians for use. The properties and characteristic behaviours of ideal biosensors include their selectivity, response time, linear range, limit of detection, reproducibility, stability and lifetime, in accordance with standard International Union of Pure and Applied Chemistry (IUPAC) protocols or definition (Daniel *et al.*, 1999; Bilitewski and Turner, 2000; Andreescu and Sadik, 2004). For convenient use, biosensors are usually coupled with an electrochemical sensor. This coupling gives data in readable form (Bhadekar *et al.*, 2011). A number of electrochemical sensors are available commercially.

The biosensors may be classified into electrochemical, optical, piezoelectric and mechanical biosensors. The high sensitivity of electrochemical transducers makes them quite popular for pesticide detection (Jaffrezic-Renault, 2001; Luque de Castro and Herrera, 2003; Andreescu and Marty, 2006). Moreover, they are considered as excellent candidates for the development of portable biosensors because of their low cost, simple design and small size (Thévenot *et al.*, 1999; Freire *et al.*, 2003; Grieshaber *et al.*, 2008; Yogeswaran and Chen, 2008; Ronkainen *et al.*, 2010). The enzymatic, whole-cell, immunochemical and DNA biosensors are another class based on biorecognition element, which have been developed for pesticide detection.

Using innovative techniques, molecular imprinted polymers (MIPs) have been developed as environmental sensors. The precise and selective identification of a pesticide in a sample can be carried out using artificial neural networks coupled with a sensor array. The use of NPs and nano tubes has improved the efficiency of biosensors for the detection of pesticides, making them promising tools in nano science (Sassolas *et al.*, 2012).

4.4.1 Enzyme-based biosensor (Mostafa, 2010)

Enzyme biosensors for pesticide detection are based on measurements of inhibition of enzymes by the pesticides or compounds involved in the enzymatic reaction. Enzymes used in biosensors are highly selective to the substrates and sensitive to pH, temperature, inhibitors, denaturing and chelating agents. Two conditions must be fulfilled to adopt a specific immobilization method in a biosensor: (i) it must provide good mechanical stability; and (ii) it must supply excellent conformation and freedom to the enzymes in their catalytic activity (Vargas-Bernal *et al.*, 2012). The output of the biosensor is measured using a detector or sensor, which is generally termed as a transducer. The electrochemical transducers (potentiometry, amperometry, voltammetry, surface charge using field effect transistors and conductometry) are often used to measure the output signal from the biorecognition domain. The enzymatic biosensors represent the most attractive area of research in the biosensing field for food analysis, for their robustness, ease of immobilization and fast response when compared with antibody-based biosensors. These enzymatic biosensors may work based on direct detection of pesticides or indirect detection depending on the analyte and substrate (Tables 4.2 and 4.3).

Inhibition-based biosensors

ACETYLCHOLINESTERASE ENZYME INHIBITION Several electrochemical biosensors have been described in literature based on acetylcholinesterase (AChE) enzyme (Andreescu and Marty, 2006; Periasamy *et al.*, 2009; Arduini *et al.*, 2010a; Amine *et al.*, 2016). AChE is able to hydrolyse acetylcholine (ACh) following the equation:

$$\text{Acetylcholine} + H_2O \rightarrow \text{Choline} + \text{Acetic acid}$$

This reaction can be monitored measuring (i) the variation of pH by potentiometric transduction (Benilova *et al.*, 2006); (ii) the variation of conductibility by conductometric transduction (Soldatkin *et al.*, 2013); and (iii) the detection of hydrogen peroxide by amperometric transduction, using a bienzymatic approach (in combination with choline oxidase enzyme, ChO). Pesticides such as organophosphates and carbamates have a shape resembling the shape of their substrate, thus they block the active centre of the cholinesterase (ChE) enzyme, inhibiting its activity, which is involved in the proper functioning of the central nervous system of humans. Due to inhibition of the acetylcholinesterase enzyme AChE, ACh (neurotransmitter) accumulates in the body and interferes with the muscular responses and finally leads to respiratory and other problems, and even death (Donarski *et al.*, 1989; Chapalamadugu and Chaudhry, 1992). In the case of AChE inhibition-based OP biosensors, the signal generated is inversely proportional to the concentration of OP compound or, in other words, increased concentration of OP compound leads to weak signals.

Numerous reports are available on the development of biosensors based on ChE inhibition by OPs and carbamate pesticides (Skladal, 1996; Trojanowitz and Hitchman, 1996). The principle of these biosensors is based on the inhibition of ChE and the degree of inhibition is dependent on the concentration of anti-ChE in contact with the enzyme.

Monoenzymatic biosensors AChE and butyrylcholinesterase (BChE) are two types of natural ChE enzymes. These enzymes

Table 4.2. Enzymatic biosensors for direct detection of pesticides. (From Mostafa, 2010.)

Analyte	Enzyme	Detection limit	System	References
Di-isopropyl fluorophosphates	OPAA	20 and 12.5 μM for batch and flow injection	Amperometry	Simonian *et al.*, 2001
Parathion	Parathion hydrolase	1 ng/ml	Amperometry	Sacks *et al.*, 2000
Parathion/paroxon	Parathion hydrolase	Parathion 15/20 nM	Amperometry	Chough *et al.*, 2002
Paraoxon/methyl parathion	OPH	20 nM	Amperometry	Mulchandani *et al.*, 2001a
Organophosphorus neurotoxin	OPH	2 μM and 6 μM for paraoxon and dichlorvos, respectively (potentiometry), and 70 nM for paraoxon (amperometry)	Amperometry/ potentiometry	Wang *et al.*, 2002

OPAA, organophosphorus acid anhydrolase; OPH, organophosphorus hydrolase.

have different substrates: AChE preferentially hydrolyses acetyl esters, such as ACh; whereas BChE hydrolyses butyrylcholine, to give reaction products, for example, choline (Ch) and corresponding organic acids (Andreescu and Marty, 2006).

pH variation due to the formation of acids can be measured using potentiometry methods (Reybier *et al.*, 2002). pH variation produced by acid formation can be measured using electrochemical methods, such as potentiometry (Liu *et al.*, 2005). pH-sensitive spectrophotometric indicators (Andreou and Clonis, 2002; Wong *et al.*, 2006) or pH-sensitive fluorescence indicators (Tsai and Doong, 2005) can also be used for measuring pH variation. Artificial substrates, acetylthiocholine for AChE and

Table 4.3. Enzymatic biosensors for the indirect detection of pesticides.

Analyte	Substrate	Enzyme	Detection limit	References
Trichlorfon	BuCh	BuTChE	<0.1 μM	Reybier *et al.*, 2002
Coumaphose/ trichlorofon/aldicarb/ methiocarb	TCh	TChE	0.002/0.4/0.3/ 0.08 μg/ml	Ivanov *et al.*, 2002
Carbaryl	ATCh	ATChE	5×10^{-5} mol/l	Josiane and Sergio, 2008
Carbafuran	ATCh	ATChE	at nM	Olga and Jon, 2007
Malathion	ATCh	ATChE	0.03 ng/ml	Dan *et al.*, 2008
Triazophos	ATCh	ATChE	0.01 μM	Dan *et al.*, 2007
Paraoxon/chlorpyrifos ethyloxone	P-aminophenol	ChE	19.1/1.24 nM	Andreescu *et al.*, 2002
Chloropyrifosmethyl/ coumaphos/ carbofuran	TCh	TChE	2×10^{-8}, 5×10^{-8} and 8×10^{-8} M	Ivanov *et al.*, 2003
Carbofuran/paraoxon	ACh/BuCh	AChE/BuTChE	0.2 and 0.6 nM	Albareda-Sirvent *et al.*, 2001
Chlorpyrifos ethyl oxon	TCh	ChE	1 ppb	Montesinos *et al.*, 2001
Dichlorvos/carbofuran	ACh/Ch	ChE/ChO	1.3×10^{-3}/ 0.01 ppb	Snejdarkova *et al.*, 2003
Carbaryl/carbofuran	ACh/Ch	AChE/ChO	2 μg/ml	Palchetti *et al.*, 1997
Trichlorofon	ACh/Ch/H_2O_2	ChE/ChO/ peroxidase	5 ng/ml	Mazzei *et al.*, 1996
Aldicarb	Glucose-6-phosphate/ glucose	AP/GOD	40 μg/ml	Mazzei *et al.*, 1996
Methyl parathion/ diazinon/carbofuran/ carboy	Monophenol	Tyrosinase	6/19/5/10 ppb	Yaico *et al.*, 2007
Triazine	Monophenol	Tyrosinase	0.5×10^{-9} mol/l	Campanella *et al.*, 2006
Diazinon/dichlorvos	Monophenol	Tyrosinase	5 and 75 μM	Everett and Rechnitz, 1998
Ziram/diram/zinc diethydithiocarbamate	Monophenol	Tyrosinase	0.074/1.3/ 1.7 μM	Perez-Pita *et al.*, 1997
Zineb	Propinoaldehyde	Aldehyde dehydrogenase	8 ppb	Noguer *et al.*, 1999
Herbicide	Pyruvate	Acetolactate synthase	1 μM	Seki *et al.*, 1996
Ethyl paraoxon	Ascorbate	oxidase Ascorbate	1 ppm	Rekha *et al.*, 2000

ACh, acetylcholine; ATCh, acetylthiocholine; BuCh, butyrylcholine; TCh, thiocholine; Ch, choline; ATChE, acetyl thiocholine esterase; BuTChE, butyryl thiocholine esterase; ChO, choline oxidase; ChE, choline esterase; AP, acid phosphates; GOD, glucose oxidase.

butyrylthiocholine for BChE, have also been used. The enzymatic hydrolysis of these substrates produces electro-active thiocholine (Equations 4.1 and 4.2).

Thiocholine can be oxidized anodically using platinum electrodes or modified electrodes. For a review article on ChE biosensors from basic research to its practical applications, see Silvana and Jean-Louis (2006). The monoenzyme-based biosensors have a simpler design, and the detection potential is lower than the one used for the oxidation of H_2O_2.

Bienzymatic biosensors AChE hydrolyses its natural substrate to Ch and acetic acid. In this system, ChE is coupled to a second enzyme, ChO (Equations 4.3 and 4.4), since Ch is not electrochemically active (Thévenot *et al.*, 1999).

In the reaction, ChO catalyses the oxidation of Ch to produce H_2O_2, where oxygen is consumed. Hence, a change of concentration of one of these can be the basis for the bienzymatic response. Oxygen detection is achieved by Clark electrodes and H_2O_2 with platinum electrodes, graphite or screen print electrodes or other electrodes (Trojanowicz and Hitchman, 1996).

For the detection of anti-ChE compounds, a biosensor was developed by coimmobilizing AChE and ChO on poly (2-hydroxyethyl methacrylate) membranes (Kok and Hasirci, 2004). The authors used

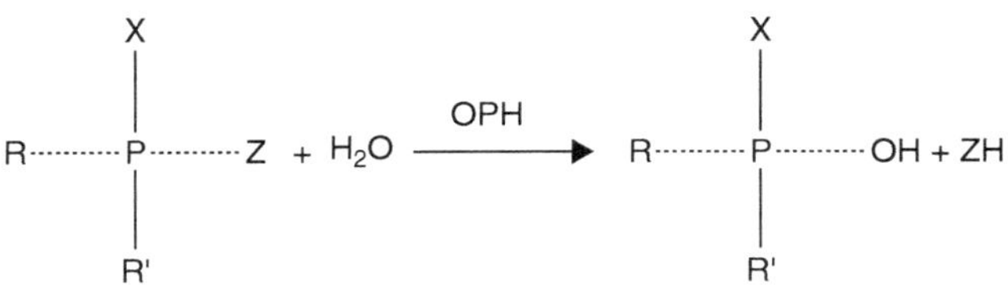

(Eqn 4.1)

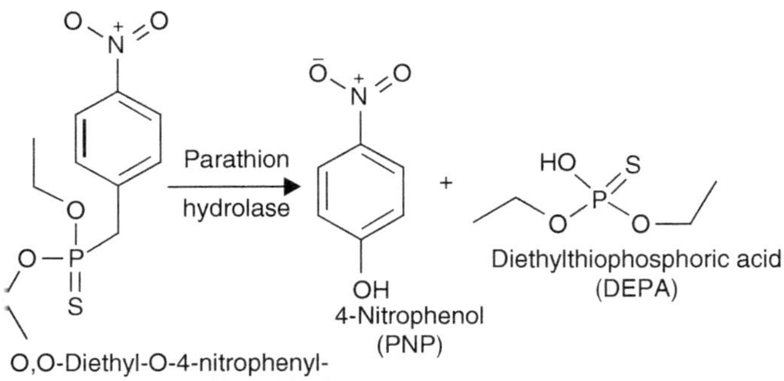

(Eqn 4.2) (Mostafa, 2010)

$$\overset{Cl^-}{\overset{+}{(H_3C)_3N}}—CH_2CH_2—O—COR \xrightarrow[ChE]{H_2O} \underset{\text{Choline}}{\overset{Cl}{\overset{+}{(H_3C)_3N}}}—CH_2CH_2—OH + RCO\overset{-}{O} + \overset{+}{H}$$

R = CH_3; acetylcholine

R = $(CH_2)_2CH_3$; butylarylcholine

(Eqn 4.3)

$$\underset{\text{Choline}}{\overset{Cl}{\overset{+}{(H_3C)_3N}}}—CH_2CH_2— OH + H_2O + 2O_2 \xrightarrow{ChO} \underset{\text{Choline}}{\overset{Cl}{\overset{+}{(H_3C)_3N}}}—CH_2COOH + 2H_2O_2$$

(Eqn 4.4)

the biosensor for the detection of anti-ChE activity of aldicarb (AC), carbofuran (CF) and carbaryl (CL), as well as two mixtures, (AC + CF) and (AC + CL). The total anti-ChE activity of binary pesticide mixtures was found to be lower than the sum of the individual inhibition values.

Trienzymatic biosensors For a trienzyme device, peroxidase (POD) may be added to the bienzyme system (Equation 4.5). The H_2O_2 is produced in the second reaction, provoking a potential change in the electrode due to its bioelectrocatalysis, where POD is regenerated without the presence of a mediator. The potential change in the electrode is due to direct electron transfer to POD. A quartz crystal microbalance (QCM) has been used to develop a trienzymatic biosensor for the determination of organophosphorus and carbamate pesticides (Karousos *et al.*, 2002). AChE converted ACh to Ch, which in turn was converted to H_2O_2 by ChO. AChE inhibition caused by pesticides reduced the amount of QCM-detectable precipitate produced. This QCM–enzyme sensor system allowed detection of carbaryl and dicholorvos up to 1 ppm concentrations.

$$2H_2O_2 \xrightarrow{\text{Oxidation}} O_2 + 2H^+ + 2e^- \quad \text{(Eqn 4.5)}$$

A few carbamates and herbicides like atrazine are capable of inhibiting tyrosinase enzyme, leading to reports of the development of several electrochemical biosensors (Vidal *et al.*, 2006; de Albuquerque and Ferreira, 2007; Campanella *et al.*, 2007; Kim *et al.*, 2008a, 2008b; De Lima *et al.*, 2010). However, due to interference of various substrates, they have poor specificity. Although the tyrosinase enzyme can withstand high temperatures and organic solvents used to dissolve the pesticides, it is inherently unstable, reducing the lifetime of the tyrosinase-based biosensors. The biosensors based on this enzyme, therefore, have reduced lifetime.

Biosensors, based on another enzyme, alkaline phosphatase (ALP), have also been reported to detect various pesticides. Paraoxon, a metabolite of parathion, has been dephosphorylated and detected using a chemiluminescent ALP-based biosensor. It is based on the measurement of the intensity of the light generated in the reaction (Ayyagari *et al.*, 1995). Similarly organochlorine, carbamate and one of the organophosphates, fenitrothion, along with heavy metals have been detected using a fluorescent ALP-based biosensor, which catalyses the reaction to produce fluorescent molecules (Garcia Sanchez *et al.*, 2003). A study by Mazzei *et al.* (2004) reported on the use of an ALP-based electrochemical biosensor for the detection of malathion and 2,4-dichlorophenoxyacetic acid (2,4-D).

ACID PHOSPHATASE A biosensor based on a bienzymatic system, consisting of two enzymes, namely AP and glucose oxidase (GOD) has been reported for the determination of organophosphorus and carbamic acid pesticides (Mazei *et al.*, 1996). These pesticides are reported to inhibit the hydrolysis of glucose-6-phosphate reaction reversibly in the presence of AP as a catalyst (see Equations 4.6 and 4.7).

$$\text{Glucose-6-phosphate} + H_2O \xrightarrow{AP} \text{Glucose} + \text{inorganic phosphate} \quad \text{(Eqn 4.6)}$$

$$\text{Glucose} + O_2 \xrightarrow{GOD} \text{Gluconolactone} + H_2O_2 \quad \text{(Eqn 4.7)}$$

The mixture of pesticides can be evaluated using a biosensor consisting of three pH electrodes according to Danzer and Schwedt (1996). The ion-sensitive area of each electrode was covered with a cellulose acetate membrane incorporating AChE, ALP or acid phosphatase. The relative inhibition of each test substance was obtained by potentiometrically measuring.

Catalytic biosensors

The catalysts used in these biosensors are: (i) enzymes; (ii) microorganisms; (iii) organelles; and (iv) tissue samples (see Lecture 17; http://ocw.mit.edu/courses/materials-science-and-engineering/3-051j-materials-for-biomedical-applications-spring-2006/lecture-notes/lecture17.pdf).

The organophosphorus hydrolase (OPH) enzyme is used to hydrolyse organophosphorus pesticides (Mulchandani *et al.*, 2001b), such as parathion, methyl parathion (Du *et al.*, 2010) or paraoxon (Lee *et al.*, 2010; Pedrosa *et al.*, 2010). This enzyme can hydrolyse P–O, P–S and P–CN bonds generating two protons, which can be electrochemically detected.

However, these biosensors show lower sensitivity values and higher detection limits than ChE-based biosensors. Moreover, they can only detect some organophosphorus compounds. A fibre optic biosensor has been developed, using glutathion-s-transferase, for detection of a herbicide, atrazine, as reported by Nunes and Barcelo (1998).

Amperometric biosensing devices have been used to measure thiocholine produced by the hydrolysis of butyrylthiocholine, or hydrogen peroxide generated as a result of the oxidation of Ch produced from ACh hydrolysis in the presence of ChO (Skladal, 1991; Marty *et al.*, 1992; Palleschi *et al.*, 1992; Skladal and Mascini, 1992; La Rosa *et al.*, 1994; Martorell *et al.*, 1994; Mionetto *et al.*, 1994; Trojanowicz and Hitchman, 1996; Martorell *et al.*, 1997; Palchetti *et al.*, 1997); potentiometric devices for measuring the pH change as a result of acetic acid production (Tran-Minh *et al.*, 1990; Chuna Bastos *et al.*, 1991; Kumaran and Tran-Minh, 1992; Kumaran and Morita, 1995; Danzer and Schwedt, 1996), conductometric (Dzyadevich *et al.*, 1994) or fibre optic for monitoring the pH change using pH indicators or the formation of chromophoric product transducers have also been reported (Rogers *et al.*, 1991; Hobel and Polster, 1992; Trettnak *et al.*, 1993; Garcia de Maria *et al.*, 1994; Moris *et al.*, 1995; Andres and Narayanaswamy, 1997).

These biosensors are sensitive and useful as single-use disposable sensors for environmental monitoring; however, they require long incubation with inhibitors prior to analysis for good sensitivity, and treatment with pyridine-2-aldoxime (2-PAM) after analysis partly to regenerate/recover the enzyme activity, which is inhibited irreversibly by organophosphate (Marty *et al.*, 1992; Dzyadevich *et al.*, 1994; La Rosa *et al.*, 1994). Lee *et al.* (2002) reported the ability of 2-PAM to reactivate the enzyme inhibited by organophosphates. The enzyme was treated with 20 μM 2-PAM in a HEPES (4-(2-hydroxyethyl)-1-piperazineethanesulfon ic acid is a zwitterionic organic chemical buffering agent) working buffer for 20 min, followed by the passage of HEPES working buffer for 20 min. AChE is inhibited by neurotoxins such as organophosphates and carbamates, making these analytical tools not selective enough to quantify either an individual or a class of pesticides that may be required for monitoring detoxification processes.

4.4.2 Portable biosensors

Portable biosensors, with the advantage of enabling on-site testing, have the potential to revolutionize medical diagnostics and environmental monitoring (Xiang and Lu, 2011). They are highly useful for the analysis of food contaminants Dhull *et al.* (2013) developed a portable paper-based biosensor for the detection of the degradation products of organophosphorus pesticides and their metabolites. The biosensor is reported to quantify AChE inhibitors in a fast, disposable, cheap and accurate format. To achieve longer shelf-life, sugar or protein were used as a stabilizer. The biosensor could detect malathion with a detection limit of 2.5 ppm in 5 min incubation time.

The limitation on the use of portable biosensors is the source of power. Another limitation is the necessity for sample preparation. Contaminants in liquid samples can be analysed easily, but pesticides in a matrix require more complicated sample preparation, for example, the homogenization of fruit or vegetable or meat samples. If there is a requirement for enrichment steps or other forms of sample concentration, even the most portable biosensor will be restricted by the need for the pre-analysis sample preparation to be performed in a laboratory.

4.4.3 Nano particle based immunosensors

Immunosensors are the most sophisticated immunoassay format to date. They are designed so that the formation of the immune complex on the transducer surface is directly determined by measuring the physical changes (electrical or optical) induced by the binding reaction. These biosensors have two immunological molecules, namely antigen and antibodies, which bind specifically. Either the antibody or the antigen is immobilized on the transducer carrier surface by forming a sensing device reacting with the complementary antigen or antibody to form an immune complex. They are characterized by sensitivity, rapidity, specificity, low cost and ability to analyse large number of samples, and on-site monitoring of pesticide pollutants.

Recent developments in immunosensors for pesticide detection have been summarized by Suri *et al.* (2009). Only the specific antigen–antibody reactions of pesticides are detected, and the reactions coupled with enzyme labels can be quantified. There can be: (i) labelled; and (ii) label-free types of immunosensors. The first type makes use of different enzymes like glucose oxidase, horseradish peroxidase, ALP, etc. The sandwich assay and competitive assays are two methods that use the labelled type. Similarly, label-free types of sensors are grouped into direct and indirect types. The applications for immunoassay as a pesticide detection method have been reviewed in many papers (Sherry, 1992; Meulenberg *et al.*, 1995; Sadik and Emon, 1996). Commercial immunoassay kits are also available on the market. Valera *et al.* (2014) reported a quantum dot-based immunosensor in place of enzymes. The quantum dots in the immunosensor allowed a fast detection of the immunocomplex, avoiding the time lapse needed for the reaction between enzyme and substrate. In an immunosensor, paraquat-specific antibodies labelled with CdS nanoparticles are used with an antigen, biofunctionalized with magnetic μ-particles, confined to the working electrode surface, for the paraquat pesticide. The metal ions released in an acid solution from CdS nanoparticles quantify the formation of the immuno-complex. The paraquat was detected with an IC_{50} equal to 0.18 ± 0.31 μg/l (7×10^{-10} M), a very high detectability level, due to the amplification effect produced by CdS nanoparticles on the electrochemical signal. A detection limit of 1.4 μg/kg was observed for paraquat detection in potatoes, which is far below the MRL (20 μg/kg) established by the EU for this pesticide.

The nanomaterials in immunoassay are able to overcome the drawbacks of conventional immunosensors, including multiple washing steps, extended enzymatic reaction time and single analyte detection. Due to the intrinsic advantages of using nanomaterials in immunoassay development, the results attained a rapid evaluation of the immunocomplex leading to smart detection of the target analyte.

Depending upon the transducer technology employed, immunosensors are broadly classified (Table 4.4) into optical, piezoelectric, electrochemical and micromechanical.

- *Piezoelectric immunosensors*: these are more common due to label-free detection of atrazine, parathion, etc. (Halamek *et al.*, 2001; 2005; Pribyl *et al.*, 2003). A piezoelectric crystal can be coated with an antigen or antibody and the change in the mass by the binding of the analyte can be correlated to the concentration of the analyte (Guilbault *et al.*, 1992).
- *Optical immunosensors*: the main optical immunosensors developed are based on a surface plasmon resonance (SPR) device. In another type of optical immunosensor, the antibody is coated on the metal sheet and causes a minute change in the refractive index when bound with the analyte, and this change can be detected by the SPR device. Another optical immunosensor is based on total internal reflection fluorescence. These biosensors are used to detect terbutryn, atrazine, parathion, polychlorophenol, etc (Erickson *et al.*, 2008).

The biological immunoassay was successfully emulated by Mosbach's group in 1993

Table 4.4. Immunosensor detection of pesticides.

Analyte	Immunosensor	System	Detection limit	References
Simazine	Peroxidase label antibody	Potentiometry	3 ng/ml	Yulaev *et al.*, 2001
Paraoxon	Paraoxon antibodies	Amperometry	12 ng/ml	Hu *et al.*, 2003
Chlorsulfuran	Anti-chlorsulfuron antibodies	Amperometry	0.01 ng/ml	Dzantiev *et al.*, 2004
2,4-D /2,4,5-T	Monocolonal/ polyclonal antibodies	Amperometry	5/10 PM	Medyantseva *et al.*, 1997
2,4-D	2,4-D	Amperometry	10 PM	Medyantseva *et al.*, 1995
Atrazine	Biotinylated-fabfragment K 47 antibody	Conductommetry	10 ng/ml	Hleli *et al.*, 2006
Paraoxon/methyl parathion/ parathion	(Cell-based biosensor) microbial (*Pseudomonas putida* JS444)	Amperometry	0.28/0.26/ 0.29 ppb	Lei *et al.*, 2005
Paraoxon/parathion, methyl parathion/ diazinon	Culture of *Escherichia coli* cells	Potentiometry	3 μM	Mulchandani *et al.*, 1998
Paraoxon	Amicrobial (PNP-degrader *Moraxella*)	Amperometry	27.5 ppb	Priti *et al.*, 2006
Chlorpyrifos/ malathion	(DNA) (calf thymus DNA)	Amperometry	0.0016/ 0.17 ppm	Nirmal *et al.*, 2007
Chlorpyrifos/ malathion	Double-stranded calf thymus DNA	Voltammetry, FTIR, SEM and electrochemical impedance	0.5 ppb and 0.01 ppm	Nirmal *et al.*, 2008

2,4-D, 2,4-dichlorophenoxyacetic acid; 2,4,5-T, 2,4,5-trichlorophenoxyacetic acid; PNP, p-nitropheny; DNA, deoxyribonucleic acid; FTIR, Fourier transform infrared spectroscopy; SEM, scanning electron microscope.

(Vlatakis *et al.*, 1993) by using MIP (pseudo-immunoassay). MIP-based immunoassays have been successfully applied for the detection of pesticides, such as 2,4-D (Lu *et al.*, 2007; Vandevelde *et al.*, 2007), atrazine (Prasad *et al.*, 2007), parathion (Li *et al.*, 2005), etc. It is believed that these alternatives to antibodies will be critical materials for the rapid detection of pesticides in the future.

4.4.4 Aptamer-based biosensor

Aptamers are oligonucleotide or peptide molecules with specific binding affinity for a specific target molecule. They are often termed as chemical antibodies, as they can be isolated from chemically synthesized combinatorial libraries by selection evolution of ligands by exponential enrichment (SELEX), produced by chemical synthesis, and purified to a very high degree (Scognamiglio *et al.*, 2015).

Aptamers are easily selected against small, non-immunogenic molecules and molecules that are toxic to antibody-producing cells. They possess several competitive advantages over antibodies, such as accurate and reproducible chemical production, their stability in extreme conditions and the ease with which they can be modified with functional groups. Through chemical modification, a vast range of additional functional groups can be added at any desired position in the oligonucleotide sequence; therefore, the best features of small molecule drugs, proteins and antibodies can be brought together into aptamers, making aptamers the most versatile reagent in proteomics. Moreover, immunization and animal hosts are not required for their production. Lim *et al.* (2010) have devised numerous aptamer-based biosensors for the detection of a wide range of target analytes.

In one review article, Sassolas *et al.* (2012) reported and described a DNA aptamer specific for acetamiprid. The potential of aptamers for pesticide detection has not been exploited but aptamer-based biosensors could be an alternative to the conventional methods for pesticide analysis. He *et al.* (2011) demonstrated a high potential for pesticide detection, and highlighted the importance of the discovery of the aptamer. Fan *et al.* (2013) constructed a biosensor by immobilizing the aptamer onto a gold electrode modified with gold nanoparticles, which could detect acetamiprid at low levels. The resulting acetamiprid–aptamer complex was evaluated by using electrochemical impedance spectroscopy (EIS), a label-free technique widely used in the biosensor field. It can provide the desired sensitive measurement platform. Here, the analyte binds to the capture molecules immobilized on the surface and increases the charge transfer resistance (Rct) proportional to the analyte concentration. The charge transfer resistance describes the resistance that a redox probe experiences on an electrode surface with respect to an electron transfer. Rct is usually the calculated parameter, which is estimated according to the diameter of the semicircle present at the high frequency region in the Nyquist plot. With this configuration, the authors evaluated the formation of acetamiprid–aptamer complex following the increasing Rct value. A wide linear range was attained from $5–600 \times 10^{-9}$ M with a low detection limit of 1×10^{-9} M. The applicability of this aptasensor was successfully evaluated by determining acetamiprid in tomatoes (Fan *et al.*, 2013).

4.4.5 Cell-based biosensors

Living cells or microorganisms (algae, bacteria, yeast and fungi) have been used to develop a microbial biosensor, which is low in cost. These microbials can be used as the biocatalytic elements for biosensors. A biosensor comprising *Pseudomonas putida* JS444, as a p-nitrophenol degrader, genetically engineered to express organophosphorus hydrolase (OPH) on the cell surface, was developed by immobilizing it on the carbon paste electrode (Lei *et al.*, 2005). The detection limits were found to be comparable to ChE inhibition-based biosensors. The biosensor could measure as low as 0.28, 0.26 and 0.29 ppb of paraoxon, methyl parathion, and parathion, respectively. Another conductometric enzymatic biosensor was developed using immobilized *Chlorella vulgaris* microalgae (an algae), immobilized inside bovine serum albumin membranes, as bioreceptors (Chouteau *et al.*, 2005). It was observed with initial experiments that paraoxon-methyl inhibits *C. vulgaris* AChE contrary to parathion-methyl and carbofuran.

Priti *et al.* (2006) reported a microbial biosensor consisting of a dissolved oxygen electrode modified with the genetically engineered p-nitrophenyl degrader *Moraxella* sp. OPH was used on the cell surface for sensitive, selective, rapid and direct determination of p-nitrophenyl substituted organophosphates (Priti *et al.*, 2006). The biosensor was able to measure as low as 27.5 ppb of paraoxon and had excellent selectivity against triazines, carbamates and organophosphates without p-nitrophenyl substitution.

4.4.6 DNA-based biosensors

These biosensors utilize the interaction of DNA with various molecules either by monitoring changes in the DNA redox reactions (i.e. oxidation of guanine) or with an electroactive analyte intercalated on a DNA layer (Mascini *et al.*, 2001; Wang, 2002). DNA biosensors are based on polyaniline (PANI)-polyvinyl sulfonate (PVS) and fabricated using an electrochemical entrapment technique into indium-tin-oxide (ITO) for detection of organophosphorus pesticides (chlorpyrifos and malathion) (Nirmal *et al.*, 2008). These double-stranded calf thymus bioelectrodes were characterized using square wave voltammetry, Fourier transform infrared spectroscopy, scanning electron microscopy and electrochemical impedance techniques, respectively. The dsCT-DNA entrapped PANIPVS/ITO bioelectrode was found to have a response time of 30 s, with a stability of about 6 months and detection limit of 0.5 ppb and 0.01 ppm for chlorpyrifos and malathion, respectively.

4.4.7 Nanoscience-based biosensors

Nowadays, nanomaterials are an attractive proposition as biosensing tools because of their unique electrical, chemical, mechanical, magnetic and physical properties (i.e. size, composition, conductivity, magnetism, mechanical strength, light absorbing and emitting properties). Carbon nanotube (CNT), graphene, metal nanoparticles, gold nanoparticles (AuNP), nanowires, nanorods, nanofibres, as well as nanocomposites, nanofilms, nanopolymers, nanoplates and quantum dots are common nanomaterials, especially targeted for developing novel biosensors (Zhang *et al.*, 2017).

For instance, graphene and its derivatives have attracted considerable attention in the application of biosensors due to their super large surface area, biocompatibility, excellent electrocatalytic activity, fast electron transfer, strong mechanical strength and high chemical stability. Nanotechnology can be expected to be an effective approach to develop more sophisticated multianalyte detection systems with low costs.

Optical biosensors have been developed using nanoparticles (Lin *et al.*, 2006) for the detection of monocrotophos using CdTe as a fluorescence probe (Sun *et al.*, 2011). In the absence of the pesticide, ACh was biocatalytically hydrolysed inducing the production of Ch and acetic acid, resulting in a pH decrease due to release of acid. The pH variation was sensed by the immobilized pH indicator (CdTe). The presence of monocrotophos induced a change in the fluorescence intensity that was related to the pesticide concentration.

The AuNPs have great potential for the development of a localized SPR (LSPR) sensor due to their optic properties (Chau *et al.*, 2006; Fu *et al.*, 2009). The absorption band of AuNPs results when the incident photon frequency is resonant with the collective oscillation of the conduction electrons and is known as the LSPR. The LSPR was used to develop a biosensor for the detection of paraoxon by immobilizing AChE onto the AuNPs layer using a self-assembling technique (Lin *et al.*, 2006). Enzyme inhibition, due to the presence of pesticides, caused a change in light attenuation. The detection limit with optimal conditions was 0.2 ppb. The biosensor retained 94% of its original activity after six cycles of inhibition with 500 ppb paraoxon followed by reactivation of AChE with 0.5 mM 2-pyridine-adoxime methoiodide. In addition, the sensor retained its activity after two months storage in dry state at 4°C.

Due to the increase in carbon-based nanomaterials, and their large use in this field, this section has been divided into two main areas of interest: biosensors obtained by using carbonaceous-based modifiers (both pristine and composites) and biosensors modified with non-carbonaceous nanomaterials (i.e. noble metal nanoparticles, metal oxide nanoparticles, quantum dots).

4.4.8 Carbon nanotube-based biosensors

Carbon nanotubes (CNTs) consist of cylindrical graphene sheets with diameters in nanometres. Nanotubes are categorized as single-walled nanotubes (SWNTs) and multi-walled nanotubes (MWNTs). Since their discovery, CNTs have been used in nanoelectronics, biomedical engineering, biosensing and bioanalysis. A CNT-based biosensor using an organophosphorus hydrolase (OPH) layer on CNT film has high sensitivity and stability according to Deo *et al.* (2005). Under optimal conditions the biosensor could measure as low as 0.15 µM paraoxon and 0.8 µM methyl parathion with sensitivities of 25 and 6 nA/µM, respectively. Yuehe *et al.* (2004) developed a disposable CNT-based biosensor for the detection of organophosphorus pesticides and nerve agents. The biosensors provided a high sensitivity, large linear range and low detection limits for the analysis of organophosphate compounds when AChE/ChO enzymes are used, indicating catalytic activity of CNT with a large surface area.

A novel bienzyme biosensor incorporating electrostatically interacted enzyme-armoured MWCNT-OPH and MWCNT-AChE along with a set of cushioning bilayers consisting of MWCNT-polyethyleneimine and MWCNT-DNA on glassy carbon electrode for the discriminative detection of organophosphate and non-organophosphate pesticides has been reported. The detection limit was found to be ~0.5 for organophosphate pesticide paraoxon and 1 μM for non-organophosphate pesticide carbaryl, in a wide linear range. The use of biosensor resulted in remarkable discrimination between and straightforward detection of organophosphate and non-organophosphate neurotoxins, and was successfully achieved with cyclic voltammetry and UV-vis methods on the MWCNT-(PEI/DNA) 2/OPH/AChE biosensor, showing great potential for large screening of organophosphate and non-organophosphate pesticides in practical applications.

Pattern-oriented antibodies have been assembled on a glassy carbon substrate using SWCNTs. The assembled SWCNTs were able to furnish a well-arranged orientation of the bioelements as well as an efficient electron transfer between biomolecules and the electrode. Liu *et al.* (2012) assembled a forest of SWCNTs on glassy carbon substrate using a microcontact-printing technique. The amide bonds from the reaction between the amines located on the modified substrate and the carboxylic groups at the ends of the nanotubes led to SWCNTs covalently and vertically anchored on the electrode surface. Ferrocenedimethylamine was attached to the ends of SWCNTs through amide bonding followed by the attachment of an epitope, endosulfan hapten, to which an antibody binds. This structure led to a well assembled orientation of the haptens and consequently of the antibody. The use of such a SWCNT-modified sensing platform allowed the detection of endosulfan in water samples, at 0.01 μg/l (2×10^{-11} M), with recovery values of 93%.

4.4.9 Graphene-based biosensors

The structure of graphene and graphene-based nanocomposites, such as pristine graphene, is characterized as an array of 2D and sp^2-hybridization of pure carbon atoms arranged in a hexagonal lattice with covalent bonds. This kind of structure led to the development of a plethora of biosensors to be used for the detection of pesticides in food samples. Zheng *et al.* (2015) modified a glassy carbon electrode by using graphene functionalized with ionic liquid (IL-GR) and gelatine. The IL-GR–Gel nanocomposite had excellent conductivity and biocompatibility, offering an extremely hydrophilic surface for AChE adhesion, and promoted the electron transfer rate of the electrode interface and facilitated the access of substrates to the active centres. The biosensor could detect carbaryl and monocrotophos with detection limits of 5.3×10^{-15} M and 4.6×10^{-14} M, respectively. Recoveries for carbaryl and monocrotophos, spiked in tomato juice samples, were found to be in the range of 92.5–105.0% and 91.2–110.0% for carbaryl and monocrotophos, respectively, indicating acceptable accuracy for the analysis of food samples. Owing to the excellent properties of IL-GR–Gel composite, such as excellent electrical conductivity, high accessible surface area and good biocompatibility, the prepared GA/IL-GR–Gel/GCE biosensor showed higher sensitivity, lower detection limit, good fabrication reproducibility and acceptable stability for the purpose of organophosphate detection.

Oliveira *et al.* (2014) developed a sensitive bienzymatic biosensor based on polyphenoloxidases-gold nanoparticles-chitosan hybrid film-graphene doped carbon paste electrode for carbamate detection. The authors enhanced the conductivity of biosensor using carbon paste electrodes doped with 20% (w/w) of graphene as substrate to electrodeposit hybrid film composed of laccase, tyrosinase and gold nanoparticles entrapped in a chitosan polymeric matrix. They observed low detection limits in the range of ($1.68 \times 10^{-9} - 2.15 \times 10^{-7}$ M) for the determination of carbaryl, formetanate hydrochloride, propoxur and ziram in citrus fruits based on their inhibitory capacity on the polyphenoloxidases' activity. Recoveries at two fortified levels ranged from $93.8 \pm 0.3\%$ (lemon) to $97.8 \pm 0.3\%$ (orange). Zhang *et al.* (2015) developed an AChE biosensor based on a reduced graphene oxide/silver nanocluster/chitosan as glassy carbon electrode modifier. The developed biosensor, using 2D configuration of graphene, exhibited good sensitivity, stability and reproducibility, thus providing a promising tool for the analysis of enzyme inhibitors and direct analysis of practical samples. Chitosan was used as a protective membrane to improve the stability of the biosensor. The detection limits of phoxim were observed as 0.2 to 250×10^{-9} M with a detection limit of 81×10^{-12} M. Using the standard addition method, the accuracy of the biosensor for organophosphates was evaluated in tap water samples observing recoveries in the range of 89.7–96.6%. Yang *et al.* (2014) developed a nanohybrid of gold nanoparticles, polypyrrole and reduced graphene oxide coupled with AChE for the detection of paraoxon-ethyl. Polypyrrole helped to avoid aggregation of reduced graphene oxide and, together with gold nanoparticles, led to a significant increase in the surface area of the modified electrode, from 0.15 cm^2 (glassy carbon electrode) to 1.66 cm^2 (glassy carbon electrode modified with hybrid nanocomposite). The nanocomposite showed excellent conductivity for thiocholine detection at around 0.7 V vs. SCE, with an overpotential decrease of around 200 mV compared with that of the bare electrode. Paraoxon-ethyl was detected from 1.0×10^{-9} M to 5×10^{-3} M with a detection limit of 0.5×10^{-9} M.

Wang *et al.* (2014) developed a novel AChE biosensor based on homogeneous dispersion of zinc oxide nanoparticles and carboxylic graphene in nafion hybrid-modified glassy carbon electrode. The oxidation peak potential of thiocholine shifted to lower potentials (0.47 V vs. SCE). The detection limits for chlorpyrifos and carbofuran were observed as 5×10^{-14} M and 5.2×10^{-13} M, respectively. To further demonstrate the reasonableness of the proposed method, recovery tests were performed on tap water by the standard addition method, observing recovery values in the range of 93.2–104.8%. The low relative standard deviation (RSD = 5.9%) for chlorpyrifos demonstrated the high precision of analysis (Arduini *et al.*, 2016). Carboxylic graphene has also been chemically modified with nickel oxide nanoparticles. Yang *et al.* (2013) developed an AChE biosensor based on nickel oxide nanoparticles, carboxylic graphene and nafion-modified glassy carbon electrode. The oxidation peak potential of thiocholine shifted to lower potentials (0.47 V vs. SCE), allowing the detection of carbofuran and chlorpyrifos at 10^{-13} M level, and methyl parathion at 10^{-14} M level. The same research group reported similar work, employing SnO_2 instead of NiO (Zhou *et al.*, 2013), reaching in this case detection limits at the $10^{-13}/10^{-14}$ M level (Arduini *et al.*, 2016). The developed biosensor exhibited good sensitivity, stability, reproducibility and low cost, providing a promising tool for analysis of enzyme inhibitors.

Wang *et al.* (2011) developed a rapid and sensitive AChE amperometric biosensor based on the quantum dots made of CdS-decorated graphene nanocomposite. The presence of CdS/graphene nanocomposite on glassy carbon electrodes provided a sensitivity towards organophosphates about 3-fold higher than the unmodified glassy carbon electrodes. However, the applied potential was quite high, being equal to 680 mV vs. SCE (Arduini *et al.*, 2016). The detection limit of biosensor to detect carbaryl was observed as 0.7 ng/ml (3×10^{-10} M). A nanocomposite formed by graphene and quantum dots was also subsequently adopted by Li and colleagues (2015). The authors deposited a nanocomposite based on CdSe/ZnS nanoparticles and graphene onto an indium tin oxide glass electrode in order to develop a novel photoelectrochemical AChE biosensor for paraoxon and dichlorvos detection at 10^{-14} M and 10^{-12} M. To demonstrate the feasibility of this biosensor, recovery studies in apple were performed, reaching a recovery from 93.6±12.1% to 103.3±12.8%, demonstrating its successful application in detection of organophosphates in real fruit samples.

4.4.10 Carbon-based biosensors

In spite of the predominant use of CNT- and graphene-based nanocomposites since the 1990s (Iijima, 1991), the interest in conductive carbon nanostructured materials has been increasing in different areas; for example, medicine, food and environment, owing to its outstanding properties (Arduini *et al.*, 2010b; Lo *et al.*, 2012; Cinti *et al.*, 2015; Vicentini *et al.*, 2015). Nanostructured carbonaceous materials present characteristics of high surface area and electrocatalysis. Several research groups have demonstrated the high potential of carbon black (CB) to develop smart electrochemical biosensors for pesticide detection. CB is a conductive low-cost nanomaterial (Silva *et al.*, 2017), characterized by carbon nanoparticle aggregates with diameters between 17.95 and 32.5 nm (Arduini *et al.*, 2010b). Some research groups have also used CB in the fabrication of electrodes based on thin films and composites. CB is comparable to other well-explored carbon nanomaterials, including graphene, CNTs (SWCNTs and MWCNTs), fullerene (C60) and single-walled carbon nanohorns (SWCNHs). When considering the electrochemical sensing of the neurotransmitter dopamine by using different carbon nanomaterials, CB has been highlighted as an excellent modifier (Silva *et al.*, 2017).

Arduini *et al.* (2015a) developed a BChE inhibition biosensor based on screen-printed electrochemical electrodes (SPEs) for paraoxon detection. The SPEs were modified by drop casting with carbon black nano particles (CBNP) dispersion, and BChE immobilized on the surface by cross-linking. Compared with the bare SPE, the use of CB allowed an increase in the sensitivity and a decrease in the overpotential, from 0.7 to 0.3 V (vs. Ag/AgCl). Indeed, the enzymatic hydrolysis of butyrylthiocholine over time was determined by measuring the enzymatic product thiocholine at a working voltage of +300 mV, demonstrating the electrocatalytic properties of CB (Arduini *et al.*, 2016). The improved electrochemical performances of CBNPs can be attributed to the high number of defect sites as well as to the better uniform coverage of the working electrode when compared with working electrode surface modified with other nanomaterials like CNTs or graphene (Arduini *et al.*, 2010b; Cinti *et al.*, 2015). Enzyme inhibition was linearly related to the concentration of paraoxon up to 30 μg/l, and the detection limit was 5 μg/l (2×10^{-8} M). The biosensor was stable for up to 78 days of storage at room temperature under dry conditions, and was able to determine paraoxon in drinking water samples with a recovery value of 96±2%, demonstrating its accuracy in drinking water matrices. Considering the results of Arduini *et al.* (2015a), it is important to highlight that the use of only CB as an electrode modifier allows thiocholine detection at lower applied potential than the use of nanocomposite consisting of several nanomaterials, such as the nanocomposite consisting of Prussian blue nanocube/reduced graphene oxide nanocomposite reported by Zhang *et al.* (2012), or the nanocomposite assembled with iron oxide and chitosan described by Jeyapragasam and Saraswathi (2014).

Evtugyn's group (Evtugyn *et al.*, 2014) also used CB as an electrode modifier. The AChE was immobilized on a CB layer deposited on a glassy carbon electrode, along with Ag nanoparticles decorated with a macrocyclic ligand known as thiacalix[4]arene. The working potential of the biosensor was reduced to 50 mV, and could detect malaoxon, paraoxon, carbofuran and aldicarb, respectively, down to 0.1, 0.05, 0.1 and 10×10^{-9} M (detection limits) with 10 min incubation. The AChE sensor was tested for the detection of residual amounts of pesticides in spiked samples of groundnut and grape juice. The limit of quantification in peanut extract was found to be 10×10^{-9} M for carbofuran and 20×10^{-9} M for aldicarb. This was much lower than the tolerance level established for groundnut by the Code of Federal Regulations (1 ppm = 4.5×10^{-6} M of carbofuran, and 0.05 ppm = 0.26×10^{-6} M of aldicarb). The pesticide recovery in groundnut showed values equal to 98–105% for 10×10^{-9} M – 1.0×10^{-6} M carbofuran and 110–130% for 20×10^{-9} M – 1.0×10^{-6} M aldicarb. The recovery values in grape juice for two concentrations of malaoxon (10 and 100×10^{-9} M) were equal to 95–98%. In further studies by Shamagsumova *et al.* (2015), a similar AChE biosensor was developed based on unsubstituted pillar[5]arene (P5A) as an electron mediator binding on CB placed on a glassy carbon electrode. Using P5A electrodes, the biosensor could detect thiocholine enzymatic product at 200 mV in chronoamperometric mode, with detection limits 4×10^{-12} M, 5×10^{-9} M, 0.02×10^{-9} M and 0.6×10^{-9} M, respectively, for malaoxon, methyl-paraoxon, carbofuran and aldicarb. These limits were lower than what was obtained in previous work (Evtugyn *et al.*, 2014). The AChE biosensor was tested in the analysis of pesticide residuals in spiked samples of groundnut and beetroot. As highlighted by the authors, the advantage of P5A lies in its high solubility in polar organic solvents (unlike cobalt phtalocynine, for example), giving a homogeneous mixture with CB and hence allowing high reproducibility of the results.

Wei and Wang (2015) developed an AChE biosensor based on honeycomb-like porous carbon, gold nanoparticles and ionic liquid on a boron-doped diamond electrode, for dichlorvos detection. The presence of the nanocomposite provided a better electric linkage and an acceleration of the electron transfer. By utilizing the porous carbon/gold nanoparticles/ionic liquid–boron-doped diamond, sensitivity towards thiocholine was higher than 1.5- and 4.5-fold, if compared with porous carbon/gold nanoparticles/boron-doped diamond and bare boron-doped diamond. The limit of detection of dichlorvos was 2.99×10^{-13} M (calculated for 10% inhibition). The projected biosensor was employed to detect the content of dichlorvos in a lettuce leaf sample, giving recoveries between 80.8 and 93.1%. The same research group (Liu and Wei, 2014) prepared platinum–carbon aerogels (Pt-CAs) composite using the sol-gel polymerization and the ethylene glycol liquid-phase reduction method to modify a boron-doped diamond electrode for the development of a AChE biosensor for the detection of methamidophos. The presence of Pt–CAs composite on the electrode surface could improve the reactive site, reduce the interfacial resistance and make the electron transfer easier. The biosensor showed satisfactory recovery values (between 91 and 106%) in spiked apple juice samples, and a limit of detection equal to 3.1×10^{-13} M.

4.4.11 Non-carbon-based biosensors

Many platforms have also been developed avoiding the use of carbon-based nanomaterials. Wang (2012) employed surface modification strategies, using noble metal nanoparticles (NMNPs) to develop novel electrochemical sensing platforms and improving their performances with extraordinary conductivity, large surface-to-volume ratio and biocompatibility. These surface modifications improve biorecognition processes as in the immunoreactions, to be conjugated with biological elements, i.e. antibodies, nucleic acids, enzymes, to facilitate the electron transfer between the redox centre of biomolecules and the electrode surface. Moreover, as previously reported (Vertegel *et al.*, 2004; Roach *et al.*, 2006), the size of the nanoparticles represents a contributing factor in the orientation and activity of bioelements. For instance, Vertegel *et al.* (2004) observed adsorption patterns and found that lysozyme's secondary structure and function are strongly dependent on the size of the nanoparticles. When the protein was adsorbed onto <20-nm-diameter silica nanoparticles, lysozyme's secondary structure was not significantly perturbed. Moreover, a decrease in the solution pH results in a decrease in lysozyme adsorption. A change of protein structure upon adsorption is observed, as characterized by a loss in α-helix content, and this is strongly dependent on the size of the nanoparticle and the solution pH. Many biosensors involving the use of different kind of nanoparticles have been adopted in pesticide monitoring research. The high mobility of atrazine in soil leads to trace contamination of groundwater. Pichetsurnthorn *et al.* (2012) designed an electrochemical impedance spectroscopy (EIS)-based biosensor, comprising a nanoporous alumina membrane integrated with printed circuit electrode, to detect trace amounts of atrazine. The nanoporous alumina generates a high density array of nanoscale-confined spaces. Binding of the pesticide to the anti-atrazine antibody has been measured by changes of the impedance associated with the perturbation to the capacitance of the electrical double layer. The limit of detection was observed in the femtogram/ml (fg/ml) (parts-per-trillion) regime with a dynamic range spanning from 10 fg/ml (5×10^{-14} M) to 1 ng/ml (5×10 M) (0.01 ppt to 1 ppm).

Wei *et al.* (2012) fabricated an amperometric immunosensor by modifying a screen-printed carbon electrode with a nanocomposite made by skilful doping of bovine serum albumin conjugated chlorpyrifos-methyl antigen (BSA-Ag) and platinum colloid into silica sol-gel. Chlorpyrifos-methyl antigen and horseradish peroxidase as an antibody label were used for determination of chlorpyrifos-methyl. The use of Pt colloid led to an increase of the electrochemical response of five times, due to the electron transfer properties of Pt NPs. The dual signal amplification provided a limit of detection equal to 22.6 ng/l (6.45×10 M). Nesakumar *et al.* (2015) developed an AChE electrochemical biosensor by immobilizing it on a ZnO nanorod for detection of captan in apple samples. The Pt/ZnO/AChE bioelectrode showed a high sensitivity of 0.538×10^{-6} A/cm^2/μM in the linear range from 0.05 to 25.0×10^{-6} M with a limit of detection of 107×10^{-9} M, demonstrating that metallic nanoparticles improved the sensitivity. The recovery results were observed between 98.4 and 102.4% from the apple sample. The same research group developed a pesticide biosensor by exploiting the favourable properties of ZnO in a brand new configuration. Nesakumar *et al.* (2016) developed AChE cyclic voltammetric biosensor by modifying a platinum electrode with ZnO nanocuboids. A Pt/ZnO/AChE/Chitosan bioelectrode was employed for the electrochemical detection of carbosulfan in the rice sample. ZnO cuboids are characterized by a mean length of 398.24 ± 57.0 nm and a mean height of 163.28 ± 28.86 nm. The ZnO-based matrix was observed to assist the electronic communication between AChE and the electrode, and to provide an effective microenvironment for boosting the enzyme–substrate interaction. The limit of detection for carbosulfan in the rice sample was found to be 0.24×10^{-9} M with satisfactory recoveries (99.06–100.96%).

Arduini *et al.* (2015b) reported development of an electrochemical biosensor based on BChE enzyme immobilized on SPEs modified with Prussian blue nanoparticles (PBNP), for continuous flow measurements of organophosphorus pesticides. This readily detects the electro-active by-product (thiocholine) of the enzymatic reaction between butyrylthiocholine and BChE. The PB is commonly used to catalyse both the reduction of hydrogen peroxide and the oxidation of thiol-containing molecules. The effectiveness of PB was also exploited by Wu *et al.* (2011a) in the development of an amperometric biosensor to detect monocrotophos based on AChE inhibition. The AChE was immobilized on AuNPs/poly (dimethyldiallylammonium chloride) protected Prussian blue (Au–PDDA–PB) nanocomposite-modified electrode surface for the detection of organophosphate pesticide. The Au–PDDA–PB membrane possessed a relatively large surface area and served as an excellent matrix for immobilization of the enzyme, by enhancing electron transfer. The authors controlled the surface hydrophilicity by modulating the ratio of gold nanoparticles to PDDA–PB; indeed the static water contact angle of the modified surface was tailored from 25.6° to 78.1°. AChE was finely oriented onto this surface with significantly improved stability and activity, leading to significantly improved sensitivity to monocrotophos, with an ultralow detection limit of 0.8×10^{-9} g/ml.

On similar lines, AChE was immobilized on a MWNT–chitosan (MC) composite to develop a sensitive, fast and stable amperometric sensor for quantitative determination of triazophos (Du *et al.*, 2007b). The MWNTs, as stated above, promoted electron transfer reactions at a lower potential and catalysed the electro-oxidation of thiocholine, thus increasing detection sensitivity, with a detection limit of 0.8 pg/mll (4×10^{-12} M) and this biosensor was challenged in garlic samples with recovery values from 92.8–106 % (Du *et al.*, 2007a). Turan *et al.* (2016) utilized a conducting polymer and silver nanowires for fabricating a rapid, highly sensitive and stable amperometric biosensor for the quantitative determination of paraoxon in milk and tap water. A graphite electrode was modified with poly(5,6-bis (octyloxy)-4,7-di (thieno[3] [3,2-b] thiophen-2-yl) benzo[c] [1,2,5] oxoadiazole) (PTTBO), which provided a hydrophobic microenvironment to enhance BChE activity. Since biomolecules contain both hydrophobic and hydrophilic parts in their structure, alkyl chains interact with the proteins that provide an enhanced stability. Biosensor performance was improved through the deposition of silver nanowires on the polymer-coated surfaces, which enhances the charge transfer rate. After 5 min of incubation, paraoxon was able to inhibit the enzyme activity allowing it to be detected down to 0.212×10^{-6} M, when butyrylthiocholine iodide was used as the substrate. Kumar *et al.* (2013) observed, during their preliminary lab studies for developing biosensor, that bare ZnO NPs are insensitive to pesticides while GSH-coated ZnO NPs exhibit monotonous increase in sensitivity as a function of organophosphate concentration over a wide range (10 ppb to 1000 ppb) with a sensitivity of 2.1 μA/ppm/cm^2, detection limit of 3 ppb and regression coefficient of 0.92. Electrochemical measurement indicated that sensors can be used for a few times only due to the known inhibitory nature of the organophosphate and AChE reaction.

In order to develop a sensitive biosensor for the determination of methomyl in carrot and tomato, Zapp *et al.* (2011) immobilized laccase (from GM *Aspergillus oryzae*) on a supported ionic liquid phase based on platinum nanoparticles and 1-butyl-3-methylimidazolium tetrafluoroborate ionic liquid supported in montmorillonite (clay mineral). This composite was subsequently used to fabricate a carbon paste electrode. The ionic liquid into the electrode led to a decrease in the resistance and increase in the rate of charge transfer due to the high conductivity of this material. Furthermore, the platinum nanoparticles facilitated the electron transfer. The association of supported catalysts with a biocatalyst has proved a promising field of research for quantification of methomyl, with a detection limit of 2.35×10^{-7} M and a quantification limit of 7.8×10^{-7} M. For quantification of the pesticide formetanate hydrochloride (FMT) in mango and grapes, Ribeiro *et al.* (2014) used a gold electrode modified with AuNPs to design a laccase-based biosensor. Using impedimetric experiments, the authors evaluated the improvement of the conductivity of the platform upon AuNP integration, by effectively lowering the resistance to charge transfer to 350 Ω (starting from 960 Ω when AuNPs were absent). Recoveries for the five tested spiking levels ranged from 95.5±2.9% (grapes) to 108.6±2.5% (mango). A sensitive biosensor was fabricated using mesocellular silica foam (MSF, a mesoporous material), pores of which entrapped AChE, preventing the enzyme from unfolding and enhancing its activity, for detection of monocrotophos (Wu *et al.*, 2011b). MSF with a large surface area could adsorb monocrotophos, via the hydrogen bond or physical adsorption interaction. MSF containing AChE was attached to a glassy carbon electrode by using poly (vinyl alcohol) (PVA) as a polymeric membrane. The use of MSF-AChE-PVA/ glassy carbon electrode allowed the increase of the monocrotophos adsorption up to three times with respect to AChE-PVA/ glassy carbon electrode. The combination of the SPE and the improved enzyme immobilization ability in MSF significantly promoted the sensitivity of the biosensor, and the limit of detection was lowered to 0.05 ng/ml (2×10^{-11} M), expanding the linearity up to 10 ppb (4×10^{-8} M).

A gold substrate modified with AgNPs was used to immobilize guanine-rich single strand DNA-labelled monoclonal antibody (mAb) specific for coumaphos and the coating antigen, in order to develop an immunosensor for detection of ultrasensitive coumaphos, in milk samples with good recoveries (Dai *et al.*, 2012). Cesarino *et al.* (2012) developed an AChE biosensor using polyaniline (PANI) and MWCNT core-shell modified glassy carbon electrode (GC), to detect carbaryl and methomyl in fruit and vegetables (apple, broccoli and cabbage) using chronoamperometry (CA). The detection limits for carbaryl and methomyl (1.4 and 0.95 μmol/l, respectively) were below the Brazilian permissible standards. The results were in full agreement with those from the HPLC procedure. In the same context, Liu *et al.* (2014) developed a simple, selective, highly sensitive and label-free electrochemical immunosensor for atrazine detection in maize samples, with recovery of 95.5–119.8%. The gold nanoparticles were immobilized on the gold electrode surface by means of a self-assembled monolayer strategy. The high sensitivity of detection was achieved by increasing the surface area of the electrode to capture more anti-atrazine monoclonal antibodies. The signal was triggered based on the interaction between anti-atrazine monoclonal antibody and atrazine, which interrupted the electron flow from the redox centre of anti-atrazine and the electrode surface. Following the addition of atrazine, the differential pulse voltammetric peak rapidly decreased with increasing concentration of the pesticide. The hindering of the electron transfer allowed the immunosensor to detect atrazine with a limit of detection of 0.016 ng/ml (7.4×10^{-11} M). Jia *et al.* (2015) exploited the sensing properties of AuNPs in developing an impedance immunosensor, by embedding a gold interdigitated array microelectrode (IDAM) in the microchannel of a microfluidic chip to detect chlorpyrifos residue in vegetable samples, i.e. cucumber, lettuce. The microelectrode surface was modified with PDDA and a layer of AuNPs, followed by immobilization of the anti-chlorpyrifos monoclonal antibody onto the IDAM surface through the use of protein A. The impedance measurements were conducted in the presence of ferri/ferrocyanide mixture, allowing the detection of the pesticide with a linear relationship between the relative impedance change and logarithm of chlorpyrifos concentration in the range of 0.5–500 ng/ml, reaching a detection limit of 0.5 ng/ml (1×10^{-9} M).

4.5 Conclusion

Pesticides are used to reduce crop losses both before and after harvest to ensure the production of adequate food supplies for the increasing world population; however, their injudicious use has led to environmental contamination, including severe risk of adverse consequences for human health. Above the prescribed MRL, the residues may cause serious health problems, which makes it necessary to detect even the minutest levels of contamination. Various diagnostic tools are available to detect and estimate the presence of pesticides in food and environmental samples to know the level of pesticides. Among these, the analytical method is very common and is categorized under monitoring studies, involving three major steps: extraction, clean-up and estimation. Various extraction techniques involve SFE, MAE and SFC. The QuEChERS method has gained significant popularity, as it extends the range of analytes recovered compared with older, tedious extraction methods. An organic solvent is used for the extraction of the pre-homogenized sample, containing water, followed by partitioning and clean-up of the extract using SPE. The final extract is analysed for pesticides by chromatographic methods, using MS techniques. Among other diagnostic tools, immunoassays, using the ELISA technique based on antibody binding to an antigen, are quite popular. Available commercial kits include sandwich and dot ELISA. A very important application of pesticide immunoassay is biomonitoring, involving measurement of a parent chemical and/or metabolites with cellular components in selected tissues and body fluids such as blood, milk, urine or sweat, or the expired breath of an exposed individual. Sample preparation is generally not required for most of the analytes. The biosensor detection of pesticides is considered as a promising alternative to the conventional methods. These may be classified into electrochemical, optical, piezoelectric and mechanical biosensors. Enzyme biosensors, another option for pesticide detection, are based on measurements of inhibition of enzyme, especially ChE, by the pesticides involved in the enzymatic reaction. ChE-based biosensors have been shown to be powerful tools when a rapid toxicity screening is required. Enzymes used in biosensors are highly selective to the substrates and sensitive to pH, temperature, inhibitors, denaturing and chelating agents. Organophosphate and carbamate insecticides are the main ChE inhibitors. Immunosensors are the most sophisticated immunoassay format to date. The immune complex is formed on the transducer surface, and is directly determined by measuring the physical changes (electrical or optical) induced by the binding reaction. In another interesting tool, microbials (algae, bacteria, yeast and fungi) can be used as the biocatalytic elements for microbial biosensors. The DNA-based biosensors utilize the interaction of DNA with various molecules either by monitoring changes in the DNA redox reactions (i.e. oxidation of guanine) or with an electroactive analyte intercalated on a DNA layer. Nowadays, emerging nanomaterials have been employed to develop biosensing tools with custom-made properties and controlled nanoscale including nanomaterials. NPs have been used for developing optical biosensors.

References

Abad, A. and Montoya, A. (1995) Application of a monoclonal antibody-based ELISA to the determination of carbaryl in apple and grape juices. *Analytica Chimica Acta* 311(3), 365–370.

Abad, A., Moreno, M.J. and Montoya, A. (1997) A monoclonal immunoassay for carbofuran and its application to the analysis of fruit juices. *Analytica Chimica Acta* 347(1–2), 103–110.

Aga, D.S., Thurman, E.M. and Pomes, M.L. (1994) Determination of alachlor and its sulfonic acid metabolite in water by solid-phase extraction and enzyme-linked immunosorbent assay. *Analytical Chemistry* 66(9), 1495–1499.

Aguilera, A., Brotons, M., Rodríguez, M. and Valverde, A. (2003) Supercritical fluid extraction of pesticides from a table-ready food composite of plant origin (gazpacho). *Journal of Agricultural Food and Chemistry* 51(19), 5616–5621.

Aguilera, A., Rodríguez, M., Brotons, M., Boulaid, M. and Valverde, A. (2005) Evaluation of supercritical fluid extraction/aminopropyl solid-phase 'In-Line' cleanup for analysis of pesticide residues in rice. *Journal of Agricultural Food and Chemistry* 53(24), 9374–9382

Albareda-Sirvent, M., Merkoci, A. and Alegret, S. (2001) Pesticide determination in tap water and juice samples using disposable amperometric biosensors made using thick-film technology. *Analytica Chimica Acta* 31, 35–44.

de Albuquerque, Y.D. and Ferreira, L.F. (2007) Amperometric biosensing of carbamate and organophosphate pesticides utilizing screen-printed tyrosinase-modified electrodes. *Analytica Chimica Acta* 596(2), 210–221.

Amine, A., Arduini, F., Moscone, D. and Palleschi, G. (2016) Recent advances in biosensors based on enzyme inhibition. *Biosensors and Bioelectronics* 76, 180–194.

Anastassiades, M., Lehotay, S.J., Stajnbaher, D. and Schenck, F.J. (2003) Fast and easy multiresidue method employing acetonitrile extraction/partitioning and 'dispersive solid-phase extraction' for the determination of pesticide residues in produce. *Journal of AOAC International* 86, 412–431.

De Andréa, M.M., Papini, S. and Nakagawa, L.E. (2001) Optimizing microwave-assisted solvent extraction (MASE) of pesticides from soil. *Journal of Environmental Science and Health, Part B Pesticides, Food Contaminants, and Agricultural Wastes* 36(1), 87–93.

Andreescu, S. and Marty, J.L. (2006) Twenty years research in cholinesterase biosensors: from basic research to practical applications. *Biomolecular Engineering* 23(1), 1–15.

Andreescu, S. and Sadik, O.A. (2004) Trends and challenges in biochemical sensors for clinical and environmental monitoring. *Pure and Applied Chemistry* 76, 861–878.

Andreescu, S., Noguer, T., Magearu, V. and Marty, J.L. (2002) Screen-printed electrode based on acetylcholinesterase for the detection of pesticides in presence of organic solvents. *Talanta* 22, 169–176.

Andreou, V.G. and Clonis, Y.D. (2002) A portable fiber-optic pesticide biosensor based on immobilized cholinesterase and sol-gel entrapped bromocresol purple for in-field use. *Biosensors and Bioelectronics* 17(1–2), 61–69.

Andres, R.T. and Narayanaswamy, R. (1997) Fiber-optic pesticide biosensor based on covalently immobilized acetylcholinesterase and thymol blue. *Talanta* 44, 1335–1352.

Antunes, P., Gil, O. and Gil, M.G. (2003) Supercritical fluid extraction of organochlorines from fish muscle with different sample preparation. *Journal of Supercritical Fluids* 25, 135–142.

Arduini, F., Amine, A., Moscone, D. and Palleschi, G. (2010a) Biosensors based on cholinesterase inhibition for insecticides, nerve agents and aflatoxin B1 detection (review). *Microchimica Acta* 170, 193–214.

Arduini, F., Amine, A., Majorani, C., Di Giorgio, F., De Felicis, D., Cataldo, F., Moscone, D. and Palleschi, G. (2010b) High performance electrochemical sensor based on modified screen-printed electrodes with cost-effective dispersion of nanostructured carbon black. *Electrochemistry Communications* 31, 346–350.

Arduini, F., Forchielli, M., Amine, A., Neagu, D., Cacciotti, I., Nanni, F., Moscone, D. and Palleschi, G. (2015a) Screen-printed biosensor modified with carbon black nanoparticles for the determination of paraoxon based on the inhibition of butyrylcholinesterase. *Microchimica Acta* 182, 643–651.

Arduini, F., Neagu, D., Scognamiglio, V., Patarino, S., Moscone, D. and Palleschi, G. (2015b) Automatable flow system for paraoxon detection with an embedded screen-printed electrode tailored with butyrylcholinesterase and Prussian blue nanoparticles. *Chemosensors* 3, 129–145.

Arduini, F., Cinti, S., Scognamiglio, V. and Moscone, D. (2016) Nanomaterials in electrochemical biosensors for pesticide detection: advances and challenges in food analysis. *Microchimica Acta* 183, 2016–2083.

Ayyagari, M.S., Kamtekar, S., Pande, R., Marx, K.A., Kumar, J., Tripathy, S.K. and Kaplan, D.L. (1995) Biosensors for pesticide detection based on alkaline phosphatase-catalyzed chemiluminescence. *Materials Science and Engineering C* 2(4), 191–196.

Baez, M.E., Aponte, A. and Sanchez-Rasero, F. (2003) Microwave-assisted solvent extraction of the herbicide methabenzthiazuron from solid and some soil natural organic and inorganic constituents. Influence of environmental factors on its extractability. *Analyst* 128, 1478–1484.

Barnabus, I.J., Dean, J.R., Fowlis, I.A. and Owen, S.P. (1995) Extraction of polycyclic aromatic hydrocarbons from highly contaminated soils using microwave energy. *Analyst* 120, 1897–1903.

Barriada-Pereira, M., Concha-Grana, E., Gonzalez-Castro, M.J., Muniategui-Lorenzo, S., Lopez-Mahia, P., Prada-Rodriguez, D. and Fernandez-Fernandez, E. (2003) Microwave assisted extraction versus Soxhlet extraction in the analysis of 21 organochlorine pesticides in plants. *Journal of Chromatography A* 1008, 115–122.

Benilova, I., Arkhypova Dzydevych, S.V., Jaffrezic-Renault, N., Martelet, C. and Soldatkin, A.P. (2006) Kinetics of human and horse sera cholinesterases inhibition with solanaceous glycoalkaloids: study by potentiometric biosensor. *Pesticide Biochemistry and Physiology* 86, 203–210.

Bhadekar, R., Pote, S., Tale, V. and Nirichan, B. (2011) Developments in analytical methods for detection of pesticides in environmental samples. *American Journal of Analytical Chemistry* 2, 1–15.

Bilitewski, U. and Turner, A.P.F. (2000) *Biosensors for Environmental Monitoring*. Harwood Academic Publishers, London.

Bouaid, A., Martín-Esteban, A., Fernández, P. and Cámara, C. (2000) Microwave-assisted extraction method for the determination of atrazine and four organophosphorus pesticides in oranges by gas chromatography (GC). *Fresenius' Journal of Analytical Chemistry* 367(3), 291–294.

Cajka, T., Hajslova, J., Lacina, O., Mastovska, K. and Lehotay, S.J. (2008) Rapid analysis of multiple pesticide residues in fruit-based baby food using programmed temperature vaporiser injection-low-pressure gas chromatography-high-resolution time-of-flight mass spectrometry. *Journal of Chromatography A* 1186, 281–294.

California Department of Food and Agriculture (CDFA) (1988) Pesticide analytical methods. In *Pesticide Residues in Food: Technologies for Detection*. OTA workshop held on 14–16 March 1988. The California Department of Food and Agriculture. Office of Technology Assessment, p. 60.

Campanella, L., Dragone, R., Lelo, D., Martini, E. and Tomassetti, M. (2006) Tyrosinase inhibition organic phase biosensor for triazinic and benzotriazinic pesticide analysis (part two). *Analytical and Bioanalytical Chemistry* 384, 915–921.

Campanella, L., Lelo, D., Martini, E. and Tomassetti, M. (2007) Organophosphorus and carbamate pesticide analysis using an inhibition tyrosinase organic phase enzyme sensor; comparison by butyrylcholinesterase + choline oxidase opee and application to natural waters. *Analytica Chimica Acta* 587(1), 22–32.

Cesarino, I., Moraes, F.C., Lanza, M.R. and Machado, S.A. (2012) Electrochemical detection of carbamate pesticides in fruit and vegetables with a biosensor based on acetylcholinesterase immobilised on a composite of polyaniline–carbon nanotubes. *Food Chemistry* 135, 873–879.

Chapalamadugu, S. and Chaudhry, G.R. (1992) Microbiological and biotechnological aspects of metabolism of carbamates and organophosphates. *Critical Reviews in Biotechnology* 12 (5–6), 357–389.

Chau, L.K., Lin, Y.F., Cheng, S.F. and Lin, T.J. (2006) Fiber-optic chemical and biochemical probes based on localized surface plasmon resonance. *Sensors and Actuators B* 113(1), 100–105.

Chen, Y.I., Su, Y.S. and Jen, J.F. (2002) Determination of dichlorvos by on-line microwave assisted extraction coupled to headspace solid phase micro-extraction and gas chromatography-electron capture detection. *Journal of Chromatography A* 976, 349–355.

Cho, Y.A., Seok, J.A., Lee, H.S., Kim, Y.J., Park, Y.C. and Lee, Y.T. (2004) Synthesis of haptens of organophosphorus pesticides and development of immunoassays for fenitrothion. *Analytica Chimica Acta* 522(2), 215–222.

Chough, S.H., Mulchandani, A., Mulchandani, P., Chen, W., Wang, J. and Rogers, K.R. (2002) Organophosphorus hydrolase-based amperometric sensor: modulation of sensitivity and substrate selectivity. *Electroanalysis* 14, 273–276.

Chouteau, C., Dzyadevych, S., Durrieu, C. and Chovelon, J.M. (2005) A bienzymatic whole cell conductometric biosensor for heavy metal ions and pesticide detection in water samples. *Biosensors and Bioelectronics* 21, 273–281.

Chowdhury, M.A.Z., Jahan, I., Karim, N., Alam, M.K., Rahman, M.A., Moniruzzaman, M., Gan, S.H. and Fakhruddin, A.N.M. (2014) Determination of carbamate and organophosphorus pesticides in vegetable samples and the efficiency of gamma-radiation in their removal. *BioMed Research International* Article ID 145159, p. 9.

Chuang, J.C., Van Emon, J.M., Durnford, J. and Thomas, K. (2005) Develoment and evaluation of an enzyme-linked immunosorbent assay (ELISA) method for the measurement of 2,4-dichlorophenoxyacetic acid in human urine. *Talanta* 67(3), 658–666.

Chuna Bastos, V.L.F., Chuna Bastos, J., Lima, J.S. and Castro Faria, M.V. (1991) Brain acetylcholinesterase as an in vitro detector of organophosphorus and carbamate insecticides in water. *Water Research* 25, 835–840.

Cinti, S., Arduini, F., Carbone M., Sansone, L., Cacciotti, I., Moscone, D. and Palleschi, G. (2015) Screen-printed electrodes modified with carbon nanomaterials: a comparison among carbon black, carbon nanotubes and graphene. *Electroanalysis* 27, 2230–2238.

Cserhati, T., Forgacs, E., Deyl, Z., Miksik, I. and Eckhardt, A. (2004) Chromatographic determination of herbicide residues in various matrices. *Biomedical Chromatography* 18(6), 350–359.

Cunha, S.C., Lehotay, S.J., Mastovska, K., Fernandes, J.O. and Oliveira, M.B.P.P. (2010) Sample preparation approaches for the analysis of pesticide residues in olives and olive oils. In: Preedy, V.R. and Watson, R.R. (eds) *Olives and Olive Oil in Health and Disease Prevention.* Academic Press, Abingdon, UK.

Dai, Z., Liu, H., Shen, Y., Su, X., Xu, Z., Sun, Y. and Zou, X. (2012) Attomolar determination of coumaphos by electrochemical displacement immunoassay coupled with oligonucleotide sensing. *Analytical Chemistry* 84, 8157–8163.

Dan, D., Xi, H., Jie, C. and Aidong, Z. (2007) Amperometric detection of triazophos pesticide using acetylcholinesterase biosensor based on multiwall carbon nanotube-chitosan matrix. *Sensors and Actuators B* 127, 531–535.

Dan, D., Jiawang, D., Yuan, T. and Xi, C. (2008) Application of chemisorption/desorption process of thiocholine for pesticide detection based on acetylcholinesterase biosensor. *Sensors and Actuators B* 134, 908–912.

Daniel, R.T., Klara, T., Richard, A.D. and George, S.W. (1999) Electrochemical biosensors: recommendation definitions and classification, IUPAC. *Pure and Applied Chemistry* 71, 2333–2348.

Dankwardt, A. (2001) Immunochemical assays in pesticide analysis, In: Meyers, R.A. (ed.) *Encyclopedia of Analytical Chemistry.* Wiley, Chichester, UK, pp. 1–26.

Danzer, T. and Schwedt, G. (1996) Chemometric methods for the development of a biosensor system and the evaluation of inhibition studies with solutions and mixtures of pesticides and heavy metals: Part 1. Development of an enzyme electrodes system for pesticide and heavy metal. *Analytica Chimica Acta* 318, 275–286.

Deo, R.P., Wang, J., Block, I., Mulchandani, A., Josh, K.A., Trojanowicz, M., Scholz, F., Chen, W. and Lin, Y. (2005) Determination of organophosphate pesticides at a carbon nanotube/organophosphorus hydrolase electrochemical biosensor. *Analytica Chimica Acta* 530(2), 185–189.

Dhull, V., Gahlaut, A., Dilbaghi, N., and Hooda, V. (2013) Acetylcholinesterase biosensors for electrochemical detection of organophosphorus

compounds: a review. *Biochemistry Research International* Article ID 731501, 18.
Diagne, R.G., Foster, G.D. and Khan, S.U. (2002) Comparison of Soxhlet and microwave assisted extractions for the determination of fenitrothion residues in beans. *Journal of Agricultural and Food Chemistry* 50, 3204–3207.
Dogheim, S.M., Gad Alla, S.A., El-Syes, S.M.A., Almaz, M.M. and Salama, E.Y. (1996) Organochlorine and organophosphorus pesticide residues in food from Egyptian local markets. *Journal of AOAC International* 79(4), 949–952.
Donarski, W.J., Dumas, D.P., Heitmeyer, D.P., Lewis, V.E and Raushel, F.M. (1989) Structure-activity relationships in the hydrolysis of substrates by the phosphotriesterase from *Pseudomonas diminuta*. *Biochemistry* 23(11), 4650–4655.
Dost, K., Jones, D.C., Auerbach, R. and Davidson, G. (2000) Determination of pesticides in soil samples by supercritical fluid chromatography atmospheric pressure chemical ionization mass spectrometric detection. *Analyst* 125, 1751–1755.
Du, D., Chen, S., Cai, J. and Zhang, A. (2007a) Immobilization of acetylcholinesterase on gold nanoparticles embedded in sol-gel film for amperometric detection of organophosphorous insecticide. *Biosensors and Bioelectronics* 23, 130–134.
Du, D., Huang, X., Cai, J. and Zhang, A.D. (2007b) Amperometric detection of triazophos pesticide using acetylcholinesterase biosensor based on multiwall carbon nanotube-chitosan matrix. *Sensors and Actuators B: Chemical* 127(2), 531–535.
Du, D., Chen, W., Zhang, W., Liu, D., Li, H. and Lin, Y. (2010) Covalent coupling of organophosphorus hydrolase loaded quantum dots to carbon nanotube. Au nano-composite for enhanced detection of methyl parathion. *Biosensors and Bioelectronics* 25(6), 1370–1375.
Dzantiev, B.B., Yazynina, E.V., Zherdev, A.V., Plekhanova, Y.V., Reshetilov, A.N., Chang, S.C. and McNeil, C.J. (2004) Determination of the herbicide chlorsulfuron by amperometric sensor based on separation-free bienzyme immunoassay. *Sensors and Actuators B* 98, 254–261.
Dzyadevich, S.V., Soldatkin, A.P., Shul'ga, A.A., Strikha, V.I. and El'skaya, A.V. (1994) Conductometric biosensor for determination of organophosphorus pesticides. *Journal of Analytical Chemistry* 49, 874–878.
El-Saeid, M.H. (2003) Pesticide residues in canned foods, fruits, and vegetables: the application of supercritical fluid extraction and chromatographic techniques in the analysis. *Scientific World Journal* 3, 1314–1326.
El-Saeid, M.H., Kanu, I., Anyanwu, E.C. and Saleh, M.A. (2005) Impact of extraction methods in the rapid determination of atrazine residues in foods using supercritical fluid chromatography and enzyme linked immunosorbent assay: microwave solvent vs supercritical fluid extractions. *Science World Journal* 5, 11–19.
Ercegovich, C.D. (1971) Analysis of pesticide residues: Immunological techniques, In: Biros, F.J. (ed.) *Pesticides Identification*. American Chemical Society, Washington, DC, pp. 162–178.
Erickson, D., Mandal, S., Allen, H., Yang, J. and Cordovez, B. (2008) Nanobiosensors: optofluidic, electrical and mechanical approaches to biomolecular detection at the nanoscale. *Microfluid Nanofluidics* 4(1–2), 33–52.
Everett, W.R. and Rechnitz, G.A. (1998) Mediated bioelectrocatalytic determination of organophosphorus pesticides with a tyrosinase-based oxygen biosensor. *Analytical Chemistry* 70, 807–810.
Evtugyn, G.A., Shamagsumova, R.V., Padnya, P.V., Stoikov, I.I. and Antipin, I.S. (2014) Cholinesterase sensor based on glassy carbon electrode modified with Ag nanoparticles decorated with macrocyclic ligands. *Talanta* 127, 9–17.
Fan, M. and He, J. (2011) Pesticide immunoassay. In: Stoytcheva, M. (ed.) *Pesticides: Strategies for Pesticides Analysis*. InTechOpen, London, pp. 293–314. Available at: http://cdn.intechopen.com/pdfs-wm/12958.pdf
Fan, L., Zhao, G., Shi, H., Liu, M. and Li, Z. (2013) A highly selective electrochemical impedance spectroscopy-based aptasensor for sensitive detection of acetamiprid. *Biosensors and Bioelectronics* 43, 12–18.
Fenik, J., Tankiewicz, M. and Biziuk, M. (2011) Properties and determination of pesticides in fruits and vegetables. *Trends in Analytical Chemistry* 30(6), 814–826.
Fernandez-Alba, A.R., Valverde, A., Aguera, A., Contreras, M. and Rodriguez, D. (1995) Determination of procymidone in vegetables by a commercial competitive inhibition enzyme immunoassay. *Analytica Chimica Acta* 311(3), 371–376.
Fillion, J., Sauvé, F. and Selwyn, J. (2000) Multiresidue method for the determination of residues of 251 pesticides in fruits and vegetables by gas chromatography/mass spectrometry and liquid chromatography with fluorescence detection. *Journal of AOAC International* 83(3), 698–713.
Food Safety and Inspection Service (FSIS) (1987) US Department of Agriculture, Compound Evacuation and Analytical Capability National Residue Plan, 1987. FSIS, Washington, DC.

Freire, R.S., Pessao, C.A., Mello, L.D. and Kubota, L.T. (2003) Direct electron transfer: an approach for electrochemical biosensors with higher selectivity and sensitivity. *Journal of Brazilian Chemical Society* 14(2), 230–243.

Fu, J., Park, B and Zhao, Y. (2009) Limitation of a localized surface plasmon resonance sensor for salmonella detection. *Sensors and Actuators B* 141(1), 276–283.

Gabaldón, J.A., Maquieira, A. and Puchades, R. (1999) Current trends in immunoassay-based kits for pesticide analysis. *Critical Reviews in Food Science and Nutrition* 39(6), 519–538.

Ganzler, K., Salgo, A. and Valko, K. (1986) Microwave extraction: a novel sample preparation method for chromatography. *Journal of Chromatography* 371, 299–306.

Garcia De Maria, C., Munoz, T.M. and Townhend, A. (1994) Reactivation of an immobilized enzyme reactor for the determination of acetylcholinesterase inhibitors. Flow injection determination of paraoxon. *Analytica Chimica Acta* 295, 287–296.

Garcia Sanchez, F., Navas Diaz, A., Ramos Peinado, M.C. and Belledone, C. (2003) Free and sol-gel immobilized alkaline phosphatase-based biosensor for the determination of pesticides and inorganic compounds. *Analytica Chimica Acta* 484(1), 45–51.

Gfrerer, M., Chen, S., Lankmayr, E.P., Quan, X. and Yang, F. (2004) Comparison of different extraction techniques for the determination of chlorinated pesticides in animal feed. *Analytical and Bioanalytical Chemistry* 378, 1861–1867.

Goel, P. (2013) Immunodiagnosis of pesticides: a review. *African Journal of Biotechnology* 12(52), 7158–7167.

Grieshaber, D., MacKenzie, R., Vöros, J. and Reimhult, E. (2008) Electrochemical biosensors – sensor principles and architectures. *Sensors* 8(3), 1400–1458.

Guan, H., Brewer, W.E., Garris, S.T. and Morgan, S.L. (2010) Disposable pipette extraction for the analysis of pesticides in fruit and vegetables using gas chromatography/mass spectrometry. *Journal of Chromatography A* 1217(12), 1867–1874.

Guilbault, G.G., Hock, B. and Schmid, R. (1992) A piezoelectric immunosensor for atrazine in drinking water. *Biosensors and Bioelectronics* 7(6), 411–419.

Halamek, J., Hepel, M. and Skladal, P. (2001) Investigation of highly sensitive piezoelectric immunosensors for 2,4-dichlorophenoxyacetic acid. *Biosensors and Bioelectronics* 16(4–5), 253–260.

Halamek, J., Pribyl, J., Makower, A., Skladal, P. and Scheller, F.W. (2005) Sensitive detection of organophosphates in river water by means of a piezoelectric biosensor. *Analytical and Bioanalytical Chemistry* 382(8), 1904–1911.

Hammock, B., Gee, S., Cheung, P. *et al.* (1988) Utility of immunoassay in pesticide trace analysis, In: *Pesticide Science and Biotechnology*, vol. 29(2–3), Blackwell Scientific Publications, London, pp. 105–129.

Hawthorne, S.B. (1990) Analytical scale supercritical fluid extraction. *Analytical Chemistry* 62, 633A–642A.

He, J., Liu, Y., Fan, M. and Liu, X. (2011) Isolation and identification of the DNA aptamer target to acetamiprid. *Journal of Agricultural and Food Chemistry* 59, 1582–1586.

Hennion, M.C. and Barcelo, D. (1998) Strengths and limitations of immunoassays for effective and efficient use for pesticide analysis in water samples: a review. *Analytica Chimica Acta* 362(1), 3–34.

Herdman, R.C., Parham, W.E. and Shen, S. (1988a) Pesticide analytical methods. Chapter-6. In: *Pesticide Residues in Food: Technologies for Detection.* OTA workshop held on 14–16 March 1988, Washington, DC.

Herdman, R.C., Parham, W.E. and Shen, S. (1988b) Federal Pesticide Residues in Food Monitoring Programs. Chapter-2. In: *Pesticide Residues in Food: Technologies for Detection.* OTA workshop held on 14–16 March 1988.

Herdman, R.C., Parham, W.E. and Shen, S. (1988c) Immunoassay: An Emerging Technology. Chapter 4. In: *Pesticide Residues in Food: Technologies for Detection.* OTA workshop held on 14–16 March 1988.

Herzog, D.P. (1997) Immunoassays for environmental contaminants (pesticides) in food and water, *Proceedings of Food Contaminants Workshop*, AOAC, Washington, DC, pp. 1–26

Hiemstra, M. and de Kok, A. (2007) Comprehensive multi-residue method for the target analysis of pesticides in crops using liquid chromatography-tandem mass spectrometry. *Journal of Chromatography A* 1154(1–2), 3–25.

Hleli, S., Martelet, C. and Abdelghani, A. (2006) Atrazine analysis using an impedimetric immunosensor based on mixed biotinylated self-assembled monolayer. *Sensors and Actuators B* 113, 711–717.

Hobel, W. and Polster, J. (1992) Fiber optic biosensor for pesticides based on acetylcholine esterase. *Fresenius' Journal of Analytical Chemistry* 343, 101–102.

Hu, S.Q., Xie, J.W., Xu, Q.H., Rong, K.T., Shen, G.L. and Yu, R.Q. (2003) A label-free electrochemical immunosensor based on gold nanoparticles for detection of paraoxon. *Talanta* 61, 769–777.

Iijima, S. (1991) Helical microtubules of graphitic carbon. *Nature* 354(6348), 56–58.

Ivanov, A.N., Lukachova, L.V., Evtugyn, G.A., Karyakina, E.E., Kiseleva, S.G., Budnikov, H.C., Orlov, A.V., Karpacheva, G.P. and Karyakin, A.A. (2002) Polyaniline-modified cholinesterase sensor for pesticide determination. *Bioelectrochemistry* 55, 75–78.

Ivanov, A., Evtugyn, G., Budnikov, H., Ricci, F., Moscone, D. and Palleschi, G. (2003) Cholinesterase sensors based on screen-printed electrodes for detection of organophosphorus and carbamic pesticides. *Analytical and Bioanalytical Chemistry* 377, 624–631.

Jaffrezic-Renault, N. (2001) New trends in biosensors for organophosphorus pesticides. *Sensors* 1(2), 60–74.

Jayaraman, S., Pruell, R.J. and McKinney, R. (2001) Extraction of organic contaminants from marine sediments and tissues using microwave energy. *Chemosphere* 44, 181–191.

Jeyapragasam, T. and Saraswathi, R. (2014) Electrochemical biosensing of carbofuran based on acetylcholinesterase immobilized onto iron oxide–chitosan nanocomposite. *Sensors and Actuators B* 191, 681–687.

Jia, H., Guo, Y., Sun, X. and Wang, X. (2015) An electrochemical immunosensor based on microfluidic chip for detection of chlorpyrifos. *International Journal of Electrochemical Science* 10, 8750–8758.

Johnson, J.C., Van Emon, J.M., Clarke, A.N. and Wamsley, B.N. (2001) Quantitative ELISA of polychlorinated biphenyls in an oily soil matrix using supercritical fluid extraction. *Analytica Chimica Acta* 428(2), 191–199.

Josiane, C. and Sergio, A.S.M. (2008) Determination of carbaryl in tomato 'in natura' using an amperometric biosensor based on the inhibition of acetylcholinesterase activity. *Sensors and Actuators B* 129, 40–46.

Karousos, N.G., Aouabdi, S., Way, A.S. and Reddy, S.M. (2002) Quartz crystal microbalance determination of organophosphorus and carbamate pesticides. *Analytica Chimica Acta* 469(2), 189–196.

Kim, G.Y., Kang, M.S. and Moon, S.H. (2008a) Substrate-bound tyrosinase electrode using gold nanoparticles anchored to pyrroloquinoline quinone for a pesticide biosensor. *Sensors and Actuators B* 133(1), 1–4.

Kim, G.Y., Shim, J., Kang, M.S. and Moon, S.H. (2008b) Optimized coverage of gold nanoparticles at tyrosinase electrode for measurement of a pesticide in various water samples. *Journal of Hazardous Materials* 156(1–3), 141–147.

King, J.W. and Zhang, Z. (2002) Derivatization reactions of carbamate pesticides in super critical carbon dioxide. *Analytical and Bioanalytical Chemistry* 374, 88–92.

Kok, F.N. and Hasirci, V. (2004) Determination of binary pesticide mixtures by an acetylcholinesterase-choline oxidase biosensor. *Biosensors and Bioelectronics* 19, 661–665.

Kovacs, M.F. Jr and Trichilo, C.L. (1987) Regulatory perspective of pesticide analytical enforcement methodology in the United States. *Journal of Association of Official Analytical Chemists* 70(6), 937–940.

Kumar, A., Arora, S., Mogha, N., Al-Deyab, S.S., Ansari, Z.A. and Ansari, S.G. (2013) Glutathione coated zinc oxide nanoparticles: a promising material for pesticide detection. *Energy and Environment Focus* 2, 101–107.

Kumaran, S. and Morita, M. (1995) Application of a cholinesterase biosensor to screen for organophosphorus pesticides extracted from soil. *Talanta* 42, 649–655.

Kumaran, S. and Tran-Minh, C. (1992) Determination of organophosphorus and carbamate insecticides by flow injection analysis. *Analytical Biochemistry* 200, 187–194.

La Rosa, C., Pariente, F., Hernandez, L. and Lorenzo, E. (1994) Determination of organophosphorus and carbamic pesticides with an acetylcholinesterase amperometric biosensor using 4-aminophenyl acetate as substrate. *Analytica Chimica Acta* 295, 273–282.

Lee, H.S., Kim, Y.A., Cho, Y.A. and Lee, Y.T. (2002) Oxidation of organophosphorus pesticides for the sensitive detection by a cholinesterase based biosensor. *Chemosphere* 46, 571–576.

Lee, W.Y., Lee, E.K., Kim, Y.J., Park, W.C., Chung, T. and Lee, Y.T. (2006) Monoclonal antibody-based enzyme-linked immunosorbent assays for the detection of the organophosphorus insecticide isofenphos. *Analytica Chimica Acta* 557(1–2), 169–178.

Lee, J.H., Park, J.Y., Min, K., Cha, H.J., Choi, S.S. and Yoo, Y.J. (2010) A novel organophosphorus hydrolase-based biosensor using mesoporous carbons and carbon black for the detection of organophosphate nerve agents. *Biosensors and Bioelectronics* 25, 1566–1570.

Lehotay, S.J. (2000) Multiclass, multiresidue analysis of pesticides, strategies for. In: Meyers, R.A. (ed.) *Encyclopedia of Analytical Chemistry*, John Wiley and Sons, Chichester, UK, pp. 6344–6884.

Lehotay, S.J. (2007) Determination of pesticide residues in foods by acetonitrile extraction and partitioning with magnesium sulfate: collaborative study. *Journal of AOAC International* 90, 485–520.

Lehotay, S.J., Mastovska, K. and Yun, S.J. (2005a) Evaluation of two fast and easy methods for pesticide residue analysis in fatty food matrixes. *Journal of AOAC International* 88, 630–638.

Lehotay, S.J., Mastovska, K. and Lightfield, A.R. (2005b) Use of buffering and other means to improve results of problematic pesticides in a fast and easy method for residue analysis of fruits and vegetables. *Journal of AOAC International* 88, 615–629.

Lei, Y., Mulchandani, P., Wang, J., Chen, W., Chen, W. and Mulchandani, A. (2005) Highly sensitive and selective amperometric microbial biosensor for direct determination of p-nitrophenyl-substituted organophosphate nerve agents. *Environmental Science and Technology* 39, 853–857.

Letellier, M. and Budzinski, H. (1999) Microwave assisted extraction of organic compounds. *Analysis* 27, 259–271.

Li, C., Wang, C., Guan, B., Zhang, Y. and Hu, S. (2005) Electrochemical sensor for the determination of parathion based on p-tert-butylcalix [6] arene-1,4-crown-4 sol–gel film and its characterization by electrochemical methods. *Sensors and Actuators B: Chemical* 107(1), 411–417.

Li, X., Zheng, Z., Liu, X., Zhao, S. and Liu, S. (2015) Nanostructured photoelectrochemical biosensor for highly sensitive detection of organophosphorous pesticides. *Biosensors and Bioelectronics* 64, 1–5.

Lim, Y.C., Kouzani, A.Z. and Duan, W. (2010) Aptasensors: a review. *Journal of Biomedical Nanotechnology* 6, 93–105.

De Lima, F., Lucca, B., Barbosa, A.M.J., Ferreira, V.S., Moccelini, S.K., Franzoi, A.C. and Vieira, I.C. (2010) Biosensor based on pequi polyphenol oxidase immobilized on chitosan crosslinked with cyanuric chloride for thiodicarb determination. *Enzyme and Microbial Technology* 47(4), 153–158.

Lin, T.J., Huang, K.T. and Liu, C.Y. (2006) Determination of organophosphorous pesticides by a novel biosensor based on localized surface plasmon resonance. *Biosensors and Bioelectronics* 22(4), 513–518.

Lisa, M., Chouhan, R.S., Vinayaka, A.C., Manonmani, H.K. and Thakur, M.S. (2009) Gold nanoparticles based dipstick immunoassay for the rapid detection of dichlorodiphenyltrichloroethane: an organochlorine pesticide. *Biosensors and Bioelectronics* 25(1), 224–227.

Liu, Y. and Wei, M. (2014) Development of acetylcholinesterase biosensor based on platinum–carbon aerogels composite for determination of organophosphorus pesticides. *Food Control* 36, 49–54.

Liu, B., Yang, Y.H., Wu, Z.Y., Wang, H., Shen, G.L. and Yu, R.Q. (2005) A potentiometric acetylcholinesterase bio-sensor based on plasma-polymerized film. *Sensors and Actuators B* 104 (2), 186–190.

Liu, G., Wang, S., Liu, J. and Song, D. (2012) An electrochemical immunosensor based on chemical assembly of vertically aligned carbon nanotubes on carbon substrates for direct detection of the pesticide endosulfan in environmental water. *Analytical Chemistry* 84, 3921–3928.

Liu, X., Li, W.J., Li, L., Yang, Y., Mao, L.G. and Peng, Z. (2014) A label-free electrochemical immunosensor based on gold nanoparticles for direct detection of atrazine. *Sensors and Actuators B* 191, 408–414.

Lo, T.W., Aldous, L. and Compton, R.G. (2012) The use of nano-carbon as an alternative to multi-walled carbon nanotubes in modified electrodes for adsorptive stripping voltammetry. *Sensors and Actuators B* 162, 361–368.

Lu, C.H., Zhou, W.H., Han, B., Yang, H.H., Chen, X. and Wang, X.R. (2007) Surface imprinted core-shell nanoparticles for sorbent assays. *Analytical Chemistry* 79(14), 5457–5461.

Luque de Castro, M.D. and Herrera, M.C. (2003) Enzyme inhibition-based biosensors and biosensing systems: questionable analytical devices. *Biosensors and Bioelectronics* 18 (2–3), 279–294.

Martorell, D., Céspedes, F., Alegret, S. and Martínez-Fàregas, E. (1994) Amperometric determination of pesticides using a biosensor based on polishable graphite–epoxy biocomposite. *Analytica Chimica Acta* 290, 343–348.

Martorell, D., Céspedes, F., Martínez-Fàregas, E. and Alegret, S. (1997) Determination of organophosphorus and carbamate pesticides using a biosensor based on polishable, 7,7,8,8-tetrac yanoquinodimethanemodified, graphite-epoxy biocomposite. *Analytica Chimica Acta* 337, 305–313.

Marty, J.L., Sode, K. and Karube, I. (1992) Biosensor for detection of organophosphate and carbamate insecticides. *Electroanalysis* 4, 249–252.

Mascini, M., Palchetti, I. and Marrazza, G. (2001) DNA electrochemical biosensors. *Fresenius' Journal of Analytical Chemistry* 369, 15–22.

Mastovska, K. and Lehotay, S.J. (2006) Rapid sample preparation method for LC-MSIMS or GC-MS analysis of acrylamide in various food matrices. *Journal of Agricultural and Food Chemistry* 54, 7001–7008.

Mazzei, F., Botre, F. and Botre, C. (1996) Acid phosphatase/glucose oxidase based biosensors for the determination of pesticides. *Analytica Chimica Acta* 336, 67–75.

Mazzei, F., Botrè, F., Montilla, S., Pilloton, R., Podesta, E. and Botrè, C. (2004) Alkaline phosphatase inhibition based electrochemical sensors for the detection of pesticides. *Journal of Electroanalytical Chemistry* 574(1), 95–100.

McHugh, M. and Krukonis, V. (1986) *Supercritical Fluid Extraction – Principles and Practice.* Butterworths, Cambridge, Massachusetts.

McNally, M.E. and Wheeler, J.R. (1988) Supercritical fluid extraction coupled with supercritical fluid chromatography for the separation of sulfonyl urea herbicides and their metabolites from complex matrices. *Journal of Chromatography* 435, 63–71.

Medyantseva, E.P., Vertlib, M.G., Budnikov, G.K., Babkina, S.S. and Eremin, S.A. (1995) Amperometric biochemical sensor based on immobilized cholinesterase in the immunoassay of pesticides. *Zhurnal Analiticheskoi Khimii* 50, 782–786.

Medyantseva, E.P., Vertlib, M.G., Kutyreva, M.P., Khaldeeva, E.I., Budnikov, G.K. and Eremin, S.A. (1997) The specific immunochemical detection of 2,4-dichlorophenoxyacetic acid and 2,4,5-trichloro-phenoxyacetic acid pesticides by amperometric cholinesterase biosensors. *Analytica Chimica Acta* 347, 71–78.

Mercader, J.V. and Montoya, A. (1997) A monoclonal antibody-based ELISA for the analysis of azinphos-methyl in fruit juices. *Analytica Chimica Acta* 347(1–2), 95–101.

Meulenberg, E.P., Mulder, W.H. and Stoks, P.G. (1995) Immunoassays for pesticides. *Environmental Science Technology* 29(3), 553–561.

Michelangeli, F., Robson, M.J., East, J.M. and Lee, A.G. (1990) Fluorescence and kinetic studies of the interactions of pyrethroids with the (Ca2++ Mg2+)-ATPase. *Biochimica et Biophysica Acta* 1028, 58–66.

Mills, P.A., Onley, J.J. and Gaither, R.A. (1963) Rapid method for chlorinated pesticide residues in non-fatty foods. *Journal of Association of Official Agricultural Chemists* 46, 186–191.

Mionetto, N., Marty, J.L. and Karube, I. (1994) Acetylcholinesterase in organic solvents for the detection of pesticides: biosensor application. *Biosensors and Bioelectronics* 9, 463–470.

Montesinos, T., Perez-Munguia, S., Valdez, F. and Marty, J.L. (2001) Disposable cholinesterase biosensor for the detection of pesticides in water-miscible organic solvents. *Analytica Chimica Acta* 431, 231–237.

Moris, P., Alexandre, I., Roger, M. and Remacle, J. (1995) Chemiluminescence assay of organophosphorus and carbamate pesticides. *Analytica Chimica Acta* 302, 53–59.

Mostafa, G.A.E. (2010) Electrochemical biosensors for the detection of pesticides. *The Open Electrochemistry Journal* 2, 22–42.

Mukherjee, I. and Gopal, M. (1996) Insecticide residues in baby food, animal feed, and vegetables by gas liquid chromatography. *Bulletin of Environmental Contamination and Toxicology* 56(3), 381–388.

Mulchandani, A., Mulchandani, P., Chauhan, S., Kaneva, I. and Chen, W. (1998) A potentiometric microbial biosensor for direct determination of organophosphate nerve agents. *Electroanalysis* 10, 733–737.

Mulchandani, A., Chen, W., Mulchandani, P., Wang, J. and Rogers, K.R. (2001a) Biosensors for direct determination of organophosphate pesticides. *Biosensors and Bioelectronics* 16, 225–230.

Mulchandani, P., Chen, W. and Mulchandani, A. (2001b) Flow-injection amperometric enzyme biosensor for direct determination of organophosphate nerve agents. *Environmental Science and Technology* 35, 2562–2565.

Nesakumar, N., Sethuraman, S., Krishnan, U.M. and Rayappan, J.B.B. (2015) Cyclic voltammetric acetylcholinesterase biosensor for the detection of captan in apple samples with the aid of chemometrics. *Analytical and Bioanalytical Chemistry* 407(16), 4863–4868.

Nesakumar, N., Sethuraman, S., Krishnan, U.M. and Rayappan, J.B.B. (2016) Electrochemical acetylcholinesterase biosensor based on ZnO nanocuboids modified platinum electrode for the detection of carbosulfan in rice. *Biosensors and Bioelectronics* 77, 1070–1077.

Nguyen, T.D., Lee, B.S., Lee, B.R., Lee, D.M. and Lee, G.H. (2007) A multiresidue method for the determination of 109 pesticides in rice using the quick easy cheap effective rugged and safe (QuEChERS) sample preparation method and gas chromatography/mass spectrometry with temperature control and vacuum concentration. *Rapid Communications in Mass Spectrometry* 21, 3115–3122.

Nirmal, P., Kavita, A., Surinder, P.S., Manoj, K.P., Harpal, S. and Bansi, D.M. (2007) Polypyrrole-polyvinyl sulphonate film based disposable nucleic acid biosensor. *Analytica Chimica Acta* 589, 6–13.

Nirmal, P., Sumana, G., Kavita, A., Harpal, S. and Malhotra, B.D. (2008) Improved electrochemical nucleic acid biosensor based on polyaniline-polyvinyl sulphonate. *Talanta* 74, 1337–1343.

Noguer, T., Gradinaru, A., Ciucu, A. and Marty, J.L. (1999) A new disposable biosensor for the accurate and sensitive detection of ethylenebis (dithiocarbamate) fungicides. *Analytical Letters* 32, 1723–1738.

del Nozal, M.J., Toribio, L., Bernal, J.L. and Castano, N. (2003) Separation of triadimefon and triadimenol enantiomers and diastereoisomers by supercritical fluid chromatography. *Journal of Chromatography A* 986, 135–141.

Numata, M., Yarita, T., Aoyagi, Y. and Takatsu, A. (2004) Evaluation of microwave-assisted extraction technique for the determination of polychlorinated biphenyls and organochlorine pesticides in sediments. *Annals of Science* 20, 793–798.

Nunes, G.S. and Barcelo, D. (1998) Electrochemical biosensors for pesticide determination in food samples. *Analysis* 26, M156–M159.

Olga, S. and Jon, R.K. (2007) An acetylcholinesterase enzyme electrode stabilized by an electrodeposited gold nanoparticle layer. *Electrochemistry Communications* 9, 935–940.

Oliveira, T.M., Barroso, M.F., Morais, S., Araújo, M., Freire, C., de Lima-Neto, P., Correia, A.N., Oliveira, M.B.P.P. and Delerue-Matos, C. (2014) Sensitive bi-enzymatic biosensor based on polyphenoloxidases–gold nanoparticles–chitosan hybrid film–graphene doped carbon paste electrode for carbamates detection. *Bioelectrochemistry* 98, 20–29.

Osman, K.A., Al-Humaid, A.M., Al-Rehiayani, S.M. and Al-Redhaiman, K.N. (2010) Monitoring of pesticide residues in vegetables marketed in Al-Qassim region, Saudi Arabia. *Ecotoxicology and Environmental Safety* 73(6), 1433–1439.

PRMP (2014) Pesticide residue monitoring program, fiscal year 2014, pesticide report, US Food and Drug Administration. Available at: https://www.fda.gov/downloads/Food/FoodborneIllnessContaminants/Pesticides/UCM546325.pdf (accessed 1 October 2018).

Palchetti, I., Cagnini, A., Del Carlo, M., Coppi, C., Mascini, M. and Turner, A.P.F. (1997) Determination of acetylcholinesterase pesticides in real samples using a disposable biosensor. *Analytica Chimica Acta* 337, 315–321.

Palleschi, G., Bernabei, M., Cremisini, C. and Mascini, M. (1992) Determination of organophosphorus insecticides with a choline electrochemical biosensor. *Sensors and Actuators B* 7, 513–517.

Parmar, P. and Kocher, D.K. (2012) Dot Elisa: immunological technique for pesticide residue analysis. *Research Journal of Recent Sciences* 2 (ISC-2012), 5–11.

Patsias, J., Papadakis, E.N. and Papadopoulou-Mourkidou, E. (2002) Analysis of phenoxyalkanoic acid herbicides and their phenolic conversion products in soil by microwave assisted solvent extraction and subsequent analysis of extracts by on-line solid phase extraction liquid chromatography. *Journal of Chromatography A* 959, 153–161.

Paya, P., Anastassiades, M., Mack, D., Sigalova, I., Tasdelen, B., Oliva, J. and Barba, A. (2007) Analysis of pesticide residues using the quick easy cheap effective rugged and safe (QuEChERS) pesticide multiresidue method in combination with gas and liquid chromatography and tandem mass spectrometric detection. *Analytical and Bioanalytical Chemistry* 389, 1697–1714.

Pedrosa, V.A., Paliwal, S., Balasubramanian, S., Nepal, D., Davis, V., Wild, J., Ramanculov, E. and Simonian, A. (2010) Enhanced stability of enzyme organophosphate hydrolase interfaced on the carbon nanotubes. *Colloids and Surfaces B Biointerfaces* 77(1), 69–74.

Perez-Pita, M.T., Reviejo, A.J., Manuel-de-Villena, F.J. and Pingarron, J.M. (1997) Amperometric selective biosensing of dimethyl- and diethyldithiocarbamates based on inhibition processes in a medium of reversed micelles. *Analytica Chimica Acta* 340(1–3), 89–97.

Periasamy, A.P., Umasankar, Y. and Chen, S.M. (2009) Nanomaterials acetylcholinesterase enzyme matrices for organophosphorus pesticides electrochemical sensors: a review. *Sensors* 9, 4034–4055.

Pichetsurnthorn, P., Vattipalli, K. and Prasad, S. (2012) Nanoporous impedemetric biosensor for detection of trace atrazine from water samples. *Biosensors and Bioelectronics* 32, 155–162.

Plossl, F., Giera, M. and Bracher, F. (2006) Multiresidue analytical method using dispersive solid-phase extraction and gas chromatography/ion trap mass spectrometry to determine pharmaceuticals in whole blood. *Journal of Chromatography, A* 1135, 19–26.

Posyniak, A., Zmudzki, J. and Mitrowska, K. (2005) Dispersive solid-phase extraction for the determination of sulfonamides in chicken muscle by liquid chromatography. *Journal of Chromatography A* 1087, 259–264.

Prasad, K., Prathish, K.P., Gladis, J.M., Naidu, G.R.K. and Rao, T.P. (2007) Molecularly imprinted polymer (biomimetic) based potentiometric sensor for atrazine. *Sensors and Actuators B: Chemical* 123(1), 65–70.

Pribyl, J., Hepel, M., Halámek, J. and Skladal, P. (2003) Development of piezoelectric immunosensors for competitive and direct determination of atrazine. *Sensors and Actuators B: Chemical* 91(1–3), 333–341.

Priti, M., Wilfred, C. and Ashok, M. (2006) Microbial biosensor for direct determination of nitrophenyl-substituted organophosphate nerve agents using genetically *Mcraxella* sp. *Analytica Chimica Acta* 568, 217–221.

Priyadharshini, K.A. (2012) Immunochemical techniques. Chapter-XII. Department of pharmaceutical chemistry, SRM College of Pharmacy. Available at: www.srmuniv.ac.in/sites/default/files/downloads/Immunochemical_Techniques.pdf (accessed 31 July 2013).

Rekha, K., Gouda, M.D., Thakur, M.S. and Karanth, N.G. (2000) Ascorbate oxidase based amperometric biosensor for organophosphorus pesticide monitoring. *Biosensors and Bioelectronics* 15(9–10), 499–502.

Reybier, K., Zairi, S. and Jaffrezic-Renault, N. (2002) The use of polyethylenimine for fabrication of potentiometric cholinesterase biosensors. *Talanta* 56 (6), 1015–1020.

Ribeiro, F.W.P., Barroso, M.F., Morais, S., Viswanathan, S., de Lima-Neto, P., Correia, A.N., Oliveira, A.B.P.P. and Delerue-Matos, C. (2014) Simple laccase-based biosensor for formetanate hydrochloride quantification in fruits. *Bioelectrochemistry* 95, 7–14.

Rissato, S.R., Galhiane, M.S., Knoll, F.R.N. and Apon, B.M. (2004) Supercritical fluid extraction for pesticide multiresidue analysis in honey: determination by gas chromatography with electron-capture and mass spectrometry detection. *Journal of Chromatography A* 1048(2), 153–159.

Rissato, S.R., Galhiane, M.S., Apon, B.M. and Arruda, M.S. (2005) Multiresidue analysis of pesticides in soil by supercritical fluid extraction/gas chromatography with electron-capture detection and confirmation by gas chromatography-mass spectrometry. *Journal of Agricultural Food and Chemistry* 53(1),62–69.

Roach, P., Farrar, D. and Perry, C.C. (2006) Surface tailoring for controlled protein adsorption: effect of topography at the nanometer scale and chemistry. *Journal of the American Chemical Society* 128, 3939–3945.

Rogers, K.R., Cao, C.J., Valdes, J.J., Eldefrawi, A.T. and Eldefrawi, M.E. (1991) Acetylcholinesterase fiber-optic biosensor for detection of acetylcholinesterases. *Fundamental and Applied Toxicology* 16, 810–820.

Ronkainen, N.J., Halsall, H.B. and Heineman, W.R. (2010) Electrochemical biosensors. *Chemical Society Reviews* 39(11), 1747–1763.

Rosenblum, L., Garris, S.T. and Morgan, J.N. (2002) Comparison of five extraction methods for determination of incurred and added pesticides in dietary composites. *Journal of AOAC International* 85, 1167–1176.

Sacks, V., Eshkenazi, I., Neufeld, T., Dosoretz, C. and Rishpon, J. (2000) Immobilized parathion hydrolase: an amperometric sensor for parathion. *Analytical Chemistry* 72, 2055–2058.

Sadik, O.A. and Emon Van, J.M. (1996) Application of electrochemical immunosensors to environmental monitoring. *Biosensors and Bioelectronics* 11(8), 1–11.

Sassolas, A., Prieto-Simón, B. and Marty, J.L. (2012) Biosensors for pesticide detection: new trends. *American Journal of Analytical Chemistry* 3, 210–232.

Sawyer, L. (1988) The development of analytical methods for pesticide residues. Appendix B, OTA workshop papers. In: *Pesticide Residues in Food: Technologies for Detection.* OTA workshop held on 14–16 March 1988, page 112. US Food and Drug Administration, Washington, DC.

Scognamiglio, V., Antonacci, A., Lambreva, M.D., Litescu, S.C. and Rea, G. (2015) Synthetic biology and biomimetic chemistry as converging technologies fostering a new generation of smart biosensors. *Biosensors and Bioelectronics* 74, 1076–1086.

Seki, A., Ortega, F. and Marty, J.L. (1996) Enzyme sensor for the detection of herbicides inhibiting acetolactate synthase. *Analytical Letters* 29(8), 1259–1271.

Shamagsumova, R.V., Shurpik, D.N., Padnya, P.L., Stoikov, I.I. and Evtugyn, G.A. (2015) Acetylcholinesterase biosensor for inhibitor measurements based on glassy carbon electrode modified with carbon black and pillar [5] arene. *Talanta* 144, 559–568.

Shen, G. and Lee, H.K. (2003) Determination of triazines in soil by microwave-assisted extraction followed by solid phase microextraction and gas chromatography-mass spectrometry. *Journal of Chromatography A* 985, 167–174.

Sherry, J.P. (1992) Environmental chemistry: the immunoassay option. *Critical Reviews in Analytical Chemistry* 23(4), 217–300.

Silva, T.A., Cruz Moraes, F., Janegitz, B.C. and Fatibello-Filho, O. (2017) Electrochemical biosensors based on nanostructured carbon black: a review. *Journal of Nanomaterials* Article ID 4571614, p. 14.

Silvana, A. and Jean-Louis, M. (2006) Twenty years in cholinesterase biosensors: from basic research to practical applications. *Biomolecular Engineering* 23, 1–15.

Simonian, A.L., Grimsley, J.K., Flounders, A.W., Schoeniger, J.S., Cheng, T.C., DeFrank, J.J. and Wild, J.R. (2001) Enzyme-based biosensor for the direct detection of fluorine-containing organophosphates. *Analytica Chimica Acta* 442, 15–23.

Singh, S.B., Foster, G.D. and Khan, S.U. (2004) Microwave-assisted extraction for the simultaneous determination of thiamethoxam, imidacloprid, and carbendazim residues in fresh and cooked vegetable samples. *Journal of Agricultural and Food Chemistry* 52(1), 105–109.

Sinha, S.N., Rao, M.V.V. and Vasudev, K. (2012) Distribution of pesticides in different commonly used vegetables from Hyderabad, India. *Food Research International* 45(1), 161–169.

Skladal, P. (1991) Determination of organophosphate and carbamate pesticides using cobalt phthalocyanine-modified carbon paste electrode and a cholinesterase enzyme membrane. *Analytica Chimica Acta* 252, 11–15.

Skladal, P. (1996) Biosensors based on cholinesterase for detection of pesticides. *Food Technology and Biotechnology* 34, 43–49.

Skladal, P. and Mascini, M. (1992) Sensitive detection of pesticides using amperometric sensors based on cobalt phthalocyanine-modified composite electrodes. *Biosensors and Bioelectronics* 7, 335–343.

Snejdarkova, M., Svobodova, L., Nikolelis, D.P., Wang, J. and Hianik, T. (2003) Acetylcholine biosensor based on dendrimer layers for pesticides detection. *Electroanalysis* 15, 1185–1191.

Soldatkin, O.O., Burdak, O.S., Sergeyeva, T.A., Arkhypova, V.M., Dzyadevych, S.V. and Soldatkin, A.P. (2013) Acetylcholinesterase-based conductometric biosensor for determination of aflatoxin B1. *Sensors and Actuators B* 188, 999–1003.

Stahl, E., Quirin, K.W. and Gerard, D. (1988) *Dense Gases for Extraction and Refining*. Springer-Verlag, Berlin.

Sun, L. and Lee, H.K. (2003) Stability studies of propoxure herbicide in environmental water samples by liquid chromatography-atmospheric pressure chemical ionization ion-trap mass spectrometry. *Journal of Chromatography A* 1014, 153.

Sun, X., Liu, B. and Xia, K. (2011) A sensitive and regenerable biosensor for organophosphate pesticide based on self-assembled multilayer film with CdTe as fluorescence probe. *Luminescence* 26(6), 616–621.

Suri, C.R., Boro, R., Nangia, Y., Gandhi, S., Sharma, P., Wangoo, N., Rajesh, K. and Shekhawat, G.S. (2009) Immunoanalytical techniques for analyzing pesticides in the environment. *Trends in Analytical Chemistry* 28(1), 29–39.

Thévenot, D.R., Toth, K., Durst, R.A. and Wilson, G.S. (1999) Electrochemical biosensors: recommended definitions and classification. *Pure Applied Chemistry* 71(12), 2333–2348.

Tong, D.W., Yang, H.S. and Wang, J.Y. (2007) Recent advances in the synthesis of artificial antigen and its application in the detection of pesticide residue. *American Journal of Agricultural and Biological Sciences* 2(2), 88–93.

Toribio, L., del Nozal, M.J., Bernal, J.L., Jimenez, J.J. and Serna, M.L. (1998) Packed column supercritical fluid chromatography coupled with solid phase extraction for the determination of organic micro-contaminants in water. *Journal of Chromatography A* 823, 163–170.

Toribio, L., del Nozal, M.J., Bernal, J.L., Jimenez, J.J. and Alonso, C. (2004) Chiral separation of some triazole pesticides by supercritical fluid chromatography. *Journal of Chromatography A* 1046, 249–253.

Toribio, L., del Nozal, M.J., Bernal, J.L., Alonso, C. and Jimenez, J.J. (2005) Comparative study of the enantioselective separation of several antiulcer drugs by high performance liquid chromatography and supercritical fluid chromatography. *Journal of Chromatography A* 1091, 118–123.

Tran-Minh, C., Pandey, P.C. and Kumaran, S. (1990) Studies of acetylcholine sensor and its analytical application based on the inhibition of cholinesterase. *Biosensors and Bioelectronics* 5, 461–471.

Trettnak, W., Reininger, F., Zinterl, E. and Wolfbeis, O.S. (1993) Fiber optic remote detection of pesticides and related inhibitors of the enzyme acetylcholinesterase. *Sensors and Actuators B* 11, 87–93.

Trojanowitz, M. and Hitchman, M. (1996) Determination of pesticides using electrochemical biosensors. *Trends in Analytical Chemistry* 15, 38–45.

Tsai, H.C. and Doong, R.A. (2005) Simultaneous determination of pH, urea, acetylcholine and heavy metals using array-based enzymatic optical biosensor. *Biosensors and Bioelectronics* 20(9), 1796–1804.

Turan, J., Kesik, M., Soylemez, S., Goker, S., Coskun, S., Unalan, H.E. and Toppare, L. (2016) An effective surface design based on a conjugated polymer and silver nanowires for the detection of paraoxon in tap water and milk. *Sensors and Actuators B* 228, 278–286.

US Congress, Office of Technology Assessment (1988) Workshop on Technologies to Detect Pesticide Residues in Food, Washington, DC, 14–16 March 1988.

Valera, E., García-Febrero, R., Pividori, I., Sánchez-Baeza, F. and Marco, M.P. (2014) Coulombimetric immunosensor for paraquat based on electrochemical nanoprobes. *Sensors and Actuators B* 194, 353–360.

Vandevelde, F., Leichle, T., Ayela, C., Bergaud, C., Nicu, L. and Haupt, K. (2007) Direct patterning of molecularly imprinted microdot arrays for sensors and biochips. *Langmuir* 23(12), 6490–6493.

Vargas-Bernal, R., Rodríguez-Miranda, E. and Herrera-Pérez, G. (2012) Evolution and expectations of enzymatic biosensors for pesticides. In: Soundararajan, R.P. (ed.) *Pesticides: Advances in Chemical and Botanical Pesticides*. IntechOpen, London. Available at: https://www.intechopen.com/books/pesticides-advances-in-chemical-and-botanical-pesticides/evolution-and-expectations-of-enzymatic-biosensors-for-pesticides (accessed 31 July 2018).

Vertegel, A.A., Siegel, R.W. and Dordick, J.S. (2004) Silica nanoparticle size influences the structure and enzymatic activity of adsorbed lysozyme. *Langmuir* 20, 6800–6807.

Vicentini, F.C., Ravanini, A.E., Figueiredo-Filho, L.C., Iniesta, J., Banks, C.E. and Fatibello-Filho, O. (2015) Imparting improvements in electrochemical sensors: evaluation of different carbon blacks that give rise to significant improvement in the performance of electroanalytical sensing platforms. *Electrochimica Acta* 157, 125–133.

Vidal, J.C., Esteban, S., Gil, J. and Castillo, J.R. (2006) A comparative study of immobilization methods of a tyrosinase enzyme on electrodes and their application to the detection of dichlorvos organophosphorus insecticide. *Talanta* 68(3), 791–799.

Villani, J. (1995) Pesticide testing: what's the best way? In: Villani, J. (ed.) *Other Diagnostic Tools for Pesticides Detection*. Available at: https://www.naturalproductsinsider.com/labstesting/pesticide-testing-whats-best-way (accessed 21 September 2018).

Vlatakis, G., Andersson, L.I., Muller, R. and Mosbach, K. (1993) Drug assay using antibody mimics made by molecular imprinting. *Nature* 361(2), 645–647.

Vryzas, Z., Papadakis, E.N. and Papadopoulou-Mourkidou, E. (2002) Microwave assisted extraction (MAE) acid hydrolysis of dithiocarbamates for trace analysis in tobacco and peaches. *Journal of Agricultural and Food Chemistry* 50, 2220–2226.

Wang, J. (2002) Electrochemical nucleic acid biosensors. *Analytica Chimica Acta* 569, 63–71.

Wang, J. (2012) Electrochemical biosensing based on noble metal nanoparticles. *Microchimica Acta* 177, 245–270.

Wang, J., Krause, R., Block, K., Musameh, M., Mulchandani, A. Mulchandani, P., Chen, W. and Schoening, M.J. (2002) Dual amperometric potentiometric biosensor detection system for monitoring organophosphorus neurotoxins. *Analytica Chimica Acta* 469(2), 197–203.

Wang, Z., Li, S., Jonca, M., Lambros, T., Ferguson, S., Goodnow, R. and Ho, C.T. (2006) Comparison of supercritical fluid chromatography and liquid chromatography for the separation of urinary metabolites of nobiletin with chiral and non chiral stationary phases. *Biomedical Chromatography* 20, 1206–1215.

Wang, K., Liu, Q., Dai, L., Yan, J., Ju, C., Qiu, B. and Wu, X. (2011) A highly sensitive and rapid organophosphate biosensor based on enhancement of CdS-decorated graphene nanocomposite. *Analytica Chimica Acta* 695, 84–88.

Wang, G., Tan, X., Zhou, Q., Liu, Y., Wang, M. and Yang, L. (2014) Synthesis of highly dispersed zinc oxide nanoparticles on carboxylic graphene for development of a sensitive acetylcholinesterase biosensor. *Sensors and Actuators B* 190, 730–736.

Watanabe, E., Baba, K., Eun, H. and Miyake, S. (2007) Application of a commercial immunoassay to the direct determination of insecticide imidacloprid in fruit juices. *Food Chemistry* 102(3), 745–750.

Watanabe, E. and Miyake, S. (2007) Immunoassay for iprodione: key estimation for residue analysis and method validation with chromatographic technique. *Analytica Chimica Acta* 583(2), 370–376.

Wei, M. and Wang, J. (2015) A novel acetylcholinesterase biosensor based on ionic liquids-AuNPs-porous carbon composite matrix for detection of organophosphate pesticides. *Sensors and Actuators B* 211, 290–296.

Wei, W., Zong, X., Wang, X., Yin, L., Pu, Y. and Liu, S. (2012) A disposable amperometric immunosensor for chlorpyrifos-methyl based on immunogen/platinum doped silica sol–gel film modified screen-printed carbon electrode. *Food Chemistry* 135, 888–892.

Wheeler, J.R. and McNally, M.E. (1987) Comparison of packed column and capillary column super critical fluid chromatography and high performance liquid chromatography using representative herbicides and pesticides as typical moderate polarity and molecular weight range molecules. *Journal of Chromatography* 410, 343–353.

Wong, F.C., Ahmad, M., Heng, L.Y. and Peng, L.B. (2006) An optical biosensor for dichlovos using stacked sol-gel films containing acetylcholinesterase and a lipophilic chromoionophore. *Talanta* 69(4), 888–893.

Wu, S., Lan, X., Zhao, W., Li, Y., Zhang, L., Wang, H., Han, M. and Tao, S. (2011a) Controlled immobilization of acetylcholinesterase on improved hydrophobic gold nanoparticle/Prussian blue modified surface for ultra-trace organophosphate pesticide detection. *Biosensors and Bioelectronics* 27, 82–87.

Wu, S., Zhang, L., Qi, L., Tao, S., Lan, X., Liu, Z. and Meng, C. (2011b) Ultra-sensitive biosensor based on mesocellular silica foam for organophosphorous pesticide detection. *Biosensors and Bioelectronics* 26, 2864–2869.

Xiang, Y. and Lu, Y. (2011) Using personal glucose meters and functional DNA sensors to quantify a variety of analytical targets. *Nature Chemistry* 3, 697–703.

Yaico, D., Tanimoto de, A. and Lucas, F.F. (2007) Amperometric biosensing of carbamate and organophosphate pesticides utilizing screen-printed tyrosinase-modified electrodes. *Analytica Chimica Acta* 596, 210–221.

Yang, L., Wang, G., Liu, Y. and Wang, M. (2013) Development of a biosensor based on immobilization of acetylcholinesterase on NiO nanoparticles–carboxylic graphene–nafion modified electrode for detection of pesticides. *Talanta* 113, 135–141.

Yang, Y., Asiri, A.M., Du, D. and Lin, Y. (2014) Acetylcholinesterase biosensor based on a gold nanoparticle–polypyrrole–reduced graphene oxide nanocomposite modified electrode for the amperometric detection of organophosphorus pesticides. *Analyst* 139, 3055–3060.

Yogeswaran, U. and Chen, S.-M. (2008) A review on the electrochemical sensors and biosensors composed of nanowires as sensing material. *Sensors* 8(1), 290–313.

Yu, J.J. (2002) Removal of organophosphate pesticides from waste water by supercritical carbon dioxide extraction. *Water Research* 36, 1095–1101.

Yuehe, L., Fang, L. and Wang, J. (2004) Disposable carbon nanotube modified screen-printed biosensor for amperometric detection of organophosphorus pesticides and nerve agents. *Electroanalysis* 16 (1–2), 145–149.

Yulaev, M.F., Sitdikov, R.A., Dmitrieva, N.M., Yazynina, E.V., Zherdev, A.V. and Dzantiev, B.B. (2001) Development of a potentiometric immunosensor for herbicide simazine and its application for food testing. *Sensors and Actuators B* 75, 129–135.

Zapp, E., Brondani, D., Vieira, I.C., Scheeren, C.W., Dupont, J., Barbosa, A.M. and Ferreira, V.S. (2011) Biomonitoring of methomyl pesticide by laccase inhibition on sensor containing platinum nanoparticles in ionic liquid phase supported in montmorillonite. *Sensors and Actuators B* 155, 331–339.

Zhang, L., Zhang, A., Du, D. and Lin, Y. (2012) Biosensor based on Prussian blue nanocubes/reduced graphene oxide nanocomposite for detection of organophosphorus pesticides. *Nanoscale* 4, 4674–4679.

Zhang, Y., Liu, H., Yang, Z., Ji, S., Wang, J., Pang, P., Feng, L., Wang, H., Wu, Z., Yang, W. (2015) An acetylcholinesterase inhibition biosensor based on a reduced graphene oxide/silver nanocluster/chitosan nanocomposite for detection of organophosphorus pesticides. *Analytical Methods* 7, 6213–6219.

Zhang, W., Guo, Z., Chen, Y., and Cao, Y. (2017) Nanomaterial based biosensors for detection of biomarkers of exposure to OP pesticides and nerve agents: A review. *Electroanalysis* 29, 1206–1213. Wiley Online Library DOI:10.1002/elan.201600748.

Zhao, C., Hao, G., Li, H. and Chen, Y. (2002) Supercritical fluid extraction for the separation of organochlorine pesticides residue in *Angelica sinensis. Biomedical Chromatography* 16(7), 441–445.

Zhenchao, W., Dandan, Y., Li, X., Mengjiao, Z. and Zhuo, C. *et al.* (2012) The development and application of a Dot-ELISA assay for diagnosis of southern rice black streaked dwarf disease in the field. *Viruses* 4, 167–183.

Zheng, Y., Liu, Z., Jing, Y., Li, J. and Zhan, H. (2015) An acetylcholinesterase biosensor based on ionic liquid functionalized graphene–gelatin-modified electrode for sensitive detection of pesticides. *Sensors and Actuators B* 210, 389–397.

Zhou, Q., Yang, L., Wang, G. and Yang, Y. (2013) Acetylcholinesterase biosensor based on SnO_2 nanoparticles–carboxylic graphene–nafion modified electrode for detection of pesticides. *Biosensors and Bioelectronics* 49, 25–31.

5 Toxicity of Pesticides in Mixtures with the Same Mode of Action

5.1 Introduction

Pesticides are likely to co-occur in the environment (water, air, soil), in food or in consumer products as mixtures, posing a risk to human beings and all other organisms present in the surrounding environment. However, most chemical risk assessment reports that are available, with a few exceptions, consider the effects of single substances in isolation, as if mixtures of substances do not bear the risk of an increased toxicity. This would be the case, for example, if only one chemical in a mixture is toxic while the others are biologically inert, or if empirical evidence showed that the joint action of chemicals is typically not greater than the effect of the most toxic compound. Richardson *et al.* (2001) pointed out that even though individual pesticides may be at a 'no adverse effect' level, their presence in a mixture may cause a cumulative adverse effect.

A marked example of malathion (MLT) toxicity by ethyl p-nitrophenyl thionobenzenephosphonate (EPN) occurred in 1957 (Gupta, 2006), when Frawley *et al.* (1957) investigated mixture toxicity on rats. They observed 100% mortality of rats and dogs when fed with non-lethal dosages of these pesticides. The mechanism underlying the potentiation of toxicity of MLT by EPN was studied (Murphy and Dubois, 1957; Seume and O'Brien, 1960a), and it was found that EPN inhibited hydrolysis of MLT by carboxylesterase. Potentiation of toxicity was also observed between EPN and other organophosphate compounds (Rosenberg and Coon, 1958; Seume and O'Brien, 1960b). Similar studies revealed more than additive interaction between other OP compounds, which were potent inhibitors of carboxylesterase activity (Murphy *et al.*, 1959; Lauwerys and Murphy, 1969; Polak and Cohen, 1969; Cohen and Murphy, 1971). Casida *et al.* (1963) assessed 112 organophosphate compounds for their ability to lower the LD_{50} of MLT in mice, and observed a vide variation in the degree of potentiation of toxicity of MLT.

Mixture toxicity of similar- and dissimilar-acting toxicants can be predicted by the model's concentration addition and independent action using single substance toxicity data. Knowledge of the toxicant's mode of action is thus required in order to use the models. Both the models predict binary and ternary mixtures of similar- and dissimilar-acting toxicants equally well (Syberg *et al.*, 2008).

The United States Food Quality Protection Act (US FQPA) introduced the terms aggregate and cumulative risk assessment. The former term means exposure to a single chemical from multiple sources, while the latter means taking account of exposure to multiple chemicals with a common mechanism of toxicity. Descriptions of the various ways in which combined actions may occur are presented in Table 5.1.

For mixtures of similarly acting chemicals the effects can be estimated directly from the sum of the doses/concentrations, scaled for relative toxicity (dose/concentration addition).

Many methods have been proposed for the expression of the overall effects of combinations of chemicals (European Commission, 2011). These include, for example:

- relative potency factors;
- toxicity equivalent factors;
- hazard indices; and
- toxic unit.

5.2 Approaches Used for Assessing Mixture Toxicity

5.2.1 Toxic unit approach

The toxic unit (TU) model has been used extensively in mixture evaluation to determine the overall toxicity of a mixture (Spehar and Fiandt, 1986; Forget *et al.*, 1999; Woods *et al.*, 2002) added directly. TUs are used to generate a normalized scale, and in mathematical presentation of TU addition (Belden *et al.*, 2007) *sumTU* is the sum of all TUs in the mixture, c_i is the exposure concentration of the *i*th chemical in the mixture, and $c_i/EC_{x,i}$ is the TU that quantifies the contribution to toxicity of the *i*th chemical in the mixture.

TU is the sum of the toxic contributions of each component in the mixture. The TU (Eqn 5.1) for a binary mixture is given by the following equation:

$$TU = \frac{ECx\ A\ (\text{mix})}{ECx\ A\ (\text{alone})} + \frac{ECx\ B\ (\text{mix})}{ECx\ B\ (\text{alone})} \quad \text{(Eqn 5.1)}$$

Where A and B are chemicals, EC_x (mix) is the effect of each component in the binary mix and EC_x (alone) is the EC_{50} or EC_{20} of A and B applied as single components. According to this model, if TU equals one, the toxicity of the mixture is additive. If TU is greater than one, the toxicity is less than additive (antagonistic) and if the TU is less than one, the toxicity of the mixture is more than additive (synergistic) as proposed by Spehar and Fiandt (1986).

The concentration addition model is also used in the calculation of toxic equivalents (TEs) (George *et al.*, 2003). Toxic equivalent factors (TEFs) are estimates of the toxicity of a chemical typically based on the effects observed in a single species, relative to a reference chemical (Safe, 1998; 1999; George *et al.*, 2003). The reference chemical is usually the most toxic chemical in the mixture. The toxicity of a mixture can then be estimated by summing the products of individual component concentrations and their respective TEFs. This method has been used in what has been termed the probabilistic ecological risk assessment-toxic equivalents approach (PERA-TE) (George *et al.*, 2003). This application of concentration addition has the additional assumption of parallel dose (or concentration)–response curves (Birnbaum, 1999; Kortenkamp *et al.*, 2009), and is also referred to as simple similar action (Jonker *et al.*, 1996).

Table 5.1. Nomenclature used in this report for combined actions of components of mixtures.

Concept	Term used in report	Synonym(s)	Effects observed
Non-interaction	Simple similar action	Simple joint action	Concentration/dose addition
	Simple dissimilar action	Simple independent action Independent joint action	Effect/response addition
Interaction	Potentiation	Synergy, supra-additivity	Greater than additive effect
	Antagonism	Sub-additivity	Less than additive effect

5.2.2 Concentration addition model

The two most widely used conceptual models for predicting mixture toxicity are concentration addition, introduced by Hewlett and Plackett (1952), and independent action, first described by Bliss (1939). The concentration addition equation (Eqn 5.2) was implemented by Altenburger *et al.* (2004) to examine whether the concentration–response relationships of binary combinations could be predicted based on the range of effects observed in response to single compounds. The principal assumption of the concentration addition model is that all chemicals considered act as dilutions of each other, and thus the potency of the mixture is the sum of the individual potencies (Deneer, 2000). Concentration addition can be expressed as (Zhang *et al.*, 2010):

$$ECx_{mix} = (\Sigma i = 1 \text{ to } n(p_i/\mathrm{EC}x_i))^{-1} \qquad \text{(Eqn 5.2)}$$

$\mathrm{EC}x_{mix}$ is the concentration of the mixture resulting in an effect of x% relative to control, $\mathrm{EC}x_i$ is the concentration of component i resulting in an effect of x% when applied in isolation and p_i is the proportion of the total concentration of the mixture represented by component i.

Whether the toxicity of each mixture conforms to concentration addition can be determined using a two-step process. First, if the observed mixture toxicity value falls within the 95% confidence intervals of the expected value of a model, then the mixture is considered to have conformed to the model (Phyu *et al.*, 2011). If, however, an observed mixture toxicity value is outside the 95% confidence intervals of the expected value, then the mixture potentially may not conform to that model. To enable a quantitative estimation of the difference in predicted and measured toxicity, the model deviation ratio (MDR) method (Belden *et al.*, 2007) is used. For the concentration addition model, the MDR is derived by dividing the predicted toxicity value (EC_{20}) by the observed toxicity value. MDR values greater than 1.3 mean that the toxicity of the mixture conforms with synergism, while values less than 0.7 are taken to conform to antagonism (Belden *et al.*, 2007; Phyu *et al.*, 2011). According to the concentration addition model, pesticide concentrations are first normalized for potency and then the total concentration of the mixture is calculated and used to predict mixture toxicity (Faust *et al.*, 2000).

5.2.3 Combination index model

The results of mixture toxicity could be analysed using the median effect/combination index isobologram equation as described by Chou and Talalay (1984) and Chou (2006). The computation is based on the median effect principle, which assumes that there is a univocal relationship between dose and effect irrespective of the number of compounds and their mechanism of action. The median effect equation (Eqn 5.3) is as follows:

$$\frac{f_a}{1-f_a} = \left(\frac{D}{\mathrm{EC}_{50}}\right)^m \qquad \text{(Eqn 5.3)}$$

D represents the dose that affects a fraction f_a. EC_{50} represents the dose at which a 50% effect (in acetylcholine esterase inhibitors, inhibition of AChE activity) is reached. The EC_{50} (potency) and m (shape) values for each compound are determined by computing experimental f_a and D values. These parameters are then used to calculate the combined doses required to produce a given effect. For all effect levels, the combination index (CI) values are calculated according to the general CI equation proposed by Chou and Talalay (1984) where $\mathrm{CI}<1$, $\mathrm{CI}=1$ and $\mathrm{CI}>1$ indicates synergism, additive effect and antagonism, respectively.

Both independent action and dose/concentration addition concepts are based on the assumption that chemicals in a mixture do not influence each other's toxicity, that is, they do not interact with each other at the biological target site. Such chemicals can either elicit similar responses by a common or similar mode of action or they act independently and may have different endpoints and/or different target organs. However, both concepts have been suggested as default approaches in regulatory risk assessment of chemical mixtures. In reality, however, chemical mixtures are rarely composed of either only similarly or only dissimilarly acting substances. Kortenkamp *et al.* (2009) has produced a more detailed description as follows:

Dose/concentration addition

Dose/concentration addition (with similar action/similar joint action) occurs if chemicals in a mixture act by the same mechanism/mode of action, and differ only in their potencies. Different methods exist for the dose/concentration approach, which mainly differ in the required knowledge about mode of actions and toxicological similarities of the mixture components. In principle, doses or concentrations of the single components are added after being multiplied by a scaling factor that accounts for differences in the potency of the individual substances.

The mixture dose/concentration D_{mix} is the sum of the adjusted doses/concentrations (aD_i) of the individual components D_i (Eqn 5.4):

$$D_{mix} = \sum_{i=1}^{n} aDi \quad \text{(Eqn 5.4)}$$

The effect of a mixture of similarly acting compounds is equivalent to the effects of the sum of the potency-corrected (adjusted) doses/concentrations of each compound.

Dose additivity is assumed over the entire dose range, including doses/concentrations below the individual no observed adverse effect levels/concentrations (NOAEL/Cs) of the mixture components.

It is noted that the dose additivity approach relies on a correct grouping of 'similar' chemicals. Though guidance on grouping of chemicals has been issued (European Chemicals Agency, ECHA; Organisation for Economic Co-operation and Development, OECD; and European Food Safety Authority, EFSA), there is currently no general agreement on the scientifically best approach, and grouping of chemicals is most often done by expert judgement on a case-by-case basis.

As reviewed by Kortenkamp *et al.* (2009), there is evidence to show that dose/concentration addition can produce reliable estimates of combined effects, if the components share a strictly identical mechanism of action. In ecotoxicology, the concentration addition approach is also applicable to the group of so-called baseline toxicants. Faust and Altenburger published two studies on the chronic algal toxicities of binary mixtures (Faust *et al.*, 1994; Altenburger *et al.*, 1996). Altenburger studied 137 binary mixtures of different pesticides and surfactants with the result that concentration addition provided the better overall prediction for the observed toxicity data than the independent action model. A similar result was obtained by Faust, who concluded that the toxicity of 66% of the tested 38 binary pesticide mixtures was predictable by concentration addition – although all the test mixtures were composed of a herbicide and an insecticide or fungicide.

5.2.4 Toxic equivalence (TEQ) approach

A related approach based on the dose additive model is to use the concept of toxic equivalence (Eadon *et al.*, 1986; Bellin and Barnes, 1987). A possible toxic equivalence scheme for regulation could be used for a mixture of aldicarb and its transformation products aldicarb sulfoxide and aldicarb sulfone. All those compounds are cholinesterase inhibitors, and their potencies relative to that of aldicarb (as measured by 1/NOAEL) can be expressed (Table 5.2). Aldicarb has a lifetime health advisory guideline of 10 ppb (μg/l) (US EPA, 1987), and the TEs of aldicarb can be compared with this value. A litre of water containing a mixture of aldicarb at 2 μg/l, aldicarb sulfoxide at 5 μg/l and aldicarb sulfone at 30 μg/l could be expressed in TEs of aldicarb on the basis of relative potencies. The toxic equivalent concentrations for the individual compounds are obtained by multiplying the concentration by the relative potency. For example, for aldicarb sulfone, the concentration of toxic equivalence is 30 μg of aldicarb sulfone/l) (0.2) = 6 μg of aldicarb/l.

It should also be noted that the toxic equivalent concentration of the mixture is 13 *mg*/l – the sum of the concentrations of toxic equivalent – which can be compared

Table 5.2. Relative potencies and toxic equivalent concentrations of aldicarb and products.

Compound	NOAEL (mg/kg)	Relative potency	Concentration μg/l	Toxic equivalent concentration μg/l
Aldicab	0.125	1	2	2
Aldicarb sulfoxide	0.125	1	5	5
Aldicarb sulfone	0.6	0.2	30	6

with the health advisory guideline of 10 *mg*/l (US EPA, 1987). This approach assumes that these compounds do not act synergistically (National Research Council, 1989).

The known mechanisms of anticholinesterase synergism depend on interference with or competition for metabolic mechanisms of detoxification of the anticholinesterases or their precursors. Therefore, one might predict that synergism will occur only when the dosage exceeds the theshold where metabolism becomes a rate-limiting factor in toxicity. Of course, that dosage becomes smaller as critical pathways of detoxification are inhibited by other compounds.

Without specific knowledge of the mechanism of synergism and without quantitative data on the response to a range of doses of interactive chemicals, it is not possible to determine at precisely what concentrations interaction occurs. However, the interactions do not occur below some dosages, which can be predicted from data on individual compounds.

The oxygen analogues of EPN and MLT are reported (DuBois, 1961) to be strictly additive with respect to their anticholinesterase action *in vitro* and, in contrast, synergistic *in vivo*. If the compounds compete for the same active catalytic sites on the acetylcholinesterase (AChE) molecules, chemicals that are intrinsically less effective as inhibitors might sometimes occupy these sites at the expense of intrinsically more active inhibitors. This combined action will be manifested as antagonistic (Veldstra, 1956).

5.3 Interactions in Pesticide Mixtures

Kok and Hasirci (2004) found inhibition of AChE lower in the presence of the combined mixtures of carbamates, aldicarb and carbofuran compared with the sum of individual inhibition values. Tahara *et al.* (2005) investigated that the degree of inhibition increased with multiple organophosphates with an additive effect. They used low concentrations of pesticides between 0.1 and 10 μg/ml or 0.1–10 mg/l. They only studied mixtures of organophosphates. Reffstrup *et al.* (2010) provided an overview on risk assessment of mixtures of pesticides and found that mixtures will either display no interaction between each other, or display an interaction. A dose additive effect means that no interaction takes place between different pesticides and that the compounds act on the 'same biological site by the same mechanism/mode of action and differ only in their potencies' (European Food Safety Authority, 2013, p. 10). A response additive effect also involves no interaction between compounds, but in this case they act on a different biological site. Where interactions take place between compounds, such interaction can result in a stronger (synergistic) or weaker (antagonistic) effect compared with the sum of individual activities.

The number of potential pesticide mixtures available in the environment is very large, and it is impossible to study all these mixtures. Andersen *et al.* (2006) applied two models to predict the effect of pesticide mixtures and to identify pesticide combinations displaying a high risk of synergistic interactions in Denmark. The pesticide mixtures were observed to have a tendency to deviate from additivity. However, it is obvious that most of the mixtures were additive, while antagonism and synergism were rarer. The synergism attracts more attention from both the public and the regulatory authorities, due to unintended health and environmental effects, which would be larger than expected. Antagonism, on the other hand, leads to smaller effects than expected. Using the additive dose model (ADM), the effects were observed to be either equal to or lower than expected. However, in another multiplicative survival model (MSM) a lower effect of pesticide mixtures is expected than in the ADM model; that is, using MSM more mixtures can be assessed as being synergistic. However, ADM is preferable to MSM in a risk assessment context.

In principle, the interactions of mixtures can be assessed making use of mechanistic or empirical models. The empirical, or descriptive, model includes only information on doses or concentrations and effects, including the dose–response relationship, while with the mechanical approach, additional information on the sequence of reaction steps is available and quantitative parameters are known. For example, the joint action of multiple chemicals at a target receptor or enzyme can be described mechanistically by Michaelis–Menten kinetics. In general, this will result in competitive agonism and the ultimate combined effect may be less than expected on the basis of effect addition because of competition for the same receptor or active site. In fact, this type of interaction can also be considered as a special case of similar joint action (dose additivity) (http://cot.food.gov.uk/sites/default/files/cot/reportindexed.pdf).

The information gap in most combination studies prevents the application of such models, which is the main drawback of this approach. Currently empirical models play a dominant role (Groten *et al.*, 2001).

Cassee *et al.* (1998; 1999) observed widespread disagreement over the terminology and definitions used in the description of interactions of mixture components. It has been observed that a commonly used approach is to calculate the expected combined response (in the absence of interaction) from the results for single agents, and that any deviations between observed and expected response are termed an *interaction*. The terminology used to describe various interactive effects should, therefore, be clearly defined in order to avoid confusion. The International Study Group on Combination Effects attempted to implement uniform definitions for synergism, potentiation and antagonism, and for standard methodologies to assess such effects. The description of the toxicological action of mixture constituents is based on three fundamental concepts, as originally described by Bliss (1939) and referred to by Cassee *et al.* (1998; 1999). In reality, several principles of joint action may occur at the same time (partial dose and/or effect additivity), especially when mixtures consist of more than two compounds and when several target sites are involved. Hazards of mixtures cannot be characterized without the use of appropriate test designs and clear, testable hypotheses to explore joint action.

Simple similar action (or simple joint action) results in dose additivity. It should be used as the starting point if the chemicals in the mixture act in the same way, by the same mechanism(s), and differ only in their potencies. Under simple joint action, the effect (or response) for the mixture is obtained by summing the doses of the individual compounds, after adjustment for differences in their potencies. Mathematically, if $R(x)$ is the dose–response function of two similarly acting compounds A and B, the response for a mixture with standardized dose x_A of A and x_B of B is $R(x_A + x_B)$ (Figs 5.1 and 5.2).

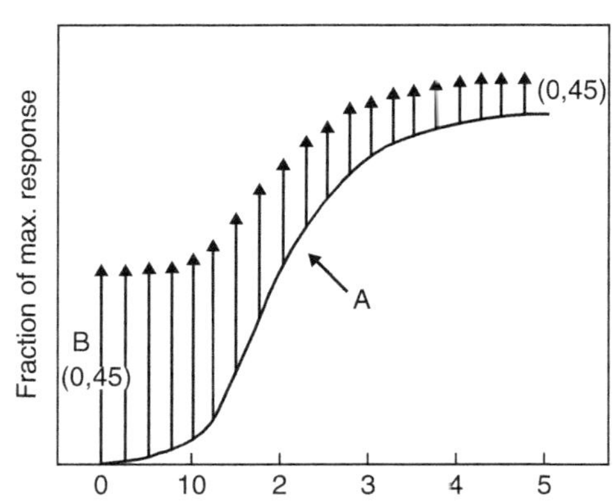

Fig.5.1. Effect of a fixed concentration of component B on the dose–response curve for A; dose vs fraction of response. (From Hughes and Woods, 2002.)

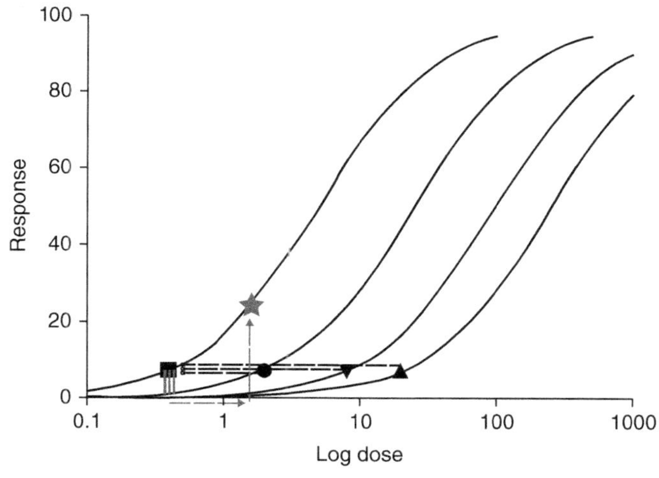

Fig. 5.2. Simple similar action: with several components. Illustration of the concept of simple similar action, with simultaneous exposure to several different compounds acting by the same mechanism. The response to a mixture of four compounds acting on the same receptor with different affinities is illustrated. The response to each compound alone is less than 10%. The response to the mixture is over 20%. (From Hughes and Woods, 2002.)

However, there is strong evidence that chemicals in mixtures with similar modes of action produce combination effects that are generally larger than the effects of an individual component applied singly. A few studies have been conducted with mixtures composed of chemicals with diverse modes of action, with results clearly pointing in the same direction: the effects of such mixtures are also higher than those of the individual components.

As per the theory of the 'same mode of action', the mixtures, for example, of two AChE-inhibiting insecticides, should clearly follow the concentration addition model (Bliss, 1939; Merino-Garcia *et al.*, 2003; Faust and Scholze, 2004), which could also be explained reasonably well using the independent action model (http://digitalcommons.unl.edu/biosciforbes). But AChE-inhibiting ability does not necessarily mean that the two pesticides share the same mode of action. A mixture of dimethoate and pirimcarb, both acetylcholine inhibitors, was observed with varying toxicities and sensitivities to target the insect *Aphis gossypii*. The increased toxic effect of dimethoate may be due to elevated cytochrome P450 mono-oxygenase (Frasco and Guilhermino, 2002; Anderson and Zhu, 2004), which could be due to oxidative activation of dimethoate into the O-analogue metabolite, omethoate, which is a more effective inhibitor of AChE (Frasco and Guilhermino, 2002; Anderson and Zhu, 2004). Benting and Nauen (2004) observed that the same mutation in AChE (S431F) led to resistance to both pirimicarb and omethoate in the insect *Aphis gossypii*. However, *A. gossypii* was many-fold less sensitive to pirimicarb than dimethoate, indicating the possibility of the presence of different residues, involved in the inhibition mechanism of pirimicarb and omethoate. It is, therefore, not wise to make a blanket statement for pesticide mixtures with a similar mode of action that bind differently to target sites, and/or differ in potency.

A question always arises of whether there are other modes of action, apart from the 'designed' effect as well as a common narcotic effect (Pope, 1999). Dimethoate is one of the insecticides that is known to have several modes of action (Casida and Quistad, 2004). Studies with organophosphorus and carbamate insecticides have shown that these compounds can induce both necrotic cell death and apoptosis (Pena-Llopis *et al.*, 2003; Caughlan *et al.*, 2004; Kim *et al.*, 2004), processes that are biochemically associated with release of cytochrome c from the mitochondria into the cytoplasm (Skulachev, 1998). Caughlan *et al.* (2004) suggested apoptosis as a novel toxic mechanism of organophosphates that could be independent of AChE inhibition. Therefore, the lower additive effect of pirimicarb and dimethoate could also be due to a similar mode of action in relation to apoptosis or necrosis as in inhibition of AChE. In the case of the mixture of pirimicarb and dimethoate predictions, using independent action would not have made a big difference to the conclusion of the study, since predictions from the two models were very similar (Syberg *et al.*, 2008). The concentration addition model could be preferred as the general reference model in risk assessment for environmental samples, consisting of both similar- and dissimilar-acting toxicants. In studies investigating the physiological interactions between chemicals, both reference models might provide valuable information despite their specific limitations.

A modulation of the acute toxicity of pesticides was observed, caused by prior exposure to relatively large doses of other pesticides, most likely through effects on their metabolism. Gaughan *et al.* (1980) observed a decrease in LD_{50} values for fenvalerate and MLT in mice, pre-treated with sulprofos, under laboratory experiments. Contrarily, animals treated with organophosphate chemicals alone exhibited no overt signs of toxicity or effects related to anti-AChE activity. Takahashi *et al.* (1987) investigated the modulating effects of a single low dose of P = S (cyanophos, fenitrothion and MLT)/P = O type organophosphate insecticide (dichlorvos) on the acute toxicities (oral LD_{50} values) of the N-methylcarbamate (fenobucarb, xylylcarb, metolcarb, carbaryl and 3,5-dimethylphenyl -N-methylcarbamate) insecticides, in male Institute of Cancer Research mice. The effects, following treatment with the N-methylcarbamates, were not altered by concurrent or subsequent treatment with organophosphate compounds. However, the acute toxicities of the carbamate compounds were potentiated. The acute toxicity of fenobucarb was markedly potentiated by each of the P = S compounds. Maximal potentiation was observed with simultaneous MLT treatment and 0.5–4-h pre-treatments with cyanophos and fenitrothion (cyanophos decreased the LD_{50} > 10-fold; fenitrothion decreased the LD_{50} > 5-fold). In contrast, fenobucarb toxicity was not significantly affected by dichlorvos (P = O compound). Potentiation of fenobucarb acute toxicity was associated with increased levels of fenobucarb found in plasma, which could be due to the changes in the respective rates of activation and detoxification.

The toxicity of pesticides also gets altered for stereoisomers, affecting not only the efficacy, but also exposure and risk to humans and the environment (Ulrich *et al.*, 2012). Substantial differences exist in the biological activity of distinct stereoisomers and their mixtures (http://digitalcommons.unl.edu/biosciforbes/23). Stereoisomers can degrade at different rates in the environment and in organisms, primarily through biological processes (Müller and Kohler, 2004). Selective degradation of stereoisomers can alter risk factors because it potentially produces differences in exposure, toxicity and bioavailability. The overall effect may be either stronger or weaker than that predicted by an additivity model. It should be noted that in this situation the term 'interaction' is used conceptually to describe deviation from an additive model. The mechanism underlying the interaction may be at the chemical, physico-chemical or biological level, for example, interactions between two chemicals in a mixture/formulation, or interactions in either the toxicokinetic or toxicodynamic phase.

5.3.1 Aquatic ecotoxicology

The evaluation of joint toxicity of pesticides in mixtures is the major challenge in aquatic ecotoxicology (Eggen *et al.*, 2004), due to entry of pesticides through various pathways (Barbash, 2003; Schulz, 2004). Moreover, the terminologies such as synergism, additivity and antagonism create ambiguity. The toxicity of individual components defines the term no interaction. The joint toxicity of a mixture is the same as would be expected from the sum of the individual components, for no interaction. Moore and Teed (2012) reported that mixtures of organophosphates and carbamates do not pose a significant direct risk to Pacific salmon. The mixture rarely occurred at levels capable of producing significant physiological and behavioural effects in Pacific salmon. Rodney *et al.* (2013) reviewed that the concentration addition model is a generally conservative and practical first-tier model for the ecological assessment of pesticide mixtures in aquatic systems.

If chemicals do not interact, it is generally assumed that the overall effect of the mixture will be additive with respect to the components, in either concentration or response (Borgert *et al.*, 2004). Interactions between mixture components may occur, however, and can either enhance or reduce toxicity relative to the predictions of an additive model (Waiser and Holm, 2005). Scholz *et al.* (2006) investigated dose additive inhibition of Chinook salmon AChE activity using insecticide mixtures. The organophosphate and carbamate insecticides were observed to be non-interactive in terms of AChE inhibition. It could be possible to estimate the cumulative neurotoxicity of mixtures by simple dose addition. Because organophosphates and carbamates are likely to have additive effects on the neurobehaviour of salmon under natural exposure conditions, ecological risk assessments that focus on individual AChEs might underestimate the actual risk to salmon in watersheds in which mixtures of these chemicals occur.

5.4 Interactive Effects of Pesticide Mixtures

The usefulness of acute study data is limited with regard to the prediction of the effects at low exposures. The metabolic pathways and receptors involved are likely to be saturated. Furthermore, the critical target for acute toxic effects may not be the same as those at lower levels of exposure. However, the mechanisms by which toxicities of compounds are modulated may have relevance to the interactions that may occur during exposure to mixtures.

For analysis of any potentially additive or interactive effects of compounds within a mixture, the dose–response curves are established for each of the individual components on its own. In the absence of such information, it is not possible to predict the effect of a single compound at doses outside the range for which the shape of the curve is known and the theoretical additive effects of the compound in combination with any other mixture component cannot be predicted. The absence of adequate dose–response characterization is a major problem in the interpretation of much of the published literature on the effects produced by mixtures of pesticides (Hughes and Woods, 2002).

The importance of dose–response characterization in determining deviation from additivity is illustrated by the effect of a fixed concentration of constituent B on the dose–response curve for constituent A (Fig. 5.1). In the absence of an interaction, when working in the region of the dose–response curve, which approximates to linear between 2 and 4 in Fig. 5.1, the effects of B will be additive with the effects of A. However, consumer exposure almost always occurs in the non-linear region of the curve, <1 in Fig. 5.1. Here, even in the absence of any interaction, the effects of B may contribute disproportionately to the response, depending where on the dose–response curve the sum of A and B appears. In the absence of information on the shape of the dose–response curve for both A and B, it is impossible to predict what the expected response would be to any combination of the two compounds, even without any interaction. The matter is further complicated in that in biological systems, for many effects, there is a threshold below which a response cannot be detected. In such cases, addition of the action of A and B can result in a detectable response where, separately, no response was detectable; this may be falsely described as potentiation.

It is necessary to ascertain all the components of the mixture before complete analysis of the effects of a mixture, while comparing with the separate effects of each of its components. This may include identification of toxicological and other relevant data (e.g. mechanism of action) regarding each of the constituents. However, experimental studies comparing the effect of the mixture with that of the individual components (and establishing a dose–response curve for each component) are required in order to evaluate effects of the mixture and potential interactive effects of components. The studies would include evaluation effects at both high (i.e. effective) and low (i.e. realistic) doses. However, the number of test groups required to evaluate all possible interactions of components within the mixture is likely to be prohibitively large on real mode, which is the major hindrance to experimental studies of mixtures. The experiments of such studies should be designed so as to minimize, as much as possible, the number of test groups, and maximize the amount of information obtained with respect to individual mixture components and combinations.

The toxicology of mixtures is studied using two basic methodological strategies, *component interaction* and *whole-mixture* analysis. For analysis of simple mixtures, where constituents are in small numbers and their composition is clearly known, component interaction analysis (bottom-up approach) can be applied. Where there are numerous components (a complex mixture) and/or no specific knowledge of the composition of a mixture, whole-mixture analysis may be more appropriate. However, such studies cannot define the extent of true interactions between components of the complex mixture without data on fractions of the mixture.

5.4.1 Component interaction analysis

In the case where a mixture comprises a small number of (e.g. two) defined components, such that sufficient experimental data can be obtained, a number of mathematical modelling approaches can be used to evaluate the data, for example, isobolographic and response-surface analyses.

The numbers of possible experimental groups required to evaluate component interactions increase exponentially with the number of compounds in the mixture, consisting of more than two components. The effects of mixtures are evaluated and compared with their individual components using available statistical designs, for example, ray, composite and factorial designs (Cassee *et al.*, 1998; 1999; Groten *et al.*, 2001; Tajima *et al.*, 2002). Another approach may be to analyse a mixture for departure from additivity, rather than for specific interactions, comparing dose–response information for individual components to the observed responses induced by specific combinations of interest.

5.4.2 Whole-mixture analysis

An alternative to the bottom-up approach is whole-mixture analysis, most frequently used in the study of complex mixtures of pesticides. With this method, the effects of the mixture, and all individual constituents are evaluated at one dose level without testing all possible combinations of components (requiring $n+1$ test groups, where n is the number of components in the mixture). However, as dose–response relationships are not characterized, this strategy does not allow for the evaluation of potential interactive effects between mixture components. It has been suggested that this approach may be appropriate for primary screening for adverse effects of a mixture.

The determination of interaction between any of the components of a complex mixture may not be possible; however, the NOAELs for all critical effects can possibly be identified. Hence, as long as exposure is at a suitable margin below these NOAELs, consumers will be protected from the effects of any potential interactions that might occur within the mixture, providing there is an adequate margin of safety. This is particularly so if the ratios of compounds in the mixture remain relatively constant.

A study on interactive effects of pesticides on serum butyryl cholinesterase (BuChE) and brain AChE, conducted by Johnston (1995), in birds, showed that red-legged partridges (*Alectoris rufa* cross) pretreated with prochloraz were more sensitive to the effects of the organophosphate MLT than controls. This was due to the induction of MLT bioactivation. Ensenbach and Nagel (1995) exposed Zebrafish (*Danio rerio*) to lindane or 3,4-dichloroaniline or mixtures of both, in tap/river water to assess their acute and chronic toxicity. The 96-h LC_{50} values for the compounds as a binary mixture in tap water were approximately half of those observed for individual chemicals. The combined toxicity of this mixture could be characterized as additive. This was presumably a consequence primarily of the induction of bioactivation by lindane.

Oskarsson and Lind (1985) investigated the effects of the dithiocarbamates (thiram and disulfiram) on levels of lead found in the blood and brain tissue of male Sprague Dawley rats ($n = 5$ per group) exposed to lead acetate (12 mM) in the drinking water for 6 weeks. Estimated intake of lead was approximately 60 mg/day. The lead levels were significantly increased in blood and brain (maximally 3-fold and 4-fold, respectively) compared with rats that received the lead treatment alone. In rats that were treated with diethyldithiocarbamate (DEDTC) or thiram by gavage after cessation of lead treatment, there was no increase in blood lead concentrations but levels of lead in the brain tissue of animals treated with thiram were increased by nearly 3-fold. Concentrations of lead were increased in both blood and brain in rats treated with DEDTC i.p. after cessation of lead treatment. However, no histopathological examination of brain tissue has been reported.

Ito *et al.* (1995a, 1995b; 1996) investigated the carcinogenic potential of combinations of 20 pesticides (19 organophosphates and one organochlorine compound) at low doses, with glutathione-s-transferase (GST) (placental or pi form) positive (GSTp + ve) hepatocyte foci as a preneoplastic endpoint marker. After tumour initiation with diethylnitrosamine (DEN) and a two-thirds partial hepatectomy, young adult rats received the test chemicals in their diets for 6 weeks. The combination of pesticides was administered at 1× the acceptable daily intake (ADI) and 100 × ADI for each individual chemical. At 1× the ADI level, there was no enhancement of the development of preneoplastic lesions initiated by DEN. However, both the number and the area of the lesions increased at 100× the ADI. The higher dose (100 × ADI) suggested several of the pesticides were acting as tumour promoters in the liver. Similar enhancing effects had been reported previously for methidathion and MLT (part of the study). A similar study with combination of 40 pesticides (high-volume compounds) and another of 20 pesticides (suspected carcinogens) was investigated at the constituents' respective ADIs. No enhancement of carcinogenesis was observed at low levels of exposure; positive effects were seen at multiples of the ADI dose. However, these studies provide little or no information regarding the nature of any interactions that may have occurred.

5.4.3 Dose additive toxicity

Reproduction and development is a continuing process, while physiological events and anatomical structures may only be present for a limited period. If a particular harmful toxic action occurs during a vulnerable period, that is, a developmental stage, it causes damage to the precursor of an adult tissue or function, which may result in permanent consequences because regeneration or repair is not feasible. It is reported that toxic actions at any stage of development follow the same dose–response relationships as do those associated with conventional target organ toxicity. However, the impact of toxicity depends on time of exposure during development, perhaps due to the possibilities of a limited period of susceptibility and of long-term consequences that may not be manifested until sexual maturation of the subsequent generation. The impact could also be observed in the quantity and quality of milk produced by the mother for her offspring, changes in the levels of immunoglobulins and other factors in her milk or abnormalities in behaviour while bringing up her young (Hughes and Woods, 2002). There may be possibilities of dose additive and effect additive toxicity, as well as potentiation and antagonism following exposure to a mixture of toxicants, if dose–response relationships are followed. Heindel *et al.* (1994; 1995) investigated the reproductive and developmental toxicity of complex mixtures contaminating groundwater sources in the field. However, these studies provide no information with regard to the nature of toxicological interactions specifically due to the chemical mixtures. Regardless of how, no adverse effects were found in rats treated with dose levels representative of 100 times the median pesticide contamination encountered in Iowa and California. There have been a few studies of the adverse effects of far less complex mixtures and their individual components on reproduction and development that have shed some light on the nature of toxicological interactions that occur.

Laetz *et al.* (2009) measured brain AChE inhibition in juvenile coho salmon (*Oncorhynchus kisutch*) exposed to sublethal concentrations of the organophosphates diazinon, MLT and chlorpyrifos (CPF); as well as the carbamates carbaryl and carbofuran, the most extensively used pesticides in the Pacific Northwest and California. It is reported that the toxic effects of pesticide mixtures on an enzyme extracted from fish (*in vitro*) are additive. Yet little is known about the effects of pesticide mixtures when live fish are exposed to pesticide 'cocktails', according to the study. Nearly every pairing of the five pesticides inhibited activity of the enzyme AChE after the salmon were exposed over a 96 h period. It was observed that for some combinations the toxic effects of reduced brain enzyme levels were much greater than predicted if each chemical were acting in isolation to the others (www.nwfsc.noaa.gov). As the doses of individual pesticides increased, the synergistic effects were almost uniformly more pronounced. For some chemical combinations, the data suggest synergistic effects are possible at concentrations below the lowest levels used in the study. They found that no salmon died when exposed to an individual pesticide, yet many died outright when exposed to combinations. The most deadly pairing was diazinon and MLT. Testing individual pesticides to determine 'safe' exposure levels may not capture the true risk to salmon, according to their study. Several combinations of organophosphates were lethal at concentrations that were sublethal in single-chemical trials. Single-chemical risk assessments are likely to underestimate the impacts of these insecticides on salmon in river systems where mixtures occur. Moreover, mixtures of pesticides that have been commonly reported in salmon habitats may pose a more important challenge for species recovery than previously anticipated (Laetz *et al.*, 2009).

AChE inhibition has been demonstrated to be useful as a biomarker for exposure to organophosphorus insecticides in many environments. Arora and Kumar (2015) investigated the response of housefly (*Musca domestica*) head AChE (HF-AChE) against five organophosphates as individual compounds and their binary mixtures under *in vitro* conditions. To examine the effects of oxidation on organophosphate potency in the HF-AChE system, bromine water was used as an oxidizing agent. The oxidized forms of CPF, MLT and triazophos (TRZ) showed increased sensitivity towards HF-AChE. Monocrotophos (MCP) and profenofos (PRF) did not exhibit any significant differences in toxicity under oxidized and unoxidized conditions. The toxicological interaction of five organophosphorus pesticides was evaluated using the concentration addition model, the CI isobologram equation and the TU approach. All three models provided similar predictions for the ten binary combinations of organophosphates under oxidized and unoxidized conditions. The antagonistic effects of the binary combination of organophosphates (CPF + PRF, CPF + MLT, MCP + MLT, PRF + MLT, MLT + TRZ and PRF + TRZ) were observed under oxidized conditions. This may be due to dispositional and/or receptor antagonism. Most of the binary combinations assayed under unoxidized conditions exhibited synergistic responses. TRZ showed very strong synergism in binary combinations with CPF, MCP and PRF in unoxidized conditions. In contrast, under oxidized conditions, only CPF + TRZ exhibited synergism. The results obtained indicate differential toxicity of binary combinations of organophosphates under oxidized and unoxidized conditions.

Kaushik *et al.* (2006) assessed the joint action of soil-applied herbicide mixtures with similar or different modes of action using the ADM. The herbicides chlorsulfuron, metsulfuron-methyl, pendimethalin and pretilachlor, applied either singly or in binary mixtures, were used on rice (*Oryza sativa* L.). The growth (shoot) response curves were described by a logistic dose–response model. The ED_{50} values and their corresponding standard errors obtained from the response curves were used to test statistically if the shape of the isoboles differed from the reference model (ADM). Results showed that mixtures of herbicides with similar molecular targets, that is, chlorsulfuron and metsulfuron (acetolactate synthase (ALS) inhibitors), and with different molecular targets, that is, pendimethalin (microtubule assembly inhibitor) and pretilachlor (very long chain fatty acids (VLCFAs) inhibitor), followed the ADM. Mixing herbicides with different molecular targets gave different results depending on whether pretilachlor or pendimethalin was involved. In general, mixtures of pretilachlor and sulfonylureas showed synergistic interactions, whereas mixtures of pendimethalin and sulfonylureas exhibited either antagonistic or additive activities. Hence, there is a large potential for both increasing the specificity of herbicides by using mixtures and lowering the total dose for weed control, while at the same time delaying the development of herbicide resistance by using mixtures with different molecular targets.

5.5 Toxicokinetic Interactions Between Pesticides

The majority of interaction studies are conducted on animals using high doses, while data on toxicokinetic interaction in humans exposed to pesticides, perhaps through residues in food, are barely available. The route of entry of a pesticide in humans and animals is different, leading to major effects on its concentration reaching the systemic circulation and nervous system. The entry of a pesticide into humans through the gastrointestinal tract usually allows rapid absorption, as compared to the skin barrier, which results in only very slow and incomplete absorption. A number of the animal studies that have reported interactions between pesticides have involved intraperitoneal injection or other routes that allow rapid and complete delivery of the dose into the general circulation. Route of administration differences can give rise to problems of interpretation with respect to the likelihood of interactions in exposed humans. Some relevant studies have already been discussed earlier in the chapter. Lauwerys and Murphy (1969) reported that the pretreatment of rats with tri-O-tolyl phosphate (which is not a pesticide) decreased the binding of paraoxon to non-vital tissue in rat liver and plasma, and increased the acute toxicity of paraoxon in rats. Although this probably represents a toxicokinetic interaction, the time course, reversibility and pharmacokinetic basis of this interaction were not defined, and the response of the combination was not compared with adequate paraoxon plasma concentration–response data. There is evidence for a protective effect of aldrin and chlordane against the acute toxicity of the carbamate carbanolate, which, at least in part, is due to the induction of cytochrome P450 (CYP)-mediated carbanolate detoxification (Williams and Casterline, 1970). Piperonyl butoxide, which inhibits CYP activities, abolished the protection afforded by these organochlorine pesticides, and this supports the theory of an underlying metabolic mechanism.

The effects of four OP insecticides on the toxicity of fenobucarb were studied by Takahashi *et al.* (1987). The study design was based on analysis of the ratios of the oral LD_{50} of fenobucarb alone compared with its LD_{50} when co-administered with 4–5% of the LD_{50} of cyanophos, dichlorvos, fenitrothion or MLT. The toxicity of fenobucarb was increased with the phosphorothioates (cyanophos, fenitrothion and MLT), and these also caused an increase in the plasma concentrations of fenobucarb. Inhibition of fenobucarb metabolism with SKF525A, which is a P450 inhibitor, enhanced plasma concentrations of fenobucarb, and enhanced its inhibition of brain acetylcholine esterase, but did not produce an equivalent shift in the LD_{50}. These findings suggest that both toxicokinetic and toxicodynamic interactions may have been occurring at these acutely toxic doses. The relevance of interactions at such high doses to human exposures is questionable. Johnston (1995) reviewed the increased sensitivity to the neurotoxicity of MLT following treatment with prochloraz (a fungicide), and concluded that the mechanistic basis was the induction of the enzymatic oxidation of MLT to its active metabolite.

5.6 Risk of Pesticides Mixtures to the Environment

The State of the Art Report on Mixture Toxicity stated in its Executive Summary (Kortenkamp *et al.*, 2009) that chemicals which present below the level of threshold doses or concentrations of regulatory concern such as ADI or predicted no effect concentrations (PNECs; or NOAELs, no observed effect concentrations (NOECs) or benchmark doses) are considered as safe. However, this does not answer the question of how to ensure sufficient protection against their combined exposures. Pesticides in combination may be risky to the environment and its various components, including human beings.

There is strong evidence that mixture effects may arise when several chemicals are combined at doses or concentrations around, or below, points of departure. In such cases, the analysis of the effect of combinations is calculated using the concept of dose or concentration addition, as discussed earlier in this chapter. The principles of dose (concentration) additivity mean that mixture effects are to be expected even when each chemical is present below zero-effect levels, because it is assumed that all toxicants in the mixture behave as if they were a dilution of one another. Hence, any concentration of any compound needs to be considered because it adds to the mixture concentration. This implies that all compounds contribute to the mixture toxicity in direct proportion to their concentration in the mixture and their individual potency. Whether the individual concentrations in the mixture are above or below the corresponding effect thresholds does not matter. This phenomenon has been termed 'something from nothing'. It has been demonstrated repeatedly for a broad range of mixtures in toxicological and ecotoxicological studies. There is unanimous agreement across all disciplines that, in the case of mixtures of similar compounds, combination effects require special consideration.

Relatively few reports have adequately investigated the nature of interactions that may be occurring between the constituents within a mixture and deviations from additivity. The mixture of pesticides seems simple and may be a single entity; however, if the mixture has many components and/or the composition is variable, the mixture is a complex mixture, as found in the environment. Unless the individual component has been studied for its interaction, there is little information regarding the nature of any interactions that may occur between the individual constituents. Toxicity of pesticides in mixtures has been observed even at very low doses in which the concentrations of the individual constituents are believed to be below independently reported ADIs (and therefore likely to be well below the individual NOAELs). However, the nature of interaction for enhancing toxicity has not been provided for by any of the studies conducted. Conversely, for a particular mixture, any interactions that may occur do not enhance toxicity at these levels of exposure. In many cases involving less complex mixtures, the effects could not be predicted by simple summation of the effects of the single constituents, leading to deviations from additivity. To study the nature of interaction involved in mixture of pesticides, dose–response curves need to be studied. For example, for simple similar action (simple joint action or concentration/dose addition), the effects of a mixture may not appear to be additive either at doses close to or just below the NOAEL, or at high doses where the response is near saturation. A conclusion that effects are summative is valid only if the dose–response is linear. Furthermore, extrapolation of high-dose data to low-dose exposures is not necessarily valid, while dose–response and the potential effects of mixtures at low doses have been considered by some workers for investigating interactions of the constituents of mixtures. For a mixture of chemicals that have the same target organ and the same mode of action, the effect induced by simultaneous administration of levels of chemicals that individually are non-toxic corresponds to what is expected on the basis of simple additivity. However, some adverse effects, albeit minor, may occur when the mixture components are present at a level equivalent to their respective NOAELs, regardless of the mechanism by which the components exert their toxic effect.

Determining the nature of interactions that may occur in simple mixtures is a challenging aspect of the study of toxicology of mixtures; however, for more complicated, complex mixtures, it is important to study how well the dose–response curves have been defined for the individual compounds. The extrapolation of effects and interactions at high doses for risk assessment of low-level exposures may not always be valid. Studies of acute effects are not thought to be helpful because the critical target for acute toxic effects may not be the same as those at lower levels of exposure, particularly over more prolonged periods. Combination effects may simultaneously include both simple addition, and/or less than and more than additive effects, depending on the interactive and non-interactive effects of mixtures. However, the precise mechanism of interaction between components of a mixture is not known. In the mixtures, there may be masking of the existence of some interactions by the effect of another interaction. In an interaction where quantitative data on absorption, distribution, metabolism and excretion and pharmacological and toxicological actions are available, it may be possible to predict the occurrence and extent of an interaction within a simple mixture purely on a mechanistic basis.

5.7 Independent Joint Action of Pesticides in Mixtures

The independent joint action of pesticides can be applicable to a mixture, where the exposure level of a pesticide is in the range of NOAELs, where no additivity and no interaction are observed. Conversely, it has been shown in *in vivo* studies targeting the same organ, that the toxicity of a mixture of pesticides with the same mode of action corresponds to the effect expected on the basis of the additivity assumption, even at levels slightly below the lowest observed adverse effect level (LOAEL) of the individual compounds (Cassee *et al.*, 1996; Jonker *et al.*, 1996; Vonier *et al.*, 1996; Stelzer and Chan, 1999; Tajima *et al.*, 2002). In these cases the dose addition model represents the basic concept to be used for hazard assessment. This model is applicable over the whole range of exposure levels from low non-toxic levels to LOAELs. Some reports have addressed the combined effect of mixtures of pesticides, while a few studies have observed clear cases of potentiation in animals exposed to effect levels of individual compounds. The chances of increased health hazards due to additivity or potentiating interaction of mixtures of pesticides at (low) non-toxic doses of the individual chemicals is very much reduced, since the dose of pesticides to which humans are exposed is generally much lower than the NOAEL. Mixtures of pesticides with similarities in their mode of action or with clear evidence of physico-chemical and/or toxicokinetic interactions, and mixtures of pesticides with no or very small margins of safety may be the exceptions to these rules. It is a well-known fact that young children are often the critical group in exposure assessment because of their relatively high intake of food on a body weight basis. There is a consensus in the field of mixture toxicology that the customary chemical-by-chemical approach to risk assessment might be too simplistic. It is in danger of underestimating the risk of chemicals to human health and to the environment.

The Panel on Plant Protection Products and their Residues (PPR Panel) was held during 2008, under European 2009 and 2013 Food Safety Authority (EFSA), for development of methodology with the purpose of performing cumulative risk assessment for pesticide residues in food. The first report (EFSA, 2008) focused on the evaluation of existing methodologies to determine their suitability for assessment of risks resulting from dietary exposure to multiple pesticide residues, and covered both aspects of hazard and exposure assessment. The active substances were proposed to be grouped into cumulative assessment groups (CAGs) based on their chemical structure, mechanism of pesticidal action, mode/mechanism of mammalian toxicity and common toxic effects. In particular, it was assumed that mixture effects for substances acting via a common mode of action might be predicted by the dose addition concept. In a next step, the PPR Panel tested the methodology proposed in the previous report by preparing a worked example of a dietary risk assessment for a group of triazole pesticides, a well-defined group in terms of structure, pesticidal mode of action and common toxic effects (EFSA, 2009).

For the third report (EFSA PPR Panel, 2013), the PPR Panel further elaborated general methodology and criteria for the establishment of CAGs of pesticides, based on their toxicological profiles. The adopted methodology follows a phenomenological approach based on organ or system toxicity, which consists of including all pesticides causing a specific effect in a CAG. Since mechanistic data supporting decisions regarding similarity of action are rarely available for toxicity endpoints on active substances, this approach takes common organ/system toxicity as an approximation of similar or interrelated modes of action. It was considered that the cumulative effects of pesticides within a CAG could then be assessed by a method derived from the dose addition concept. Nevertheless, the Panel acknowledged that CAGs based on phenomenological effects would be expected to include substances acting via different modes of action, but contributing to a common adverse outcome. Although the Panel noted that '. . . further refinement of grouping may be achieved when data on the precise toxicological mode of action may be available', it pointed out that 'information that justifies any deviation from dose addition might also be necessary to consider for such a refinement' (EFSA PPR Panel, 2013, section 2.2, p. 15).

5.8 Conclusion

Pesticides probably co-occur in the environment as mixtures, posing a risk to human beings and all other organisms present in the environment. Their number in the environment is too large to allow for all of them to be studied. However, single pesticides are assessed for their risk to the environment, ignoring the existence of their mixtures. Several models such as concentration addition, independent action and the TU approach have been used to predict the mixture toxicity of pesticides with similar and dissimilar modes of action. Pesticides with similar modes of action generally lead to additive (summation of doses of the individual compounds) effects. These combination effects are generally larger than the effects of an individual component applied singly. Pesticides with different modes of action may lead to synergistic or antagonistic effects. Pesticides enter aquifers through various pathways leading to joint toxicity of these mixtures. The overall effect of pesticide mixtures, with similar modes of action, may be either stronger or weaker than that predicted by an additivity model; in this situation the term 'interaction' is used conceptually to describe deviation from an additive model. The toxicity of pesticides is altered for stereoisomers, affecting not only the efficacy, but also exposure and risk to humans and the environment. For analysis of any potentially additive or interactive effects of compounds within a mixture, dose–response curves are established for each of the individual components on its own. The toxicology of mixtures is studied using two basic methodological strategies, component interaction and whole-mixture analysis. For analysis of simple mixtures, where constituents are in small numbers and their composition is clearly known, the component interaction analysis (bottom-up approach) can be applied. Where there are numerous components (a complex mixture) and/or no specific knowledge of the composition of a mixture, whole-mixture analysis may be more appropriate. However, such studies cannot define the true extent of interactions between components of the complex mixture without data on fractions of the mixture. The presence of chemicals in the environment below their threshold doses is considered safe; however, this does not imply sufficient protection against exposure to mixtures of these chemicals.

References

Altenburger, R., Boedeker, W., Faust, M. and Grimme, L.H. (1996) Regulations for combined effects of pollutants: Consequences from risk assessment in aquatic toxicology. *Food and Chemical Toxicology* 34, 1155–1157.

Altenburger, R., Walter, H. and Grote, M. (2004) What contributes to the combined effect of a complex mixture? *Environmental Science and Technology* 38, 6353–6362.

Andersen, H.R., Cedergreen, N., Streibig, J.C., Mathiassen, S., Møhlenberg, F. and Vinggaard, A.M. (2006) Combination effects of pesticides. *Plantekongres* (DK Version 1.5.4).

Anderson, T.D. and Zhu, K.Y. (2004) Synergistic and antagonistic effects of atrazine on the toxicity of organophosphorodithioate and organophosphorothioate insecticides to *Chironomus tentans* (Diptera: Chironomidae). *Pesticide Biochemistry and Physiology* 80(1), 54–64.

Arora, S. and Kumar, A. (2015) Binary combinations of organophosphorus pesticides exhibit differential toxicity under oxidised and un-oxidised conditions. *Ecotoxicology and Environmental Safety* 115, 93–100.

Barbash, J.E. (2003) The geochemistry of pesticides. *Treatise on Geochemistry* 9, 541–577.

Belden, J., Gilliom, R. and Lydy, M. (2007) How well can we predict the toxicity of pesticide mixtures to aquatic life? *Integrated Environmental Assessment and Management* 3, 364–372.

Bellin, J.S. and Barnes, D.G. (1987) *Interim Procedures for Estimating Risk Associated with Exposures to Mixtures of Chlorinated Dibenzo-p-dioxins and Dibenzofuran (CDDs and CDF).* US Environmental Protection Agency Report No. EPA/625/3-87/012. Washington, DC, Risk Assessment Forum, US Environmental Protection Agency.

Benting, J. and Nauen, R. (2004) Biochemical evidence that an S431F mutation in acetylcholinesterase-1 of *Aphis gossypii* mediates resistance to pirimicarb and omethoate. *Pesticide Management Science* 60(11), 1051–1055.

Birnbaum, L. (1999) TEFs: A practical approach to a real-world problem. *Human and Ecological Risk Assessment* 5, 13–24.

Bliss, C.I. (1939) The toxicity of poisons applied jointly. *Annals of Applied Biology* 26, 585–615.

Borgert, C., Quill, T.F., McCarty, L.S. *et al.* (2004) Can mode of action predict mixture toxicity for risk assessment? *Toxicology and Applied Pharmacology* 201, 85–96.

Casida, J.E. and Quistad, G.B. (2004) Organophosphate toxicology: Safety aspects of nonacetylcholinesterase secondary targets. *Chemical Research in Toxicology* 17(8), 983–998.

Casida, J.E., Baron, R.L., Eto, M. and Engel, J.L. (1963) Potentiation and neurotoxicity induced by certain organophosphates. *Biochemical Pharmacology* 12, 73–83.

Cassee, F.R., Groten, J.P. and Feron, V.J. (1996) Changes in the nasal epithelium of rats exposed by inhalation to mixtures of formaldehyde, acetaldehyde and acrolein. *Fundamental and Applied Toxicology* 29, 208–218.

Cassee, F.R., Groten, J.P., van Bladeren, P.J. and Feron, V.J. (1998) Toxicological evaluation and risk assessment of chemical mixtures. *Critical Reviews in Toxicology* 28, 73–101.

Cassee, F.R., Suhnel, J., Groten, J.P. and Feron, V.J. (1999) Toxicology of chemical mixtures. In: Ballantyne, B., Marrs, T.C. and Syversen, T. (eds) *General and Applied Toxicology*, vol. 1, Macmillan Reference Ltd, London, pp. 303–319.

Caughlan, A., Newhouse, K., Namgung, U. and Xia, Z.G. (2004) Chlorpyrifos induces apoptosis in rat cortical neurons that is regulated by a balance between p. 8 and ERK/JNK MAP kinases. *Toxicological Sciences* 78(1), 125–134.

Chou, T.C. (2006) Theoretical basis, experimental design, and computerized simulation of synergism and antagonism in drug combination studies. *Pharmacological Reviews* 58, 621–681.

Chou, T.C. and Talalay, P. (1984) Quantitative analysis of dose–effect relationships: The combined effects of multiple drugs or enzyme inhibitors. *Advances in Enzyme Regulation* 22, 27–55.

Cohen, S.D. and Murphy, S.D. (1971) Carboxylesterase inhibition as an indicator of malathion potentiation in mice. *Journal of Pharmacology and Experimental Therapeutics* 176, 733–742.

Deneer, J.W. (2000) Toxicity of mixtures of pesticides in aquatic systems. *Pest Management Science* 56, 516–551.

DuBois, K.P. (1961) Potentiation of the toxicity of organophosphorus compounds. *Advances in Pest Control Research* 4, 117–151.

Eadon, G., Kaminsky, L., Silkworth, J., Aldous, K., Hilker, D., O'Keefe, P., Smith, R., Gierthy, J., Hawley, J., Kim, N. and DeCaprio, A. (1986) Calculation of 2,3,7,8-TCDD equivalent concentrations of complex environmental contaminant mixtures. *Environmental Health Perspectives* 70, 221–227.

Eggen, R.I.L., Behra, R., Burkhardt Holm, P., Escher, B.I. and Schweigert, N. (2004) Challenges in ecotoxicology: Mechanistic understanding will help overcome the newest challenges. *Environmental Science and Technology* 35, 58A–64A.

Ensenbach, U. and Nagel, R. (1995) Toxicity of complex chemical mixtures: Acute and long-term effects on different life stages of zebrafish (*Brachydanio rerio*). *Ecotoxicology and Environmental Safety* 30, 151–157.

European Commission (2011) Toxicity and assessment of chemical mixtures. Scientific Committee on Consumer Safety, Emerging and Newly Identified Health Risks, and Health and Environmental Risks. Available at: http://ec.europa.eu/health/scientific_committees/environmental_risks/docs/scher_o_155.pdf (accessed 15 August 2018).

European Food Safety Authority (EFSA) (2008) Opinion of the Scientific Panel on Plant Protection Products and their Residues to evaluate the suitability of existing methodologies and, if appropriate, the identification of new approaches to assess cumulative and synergistic risks from pesticides to human health with a view to set MRLs for those pesticides in the frame of Regulation (EC) 396/2005. *The EFSA Journal* 704, 1–85.

European Food Safety Authority (EFSA) (2009) Panel on Plant Protection Products and their Residues (PPR Panel) Scientific Opinion on a request from EFSA on risk assessment for a selected group of pesticides from the triazole group to test possible methodologies to assess cumulative effects from exposure through food from these pesticides on human health. *The EFSA Journal* 7(9), 1167.

European Food Safety Authority (EFSA) PPR Panel (2013) Different mode of action. Scientific opinion. Scientific opinion on the relevance of dissimilar mode of action and its appropriate application for cumulative risk assessment of pesticides residues in food. Toxicity of Pesticides-mixtures. *The EFSA Journal* 11(12), 3472. Available at: www.efsa.europa.eu/efsajournal (accessed 1 October 2018).

Faust, M. and Scholze, M. (2004) Competing concepts for the prediction of mixture toxicity: do the differences matter for regulatory purposes? *Workpackage 7*. European Commission, Bremen, German, p. 64.

Faust, M., Altenburger, R., Boedeker, W. and Grimme, L.H. (1994) Algal toxicity of binary combinations of pesticides. *Bulletin of Environmental Contamination and Toxicology* 53, 134–141.

Faust, M., Altenburger, R., Backhaus, T., Boedeker, W., Scholze, M. and Grimme, L.H. (2000) Predictive assessment of the aquatic toxicity of multiple chemical mixtures. *Journal of Environmental Quality* 29, 1063–1068.

Forget, J., Pavillon, J.F., Beliaeff, B. and Bocquene, G. (1999) Joint action of pollutant combinations (pesticides and metals) on survival (LC_{50} values) and acetylcholinesterase activity of *Tigriopus brevicornis* (Copepoda: Harpacticoida). *Environmental Toxicology and Chemistry* 18, 912–918.

Frasco, M.F. and Guilhermino, L. (2002) Effects of dimethoate and betanaphthoflavone on selected biomarkers of *Poecilia reticulata*. *Fish Physiology and Biochemistry* 26(2), 149–156.

Frawley, J. P., Fuyat, H.N., Hagan, E.C., Blake, J.R. and Fitzhugh, O.G. (1957) Marked potentiation in mammalian toxicity from simultaneous administration of two anticholinesterase compounds. *Journal of Pharmacology and Experimental Therapeutics* 121, 96–106.

Gaughan, L.C., Engel, J. and Casida, J. (1980) Pesticide interactions: Effects of organophosphorus pesticides on the metabolism, toxicity, and persistence of selected pyrethroid insecticides. *Pesticide Biochemistry* 14, 18–25.

George, T.K., Liber, K., Solomon, K.R. and Sibley, P.K. (2003) Assessment of the probabilistic ecological risk assessment-toxic equivalent combination approach for evaluating pesticide mixture toxicity to zooplankton in outdoor microcosms. *Archives of Environmental Contamination and Toxicology* 45, 453–461.

Groten, J.P., Feron, V.J. and Sühnel, J. (2001) Toxicology of simple and complex mixtures. *Trends in Pharmacological Sciences* 22, 316–322.

Gupta, R.C. (2006) *Toxicology of Organophosphate and Carbamate Compound*. Elsevier Academic Press, Amsterdam.

Heindel, J.J., Chapin, R.E., Gulati, D.K., George, J.D., Price, C.J., Marr, M.C., Myers, C.B., Barnes, L.H., Fail, P.A., Grizzle, T.B., Schwetz, B.A. and Yang, R.S.H. (1994) Assessment of the reproductive and developmental toxicity of pesticide/fertilizer mixtures based on confirmed pesticide contamination in California and Iowa groundwater. *Fundamental and Applied Toxicology* 22, 605–621.

Heindel, J.J., Chapin, R.E., George, J.D., Galati, D.K., Fail, P.A., Barnes, L.H. and Yang, R.S.H. (1995) Assessment of the reproductive toxicity of a complex mixture of 25 groundwater contaminants in mice and rats. *Fundamental and Applied Toxicology* 25, 9–19.

Hewlett, P.S. and Plackett, R.L. (1952) Similar joint action of insecticides. *Nature* 169, 198–199.

Hughes, I. and Woods, H.F. (2002) Risk assessment of mixture of pesticides and similar substances. In: *Committee on Toxicity of Chemicals in Food, Consumer Products and the Environment*. Crown Copyright 2002. Food Standards Agency, London.

Ito, N., Hasegawa, R., Imaida, K., Kurata, A., Hagiwara, A. and Shirai, T. (1995a) Effect of ingestion of 20 pesticides in combination at acceptable daily intake levels on rat liver carcinogenesis. *Food and Chemical Toxicology* 33, 159–163.

Ito, N., Hagiwara, A., Tamano, S., Hasegawa, R., Imaida, K., Hirose, M. and Shirai, T. (1995b) Lack of carcinogenicity of pesticide mixtures administered in the diet at acceptable daily intake (ADI) dose levels in rats. *Toxicology Letters* 82/83, 513–520.

Ito, N., Hagiwara, A., Tamano, S., Futakuchi, M., Imaida, K. and Shirai, T. (1996) Effects of pesticide mixtures at the acceptable daily intake levels on rat carcinogenesis *Food and Chemical Toxicology* 34, 1091–1096.

Johnston, G. (1995) The study of interactive effects of pollutants: A biomarker approach. *Science of the Total Environment* 171, 205–212.

Jonker, D., Woutersen, R.A. and Feron, V.J. (1996) Toxicity of mixtures of nephrotoxicants with similar or dissimilar mode of action. *Food Chemical Toxicology* 34, 1075–1082.

Kaushik, S., Inderjit, Streibig, J.C. and Cedergreen, N. (2006) Activities of mixtures of soil-applied herbicides with different molecular targets. *Pest Management Science* 62(11), 1092–1097.

Kim, S.J., Kim, J.E., Ko, B.H. and Moon, I.S. (2004) Carbofuran induces apoptosis of rat cortical neurons and down-regulates surface alpha 7 subunit of acetylcholine receptors. *Molecules and Cells* 17(2), 242–247.

Kok, F.N. and Hasirci, V. (2004) Determination of binary pesticide mixtures by an acetylcholinesterase-choline oxidase biosensor. *Biosensors and Bioelectronics* 19, 661–665.

Kortenkamp, A., Backhaus, T. and Faust, M. (2009) *State of the Art Report on Mixture Toxicity*. Final Report, Executive Summary. Contractor number 070307/2007/485103/ETU/D.1 The School of Pharmacy, University of London. Available at: http://ec.europa.eu/environment/chemicals/effects/pdf/report_mixture_toxicity.pdf (accessed 1 October 2018).

Laetz, C.A., Baldwin, D.H., Collier, T.K., Hebert, V., Stark, J.D. and Scholz, N.L. (2009) The synergistic toxicity of pesticide mixtures: Implications for risk assessment and the conservation of endangered Pacific salmon. *Environmental Health Perspectives* 117, 348–353.

Lauwerys, R.R. and Murphy, S.D. (1969) Interaction between paraoxon and tri-o-tolyl phosphate in rats. *Toxicology and Applied Pharmacology* 14, 348–357.

Merino-Garcia, D., Kusk, K.O. and Christensen, E.R. (2003) Joint toxicity of similarly and dissimilarly acting chemicals to *Daphnia magna* at different response levels. *Archives of Environmental Contamination and Toxicology* 45(3), 289–296.

Moore, D.R.J. and Teed, S.R. (2012) Risks of carbamate and organophosphate pesticide mixtures to salmon in the Pacific Northwest. *Integrated Environmental Assessment and Management* 9(1), 70–78.

Müller, T.A. and Kohler, H.P.E. (2004) Chirality of pollutants – effects on metabolism and fate. *Applied Microbiology and Biotechnology* 64, 300–316.

Murphy, S.D. and Dubois, K.P. (1957) Quantitative measurement of inhibition of the enzymatic detoxification of malathion by EPN (ethyl p-nitrophenyl thionobenzenephosphonate). *Proceedings of the Society for Experimental Biology and Medicine* 96, 813–818.

Murphy, S.D., Anderson, R.L. and Dubois, K.P. (1959) Potentiation of toxicity of malathion by triortotolyl phosphate. *Proceedings of the Society for Experimental Biology and Medicine* 100, 483–487.

National Research Council (1989) *Drinking Water and Health, Volume 9: Selected Issues in Risk Assessment*. The National Academies Press, Washington, DC. DOI: https://doi.org/10.17226/773.

Oskarsson, A. and Lind, B. (1985) Increased lead levels in brain after long-term treatment with lead and dithiocarbamate or thiuram derivatives in rats. *Acta Pharmacologica Toxicologica* 56, 309–315.

Pena-Llopis, S., Ferrando, M.D. and Pena, J.B. (2003) Fish tolerance to organophosphate-induced oxidative stress is dependent on the glutathione metabolism and enhanced by N-acetylcysteine. *Aquatic Toxicology* 65(4), 337–360.

Phyu, Y.L., Palmer, C.G., Warne, M., Hose, G.C., Chapman, J.C. and Lim, R.P. (2011) A comparison of mixture toxicity assessment: Examining the chronic toxicity of atrazine, permethrin and chlorothalonil in mixtures to *Ceriodaphnia cf. Dubia*. *Chemosphere* 85, 1568–1573.

Polak, R.L. and Cohen, M.E. (1969) The influence of tri-orthocresylphosphate on the distribution of 32P in the body of the rat after the injection of 32P-sarin. *Biochemical Pharmacology* 18, 813–820.

Pope, C.N. (1999) Organophosphorus pesticides: do they all have the same mechanism of toxicity? *Journal of Toxicology and Environmental Health B* 2, 161–181.

Reffstrup, T.K., Larsen, J.C. and Meyer, O. (2010) Risk assessment of mixtures of pesticides. Current approaches and future strategies. *Regulatory Toxicology and Pharmacology* 56, 174–192.

Richardson, J.R., Chambers, H.W. and Chambers, J.E. (2001) Analysis of the additivity of in-vitro inhibition of cholinesterase by mixtures of chlorpyrifos-oxon and azinphosmethyl-oxon. *Toxicology and Applied Pharmacology* 172(2), 128–139.

Rodney, S.I., Teed, S.R. and Moore, D.R.J. (2013) Estimating the toxicity of pesticide mixtures to aquatic organisms: A review. *Human and Ecological Risk Assessment* 19(6), 1557–1575.

Rosenberg, P. and Coon, J.M. (1958) Potentiation between cholinesterase inhibitors. *Proceedings of the Society of Experimental Biology and Medicine* 97, 836–839.

Safe, H.S. (1998) Hazard and risk assessment of chemical mixtures using the toxic equivalency factor approach. *Environmental Health Perspectives* 106(4), 1051–1058.

Safe, S.H. (1999) Development and application of TEFs. *Human and Ecological Risk Assessment: An International Journal* 5, 9–12.

Scholz, N.L., Truelove, N.K., Labenia, J.S., Baldwin, D.H. and Collier, T.K. (2006) Dose-additive inhibition of Chinook salmon Acetylcholinesterase activity by mixtures of organophosphate and carbamate insecticides. *Environmental Toxicology and Chemistry* 25(5), 1200–1207.

Schulz, R. (2004) Field studies on exposure, effects, and risk mitigation of aquatic nonpoint-source insecticide pollution: A review. *Journal of Environmental Quality* 33, 419–448.

Seume, F.W. and O'Brien, R.D. (1960a) Metabolism of malathion by rat tissue preparations and its modification by EPN. *Journal of Agricultural and Food Chemistry* 8(1), 36–41.

Seume, F.W. and O'Brien, R.D. (1960b) Potentiation of the toxicity to insects and mice of phosphorothionates containing carboxylester and carboxyamide groups. *Toxicology and Applied Pharmacology* 2, 495–503.

Skulachev, V.P. (1998) Cytochrome c in the apoptotic and antioxidant cascades. *FEBS Letters* 423(3), 275–280.

Spehar, R.L. and Fiandt, J.T. (1986) Acute and chronic effects of water quality criteria-based metal mixtures on three aquatic species. *Environmental Toxicology and Chemistry* 5, 917–931.

Stelzer, A. and Chan, H.M. (1999) The relative estrogenic activity of technical toxaphene mixture and two individual congeners. *Toxicology* 138, 69–80.

Syberg, K., Elleby, A., Pedersen, H., Cedergreen, N. and Forbes, V.E. (2008) Mixture toxicity of three toxicants with similar and dissimilar modes of action to *Daphnia magna*. *Ecotoxicology and Environmental Safety* 69, 428–436.

Tahara, M., Kubota, R., Nakazawa, H., Tokunaga, H. and Nishimura, T. (2005) Use of cholinesterase activity as an indicator for the effects of combinations of organophosphorus pesticides in water from environmental sources. *Water Research* 39, 5112–5118.

Tajima, O., Schoen, E.D., Feron, V.J. and Groten, J.P. (2002) Statistically designed experiments in a tiered approach to screen mixtures of *Furarium* mycotoxins for possible interactions. *Food and Chemical Toxicology* 40, 685–695.

Takahashi, H., Kato, A., Yashamita, E., Naito, Y., Tsuda, S. and Shirasu, Y. (1987) Potentiations of N-methylcarbamate toxicities by organophosphorus insecticides in male mice. *Fundamental and Applied Toxicology* 8, 139–146.

Ulrich, E.M., Morrison, C.N., Goldsmith, M.R. and Foreman, W.T. (2012) Chiral pesticides: Identification, description, and environmental implications. *Reviews of Environmental Contamination and Toxicology* 217, 1–74.

US Environmental Protection Agency (EPA) (1987) Aldicarb (sulfoxide and sulfone). Health advisory (draft). Office of Drinking Water, US Environmental Protection Agency Washington, DC, p. 16.

Veldstra, H. (1956) Synergism and potentiation with special reference to the combination of structural analogues. *Pharmacological Reviews* 8, 339–387.

Vonier, P.M., Crain, D.A., McLachlan, J.A., Guillette, L.J. and Arnold, S.F. (1996) Interaction of environmental chemicals with the estrogen and progesterone receptors from the oviduct of the American alligator. *Environmental Health Perspectives* 104, 1318–1322.

Waiser, M. and Holm, J. (2005) *Report on the Cumulative and Synergistic Effects of Complex Mixtures of Herbicides on Wetland Biodiversity: Implications for Beneficial Herbicide Application Practice and Environmental Standards for Prairie Wetlands.* NAESI Technical Series Report No. 1–16 and NWRI Technical Report AEP TN-05-003. Environment Canada Burlington, ON, Canada.

Williams, C.H. and Casterline, J.L. (1970) Effects of toxicity and on enzyme activity of the interactions between aldrin, chlordane, piperonyl butoxide and banol in rats. *Experimental Biology and Medicine* 135, 46–50.

Woods, M., Kumar, A. and Correll, R. (2002) Acute toxicity mixtures of chlorpyrifos, profenefos, and endosulfan to *Ceriodaphnia dubia*. *Bulletin of Environmental Contamination and Toxicology* 68, 801–808.

Zhang, A., Xuemei, X., Ye, J., *et al.* (2010) Stereoselective toxicity of malathion and its metabolites, malaoxon and isomalathion. *Environmental Chemistry Letters* 9(3), 369–373.

6 Toxicity of Pesticides in Mixtures with Different Modes of Action

6.1 Introduction

Theoretically, mixtures of insecticides (with different modes of action) should usually prove very effective in resistance management programmes (Roush, 1993). The reason being, if the resistance mechanisms to each insecticide in the mixture are independent, the chances of resistance to both insecticides at the same time would be minimal (Curtis, 1985). Mixtures with different modes of action generally behave synergistically. According to Corbett (1974), one product or toxicant in the mixture interferes with the metabolic detoxification of the other toxicant, thereby synergizing the toxicity of the latter toxicant. Kulkarni and Hodgson (1980) investigated potentiation of toxicity in insecticide mixtures, as observed from the mixture of pyrethroid and organophosphate classes, which may be potential or competitive substrates for the same oxidase.

6.2 Pesticides in Mixtures with Different Modes of Action

Synergism may occur when one pesticide interferes with the metabolic detoxification of another pesticide (Corbett, 1974; Kulkarni and Hodgson, 1980). Certain organophosphate insecticides bind to the active site on esterase enzymes responsible for detoxification of parathyroid insecticides (Ascher *et al.*, 1986; Ishaaya *et al.*, 1987; Bynum *et al.*, 1997; Gunning *et al.*, 1999; Ahmad, 2004; Zalom *et al.*, 2005; Ahmad *et al.*, 2008; Ahmad, 2009), and so organophosphate insecticides can be useful synergists for pyrethroids (Chapman and Penman, 1980; Brattsten *et al.*, 1986; Ishaaya *et al.*, 1987; Gunning *et al.*, 1999; Martin *et al.*, 2003; Zalom *et al.*, 2005; Attique *et al.*, 2006). Moreover, it was also proposed that when pyrethroid/organophosphate mixtures were applied against resistant insects, organophosphate insecticide might bind to the monooxygenase resulting from the activation of molecules, and then prevent the binding and subsequent degradation of the pyrethroid insecticide by monooxygenase enzymes. These enzymes when binding to the organophosphate insecticide could also lead to non-toxic metabolites via the process of hydroxylation of either oxon or thioate forms, which ultimately results in their degradation via oxidative ester cleavage. In this way, the binding of monooxygenase enzymes with organophosphate insecticide would prevent or delay the degradation, and enhance the toxicity of pyrethroid insecticide via a competitive substrate inhibition mechanism (Kulkarni and Hodgson, 1980; Attique *et al.*, 2006). Previously it has been assumed that organophosphates, when used in combination with pyrethroids, inhibit the enzymes (monooxygenases and/or esterases) responsible for metabolic detoxification in different insect pests (Bryne and Devonshire, 1991; Martin *et al.*, 2003; Montella *et al.*, 2012).

Arora *et al.* (2017) evaluated toxicity of organophosphates and carbamates under oxidized and unoxidized conditions, as individual compounds and as binary mixtures, against the response of the housefly (*Musca domestica*) head acetylcholinesterase (AChE) (HF-AChE). In addition, the selected organophosphates and carbamates were evaluated for their toxicity in binary combinations with two synthetic pyrethroids, deltamethrin (DLT) and cypermethrin (CYP) at fixed concentrations of 0.1 and 10 mg/l. The toxicological interaction of five organophosphates with two carbamate pesticides was evaluated under oxidized and unoxidized conditions using a toxic unit (TU) approach and a concentration addition model. Pyrethroid combinations were assessed only under oxidized conditions. Since organophosphates and carbamates act via a similar mechanism of inhibition of AChE, a dose additive effect was expected, but not conclusively found. Triazophos (TRZ) with either carbofuran (CBF) or carbaryl (CRB) exhibited synergism under oxidized and unoxidized conditions, but the degree of synergism was stronger under unoxidized conditions. Additivity was exhibited by CBF+ monocrotophos (MCP), CRB + MCP, CRB + malathion (MLT) and CBF + MCP under unoxidized conditions and CRB + MCP and CRB + CPF under oxidized conditions. Pyrethroids in combination with organophosphates (TRZ, MLT and CPF) were highly synergistic.

Khan *et al.* (2013) investigated the use of two enzyme inhibitors, namely piperonyl butoxide (PBO) and S,S,S-tributylphosphorotrithioate (DEF), against the field population of *M. domestica* to overcome resistance to bifenthrin, cypermethrin, deltamethrin, chlorpyrifos, profenofos and emamectin. PBO significantly reduced LC values (non-overlapping of 95% CIs) for bifenthrin from 127.62 to 21.94 (6-fold), cypermethrin from 317.45 to 43.86 (7-fold), deltamethrin 50 from 209.12 to 38.76 (5-fold), chlorpyrifos from 16.31 to 6.19 (3-fold), profenofos from 39.85 to 14.77 (3-fold) and emamectin from 124.59 to 60.61 (2-fold). However, PBO did not produce a synergistic effect with fipronil. DEF also reduced LC values for bifenthrin from 127.62 to 64.35 (2-fold), cypermethrin from 317.45 to 162.50 (2-fold), deltamethrin from 209.12 to 100.09 (2-fold), chlorpyrifos from 16.31 to 3.66 (4-fold) and profenofos from 39.85 to 5.48 (7-fold). However, DEF did not show synergism with emamectin and fipronil (Porter, 2002). Moreover, PBO and DEF did not synergize any insecticide against the laboratory susceptible strain.

El-Demerdash (2011) conducted studies on lipid peroxidation and antioxidant enzyme activities and AChE, for antioxidant status and oxidative stress using male rat brain as biomarker, with an insecticide mixture of organophosphate and pyrethroids (fenitrothion 25%, lambda cyhalothrin 2.5% and piperonyl butoxide 6%). The toxic effects of fenitrothion probably occur through the generation of reactive oxygen species (ROS) causing damage to various membranous components of the cell (Goel *et al.*, 2005). The general site of action of pyrethroids is biological membranes due to their higher hydrophobicity than other classes of insecticides (Michelangeli *et al.*, 1990). The mixture was observed to cause cytotoxic effects on rat brain. It has the capability to induce oxidative damage as evidenced in terms of increased lipid peroxidation and perturbations in various antioxidant enzymes, and the effect was pronounced with high doses and long exposure time. The results indicated a risk of organ damage during exposure to a combination of insecticides.

6.3 Models for Predicting Mixture Toxicity of Pesticides

The response addition model does not always accurately predict mixture toxicity for chemicals with differing modes of action. Pape-Lindstrom and Lydy (1997) used a TU approach to test the response addition model for mixtures of pesticides with differing modes of action. Atrazine was tested in binary and ternary combinations with methoxychlor, organochlorine and methyl parathion organophosphate insecticides. It was also tested in binary combinations with additional organophosphates. The TU for atrazine was set slightly below its water solubility limits (TU = 20 ppm), which is well below its 96-h 50% effective concentration for the fourth instar of the midge, *Chironomus tentans*. The 96-h acute toxicity was found to produce synergistic (greater than additive) toxicity in a binary mixture with methyl parathion. Less than additive toxicity was found for the combination of atrazine with methoxychlor (likely due to the low TU assigned to atrazine). The ternary combination of atrazine + methyl parathion + methoxychlor was found to be marginally synergistic. The toxicity tests of atrazine in binary combinations with other organophosphates indicated more than additive toxicity for all compounds except mevinophos. A synergistic relationship was observed between atrazine and various organophosphate insecticides.

6.3.1 Aquatic ecotoxicology

The organisms in the aquatic environment are exposed to more complex mixtures of pesticides with high concentrations occurring in pulses associated with spray and rain events, in contrast to the terrestrial environment (Cedergreen and Rasmussen, 2017). The use of mixture toxicity models has been emphasized to quantify the effect of these complex mixtures. The standard is to use concentration addition (also called dose addition), where all co-occurring pesticides are transformed into one common unit, which is summed and used as the joint exposure concentration. Using this approach, both herbicides and insecticides can reach joint concentrations during storm flow, which can cause measurable effects in the plant, algae and invertebrate communities. Fungicide concentrations rarely reach the same level as predicted environmental effects, but this is likely because traditional ecotoxicological tests do not reflect aquatic fungi communities, which are likely the most sensitive to these pesticides. Except for extreme incidents, pesticide occurrences in agricultural catchments will rarely lead to acute extermination of the majority of species. There is no doubt, however, that peak pesticide concentrations can affect the species communities on sub-lethal endpoints such as growth, emergence times, feeding and drift behaviour, which ultimately may translate into altered community structure and function. In addition to the low-dose effects of pesticides, other stressors of both a chemical and physical nature also affect aquatic communities. Therefore, ensuring a diverse aquatic community requires a broad focus including both combatting excessive concentrations of pesticides and other chemicals and improving the physical habitat on a local scale as well as on a catchment scale. Only holistic approaches, including all stressors of the aquatic ecosystem, will sufficiently safeguard aquatic ecosystems and potentially recreate lost habitats, as well as increase populations of threatened species.

In more than 50% of cases of aquatic toxicity of commercial pesticide mixtures, the predicted and observed mixture toxicity deviated by less than factor 2. An indication for a synergistic interaction was only detected with regard to algal growth inhibition for mixtures of fungicides that inhibit different enzymes of ergosterol biosynthesis. The greatest degree of compliance between prediction and observation was found for the acute toxicity of fungicidal products towards *Daphnia* and fish, while the greatest degree of underestimation of product toxicity occurred for the acute toxicity of herbicidal products towards *Daphnia* and fish. Using the lowest available toxicity measures within taxonomic groups as the most conservative approach resulted in a bias towards overestimation of product toxicity, but did not eliminate cases of considerable underestimation of product toxicity (Coors and Frische, 2011).

6.3.2 Herbicide mixtures

Studies on dose–response relationships are important when determining the herbicide efficacy and mode of action. In order to improve weed control efficacy and reduce the application costs, the use of herbicide mixtures has become popular in many countries (Kudsk, 2002). This can often save on costs of application, expand the control of a diverse weed flora, slow down the selection pressure to delay development of herbicide resistant weeds, and be more efficacious than expected from the effect of single herbicides. While applying herbicide binary mixtures in a test plant, the conjoined effect of both compounds are considered. Streibig and Jensen (2000) described two reference models for assessment of the effect of herbicide mixtures, to be used in different situations. The additivity of doses, that is, horizontal assessment, uses the Additive Dose model, ADM. Another model, the Multiplicative Survival model (MSM), does not require horizontal assessment of dose–response curves. The ADM implicitly assumes that the herbicides have the same mode of action and do not interfere with each other's action at the binding site. However, we cannot expect the mixtures to follow ADM, if the compounds have the same mode of action but compete for the same binding site. The MSM, however, implicitly assumes that the herbicides elicit their effect independently of each other. Consequently, the choice of reference model could be based on the known mode of action of the mixed compounds. The choice between MSM and ADM, the two biologically reasonable models, could be based on the investigator's perception of how the herbicides act in the plants. Kaushik *et al.* (2006) assessed joint action of soil-applied herbicide mixtures with similar or different modes of action using the ADM. The herbicides chlorsulfuron, metsulfuron-methyl, pendimethalin and pretilachlor, applied either singly or in binary mixtures, were used on rice (Oryza sativa L.) (details provided in section 5.4.3 in Chapter 5). The mixtures of pretilachlor and sulfonyl ureas were observed with synergistic interactions, whereas mixtures of pendimethalin and sulfonyl ureas exhibited either antagonistic or additive activities. Hence, there is a large potential for both increasing the specificity of herbicides by using mixtures and lowering the total dose for weed control, while at the same time delaying the development of herbicide resistance by using mixtures with different molecular targets.

Two basic concepts have been established for predicting additive joint effects based on the known toxicity of the individual mixture components: the concept of concentration addition for mixtures of substances with similar modes of action and the concept of independent action (also called response addition) for mixtures of substances with dissimilar modes of action (Calamari and Vighi, 1992; Altenburger *et al.*, 2004; Backhaus *et al.*, 2004). Independent action is the appropriate conceptual assumption for predicting the joint toxicity of mixtures composed of substances with dissimilar modes of toxic action (Bliss, 1939; De Zwart and Posthuma, 2005), while the concentration addition prediction will often overestimate the toxicity of such mixtures (Altenburger *et al.*, 1996; Groten, 2000; Chèvre *et al.*, 2006; Junghans *et al.*, 2006). As deduced theoretically, the mixture toxicity prediction of the independent action and the concentration addition concept can deviate at the most by factor *n*, being the number of mixture components (Junghans *et al.*, 2006). Hence, the difference between the prediction concepts cannot explain much of the deviation between predicted and measured mixture toxicity observed here for the investigated combination products, which contained mostly two and at maximum four components (i.e. active substances).

6.4 Toxic Effects of Pesticide Mixtures in the Environment

Thiruchelvam *et al.* (2000a; 2000b) studied the combined impact of a herbicide (paraquat) and a fungicide (maneb) on male C57BL/6 mice. It was observed that decreases in motor activity immediately following injections were observed more consistently with combined exposures of maneb/paraquat. Levels of dopamine and metabolites and dopamine turnover were slightly increased immediately post-injection by combined exposures compared to maneb alone. The authors concluded that these results demonstrated potentiating effects on nigrostriatal dopamine systems of combined exposures to paraquat and maneb. Synergistic effects were observed for the combination of paraquat/maneb, expressing preferentially in the nigrostriatal dopamine system, suggesting that such mixtures could play a role in the aetiology of Parkinson's disease. Gaughan *et al.* (1980) reported the effects of organophosphate pesticides (profenofos, sulprofos, ethyl p-nitrophenyl thionobenzenephosphonate (EPN) and S,S,S-tributylphosphorotrithioate given by intraperitoneal injection) on liver pyrethroid esterase and on the LD_{50} of fenvalerate, malathion, trans-permethrin and cis-cypermethrin (given by intraperitoneal injection). The organophosphates inhibited the esterase, and at the same doses enhanced the toxicity of fenvalerate, malathion and cis-cypermethrin, but not of trans-permethrin. The studies demonstrated a toxicokinetic interaction in which inhibition of metabolism caused increased toxicity.

Das *et al.* (2007) conducted *in vitro* experiments to evaluate cytotoxicity and genotoxicity damage caused by pesticides and their mixtures in humans by culturing human peripheral blood lymphocytes. The LC_{50} values for cytotoxicity, using standard trypan blue dye exclusion were 4.18, 5.76 and 7.5 µM for endosulfan, carbofuran and monocrotophos, respectively. When combined in equimolar concentrations, the LC_{50} values for cytotoxicity were 0.7, 0.9 and 1.0 µM for monocrotophos + carbofuran, endosulfan + monocrotophos, and endosulfan + carbofuran, respectively, using the method. Using a standard alkaline comet assay procedure, high concentrations of individual pesticides (0.5–4.0 µM) caused significant DNA damage, while lower concentrations (0.05–0.5 µM) of their binary mixtures also caused the same effect. The analysis of genotoxicity may serve an important biomarker for occupational and household exposure to pesticides, especially mixtures of pesticides, with different modes of action. A number of research articles have described the toxic effects of mixtures. However, relatively few studies have adequately investigated the nature of interactions that may be occurring between the constituents within a mixture and deviations from additivity. One of the articles by Abou-Donia *et al.* (1996a) has received considerable attention in the media in relation to chemical exposures and their possible consequences during the Gulf War. The paper reports greater toxicity from a combination of pyridostigmine (a reversible AChE inhibitor given orally), N,N-diethyl-meta-toluamide (DEET; an insect repellent given by subcutaneous injection) and permethrin (a synthetic pyrethroid insecticide given by subcutaneous injection). Fixed doses were studied alone and in combinations, and their effects on toxicity in hens, and inhibition of plasma AChE were reported. However, a poor relationship appeared between the potency to inhibit plasma AChE and toxicity outcomes. No toxicokinetic data have been provided by the authors to support this hypothesis. Despite this, the authors hypothesized that 'competition for liver and plasma esterases by these compounds leads to their decreased breakdown and increased transport of the parent compound to nervous tissues' (Abou-Donia *et al.*, 1996a, p. 1).

Varga *et al.* (1999) reported the individual and combined effects of the insecticide Sumithion 50 EC (50% fenitrothion) and the herbicide Fusilade S (12.5% fluazifop-P-butyl) in a study of embryotoxicity and teratogenicity in pheasant (*Phasianus colchicus* mongolicus et torquatus) eggs. Eggs were injected with 0.1-ml volumes of various concentrations of Sumithion 50 EC or Fusilade S (0.033, 0.33, 3.33% and 0.1, 1.0, 10.0%, respectively) or mixtures (Sumithion 50 EC + Fusilade S; 0.33% + 0.1%, 0.33% + 1.0%, 0.33% + 10%) of the two chemicals, into the air space on day 12 of incubation. Sumithion 50 EC caused a dose-related decrease in embryo body mass and high embryo mortality at the top dose. Developmental abnormalities were increased at all dose levels, but their frequency was not dose-dependent. Fusilade S significantly decreased body mass at the mid-dose level but caused 100% mortality at the top dose. Malformations were observed only at the low dose. Combined treatment resulted in an apparent moderation by Sumithion 50 EC of the high dose mortality effects of Fusilade S. However, no skeletal malformations were observed in embryos receiving Fusilade S and Sumithion 50 EC together, whereas such malformations were seen with Sumithion 50 EC alone.

6.4.1 Multiple-level effects of pesticide mixtures

Jaeger *et al.* (1999) used the interacting biological systems model to interpret the data on interactive effects of mixtures of aldicarb, atrazine, and nitrate on endocrine, immune and nervous system function in context with human health and biological conservation concerns. Fractional and full factorial designs with centre replicates were used to determine multifactor effects. The chronic doses used in experiments varied in duration from 22 to 103 days. The authors tested for changes in thyroid hormone levels, ability to make antibodies to foreign proteins and aggression in wild deer mice, *Peromyscus maniculatus*, and white outbred Swiss Webster mice, *Mus musculus*, ND4 strain. Endocrine, immune and behaviour changes occurred due to doses of mixtures, but rarely due to single compounds at the same concentrations. Immune assay data suggest the possibility of seasonal effects at low doses. Figure 6.1 illustrates the current conceptual model of interactive effects of environmental chemicals across multiple levels of biological organization. The left half of the figure illustrates healthy systems, and the right side illustrates consequences of some pesticide impacts. Pesticides may undermine endocrine, immune and nervous system functions, and cellularmolecular processes that depend on mass, energy and nutrients. The availability of mass, energy and nutrients for cellularmolecular processes depends on proper functioning of the gut and the nervous system that controls gut function. The nervous, endocrine and immune systems are intimately interconnected and affect each other via extensive feedback mechanisms (Cotman *et al.*, 1987). At the individual level of organization, behaviour, growth and reproduction including developmental processes depend on cellular/molecular and organ system functions. Individual processes support population level processes of birth

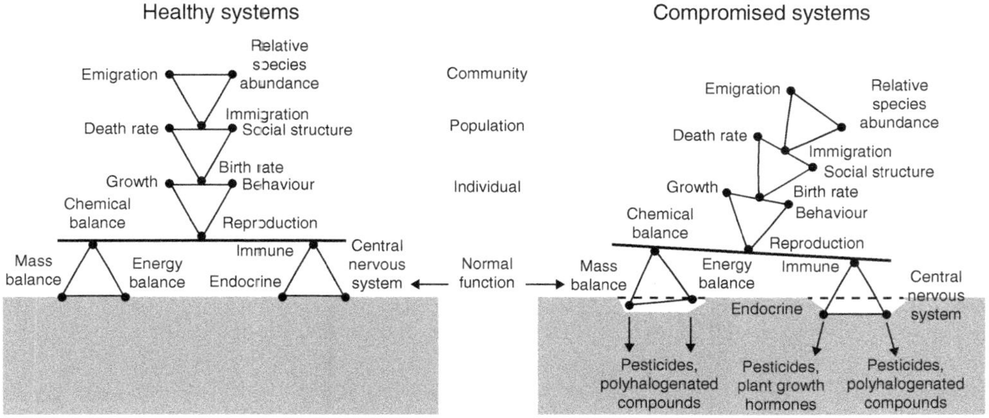

Fig. 6.1. Multiple-level effects of environmental contaminants. (From Jaeger *et al.*, 1999.)

rates and death rates and social structure. Population level processes support community level properties. These properties may include relative species abundance, immigration and emigration rates. Thus, pesticide effects on processes at the organ system level and the cellularmolecular level may affect the entire superstructure by undermining or altering functions at suborganismal levels of organization.

Professor Porter studied the contaminants aldicarb (insecticide), atrazine (herbicide) and nitrate (from fertilizer), widely detected in groundwater in the United States. Two strains of mice were used, testing for aggression, curiosity, learning speed, immune reaction to a foreign chemical, plus thyroid hormone assays. Aldicarb and atrazine alone had little or no effect, but the mixtures did. The study was conducted for concentration levels, as well as sorting by season. The mice were observed to be most vulnerable in the winter and in the spring. A reduction in spatial discrimination (nervous system), exploratory behaviour, memory, motor coordination and learning speed, and changes in aggression were observed depending on the season and the mix. The endocrine and immune systems were also affected (Porter *et al.*, 1999; Porter, 2002). In another study, developmental toxicity was investigated in mice for a common commercial formulation of herbicide containing a mixture of 2,4-dichlorophenoxyacetic acid (2,4-D), mecoprop, dicamba and inactive ingredients. A decrease in litter size was observed to be associated with a decrease in the number of implantation sites, but only at very low and low environmentally relevant doses (Cavieres *et al.*, 2002).

Pesticide exposure also leads to changes in thyroid levels. A correct level of thyroid hormone exposure is crucial to development. Children with improper thyroid levels can be difficult to teach. Unfortunately, if a mother rat's thyroid is too high or low during a crucial 2-day window early in foetal development, the foetal hind brain may not develop properly. Mouse foetuses can respond to hormonal differences in the parts per trillion. Human foetuses may, too. The mixture of pesticides, 'cocktails of contaminants' frequently detected in nature, can adversely affect the environment. The grey tree frog and leopard frog tadpoles were exposed to small amounts of the ten pesticides that are widely used throughout the world. Relyea (2008) selected five insecticides – carbaryl, chlorpyrifos, diazinon, endosulfan and malathion – and five herbicides – acetochlor, atrazine, glyphosate, metolachlor and 2,4-D. Different treatments included: each of the pesticides alone, the insecticides combined, a mix of the five herbicides or all ten of the poisons. A mixture of all ten chemicals killed 99% of leopard frog tadpoles as did the insecticide-only mixture; the herbicide mixture had no effect on the tadpoles. While leopard frogs perished, grey tree frogs did not succumb to the poisons and instead flourished in the absence of leopard frog competitors. Endosulfan, by itself, caused 84% of the leopard frogs to die. The chemical appeared to be about 1000× more lethal to amphibians than other pesticides examined, and it was observed that it was not only highly toxic to leopard frogs, but also that it served as the linchpin of the pesticide mixture that eliminated the bulk of leopard frog tadpoles. The administered concentration was far below the human-lifetime-exposure levels set by the US Environmental Protection Agency (US EPA) (2 to 16 parts per billion) and also fell short of the maximum concentrations detected in natural bodies of water. Declining amphibian populations have been recorded even with these low concentrations in pristine areas far downwind from areas of active pesticide use; the chemical cocktail could be described as a culprit. Amphibians are considered an environmental indicator species because of their unique sensitivity to pollutants. Their demise from pesticide overexposure could foreshadow the fate of less sensitive animals. Leopard frogs, in particular, are vulnerable to contamination; once plentiful across North America, including Pennsylvania, their population has declined due to an increase in pollution and deforestation. Gradual amounts of malathion – the most popular insecticide in the US – that were too small to directly kill developing leopard frog tadpoles instead sparked a biological chain of events that deprived them of their primary food source. As a result, nearly half the tadpoles in the experiment did not reach maturity and would have died in nature.

The mixture of pesticides may be responsible for declining salmon stocks, along the US Pacific Coast (Petkewich, 2008). The salmon use their sense of smell to find food and mates, detect predators and migrate seasonally from rivers to oceans.

6.4.2 Effect of mixtures on foetal implantation and litter size

What effect will a pesticide mixture have on the implantation of foetuses and on litter size? The developmental toxicity in mice of a common herbicide formulation containing a mixture of 2,4-D, mecoprop, dicamba and inactive ingredients was investigated. Pregnant mice were exposed to one of four different doses of the herbicide mixture diluted in their drinking water, either during pre-implantation and organogenesis or only during organogenesis. Litter size, birth weight and crown–rump length were determined at birth, and pups were allowed to lactate and grow without additional herbicide exposure. At weaning, dams were sacrificed, and the number of implantation sites was determined. The data, although apparently influenced by season, showed an inverted or U-shaped dose–response pattern for reduced litter size, with the low end of the dose range producing the greatest decrease in the number of live pups born (Table 6.1). In the table, the second or 'Low' dose is considered 'safe' by the US EPA. The 'very low dose' was in parts per billion – equivalent to one drop of water in 500 bathtubs. The dose is considered environmentally relevant (http://americannutritionassociation.org/newsletter/biological-effects-low-level-pesticide-mixtures-problems-solutions).

Nair *et al.* (2007) studied the individual and combined acute lethal toxicity of the common biocides – malathion, methyl parathion, endosulfan and 2,4-D on early juveniles of rohu under laboratory conditions. The 48-h LC_{50} values of individual biocides observed were: malathion 7.89 mg/l (7.28–8.61); methyl parathion 7.34 mg/l (7.25–7.43); endosulfan 0.0036 mg/l (0.0025–0.0047) and 2,4-D 962.43 mg/l (954.02–970.81). It was found that 2,4-D (chlorophenoxy herbicide) is 'moderately toxic', malathion and methyl parathion (organophosphate insecticides) are 'toxic' and endosulfan (organochlorine insecticide) is 'very toxic' to rohu juveniles under static (with renewal) 48-h LC_{50} bioassay. The combined toxicity tests indicated the 'strictly additive' nature of the insecticide–herbicide combinations and the 'more than additive' nature of the insecticide–insecticide combinations. The simultaneous or even asynchronous application of these chemicals in paddy fields and plantations thus greatly increases their potential to be toxic in freshwater and coastal ecosystems of the tropics.

6.4.3 Pulse dose exposures

Most pesticides can penetrate our immune defences by using 'pulse dose exposure'.

Table 6.1. Concentration of herbicides in drinking water (ppm) and dose (mg/kg/day) of herbicide mixture administered to mice.

	2,4-D		Mecoprop		Dicamba	
Dose	ppm	mg/kg/day	ppm	mg/kg/day	ppm	mg/kg/day
Very low	0.039	0.01	0.019	0.004	0.004	0.0009
Low	0.32	0.1	0.15	0.040	0.035	0.009
Intermediate	77	20	36.7	8.07	8	1.83
High	400	100	200	40.39	42.4	9.166

This is described by imagining a stranger approaching you smiling with his hand held out, and then suddenly he punches you in the stomach. When you put up your hands to defend yourself, he backs off smiling. After you get tired of holding up your hands and let them down, he suddenly rushes in and hits you again. Most of our defences against foreign invaders exist in our liver. The special enzymes can take from half a day to up to 5 days to develop. Pulse dose pesticides can penetrate in a few minutes (like the smiling stranger hitting you before your hands are raised in defence) (Nutrition Digest, 2015). It has been recognized that aquatic non-target organisms exposed to pesticides are typically exposed to sequential pulses with fluctuating concentrations (Handy, 1994; Reinert *et al.*, 2002), but current risk assessment relies on standard toxicity tests performed at constant concentrations and over fixed durations. The aquatic risk assessment can be improved if the effects resulting from sequential pulsed exposure to multiple compounds can be predicted quantitatively (Ashauer *et al.*, 2007). The authors evaluated two modelling approaches, both extended to suit multiple compounds, the semi-mechanistic threshold damage model (TDM), and a model based on time-weighted averages (TWA). The TDM predicted that recovery of damage to *Gammarus pulex* from exposure to chlorpyrifos took longer than that from exposure to carbaryl and consequently that the sequence of exposure matters. The survival of the freshwater invertebrate *Gammarus pulex* after sequential pulsed exposure to carbaryl and chlorpyrifos was measured. Two groups of organisms were exposed to a first pulse of either carbaryl or chlorpyrifos for 1 day and then, after a recovery period of 2 weeks, to a second pulse with the other compound. The comparison of mortalities caused by each pulse, as well as combined mortalities in both treatments, showed that the sequence of exposure to pulses of contaminants did indeed matter. Previous exposure to chlorpyrifos led to significantly increased mortality from subsequent pulses of carbaryl, but not the other way round. The TDM facilitates a process-based ecotoxicological explanation by simulating the recovery dynamics and outperforms the TWA model.

Cold and Forbes (2004) examined the effects of a short and environmentally realistic pulse exposure of different life stages of the freshwater amphipod *Gammarus pulex* to esfenvalerate, a pyrethroid insecticide. The aim was to determine the extent to which detectable effects on key life history traits persisted following cessation of a brief exposure to this pesticide. The results indicated a significant effect on the survival and reproduction of *G. pulex* with the environmentally realistic exposure concentrations of this insecticide. The *G. pulex* was found to have a similar sensitivity to esfenvalerate to the standard test invertebrate, *Daphnia magna* and was more sensitive than other common stream invertebrates. Despite 100% survival during pulses of up to 2 µg/l, mortality increased, in some cases markedly, following transfer to clean conditions. Pulse exposure to esfenvalerate at concentrations in the range 0.1–0.6 µg/l for as little as 1 h can have effects on *G. pulex* survival, pairing behaviour and reproductive output that could still be detected at least 2 weeks following the pulse. Reproductive traits were very sensitive to esfenvalerate, and exposure to 0.05 µg/l for 1 h led to immediate disruption of reproducing pairs, release of eggs or offspring from the brood pouch, and substantial delays in pair formation and subsequent reproduction following transfer to clean water.

6.5 Pesticide Mixtures and Rotations for Resistance Management

Resistance to pesticides has always been a concern to researchers and producers associated with agricultural cropping system. Resistance typically develops due to continually using pesticides with the same mode of action or chemistries. The strategies suggested to combat this issue of pest populations are the implementation of pesticide mixtures or pesticide rotations. However, neither of the strategies has been adequately demonstrated quantitatively to actually mitigate resistance. Pesticide mixtures involve exposing individuals in an arthropod population to each pesticide simultaneously, whereas pesticide rotations are the alternating use of pesticides with dissimilar modes of action. However, there are a number of important assumptions (Cloyd, 2010) dictating how successful these strategies may be in delaying the onset of resistance, including: (i) resistance associated with each pesticide is monogenic and independently genetically controlled; (ii) individuals in the arthropod pest population with doubly resistance genes or with multiple resistance mechanisms are rare; (iii) a certain frequency or proportion of individuals in the arthropod pest population are left untreated due to the presence of refugia; (iv) the pesticides used are equally persistent so that any individuals in the arthropod pest population are not exposed to just one pesticide for an extended length of time; and (v) the pesticides used have different modes of action. However, implementation of alternative pest management tactics including cultural, sanitation and biological control can reduce reliance on pesticides, which is most effective in delaying or mitigating pesticide resistance.

The non-effective insecticide, dimethoate has now become effective for cotton aphid (Alemayehu and Ababu, 1985). Shonga *et al.* (2013) conducted field experiments at Werer Agricultural Research Center, Afar region, Ethiopia, to determine the best insecticide rotation sequence and use of compatible mixtures in the insecticide resistance management strategy for cotton aphid, *Aphis gossypii*. It was observed that pesticides with reduced efficacy due to serious resistance will regain their efficacy if their use is stopped for 3–5 years. Endosulfan frequently used for bollworm control has the potential to suppress the aphid population to lower economic threshold level (ETHL or ETL). A mixture of carbosulfan and endosulfan, or furathiocarb and endosulfan is used when aphid and bollworm complexes exist together. Rotation of insecticides, carbosulfan, diafenthiuron, furathiocarb, dimethoate and endosulfan was found to be useful. Mixtures like endosulfan + carbosulfan and endosulfan + furathiocarb gave better control of the aphid. Insecticide rotations with treatment 2 (carbosulfan, diafenthiuron, furathiocarb and deltamethrin), 3 (dimethoate, carbosulfan, deltamethrin and diafenthiuron) and 5 (endosulfan, endosulfan + carbosulfan, deltamethrin and dimethoate) were found to be the best alternative combinations of insecticide groups to be used in an insecticide resistance management strategy.

6.6 Conclusion

Mixtures of insecticides with different modes of action are usually considered very effective in resistance management programmes. In pesticide mixtures with different modes of action, one product or toxicant interferes with the metabolic detoxification of the other toxicant, thereby synergizing the toxicity of the latter toxicant, leading to their synergistic behaviour. In a mixture of organophosphates with pyrethroids, binding of monooxygenase enzymes with organophosphate insecticide would prevent or delay degradation, enhancing the toxicity of pyrethroid insecticide via a competitive substrate inhibition mechanism. The response addition model does not always accurately predict mixture toxicity for chemicals with differing modes of action, while synergistic relationships have been observed using the TU approach. Two basic concepts have been established for predicting additive joint effects based on the known toxicity of the individual mixture components: the concept of concentration addition for mixtures of substances with similar modes of actions and the concept of independent action for mixtures of substances with dissimilar modes of action. Most pesticides can penetrate our immune defences by using 'pulse dose exposure' in a few minutes. Aquatic non-target organisms, exposed to pesticides, get sequential pulses with fluctuating concentrations. Implementation of pesticide mixtures or pesticide rotations can combat the issue of resistance to pesticides, due to their different modes of action.

References

Abou-Donia, M.B., Wilmarth, K.R., Jensen, K.F., Oehme, F.W. and Kurt, T.L. (1996a) Neurotoxicity resulting from coexposure to pyridostigmine bromide, DEET, and permethrin: Implications of Gulf War chemical exposures. *Journal of Toxicology and Environmental Health* 48, 35–56.

Ahmad, M. (2004) Potentiation/antagonism of deltamethrin and cypermethrins with organophosphate insecticides in the cotton bollworm, *Helicoverpa armigera* (Lepidoptera: Noctuidae). *Pesticide Biochemistry and Physiology* 80, 31–42.

Ahmad, M. (2009) Observed potentiation between pyrethroid and organophosphate insecticides for the management of *Spodoptera litura*. *Crop Protection* 28, 264–268.

Ahmad, M., Ahmad Saleem, M. and Sayyed, A.H. (2008) Efficacy of insecticide mixtures against pyrethroid and organophosphate resistant populations of *Spodoptera litura*. *Pest Management Science* 65, 266–274.

Alemayehu, R. and Ababu, D. (1985) A review of research on insect pests of fiber crops in Ethiopia. In: Tsedeke, A. (ed.) *Proceedings of the First Ethiopian Crop Protection Symposium on Review of Crop Protection Research in Ethiopia.* IAR, Addis Ababa, Ethiopia, pp. 215–222.

Altenburger, R., Boedeker, W., Faust, M. and Grimme, L.H. (1996) Regulations for combined effects of pollutants: Consequences from risk assessment in aquatic toxicology. *Food and Chemical Toxicology* 34, 1155–1157.

Altenburger, R., Walter, H. and Grote, M. (2004) What contributes to the combined effect of a complex mixture? *Environmental Science and Technology* 38, 6353–6362.

Arora, S., Balotra, S., Pandey, G. and Kumar, A. (2017) Binary combinations of organophosphorus and synthetic pyrethroids are more potent acetylcholinesterase inhibitors than organophosphorus and carbamate mixtures: an *in vitro* assessment. *Toxicology Letters* 268, 8–16.

Ascher, K.R.S., Eliyahu, M., Ishaaya, I., Zur, M. and Ben-Moshe E (1986) Synergism of pyrethroid-organophosphorus insecticide mixtures in insects and their toxicity against *Spodoptera littoralis* larvae. *Phytoparasitica* 14, 101–110.

Ashauer, R., Boxall, A.B. and Brown, C.D. (2007) Modeling combined effects of pulsed exposure to carbaryl and chlorpyrifos on *Gammarus pulex*. *Environmental Science and Technology* 41(15), 5535–5541.

Attique, M.N.R., Khaliq, A. and Sayyed, A.H. (2006) Could resistance to insecticides in *Plutella xylostella* be overcome by insecticide mixtures? *Journal of Applied Entomology* 130, 122–127.

Backhaus, T., Arrhenius, A. and Blanck, H. (2004) Toxicity of a mixture of dissimilarly acting substances to natural algal communities: Predictive power and limitations of independent action and concentration addition. *Environmental Science and Technology* 38, 6363–6370.

Bliss, C.I. (1939) The toxicity of poisons applied jointly. *Annals of Applied Biology* 26, 585–615.

Brattsten, L.B., Holyoke, C.W., Leeper, J.R. and Raffa, K.F. (1986) Insecticide resistance: Challenge to pest management and basic research. *Science* 231, 1255–1260.

Bryne, F.J. and Devonshire, A L. (1991) In vivo inhibition of esterase and acetylcholinesterase activities by profenofos treatment in the tobacco whitefly *Bemisia tabaci* (Genn): implications for routine biochemical monitoring of these enzymes. *Pesticide Biochemistry and Physiology* 40, 198–204.

Bynum, E.D., Archer, T.L. and Plapp, F.W. (1997) Comparison of banks grass mite and two spotted spider mite: Response to insecticide alone and in synergistic combinations. *Journal of Economic Entomology* 90, 1125–1130.

Calamari, D. and Vighi, M. (1992) A proposal to define quality objectives for aquatic life for mixtures of chemical substances. *Chemosphere* 25, 531–542.

Cavieres, M.F., Jaeger, J. and Porter, W. (2002) Developmental toxicity of a commercial herbicide mixture in mice: I. Effects on embryo implantation and litter size. *Environmental Health Perspectives* 110(11), 1081–1085.

Cedergreen, N. and Rasmussen, J.J. (2017) Low dose effects of pesticides in the aquatic environment. In: *Pesticide Dose: Effects on the Environment and Target and Non-Target Organisms.* Chapter 12, *ACS Symposium Series*, vol. 1249, pp. 167–187.

Chapman, B.R. and Penman, D.R. (1980) The toxicity of mixtures of a pyrethroid with organophosphorus insecticides to *Tetranychus urticae* Koch. *Pesticide Science* 11, 600–604.

Chèvre, N., Loepfe, C., Singer, H., Stamm, C., Fenner, K. and Escher, B.I. (2006) Including mixtures in the determination of water quality criteria for herbicides in surface water. *Environmental Science and Technology* 40, 426–435.

Cloyd, R.A. (2010) Pesticide mixtures and rotations: are these viable resistance mitigating strategies? *Pest Technology* 4(1), 14–18.

Cold, A. and Forbes, V.E. (2004) Consequences of a short pulse of pesticide exposure for survival and reproduction of *Gammarus pulex. Aquatic Toxicology* 67(3), 287–299.

Coors, A. and Frische, T. (2011) Predicting the aquatic toxicity of commercial pesticide mixtures. *Environmental Sciences Europe* 23, 22.

Corbett, J.R. (1974) *The Biochemical Mode of Action of Pesticides*, Academic Press, New York, p. 330.

Cotman, W.W., Brinton, R.E., Galaburda, A., McEwen, B. and Schneider, D.N. (1987) *The Neuro-Immune-Endocrine Connection.* Raven Press, New York.

Curtis, C.F. (1985) Theoretical models of the use of insecticide mixtures for the management of resistance. *Bulletin of Entomological Research* 75, 259–265.

Das, P.P., Shaik, A.P. and Jamil, K. (2007) Genotoxicity induced by pesticide mixtures: in-vitro studies on human peripheral blood lymphocytes. *Toxicology and Industrial Health* 23(8), 449–458.

El-Demerdash, F.M. (2011) Lipid peroxidation, oxidative stress and acetylcholinesterase in rat brain exposed to organophosphate and pyrethroid insecticides. *Food and Chemical Toxicology* 49(6), 1346–1352.

Gaughan, L.C., Engel, J. and Casida, J. (1980) Pesticide interactions: Effects of organophosphorus pesticides on the metabolism, toxicity, and persistence of selected pyrethroid insecticides. *Pesticide Biochemistry* 14, 18–25.

Goel, A., Dani, V. and Dhawan, D. (2005) Protective effect of zinc on lipid peroxidation, antioxidant enzymes and hepatic histoarchitecture chlorpyriphos induced hepatotoxicity. *Chemico-Biological Interactions* 156, 131–140.

Groten, J.P. (2000) Mixtures and interactions. *Food and Chemical Toxicology* 38, S65–S71.

Gunning, R.V., Moores, G.D. and Devonshire, A.L. (1999) Esterase inhibitors synergise the toxicity of pyrethroids in Australian *Helicoverpa armigera. Pesticide Biochemistry and Physiology* 63, 50–62.

Handy, R.D. (1994) Intermittent exposure to aquatic pollutants: Assessment, toxicity and sublethal responses in fish and invertebrates. *Comparative Biochemistry and Physiology Part C: Toxicology and Pharmacology at ScienceDirect* 107, 171–184.

Ishaaya, I., Mendelsson, Z., Ascher, K.R.S. and Casida, J.E. (1987) Cypermethrin synergism by pyrethroid esterase inhibitors in adults of whitefly. *Pesticide Biochemistry and Physiology* 28, 155–162.

Jaeger, J.W., Carlson, I.H. and W.P. Porter (1999) Endocrine, immune, and behavioral effects of aldicarb (carbamate), atrazine (triazine) and nitrate (fertilizer) mixtures at groundwater concentrations. *Toxicology and Industrial Health* 15, 133–151.

Junghans, M., Backhaus, T., Faust, M., Scholze, M. and Grimme, L.H. (2006) Application and validation of approaches for the predictive hazard assessment of realistic pesticide mixtures. *Aquatic Toxicology* 76, 93–110.

Kaushik, S., Inderjit, Streibig, J.C. and Cedergreen, N. (2006) Activities of mixtures of soil-applied herbicides with different molecular targets. *Pest Management Science* 62(11), 1092–1097.

Khan, H.A.A., Akram, W., Shad, S.A. and Lee, J.J. (2013) Insecticide mixtures could enhance the toxicity of insecticides in a resistant dairy population of *Musca domestica* L. *PLOS ONE* 8(4), 1–8.

Kudsk, P. (2002) Optimising herbicide performance. In: Naylor, R.E.L (ed.) *Weed Management Handbook.* Blackwell Science, Oxford, pp. 323–344.

Kulkarni, A.P. and Hodgson, E. (1980) Metabolism of insecticides by mixed function oxidase systems. *Pharmacological Therapy* 8, 379–475.

Martin, T., Ochou, O.G., Vaissayre, M. and Fournier, D. (2003) Organophosphorus insecticides synergise pyrethroids in the resistant strain of cotton bollworm from West Africa. *Journal of Economic Entomology* 96, 468–474.

Michelangeli, F., Robson, M.J., East, J.M. and Lee, A.G. (1990) Fluorescence and kinetic studies of the interactions of pyrethroids with the (Ca2++Mg2+)- ATPase. *Biochimica et Biophysica Acta* 1028, 58–66.

Montella, I.R., Schama, R. and Valle, D. (2012) The classification of esterases: An important gene family involved in insecticide resistance – a review. *Memórias do Instituto Oswaldo Cruz* 107, 437–449.

Nair, V.P., Nair, R.J. and Mercy, T.V.A. (2007) The individual and combined acute lethal toxicity of selected biocides on the juveniles of rohu (*Labeo rohita*) under tropical conditions. *Indian Journal of Fisheries* 54(3), 267–274.

Nutrition Digest (2015) American Nutrition Association. *Biological Effects of Low Level Pesticide Mixtures: Problems and Solutions* 38(2). Available at: http://americannutritionassociation.org/newsletter/biological-effects-low-level-pesticide-mixtures-problems-solutions (accessed 31 July 2018).

Pape-Lindstrom, P.A. and Lydy, M.J. (1997) Synergistic toxicity of atrazine and organophosphate insecticides contravenes the response addition mixture model. *Environmental Toxicology and Chemistry* 16(11), 2415–2420.

Porter, W.P. (2002) Biological effects of low level pesticide mixtures: Problems and solutions. *NOHA NEWS, Nutrition Digest* 38(2).

Porter, W., Jaeger, J.W. and Carlson, I.H. (1999) Endocrine, immune and behavioral effects of aldicarb (carbamate), atrazine (triazine) and nitrate (fertilizer) mixtures at groundwater concentrations. *Toxicology and Industrial Health* 15, 133–150.

Petkewich, R. (2008) Pesticide mixtures hurt salmon. *Chemical and Engineering News* 86(24), 14. DOI:10.1021/cen-v086n024.p014a

Reinert, K.H., Giddings, J.M. and Judd, L. (2002) Effects analysis of timevarying or repeated exposures in aquatic ecological risk assessment of agrochemicals. *Environmental Toxicology and Chemistry* 21, 1977–1992.

Relyea, R.A. (2008) A cocktail of contaminants: How mixtures of pesticides at low concentrations affect aquatic communities. *Oecologia* 159(2), 363–376. DOI:10.1007/s00442-008-1213-9

Roush, R.T. (1993) Occurrence, genetic and management of insecticide resistance. *Parasitology Today* 9, 174–179.

Shonga, E., Ali, K. and Azrefegne, F (2013) Effect of insecticide rotation and mixtures use for resistance management on cotton aphid, *Aphis gossypii* Glover (Hemiptera, Aphididae) in middle Awash areas of Ethiopia. *Greener Journal of Agricultural Sciences* 3(7), 569–578.

Streibig, J.C. and Jensen, J.E. (2000) Action of herbicides in mixtures. In: Cobb, A.H. and Kirkwood, R.C. (eds) *Herbicides and their Mechanisms of Action.* Sheffield Academic Press Ltd., Sheffield, pp. 153–180.

Thiruchelvam, M., Brockel, B.J., Richfield, E.K., Baggs, R.B. and Cory-Slechta, D.A. (2000a) Potentiated and preferential effects of combined paraquat and meneb on nigrostriatal dopamine systems: Environmental risk factors for Parkinson's disease. *Brain Research* 873(2), 225–234.

Thiruchelvam, M., Richfield, E.K., Baggs, R.B., Tank, A.W. and Cory-Slechta, D.A. (2000b) The nigrostriatal dopaminergic system as a preferential target of repeated exposures to combined paraquat and maneb: Implications for Parkinson's disease. *Journal of Neuroscience* 20, 9207–9214.

Varga, T., Hlubik, I., Varnagy, L., Budai, P. and Molnar, E. (1999) Embryonic toxicity of insecticide Sumithion 50 EC and herbicide Fusilade S in pheasants after individual or combined administration. *Acta Veterinaria Hungarica* 47, 123–128.

Zalom, F.G., Toscano, N.C. and Byrne, F.J. (2005) Managing resistance is critical to future use of pyrethroids and neonicotinoids. *California Agriculture* 59, 11–15.

De Zwart, D. and Posthuma, L. (2005) Complex mixture toxicity for single and multiple species: Proposed methodologies. *Environmental Toxicology and Chemistry* 24, 2665–2676.

7 Techniques for Pesticide Risk Assessment

7.1 Introduction

According to California Environmental Protection Agency (CalEPA) (www.cdpr.ca.gov/docs/dept/factshts/artic12.pdf), the toxicity of a pesticide is an inherent property. All pesticides can produce adverse health effects at some level of exposure. A hazardous substance has the potential to harm health if it is present in the environment and if people are exposed to it. Many hazards can be either contained or avoided, so not every potential hazard poses a health risk. A risk, in turn, is defined as the likelihood of the hazard occurring in a given situation. There is, therefore, a possibility of 'risk' of adverse health effects associated with exposure to a particular amount (dose) of a pesticide. Therefore, risk is a function of both toxicity and exposure. If quantifiable, risk may be calculated as the probability of an action occurring multiplied by the severity of the harm if the action occurs:

$$\text{Risk} = \text{Probability} \times \text{Severity of consequence}$$

Hazard is not risk. 'Hazard' is a descriptive term referring to the intrinsic capability of the waste to cause harm; a hazard is the source of the risk. Hazard and risk are two distinct but interrelated concepts – the first a reflection of potential effect and the second of the likelihood it will occur. In many cases the hazard cannot be completely eliminated though the probability of occurrence can be reduced with the addition of safety measures and at a financial cost.

Many pesticide risk assessment tools have been developed throughout the world over the past decade. Depending on the intended use of the resulting metrics, they have been variously called hazard rankings, yardsticks, indicators, screening benchmarks, relative risk rankings, risk assessment tools and many more. Levitan (2000) discussed different risk assessment tools depending on whether they are intended for grower decision support systems, for 'eco-labelling' purposes or to provide governments and others with an estimate of ecological damage. She distinguished between 'indicators' and 'impact assessment systems'. 'Indicators' tend to be concise summaries and integrations of various trends or highlights that evolved within a framework of policy analysis and risk communication. 'Impact assessment systems', on the other hand, retain the ecosystemic perspective as well as the depth and complexity that are appropriate for the level of knowledge of a particular environmental component. Risk assessment systems can provide insight not always available from evaluations carried out by pesticide registration authorities. The latter typically consider pesticides singly (product-specific registration procedures) and make registration decisions often under imprecise concepts of risk and benefit.

Durkin and Diamond (2002) reported the impact of and approach to three specific toxicological endpoints considered in risk assessments – neurotoxicity, immunotoxicity and endocrine disruption. The process of making these judgements involves hazard identification and dose–response assessment. Hazard identification is the process of identifying what, if any, effects a compound is likely to induce in an exposed population. Dose–response assessment is the process of estimating levels of exposure that are not likely to be associated with any adverse effects. These two steps, along with exposure and risk characterization, comprise the risk assessment process (National Research Council, 1993). In practice, risk assessments do not quantitatively address each endpoint that a chemical might cause. To do this would be extremely resource-intensive and would make each risk assessment more complicated than necessary. Instead, the hazard identification process is used to identify the *critical effect* – the adverse effect that occurs at the lowest dose level. In many US Environmental Protection Agency (US EPA) risk assessments, this is referred to as the LOAEL (lowest observed adverse effect level). Then, the dose–response assessment focuses on identifying the NOAEL (no observed adverse effect level, *the highest dose that does not cause adverse effects*) or NOEC (no observed effect concentration) for the critical effect. In human health risk assessments, NOAELs are divided by an uncertainty factor to calculated reference doses (RfDs) that are considered to be levels of exposure that are not likely to result in adverse effects in any member of the population.

It is essential to know the non-toxic-effect levels and exposure levels just below the LOAEL for pesticides, in order to assess the health risk for humans exposed to mixtures of pesticides, veterinary drugs and similar substances. Generally, when exposure levels of the chemicals within a mixture are in the range of the NOAELs, and the components of the mixture have different modes of toxic action, no additivity and no potentiating interactions are found, indicating the applicability of the basic concept of 'simple dissimilar action', which suggests that adverse reactions would be unlikely. For whole-mixture analysis, while it may not be possible to determine whether there is an interaction between any of the components of a complex mixture, it should be possible to identify NOAELs for all critical effects. Hence, as long as exposure is at a suitable margin below these NOAELs, consumers will be protected from the effects of any potential interactions that might occur within the mixture, providing there is an adequate margin of safety (Cassee *et al.*, 1998). This is particularly so if the ratios of compounds in the mixture remain relatively constant.

The risk assessment of any toxic effects of chemical mixtures is extremely difficult due to the complexity and variability of chemical mixtures that may occur in the environment. Most attention has been directed at toxic effects due to combined actions on biological systems at relatively high levels of exposure in laboratory experiments in laboratory animals or using *in vitro* systems. Moreover, direct chemical reactions can occur between the components of a mixture. Several studies claim to have identified synergistic interactions of some mixtures. However, for the most part, these studies have been inadequately designed and based on an incomplete understanding of the concepts involved, but a few well-designed studies have demonstrated the occurrence of both synergistic and antagonistic interactions, as well as additive effects in mixtures, usually at high concentrations or high experimental exposure levels, which are probably unrepresentative of exposure doses. Some interactions may not be easy to predict, such as those that may occur at the transcriptional or transductional level of the genome. The type of combined action or interaction found at clearly toxic effect levels may not predict what will happen at non-toxic effect levels, including levels only slightly lower than the LOAELs. In relation to most examples of possible human exposure to multiple residues, it will be important critically to evaluate whether any effects are likely to occur at low levels of exposure, such as those that will occur through food and water (Committee on Toxicity of Chemicals in Food, Consumer Products and the Environment, 2002).

7.2 Risk Assessment

The goal for carrying out risk analysis is required as a part of a statutory requirement, as well as emergency planning, etc. depending on the nature of the industry (Fig. 7.1).

The risk is assessed through a process designed to answer questions about how toxic a chemical is, what exposure results from its various uses, what is the probability that use will cause harm and how to characterize that risk. The US EPA has cited four basic elements in risk assessment:

- hazard identification;
- dose–response assessment;
- exposure assessment; and
- risk characterization.

7.2.1 Hazard identification

Hazard identification involves the review and evaluation of a chemical's toxic properties – the extent and type of adverse effects. Adverse effects may be acute (arising from short-term exposure), subchronic (exposures longer than a few days but less than a year) or chronic (the result of exposure of a year or more, including lifetime). It is a clear understanding of what chemicals are present at a site, their concentration and spatial

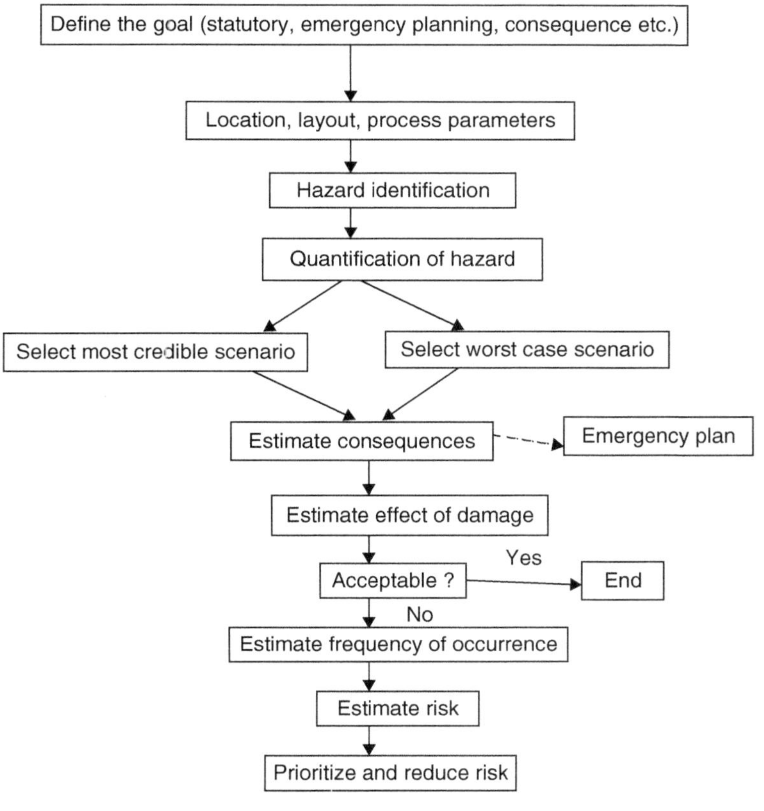

Fig. 7.1. Flow chart for risk analysis. (From BIS, 2006.)

distribution and how they could move to potential receptors. Laboratory studies on animals are generally used to define the types of toxic effects caused by a chemical, and the exposure level (doses) at which these effects may be seen. In evaluating chemicals, scientists must determine the exposure level at which adverse effects would not be expected to occur. Risk assessment commonly examines certain critical effects, including: carcinogenic (cancer), genotoxic (heritable traits or impacts), developmental (birth defects and miscarriages), reproductive (male and female fertility), endocrine (hormonal function), neurological (brain and nervous system disorder) and immunological (resistance to infection disease, occurrence of hypersensitivity disorders and autoimmune diseases). (California Department of Pesticide Regulation, chapter 5: 'Assessing pesticide risks to human health', www.cdpr.ca.gov/docs/pressrls/dprguide/chapter5.pdf).

7.2.2 Dose–response assessment

Dose–response assessment (often combined with hazard identification in a single step) documents the quantitative relationship between dose and toxic effects. The lowest dose of a chemical resulting in harmful effect is determined by considering its toxic properties. State and federal tests require that laboratory animals receive high enough doses to produce toxic effects. Animals receive a wide range of exposures, including doses that may be much higher than those to which people might be exposed. There also are doses at which no ill effects occur in the test animals. Within that range of doses, the highest tested dose that does not cause adverse effects is the 'no observed effect level' (NOEL). A dose–response curve can be developed from most chemicals. From these curves the threshold level and the relative toxicity of chemicals can be obtained to help establish safe levels of chemical exposure.

The calculation used to determine the safe human dose (SHD) is as follows:

$$\mathrm{SHD} = \frac{\mathrm{ThD}_{0.0} \times 70(\mathrm{kg})}{\mathrm{SF}} = \begin{array}{c}\text{Amount mg/day}\\ \text{of toxic substance}\end{array}$$

Where SHD = Safe Human Dose; $\mathrm{ThD}_{0.0}$, threshold dose at which no effect is observed; 70 kg, average weight of a man; SF, safety factor (ranges from 10–1000, which varies according to the type of test and data used to obtain the $\mathrm{ThD}_{0.0}$) ('The dose response relationship', Chapter 4, http://web.cjcu.edu.tw/~jdwu/toxicology/chapter04.pdf).

When complete information is not available, mathematical adjustments are used as uncertainty factors. One way they are used in risk assessments is to compensate for uncertainties in the process that estimates the dose level in humans at which there is reasonable certainty that the identified adverse effects will not occur. As a default, if the toxicity studies are based on animals, we generally use an uncertainty factor of 10 to account for assumed differences in sensitivity between humans and experimental animals to a chemical (an assumption that the least sensitive humans are 10 times more sensitive than the most sensitive animal species). An additional uncertainity factor of 10 is used to address differences in sensitivity among humans (this assumes that the most sensitive human is 10 times more sensitive than the least sensitive human). This results in a total uncertainty factor of 100.

7.2.3 Exposure assessment

Exposure is a critical connection between potentially harmful pesticides and human health effects. Exposure assessment examines what is known about the duration, frequency (continuous or intermittent) and level of contact. It is the process of finding out how people come into contact with the pesticide, how often and for how long they are in contact with the substance, and how much of the substance they are in contact with. It is the second stage of a quantitative risk assessment consisting of estimating the exposure of chemicals by the populations potentially at risk. The potentially exposed populations may include present and future populations in the vicinity of the site; subpopulations of special concern, for example, children; and on-site workers involved with remediation (Environmental Resources Management, 2001). It includes an estimate of people's potential exposure to a chemical at work, at home or in their diets.

Exposure may be of short duration (acute, occurring once or for a short time), intermediate duration (subchronic, generally 1–3 months), or long term (chronic, generally 1 year to lifetime). Rates of exposure are determined for breathing (inhalation), eating or drinking (ingestion), or contact with the skin (dermal absorption), depending on the chemical and the ways people may be exposed to it.

7.2.4 Risk characterization

A risk characterization presents qualitative and quantitative estimates of the likelihood that any of the hazards associated with the pesticide will occur in exposed people. It quantifies the results of the risk assessment. Risk characterization is the final phase of the ecological risk assessment and interpretation phase of the assessment process. It combines hazard identification and dose–response assessment (generally based on animal studies) with exposure assessment (based on estimated human exposure), and presents the major results and conclusions of the risk assessment as well as the associated uncertainties. A risk characterization provides a discussion for a diverse audience that minimizes the use of technical terms. It is an appraisal of the science that supports the risk manager in making public health decisions (OPP, 2002).

Risk characterization has two major components: *risk estimation* and risk description. Risk estimation compares exposure and effects data, considers integrated exposure and effects data in context of levels of concern (LOCs) and states the potential for risk. The risk description interprets risks based on assessment endpoints. For example, characterizing the risk to pesticide applicators requires estimating what dose of the chemical causes what effects (i.e. the dose–response assessment), and what dose workers are exposed to (the exposure assessment). The results are often expressed in one of two ways. The first is as a margin of exposure, which is calculated by dividing the NOEL by the estimated human exposure. If the NOEL is based on a study using experimental animals, the benchmark margin of exposure would be 100 to ensure that there is reasonable certainty that the effect will not occur in exposed people.

A risk characterization summary should accompany a cumulative risk assessment (CRA), and should present the conclusions of the analysis and include a discussion of the significant contributors and sources of risk. It should provide descriptions of risk for the exposed population as a whole (i.e. average levels of exposure) as well as those individuals in the high end of the distribution. Important subgroups of the population (e.g. children) who may be at disproportionate risk (e.g. through unique susceptibility to the common mechanism of toxicity, or human activity exposure patterns) should be highlighted. Methods of estimation should be described (e.g. deterministic versus probabilistic methods, or screening-level exposure methods versus highly refined monitoring data reflective of residues close to the point of consumption or contact) (OPP, 2002).

It is the final stage of the four-stage risk assessment for estimating risks. This consists in part of calculating quantitative estimates of both the carcinogenic and non-carcinogenic risks to receptors for all exposure scenarios considered. Estimates are typically done for:

- three exposure routes: inhale, ingest, skin;
- the maximally exposed individual (MEI); and
- the most probable exposed population.

7.3 Probabilistic and Deterministic Approach

The *probabilistic approach* is considered on account of variability in pesticide dosages, exposure and frequency. The postulated scenarios or events are modelled and consequences are assessed for risk related to pesticide use. The *deterministic approach* is based on certainties of causalities pertaining to risk related to pesticide use. The approach involves certainties pertaining to dosage, exposure, frequency and risk involved with use of pesticides. Deterministic risk assessments provide a single-point estimate of risk at a site of concern, while probabilistic risk assessment methods generate a range of values from probability distribution functions (PDFs) (Bruce *et al.*, 2007).

7.3.1 Risk quotient

For most risk assessments, US EPA uses a deterministic approach or the quotient method to compare toxicity to environmental exposure. In the deterministic approach, a risk quotient (RQ) is calculated by dividing a point estimate of exposure by a point estimate of effects. This ratio is a simple, screening-level estimate that identifies high- or low-risk situations. Calculation of RQs are based upon ecological effects data, pesticide use data, fate and transport data, and estimates of exposure to the pesticide. In this method, the estimated environmental concentration (EEC) is compared to an effect level, such as an LC_{50} (the concentration of a pesticide where 50% of the organisms die):

Risk quotient = Exposure/Toxicity

After the RQ(s) is calculated, it is compared to the agency's LOC. An LOC is a policy tool that the agency uses to interpret the RQ and to analyse potential risk to non-target organisms and the need to consider regulatory action.

Although the deterministic or RQ method is useful for screening purposes, it provides only a single-point estimate of environmental risk. Following the recommendations of the FIFRA Scientific Advisory Panel and the Ecological Committee on FIFRA Risk Assessment Methods (ECOFRAM), US EPA has moved towards developing more sophisticated methodologies by using probabilistic tools and methods to estimate the variability and/or uncertainty in factors that influence risk and to express risk in terms of the probability and magnitude of adverse effects. A refined or probabilistic risk assessment produces a distribution or range of values instead of one fixed value. The assessment becomes more refined as the levels increase, presenting increasingly focused effects and exposure scenarios. Because the results of the refined risk assessment show the range of possible environmental impacts and which ones are most likely to occur, they provide the risk manager with a flexible tool for making decisions regarding the level of certainty needed for a particular situation.

Monte Carlo (2001) simulations are the most common approach to estimate exposure with probability distributions. Using Monte Carlo, the probability distributions for one or more variables are sampled repeatedly. The exposure estimate is calculated each time to produce a distribution of possible values for exposure estimates. The distributions sampled in a Monte Carlo risk assessment can be either empirical data (i.e. actual ingestion rates recorded in scientific studies) or synthesized probability/statistical distributions such as ones estimated from mathematical equations that represent possible distributions of values. Monte Carlo simulations and other probabilistic approaches can provide estimates of exposure, but doing a probabilistic assessment using Monte Carlo techniques may not be necessary in situations where risk or costs of remediation are low. Typically, stressors, exposure pathways and receptors of highest potential concern are selected for a probabilistic approach to conserve resources. The *Guiding Principles for Monte Carlo Analysis* (US EPA, 1997) lays out the general approach that should be taken in all cases, beginning with defining the problem and scope of the assessment to selecting the best tools and approach. The *Guiding Principles* also describes the process of estimating and characterizing variability and uncertainty around risk estimates.

The Office of Pesticide Programs (OPP), until 1998, was using a software program, Dietary Risk Evaluation System (DRES), to conduct its acute dietary risk assessments for pesticide residues in foods. However, it was not possible to know where the pesticide exposure estimates from the DRES software fitted in the overall distribution of exposures due to the limits of the tools being used. To address these deficiencies, acute probabilistic dietary exposure guidance has been developed by OPP, to use a model that estimates exposure to pesticides in the food supply. The probabilistic Dietary Exposure Evaluation Model (DEEM) provides specific information about the range and probability of possible exposures. DEEM has been used in risk assessments to support tolerance levels for several pesticides (e.g. phosalone) and as part of CRAs for organophosphorus compounds and other pesticides (Case Study 4, 2014). In another case study (Case Study 12, 2014), US EPA analysed exposures to 30 organophosphates through food consumption, drinking water intake and exposure via pesticide application, based on distributions of human exposure factors, such as breathing rates, body weight and surface areas used in the assessment. The organophosphates CRA presented potential risk from single-day (acute) exposures across 1 year and from a series of 21 day rolling averages across the year. Generally, exposures through the food pathway dominated total margins of exposure (MOEs), and exposures through drinking water were substantially lower throughout most of the year. Residential exposures were substantially smaller than exposures through both food and drinking water.

As mentioned earlier, one of the qualitative rating systems for pesticides is the RQ. An RQ is simply the ratio of estimated or actual environmental or dietary concentration of the pesticide to a toxic effect level or threshold. Some other terms for this ratio include hazard quotient (HQ), hazard index (HI), margin of safety (MOS), toxicity exposure ratio (TER) and MOE. Peterson (2006) showed that an RQ approach is valuable for making direct comparisons of quantitative risks between pesticides. Furthermore, Peterson (2006) demonstrated that a numerical ranking of RQs for the purpose of comparing risks is valid across different levels of exposure refinement. Therefore, comparisons are equally valid whether using highly conservative exposure estimates (i.e. tier 1) or actual environmental exposures (tier 4). However, higher tiers should be used if the purpose is to accurately estimate the quantitative risk for an individual pesticide within a specific use and location scenario. A risk rating system for pesticides is attractive and has potential benefits. However, Peterson and Schleier (2014) suggested it was not advisable to use qualitative rating systems for pesticide risk assessment, management or decision making because they cannot properly discriminate between different levels of risk the way they are currently structured. They suggested use of quantitative risk models for both risk assessment and risk management of pesticides.

The US EPA approach is conservative and typically involves safety factors:

- For non-carcinogens, extrapolation from animals to humans involves two factors of 10.
- For carcinogens, the model assumes a 95% confidence limit; conservative 19 out of 20 times.

In Chapter 14 of Environmental Resources Management (2001), a few examples have been cited as follows:

Example:

Given: bis(2-ethylhexyl) phthalate (BEHP; also abbreviated as DEHP)

A = skin exposed = 20% = 0.2 × 18,150 cm^2 T14–7, p. 855 = 3630 cm^2

DA = dust adherence = 0.51 mg/cm^2

ABS = skin adsorption rate = 6%

SM = effect of soil matrix = 155 (85% is not available for skin contact)

EF = 250 exposures/year

ED = 1 year, consider using length of project

BW = 70 kg T14–7, p. 855

Find: The maximum (not average) exposure of on-site workers from dermal contact with the soil over a 1 year period.

$$I = \frac{(C \times CR \times EF \times ED \times RR \times ABS)}{(BW \times AT)}$$

units p. 854 of Chapter 14 (Environmental Resources Management, 2001).

C = concentration at exposure point;
CR = contact rate; EF = frequency;
ED = exposure duration; BW = body weight;
AT = averaging time; RR = retention rate;
ABS = adsorption into bloodstream.

$$I = \frac{(C \times 3630\ \text{cm}^2 \times .51\ \text{mg/cm}^2 \times .06 \times .15 \times 250\ \text{exposures/year} \times 1\ \text{kg}/10^6\ \text{mg})}{(70 \times 365)}$$

$I = 1.63 \times 10^{-8} \times C$ mg/kg/day

From T14-3, p. 843 (Environmental Resources Management, 2001), the maximum concentration of BEHP is 230 mg/kg, maximum soil contact.

$I = 1.63 \times 10^{-8} \times 230$ mg/kg/day

$I = 3.75 \times 10^{-6}$ mg/kg/day

Carcinogenic risk

Carcinogenic risk is defined as the chronic daily intake dose (through exposure assessment) multiplied by the carcinogenic slope factor (dose response/toxicity assessment). The risk is the probability of excess lifetime cancer from exposure to the chemical:

Risk = CDI × SF (Eqn 14-6 units p. 861)

Example:

Given: The maximum carcinogenic risk from BEHP in on-site workers from dermal contact with the soil is $I_N = 1.63 \times 10^{-8} \times C$ mg/kg/day, from previous problem.

Find: The maximum carcinogenic risk.

$I = 1.63 \times 10^{-8} \times C$ mg/kg/day where C is the exposure point concentration.

From T14-3, p. 843, C = 230 mg/kg

$I_C = 3.75 \times 10^{-6}$ mg/kg/day

SF for BEHP, use benzene 2.90×10^{-2} from app. B, p. 1056 (Environmental Resources Management, 2001)

Risk = CDI × SF units p. 861

$$\text{Risk} = 3.75 \times 10^{-6}\ \text{mg/kg/day} \times 2.90 \times 10^{-2}\ \frac{1}{\text{mg/kg/day}}$$

Risk = 1.09×10^{-7}, maximum BEHP

Characterization of carcinogenic risk first involves performing calculations similar to the example for each chemical appropriate to the exposure route and pathway under study. The risks are additive

Non-carcinogenic risk

Non-carcinogenic risk is characterized by the HI, which is the ratio of the estimated intake dose from exposure to the RfD as follows:

HI = CDI/RfD (Eqn 14-8 units p. 862)

An HI value of <1 is acceptable. For exposure to multiple non-carcinogens, the HI scores are added.

Example:

Given: chlorobenzene ex. 14-4a in which the exposure is 2.82×10^{-7} mg/kg/day

Find: HI

HI = CDI/RfD units p. 862 of chapter 14 (Environmental Resources Management, 2001)

RfD = 2×10^{-2} mg/kg/day for chlorobenzene, app. B, p. 1059

HI = 2.82×10^{-7} mg/kg/day/2×10^{-2}

HI = $1.41 \times 10^{-5} < 1$ ∴ OK

Uncertainties inherent in calculated risks

Uncertainties are inherent for the following reasons:

- The methodology may be deficient.
- Exposure assessments are uncertain.
- With the toxicity assessment, the slope factors and RfDs are uncertain.
- The computation of risk is an exercise in applied probability of extremely rare events.
- In summary, the process is very protective of human health by ensuring that

the risks are not understated – when in doubt, overdesign.

For cancer effects, risk is often expressed another way, as how much more likely it is that cancer will result from exposure to a chemical. Often, this is simplified in a kind of scientific shorthand; for example, a cancer risk of 'one in a million' in a given population. This can give the inaccurate impression that science can determine that exactly one person in a million will develop cancer; that we can determine and measure the causes of all cancers. The inherent uncertainty in risk assessment means that risk assessors can only predict the probability of risk.

The goal of any cumulative analysis should be to produce estimates of exposure that are health protective but use the available data to the greatest extent possible. Uncertainty in the cumulative assessment is also an important consideration and should be discussed in at least a qualitative sense whenever possible in the assessment. It is only when the uncertainties about the cumulative risk estimates can be adequately evaluated and conveyed to risk managers, interested parties and the general public that productive dialogue on potential refinements, responses and mitigation actions can take place.

Uncertainty should be described qualitatively, as well as quantitatively to the extent possible. No definitive standard agency policy or detailed procedure on how to perform uncertainty analysis is available. In evaluating uncertainty, unique issues for both the toxicological and exposure inputs into calculating risks for each pathway of exposure should be highlighted (Hattis and Burmaster, 1994; Baird *et al.*, 1996; Hattis and Minkowitz, 1996; Hattis and Anderson, 1999). The precision of each pathway analysis (e.g. food, drinking water, residential and other non-occupational sources) should help to determine the uncertainty in combining the different pathways in the cumulative assessment. The following two sections address uncertainty issues with respect to CRA.

OPP has developed guidance entitled 'Determination of the Appropriate FQPA Safety Factor(s) for Use in the Tolerance-Setting Process', which describes the policies employed in making determinations regarding the Food Quality Protection Act (FQPA) safety factor for single chemical risk assessments (US EPA, 1999a, b; 2002).

7.4 Cumulative Risk Assessment

In 1996, the FQPA imposed upon the OPP (OPP, 2002) the requirement to consider potential human health risks from all pathways of dietary and non-dietary exposures to more than one pesticide acting through a common mechanism of toxicity. Under the OPP program the potential human risks associated with such multi-chemical and multi-pathway exposures to pesticides are evaluated and estimated using a process termed as CRA. The CRA, therefore, enables OPP to make regulatory decisions, which protects public health and sensitive subpopulations, including infants and children.

As per OPP (2002), the CRA is conducted using the following steps:

1. Identify the common mechanism group (CMG).
2. Identify potential exposures.
3. Characterize and select common mechanism endpoint(s).
4. Determine the need for a comprehensive cumulative risk.
5. Determine candidate cumulative assessment group.
6. Conduct dose response analyses and determine relative toxic potencies of pesticides.
7. Develop detailed exposure scenarios for all routes and durations.
8. Establish exposure input parameters.
9. Conduct final CRA of pesticides.
10. Conduct characterization of cumulative risk.

7.4.1 Identify common mechanism group

A CRA begins with the identification of a group of chemicals (a common mechanism group, CMG), which induce a common toxic effect by a common mechanism of toxicity; for example, the organophosphate cumulative hazard assessment. OPP has developed a general framework for identifying the chemicals that belong in that group (US EPA, 1999b).

7.4.2 Identify potential exposures

Once a CMG has been established, the next step is to evaluate registered and proposed uses for each CMG member in order to identify potential exposure pathways (e.g. through food, drinking water and/or residential) and routes (i.e. oral, inhalation, dermal).

7.4.3 Characterize and select common mechanism endpoint(s)

During the hazard characterization phase, the various endpoints associated with the common mechanism of toxicity are identified, as well as the test species/sex that might serve as a uniform basis for determining relative potencies among the chemicals of interest. The common effect is also evaluated to determine if it is expressed across all exposure routes and durations of interest for each CMG member. The temporal aspects (e.g. time to peak effects, time to recovery) of the common mechanism toxicity are characterized to determine the critical window of its expression.

7.4.4 Determine the need for a comprehensive cumulative risk

Not all cumulative assessments need to be of the same depth and scope. Thus, early in the cumulative assessment process, it is important to determine the need for, or the capability to perform, a comprehensive risk assessment. This is done by considering the number and types of possible exposure scenarios in conjunction with the associated residue values available.

7.4.5 Determine candidate cumulative assessment group

The exposures from minor pathways are considered qualitatively. Thus, a subset of common mechanism chemicals to be included in the quantification of cumulative risk needs to be identified from the CMG. This subgroup is called the cumulative assessment group (CAG). The identification of the CAG is done throughout the process as a detailed understanding of each group member's hazard and exposure potential emerges from the analysis.

Although a chemical(s) may be removed from the quantification of risk, the rationale for such decisions should be explained. Thus, all chemicals that were grouped by a common mechanism of toxicity should be accounted for (qualitatively or quantitatively) in the final assessment. As the risk assessor proceeds with the cumulative assessment, it is important to determine candidate chemicals and uses, routes and pathways from the CMG that may cause cumulative effects. Cumulative assessments should not attempt to quantify risk resulting from those common mechanism chemicals that will have a minimal toxic contribution to the cumulative hazard, or from minor exposure pathways, routes or uses.

7.4.6 Conduct dose–response analyses and determine relative toxic potencies of pesticides

A dose–response analysis is performed on each CAG member to determine its toxic potency for the common toxic effect. The determination of toxic potency should, to the extent feasible with available data, be conducted on a uniform basis (i.e. same measure of potency, for the same effect, from the same test species/sex using studies of comparable methodology). For multi-chemical assessments involving chemicals sharing common toxic effect, assessment of combined risk of the CAG, is recommended using dose addition approach. Departures from the dose addition approach are appropriate if there are data available to support an alternative method.

The initial toxicological and exposure information is collected. A screening-level assessment may be conducted that applies more conservative approaches than would a comprehensive and refined CRA.

For example, MOEs may be based on NOAELs for the common toxic effect rather than modelling dose–response curves of each chemical member to derive more refined relative potencies and points of departure. For exposure to food, treatment of 100% of crops and tolerance level residues may be assumed for each CMG chemical registered for use on a crop. If a screening-level analysis including such overestimates of exposure indicates that there is no risk concern, then no further detailed assessment may be necessary. But if this conservative approach indicates a potential for unacceptable risk, then a refined assessment should be conducted. This may engender the need for additional data.

Once the toxic potency of each common mechanism chemical is determined, the relative potencies of the CAG members are established. To determine relative potency, a chemical from the CAG is selected to serve as the index chemical. The index chemical is used as the point of reference for standardizing the common toxicity of the other chemical members of the CAG. Once the index chemical is selected, relative potency factors (RPFs) are calculated (i.e. the ratio of the toxic potency of a given chemical relative to that of the index chemical). RPFs are used to convert exposures of all chemicals in the CAG into exposure equivalents of the index chemical. Given that the RPF method portrays risk as exposure equivalents to one chemical (the index compound), it is preferred that the index chemical: (i) have high-quality dose–response data; (ii) have a toxicological/biological profile for the common toxicity that is representative of the common toxic effect(s); and (iii) be well characterized for the common mechanism of toxicity. The last step in the dose–response assessment is to calculate a point of departure(s) for the index chemical so that the risk of the CAG can be extrapolated to anticipated human exposures.

7.4.7 Develop detailed exposure scenarios for all routes and durations

Detailed exposure scenarios for all of the uses remaining for each pesticide in the CAG should be developed. This includes the determination of potential human exposures by all relevant pathways, durations and routes that may allow simultaneous exposures, or any sequential exposures among the CAG members that could contribute to the same joint risk of the common toxic effect (i.e. either by overlapping internal doses or by overlapping toxic effects). The framework for estimating combined exposures is based on exposure to individuals, representing differing attributes of the population (e.g. human activity patterns, place of residence, age) that link pathways/route of exposure through scenario building. Cumulative risk values for a given common toxic effect are calculated separately for each exposure route and duration and then combined. To the extent data permit, the temporal and spatial linkages should be maintained for the many factors defining a possible individual exposure. A decision must be made on the relative importance of scenarios and the need for their inclusion in a quantitative assessment, as well as on the populations of interest and locations for evaluation in the assessment. The potential for co-occurrence of possible exposure scenarios is evaluated.

7.4.8 Establish exposure input parameters

Spatial, temporal and demographic considerations are major factors in determining whether a concurrent exposure is likely to occur. In other words, all exposure events need to occur over a specific interval of time; events need to agree in time, place and demographic characteristics; and an individual's dose needs to be matched with relevant toxicological values in terms of route and duration. Exposure input parameters are established.

7.4.9 Conduct final CRA of pesticides

The magnitude, frequency and duration for all pertinent exposure pathway/route combinations are determined. These parameters take into account appropriate sources of use/usage information, and residues in all appropriate media. Any modifying factors necessary for inclusion in the assessment are also identified. Where necessary, any appropriate surrogate datasets from other chemical-specific data, published literature or generic datasets are considered. A trial run of a quantitative cumulative risk is conducted by assigning route- and duration-specific risk metrics. The outputs of this trial run are evaluated and a sensitivity analysis is conducted. Subpopulations or life stages of concern are assessed. A final quantitative CRA can then be conducted.

7.4.10 Conduct characterization of cumulative risk

The last step of the assessment process is to characterize the risk. The results and conclusions of the CRA are clearly described, including the relative confidence in toxicity and exposure data sources and model inputs. The risk characterization also includes a description of the variability. Major areas of uncertainty should be described both qualitatively and quantitatively. The magnitude and direction of likely bias and the impact on the final assessment are discussed. Risk contributors are identified with regard to pesticide(s), pathway, source, time of year and impacted subpopulations (with particular attention to children). The basis for group uncertainty and FQPA safety factors should be explained.

As a guidance document and not a rule, the policy in this guidance is not binding on either US EPA or any outside parties. Although this guidance provides a starting point for OPP's risk assessments, OPP will depart from its policy where the facts or circumstances warrant. In such cases, OPP will explain why a different course was taken. Similarly, outside parties remain free to assert that a policy is not appropriate for a specific pesticide or group of pesticides, or that the circumstances surrounding a specific risk assessment demonstrate that a policy should be modified or abandoned. The CRA process will continue to evolve after this guidance is published. Thus, the agency may update this guidance or provide supplementary materials as appropriate.

7.5 Environmental Risk Assessment of Pesticides in Soil

In addition to many other factors, pesticides may lead to spatial and temporal changes in soil biological communities in the agricultural landscape. Based on the objectives of Society of Environmental Toxicology and Chemistry (SETAC) Europe, a workshop on 'Semi-Field Methods for the Environmental Risk Assessment of Pesticides in Soil' (PERAS), was convened in Coimbra, Portugal, in October 2007. During the workshop, various potential effects of pesticides on soil communities were discussed in the tiered approach of pesticide risk assessment within the European Union.

Soil provides a medium for an astounding variety of organisms that use the soil as a habitat and a source of energy and contribute to the formation of soil by influencing the soil's physical and chemical properties and the nature of vegetation that grows on it. There is a lot of difference between the structure and functions of the soil community at ecosystem level. The structure is the composition of the soil biodiversity described at the species level (i.e. abundance, biomass, diversity and dominance), while function refers to biologically determined processes in the soil ecosystems, based on the interaction of its different components (i.e. nutrient cycling, community respiration, organic matter breakdown, stabilization and structuring of soil aggregates).

There is increased recent focus on discussing semi-field methods. A tiered testing and risk assessment framework is followed for ecotoxicological data covering both acute and chronic effects, starting with relatively simple acute laboratory tests and moving on to more complex chronic laboratory, semi-field and field tests. The semi-field methods focus on those non-target organisms that predominantly inhabit the soil profile: earthworms, microarthropods, soil microorganisms and other soil non-target macroorganisms (which are used in the directive to predict effects on the breakdown of soil organic matter), as well as assessing surface active arthropods and plants.

7.5.1 Evaluation of soil semi-field methods

Schaeffer *et al.* (2011) defined semi-field tests as controlled, reproducible systems that attempt to simulate the processes and interactions of components in a portion of the terrestrial environment, either in the laboratory (small-scale) or in the field, or somewhere in between. Under these methods, the advantages of laboratory tests (e.g. standardization, controlled conditions) are combined with the advantages of field studies (natural variability, complex interactions), avoiding their disadvantages at the same time. The authors proposed various semi-field tests, as follows:

Group A methods comprise artificially assembled units with added organisms under controlled environmental conditions (A1) or under field conditions (A2). Group B tests are terrestrial model ecosystems (TMEs) using intact soil cores with natural communities under controlled environmental conditions (B1) or under field conditions (B2). Field enclosures on undisturbed soil belong to group C, in which immigration of species is prevented by barriers; tests can be performed relying on natural communities (C1) or on added organisms (C2), respectively.

In addition, 'grey zones' exist between the laboratory and the semi-field level, and between the semi-field level and the field level.

The community recovery principle (CRP) protects the life support functions of the soil for crop rotation and sustainable agriculture, with overall protection of the structure and functioning of soil community characteristics for agroecosystems, while the ecological threshold principle (ETP) allows changes in land use, with overall protection of the structure and functioning of soil community characteristics for nature reserves. The procedure provides trigger values for the half-life for dissipation (DT_{50}) from soil. The tiered approaches can determine both the predicted environmental concentrations (PECs) and the ecotoxicological endpoints.

For risk assessment a first-tier approach is proposed to base the permissible concentration on the long-term TER for a basic set of standard soil organisms, which is different for compounds with different modes of action. The species sensitivity distribution (SSD) approach is proposed for a second tier, for calculating the hazardous concentration for 5% of the species (HC5). At the highest tier of the refined effects assessment, the performance of semi-field tests is proposed.

Schaeffer *et al.* (2008) assessed structural effects on the soil community using TMEs considering it a suitable tool at the semi-field level. It represents an appropriate higher-tier test system to investigate the impact of pesticides and chemicals on the structure of the soil community and biodiversity. TMEs comprise various trophic levels of the soil community reflecting both direct and indirect effects of chemicals. The TME should contain undisturbed soil cores, for example, from an established grassland, containing microfauna, for example, microarthropods, enchytraeids, nematodes and microorganisms. Efforts should be made to link and quantify the exposure, for example, by chemical analyses and modelling, and the effects of pesticides in the TME systems. TMEs allow for the investigation of the intrinsic recovery of the soil mesofauna. However, field studies are ecologically even more relevant than TME studies, allowing for the investigation of recolonization.

For proper use of TME

Species homogeneity and sufficient abundance of sensitive organisms should be ensured for pre-screening of the soils. The soil moisture is of key importance for the biological activity of the organisms. The dimensions and size of the soil column should be optimized in terms of number and size of samples to be withdrawn along with minimization of boundary effects. The sampling frequency is dependent on dissipation rate and fate of the test substance. For statistical analysis of the experimental results, appropriate uni- and multivariate methods should be used (principal response curves); a dose–response design should be applied to derive EC_x and NOEC at the community and population levels; the statistical power should be increased by an appropriate number of replicates; and the minimum detectable difference (MDD) should be calculated.

Generally, the systems to promote and support more environmentally benign use of pesticides have relied either on hazard ranking or on very simplified indicators of risk. Levitan *et al.* (1995) highlighted issues on: (i) creating an integrated assessment (most were groundwater-human toxicity-based); (ii) integrating environmental and economic information; and (iii) balancing ease of use and complexity of process. Hazard-based systems rank pesticides solely on their potential to cause harm with no consideration of the intensity or probability of environmental exposure. This is the approach used to date within EMA (a computer-based environmental management tool).

According to Lewis *et al.* (1997) in EMA, a risk assessment system, the toxicity of pesticides towards humans, animals, birds, aquatic life and bees is indicated on labels of their containers, thereby helping the user select more benign chemicals. The fate of the pesticides was considered through simple rules such as weighting factors that varied with the distance between sprayed area and surface water. Potential for groundwater contamination was again hazard-based, associated with strength of sorption to soil and rate of degradation in soil (Gustafson, 1989), but not considering rate of application, soil type, depth to groundwater, etc. Wijnands (1997) describes a whole-farm system to support integrated pest management where environmental exposure to pesticides for air, soil and groundwater compartments is calculated as a function of the application rate of the chemical and its vapour pressure, degradation rate in soil and mobility (strength of sorption), respectively. Roussel *et al.* (2000) have developed a system based on fuzzy logic for pesticide use in Brittany, France (IPEST-B). Simple decision rules are used to assign an environmental effect score (0 to 1) based on whether a number of input parameters are 'favourable' or 'unfavourable'. For example, risk of contamination of groundwater and potential for harm to the population requires input related to pesticide properties, type of application, soil/geological conditions and acceptable daily intake for human consumption. System for Predicting the Environmental impact of Pesticides (SyPEP) (Pussemier, 1999) adopts an alternative approach where exposure is explicitly considered via the System for the Evaluation of Pesticide Transport to Waters (SEPTWA) model to estimate pesticide emissions to surface and groundwater. However, a single scenario is adopted to represent Belgian conditions and the system reverts to a standardized ranking index for pesticides with no consideration of local conditions. The environmental yardstick assesses the impact from pesticide use in The Netherlands and has a range of uses from a management tool for farmers and agricultural consultants to a policy evaluation tool (Reus and Leendertse, 2000). The calculation methods are risk-based, so that groundwater concentrations are calculated using the Pesticide Leaching and Accumulation (PESTLA) model and risk to aquatic organisms combines losses to surface water via drift with acute toxicity data for the compound. At the farm scale, the system comprises a number of look-up tables with farmer inputs restricted to choice of chemical, dose rate, soil organic matter content, time and method of application, and distance from sprayed area to surface water.

Fate and exposure considerations

The readily degradable pesticides (DT90 <100 days) should be applied to soil semi-field systems according to good agricultural practice (GAP); however, while applying persistent pesticides, formation of an accumulation plateau of their concentration in the soil should be taken into account. Soils in semi-field tests should always be covered with plants (at least following application), and irrigation of the soil should be adapted to regional circumstances and as per the water demands of the cover crop and soil fauna. The risk assessment for in-crop and off-crop concerns may need a different approach because of differences in soil cover. In order to enable extrapolation of the test results to other environmental conditions, on both a spatial and a temporal scale, different soils, soil moistures, seasons, etc., should be tested (ideally during method development). The recommendations should be established for the soils for semi-field studies. To evaluate the suitability of semi-field test systems with respect to the sensitivity of the soil biota, toxic (positive) controls should be applied to separate test systems. Ideally, analytical confirmation of the added pesticide concentrations should be achieved in the same soil strata that will be sampled for the ecotoxicological effect assessment (Schaeffer *et al.*, 2008).

Effect considerations

The TME can mirror field situations for the sensitive soil organism groups present in sufficient abundance within the cores and that are exposed to the test substance. The soil micro- and mesofauna (e.g. nematodes, microarthropods, enchytraeids) and micro-organisms are observed to be the main groups targeted by this particular system. A pre-application incubation time is needed to allow communities to adapt and stabilize within the TME after the disturbance caused by the extraction of soil cores. The TME shows strong preference for grasslands due to their higher diversity and stability of soil organism communities. Existing experience with TMEs supports sizes between 10 and 50 cm in diameter, and about 40 cm in depth. The soil moisture content should be measured and compensated regularly. Micro- and mesofauna community structure is often the most relevant effect parameter. Usually, the duration of a TME study is 16 weeks, and current experience indicates that they have a limited life span of up to 1 year. Functional endpoints as well as microbial activity (e.g. soil enzymes), respiration and functional diversity can also be assessed in TME studies. Moreover, integrative functional parameters like litter decomposition or the feeding rate via bait-lamina can also be measured.

The semi-field tests are the higher-tier methods for the evaluation of soil. Risk managers and decision makers may determine the appropriate protection goals for agricultural soil. Both literature research and field work are necessary to describe the ecology, sensitivity against plant protection products (PPPs) and recovery timeframes for mesofauna for both in-crop and off-crop communities. Data from semi-field tests needs to be extrapolated for different soil types (texture, pH, cation exchange and water-holding capacities, organic matter content, nutrient status), as well as temperature, water (irrigation vs. natural climatic conditions) and light regimes. Comparative experimental studies and models are considered necessary to determine first the variability of TME data, both within TMEs with respect to subsampling and between independent TME cores, and second, to compare the variability of TME data with that of field tests. The published data can be utilized to establish the literature database with regard to semi-field and field tests (Jänsch *et al.*, 2006). It is necessary to investigate whether subsampling of a soil core at the defined sampling intervals will affect the soil communities of the residual soil columns compared to sacrificing individual soil cores at each sampling interval. Subsampling will considerably reduce the number of soil cores to be established for testing pesticide side effects on soil communities, and therefore will enhance the practicability of semi-field tests. Research is also needed to distinguish whether techniques like principal response curves are sufficient to explore community level effects, or whether diversity indices can play an additional role. This has been done for the aquatic compartment (Van den Brink and Ter Braak, 1998), where the principal response curves (PRC) approach was considered to be most appropriate. The duration of the test may be extended to over the next season or the next year, if recovery of the organisms from pesticide effects after crop harvest is not achieved; the long-term recovery of soil communities may be followed in that case, if necessary. Considering the properties of TMEs, in particular the limited size, other methods might be more appropriate when studying recovery (e.g. field studies). From a regulatory point of view, a need exists for suitable triggers for conducting semi-field studies, including TMEs, and guidance on summarizing and evaluating results from semi-field methods, including TME studies.

7.6 Risk Management

Risk management is the evaluation and selection of mitigation options. Risk managers use risk assessment as an important tool to determine the acceptability of a level of exposure and then reduce exposures to that level. Unlike risk assessment, risk management is not based solely on scientific considerations, since it also involves social, economic and legal considerations to make regulatory and policy decisions.

The process of risk assessment is separate from risk management. Risk assessment often drives risk management, but risk management cannot and does not drive risk assessment at the Department of Pesticide Regulation. Risk assessments and risk management options are developed by separate Department of Pesticide Regulation branches and are described in separate formal documents.

The organophosphate pesticides are known to inhibit acetylcholinesterase (AChE) leading to various adverse health effects including smooth muscle contractions (e.g. abdominal cramps), glandular secretions, skeletal muscle twitching and flaccid paralysis (US EPA, 2000). These pesticides can reach humans through the ambient air, food chain and drinking water, or by direct contact. There is, therefore, a need to evaluate cumulative doses from multiple pathways for pesticide risk assessments in humans. The US EPA has suggested combination of pesticide intakes from food, residential and water pathways for cumulative assessment of organophosphate pesticide. For human risk assessment, the chemical concentration in the environment is usually estimated through modelling. Regional mass balance models, such as fugacity-based multimedia environmental fate models, have been introduced in the assessments of human exposure to various organic compounds including pesticides (Bennett *et al.*, 1999; Hertwich *et al.*, 1999; Dunn, 2004; Morra *et al.*, 2006; Strang, 2006; McKone *et al.*, 2007). Analytically simpler models with fewer parameters provide explicit and reliable predictions of chemical distributions in various environmental media. Both temporal trends and spatial variability are critical in the risk assessment of organophosphate pesticides, especially when the assessment results are used as guidance for future pesticide management practices. To properly describe the relationship between the use and resultant risks of organophosphate pesticides, a risk assessment framework is required to account for spatiotemporal variations in pesticide distribution and potential risk to human health.

Luo and Zhang (2009) studied CRA of human exposure to 13 widely used organophosphate pesticides, assumed to represent the total organophosphate usage, in the northern San Joaquin Valley of California. The transport and fate of organophosphate pesticides was investigated utilizing various models; that is, the multimedia environmental fate model, estimation of exposure media concentrations and multi-pathway pesticide intakes, and determination of the temporal trends and spatial variability in the cumulative risk characterizations.

7.6.1 Multimedia environmental fate model

According to the studies of Luo *et al.* (2007a), the multimedia environmental fate model was designed for analysing unsteady-state dispersion and distribution of chemicals in multimedia environmental systems at the watershed scale. The transport and fate processes of chemicals were considered within environmental compartments of air, canopy, soil layers, surface water and sediment. By assuming environmental compartments were completely mixed and chemical equilibrium was established instantaneously between the phases within each compartment, a fugacity approach was utilized to formulate the mechanisms of diffusion, advection and transformation reactions. Mass exchange across compartmental boundaries was represented mathematically by first-order equations as follows:

$$Q_{ij} = M_{ij} \times N_{ij} = \left(\frac{D_{ij}}{V_i Z_i}\right) N$$

Where Q_{ij} (mol/day) is the unidirectional chemical flux from compartment i to j, M_{ij} (day) is the transfer rate constant from i to j, N_{ij} (mol) and Z_i (mol/Pa/m^3) are the inventory and fugacity capacity of the chemical in i, respectively, D_{ij} (mol/Pa/day) is the overall transport coefficient (Mackay-type D value) from i to j, and V_i (m^3) is the compartment volume. The transport coefficients were obtained from hydrometeorological simulation results or estimated based upon empirical and semi-theoretical methods. Mass balance of the chemical could be described by a set of differential equations developed for environmental compartments connected in an Eulerian system:

$$\frac{dN_i}{d_t} = S_i + \sum_i (M_{ij} N_j - M_{0i} N_i)$$

Where j is a running index for all compartments connected to compartment i, S_i (mol/day) is a source term of chemical application or emission in i and M_{0i} (day) is the total loss rate constant of the chemical in i. The model has been successfully applied in the field conditions of the Connecticut River Basin and the Great Lakes region of the northern San Joaquin Valley watershed, using organic compounds with various physiochemical properties as test agents (Luo and Yang, 2007; Luo *et al.*, 2007b). The model performance yielded reasonable agreement with measured data.

The organophosphate pesticide application to plant, surface soil and air can be converted into effective environmental release levels, where the release fraction is determined by pesticide products and suggested ratios in the literature (Strang, 2006; McKone *et al.*, 2007). The available data on pesticide loads and concentrations in the surface water, in the air and in precipitation act as the primary elements in the model evaluation. Luo and Zhang (2009) compared the predicted dissolved loads to observed data at the watershed outlet by the US Geological Survey (USGS gauge #11303500, San Joaquin River at Vernalis). The authors compared simulation results to the measurements by percentiles at 50%, 75% and 95%, due to low detection frequency of pesticides. A similar method has been used in other studies in evaluating model prediction with limited observation data (US EPA, 2006).

7.6.2 Multimedia and multi-pathway risk assessment

Human exposure to organophosphate pesticides could be estimated from the mass exchange between humans and contaminated exposure media via inhalation, ingestion and dermal contact. According to the US EPA OPP, the steady state for organophosphate pesticides is reached after approximately 21–28 days of exposure (US EPA, 2006). Through this model, subchronic exposure at monthly intervals of a residential population exposed to organophosphate pesticides via everyday activities, like breathing, dining, drinking, etc., can be studied. For an individual exposure pathway, the generic equation for a specific organophosphate pesticide intake (I, g/kg [body weight]/day) is calculated as:

$$I = \frac{C^E \times CR \times EF}{BW}$$

Where C^E is concentration of the specific chemical in exposure media (g/kg [water]), CR is the daily consumption rate (kg [drinking water]/day), EF is the exposure frequency that is set to one for inhalation and ingestion exposures, and BW is the average body weight (kg). The intake calculations performed for all potential exposure pathways are summed up as total intake of the chemical for a receptor. The highest monthly total intake during a year is used to estimate subchronic exposure for that year under this study.

Concentration in exposure media

Under this study, only the exposure pathways related to residential inhalation and dietary ingestion are considered, and not the dermal exposure to water, soil or air. The US EPA (2005) suggests contribution of dermal exposure to soils to overall risk, as typically small. McKone *et al.* (2007) investigated risk assessments in the Salinas Valley of California indicating that human exposure by dermal contact accounted for less than 5% of total exposure to the major organophosphate pesticides via everyday activities. Using the multimedia environmental fate model, the pesticide concentrations in the exposure media can be calculated based on the simulation results of the model. Following the procedures of US EPA (2005), concentration of pesticides in exposure media could be converted from those in environmental media by applying transfer factors (TFs),

$$C^E = \sum_i TF \times C_i$$

Where C_i (g/m^3) is the concentration of a chemical in environmental medium i, with i as a running index for all involved environmental media. The gram unit is chosen for chemical concentration in order to simplify the conversion from chemical inventory (mol) and fugacity (Pa) used in multimedia simulation results. Burow *et al.* (1998) observed organophosphate pesticide concentrations in groundwater were significantly lower than those in surface water, following model simulation and the groundwater sampling programme. There may be overestimation of exposure by excluding underground drinking water; however, it may be helpful in establishing an upper limit of risk through water ingestion.

Luo and Zhang (2009) calculated pesticide concentrations in above ground products directly, using multimedia environmental simulation, while pesticides in the root crops were assumed to be in equilibrium with the concentration in the root zone. Similarly, pesticide residues in animal products may be estimated based on an evaluation of the animal's ingestion and accumulation rate in tissues. US EPA (2005) recommended a simplified calculation for chemical concentration in animal products ($C^E{}_{\mathrm{Animal}}$) from plant (forage, silage and/or grain) and surface soil ingestion:

$$C^E{}_{Animal} = BTF \times (C_p \times CR_p + C_g \times CR_g)$$

Where C_p and C_g (g/m^3) are chemical concentrations in plants and ground soil, respectively, CR_p and CR_g (kg[dry weight]/day) are the corresponding consumption rates and BTF (day/kg [tissue]) is the biotransfer factor for the animal. The BTF was estimated based on a regression equation developed by the Research Triangle Institute (RTI, 2005):

$$\log\left(\frac{BTF}{FC}\right) = -0.099\ (\log K_{ow})2 - 1.07\ \log K_{ow} - 3.56$$

Where K_{ow} is the octanol–water partition coefficient, and FC is the average fat content of the animal product, as suggested by US EPA (2005). The above equation is developed for organic chemicals having a log (K_{ow}) between −0.67 and 8.20. However, the equation was applied to acephate, methamidophos and oxydemeton-methyl with log (K_{ow}) values slightly lower than −0.67.

A number of environmental risk assessment tools are reported for pesticides and are listed in Table 7.1.

Table 7.1. List of the reviewed systems, in approximate order of publication. Numbering of systems is arbitrary. (From NAESI, 2005.)

System name	Country	Authors	System ID
Environmental impact quotient (EIQ)	USA	Kovach *et al.* (1992)[a] in Van der Werf (1996) and Levitan (1997)	2
Dutch environmental yardstick for pesticides (EYP)	Netherlands	Reus (1996)[a] and Reus and Pak (1993)[a] in appendix to Reus *et al.* (1999); Reus and Leendertse (2000)	6
Environmental economic injury levels		Higley and Wintersteen (1992)[a] in Van der Werf (1996)	–
Adverse water quality impact		Hornsby (1992)[a] in Van der Werf (1996)	–
Stemilt growers integrated fruit production responsible choice point summary	USA	Stemilt Growers (1993)[a] in Levitan (1997)	3
Environmental potential risk indicator for pesticides (EPRIP)	Italy	Based on Del Re *et al.*[a] as described by Trevisan *et al.* (1999) in appendix to Reus *et al.* (1999)	13
PESTDECIDE	Australia	Penrose *et al.* (1995). Supersedes 'Rating Index' of Penrose *et al.* (1994)	4
Consumer Union's Indices of Trends in Agricultural Pesticide Risk	USA	Benbrook *et al.* (1996), also reviewed by Levitan (1997)	7
Use of pesticides in California Environmental Health Policy Program Ranking System	USA	Pease *et al.* (1996)[a] in Levitan (1997)	10
CHEMS	USA	Swanson *et al.* (1997)	
SYNOPS	Germany	Version 1.1 Gutsche and Rossberg (1997); Version 2 Gutsche and Rossberg (1998)	9
Environmental performance indicators for pesticides (p-EMA)	UK	Lewis *et al.* (1997a; b)[a] in appendix to Reus *et al.* (1999); Brown *et al.* (2003); Hart *et al.* (2003)	12
USDA ERS Chronic and Acute Risk Indicators of Pesticide Use	USA	Barnard (1997)[a] in Levitan (1997)	8
Ipest	France	Van der Werf and Zimmer (1998)	5
The Hasse Diagram (HD)	Denmark	In appendix to Reus *et al.* (1999); also in Sørensen *et al.* (1998)	11
System for predicting the environmental impact of pesticides (SyPEP)	Belgium	Pussemier (in appendix to Reus *et al.*, 1999)	14
Pesticide Environmental Risk Indicator (PERI)	Sweden	Nilsson (in appendix to Reus *et al.*, 1999)	15
Bees Risk Indicator (BRI)	Italy	Villa *et al.* (2000)	16
The RIVM Indicator: AARI (and ATRI)	Netherlands	In the OECD (1999–2000); also Luttik, pers. comm.[a]	17
Danish Load Index (DLI)	Denmark	Not reported[a]; discussed in OECD (2004)	18
European Standard Characteristic Of non-target arthropod Regulatory Testing (ESCORT_2)	Europe	Candolfi *et al.* (2000)	19
Frequency of Application (FA)	Denmark	Not reported[a]; discussed in OECD (2004)	20
Norwegian Indicator (NARI)	Norway	Norwegian Agricultural Inspection Services (2000)	22
Short-term Pesticide Risk Index for Hypogean Soil systems (PRIHS-1)	Italy	Finizio *et al.* (2001)	24

continued

Table 7.1. Continued.

System name	Country	Authors	System ID
Long-term Pesticide Risk Index for Hypogean Soil systems (PRIHS-2)	Italy	Finizio *et al.* (2001)	24
Short-term Pesticide Risk Index for Epygean Soil systems (PRIES-1)	Italy	Finizio *et al.* (2001)	24
Long-term Pesticide Risk Index for Epygean Soil systems (PRIES-2)	Italy	Finizio *et al.* (2001)	24
Short-term Pesticide Risk Index for Surface Water Systems (PRISW-1)	Italy	Finizio *et al.* (2001)	24
Long-term Pesticide Risk Index for Surface Water Systems (PRISW-2)	Italy	Finizio *et al.* (2001)	24
Environmental Risk for pesticides (ERIP); for Hypogean or Epygean Soil Systems and Surface Water Systems	Italy	Finizio *et al.* (2001)	24
Multi Attribute Toxicity Factors (MATF)	USA	Benbrook *et al.* (2002)	21
Pesticide Occupational and Environmental Risk (POCER)	Belgium	Vercruysse and Steurbaut (2002)	23
Ecological Relative Risk (EcoRR)	Australia	Sánchez-Bayo *et al.* (2002)	26
Probability of bird mortality	Canada	Mineau (2002; 2005)	25
SCRAM	USA	Mitchell *et al.* (2002)	29
Recommendations for Prioritizing Risk Reduction Strategies Based on Toxicity Loading by Commodity, By Region, and by Pesticide	Canada	WWF (2003)	28
PEI Relative Ranking System	Canada	Dunn (2004)	1
A Pesticide Priority List Evaluation Scheme: APPLES	Canada	Teed (2004); draft	27

[a]Original publication not retrieved; information from reviews.

7.7 Conclusion

Pesticides, with their inherent toxicity properties, produce adverse health effects at some levels of exposure. However, it is not true that every potential hazard poses a health risk. Hazard is a reflection of potential effect and risk is the likelihood it will occur. A number of 'indicators' and 'impact assessment systems' have been reported as useful in pesticide risk assessment. Indicators are confined to a framework of policy analysis and risk communication, while 'impact assessment systems' retain the ecosystemic perspective. Risk assessment of any toxic effects of chemical mixtures is extremely difficult due to the complexity and variability of chemical mixtures that may occur in the environment. Hazard identification, dose–response assessment, exposure assessment and risk characterization are four basic elements in risk assessment. Hazard identification involves the review and evaluation of a chemical's toxic properties, and dose–response assessment documents the quantitative relationship between dose and toxic effects. Exposure assessment examines what is known about the duration, frequency and level of exposure to potentially harmful pesticides and the associated human health effects. Risk characterization presents qualitative and quantitative estimates of the likelihood that any of the hazards associated with the pesticide will occur in exposed people. It is estimated using probabilistic and deterministic approaches. The probabilistic approach takes account of variability in pesticide dosages, exposure and frequency, whereas the deterministic approach is based on certainties of causalities pertaining to risk related to pesticide use. US EPA uses the deterministic approach, also known as the RQ method, to compare toxicity to environmental exposure. Under OPP programmes, the potential human risks associated with multi-chemical and multi-pathway exposures to pesticides are evaluated and estimated using CRA. CRA helps in making regulatory decisions in order to safeguard public health and sensitive subpopulations, including infants and children. Pesticides may lead to spatial and temporal changes in soil biological communities in the agricultural landscape. Soil acts as a habitat for a variety of organisms and a source of energy. Semi-field methods focus on those non-target organisms that predominantly inhabit the soil profile. Species homogeneity and sufficient abundance of sensitive organisms should be ensured for pre-screening of the soils. Under the risk assessment system, the toxicity of pesticides towards humans, animals, birds, aquatic life and bees should be indicated on the labels of their containers, thereby helping the user to select more benign chemicals. Risk managers use risk assessment as an important tool to determine the acceptability of a level of exposure and then reduce exposures to that level. Unlike risk assessment, risk management is not based solely on scientific considerations, since it also involves social, economic and legal considerations in making regulatory and policy decisions.

References

Baird, S.J.S., Cohen, J.T., Graham, J.D., Shlyakhter, A.I. and Evans, J.S. (1996) Noncancer risk assessment: A probabilistic alternative to current practice. *Human and Ecological Risk Assessment* 2, 78–99.

Barnard, C. (1997) Pesticide use. In: Anderson, M. (ed.) *Agricultural Resources and Environmental Indicators*. USDA ERS, Washington, DC.

Benbrook, C., Groth, E. III, Halloran, J.M., Hansen, M.K. and Marquardt, S. (1996) *Pest Management at the Crossroads*. Consumers Union, Yonkers, NY, p. 272.

Benbrook, C.M., Sexson, D.L., Wyman, J.A., Stevenson, W.R., Lynch, S., Wallendal, J., Diercks, S., Van Haren, R. and Granadino, C.A. (2002) Developing a pesticide risk assessment tool to monitor progress in reducing reliance on high-risk pesticides. *American Journal of Potato Research* 79, 183–199.

Bennett, D.H., Kastenberg, W.E. and McKone, T.E. (1999) A multimedia, multiple pathway risk assessment of atrazine: The impact of age differentiated exposure including joint uncertainty and variability. *Reliability Engineering and System Safety* 63, 185–198.

BIS (2006) Indian Standard. Hazard identification and risk analysis – code of practice. ICS 13.100. Occupational Safety and Health and Chemical Hazards Sectional Committee, CHD 8. Bureau of Indian Standards. May 2006. Available at: https://ia601901.us.archive.org/12/items/gov.law.is.15656.2006/is.15656.2006.pdf (accessed 30 July 2018).

Brown, C.D., Hart, A., Lewis, K.A. and Dubus, I.G. (2003) p-EMA (I): Simulating the environmental fate of pesticides for a farm-level risk assessment system. *Agronomie* 23, 67–74.

Bruce, E.D., Abusalih, A.A., McDonald, T.J. and Autenrieth, R.L. (2007) Comparing deterministic and probabilistic risk assessments for sites contaminated by polycyclic aromatic hydrocarbons (PAHs). *Journal of Environmental Science and Health. Part A, Toxic/hazardous Substances and Environmental Engineering* 42(6), 697–706.

Burow, K.R., Shelton, J.L. and Dubrovsky, N.M. (1998) Occurrence of nitrate and pesticides in ground water beneath three agricultural land-use settings in the eastern San Joaquin Valley, California, 1993–1995. Water-Resources Investigations Report 97-4284, US Geological Survey, National Water-Quality Assessment Program, Sacramento, California.

Candolfi, M.P., Bakker, F., Cañez, V., Miles, M., Neumann, C., Pilling, E., Primiani, M., Rominj, K., Schmuck, R., Storck-Weyhermüller, S., Ufer, A. and Waltersdorfer, A. (2000) Sensitivity of non-target arthropods to plant protection products: could *Typhlodromus pyri* and *Aphidus* spp. be used as indicator species? *Chemosphere* 39, 1357–1370.

Case Study 12 (2014) Probabilistic analysis in cumulative risk assessment of organophosphorus pesticides. In: *Risk Assessment Forum White Paper: Probabilistic Risk Assessment Methods and Case Studies*. EPA/100/R-14/004 July 2014. US EPA, Washington, DC.

Case Study 4 (2014) Probabilistic analysis of dietary exposure to pesticides for use in setting tolerance levels. In: *Risk Assessment Forum White Paper: Probabilistic Risk Assessment Methods and Case Studies*. EPA/100/R-14/004 July 2014. US EPA, Washington, DC.

Cassee, F.R., Groten, J.P., van Bladeren, P.J. and Feron, V.J. (1998) Toxicological evaluation and risk assessment of chemical mixtures. *Critical Reviews in Toxicology* 28, 73–101.

Committee on Toxicity of Chemicals in Food, Consumer Products and the Environment (2002) Risk assessment of mixtures of pesticides and similar substances. Food Standards Agency, London. Available at: https://cot.food.gov.uk/sites/default/files/cot/reportindexed.pdf (accessed 31 July 2018).

Dunn, A.M. (2004) A *Relative Risk Ranking of Pesticides Used in Prince Edward Island.* Environment Canada, Environmental Protection Branch, Atlantic Region, Dartmouth, Nova Scotia.

Durkin, P. and Diamond, G. (2002) Neurotoxicity, immunotoxicity, and endocrine disruption with specific commentary on glyphosate, triclopyr, and hexazinone: Final report. Forest Service Leslie Rubin, COTR Animal and Plant Health Inspection Service (APHIS) Biotechnology, Biologics and Environmental Protection Environmental Analysis and Documentation United States Department of Agriculture, New York.

Environmental Resources Management (2001) *Hazardous Waste Management* (2nd edn). Instructors' Manual, chapter 14, Quantitative risk assessment, p. 837. Available at: https://www.coursehero.com/file/9758271/Quantitative-Risk-Assessment-Review (accessed 17 August 2018).

Finizio, A., Calliera, M. and Vighi. M. (2001) Rating systems for pesticide risk classification on different ecosystems. *Ecotoxicology and Environmental Safety* 49, 262–274.

Gustafson, D.I. (1989) Groundwater Ubiquity Score: A simple method for assessing pesticide leachability. *Environmental Toxicology and Chemistry* 8, 339–357.

Gutsche, V. and Rossberg, D. (1997) SYNOPS 1.1 – a model to assess and compare the risk potential of active ingredients of plant protection products. *Agriculture, Ecosystems and Environment* 64(2), 181–188.

Gutsche, V. and Rossberg, D. (1998) SYNOPS 2. In: Day, E. (ed.) *Integrated Pest Management Measurement Systems Workshop.* Workshop held 12–13 June 1998, Chicago, IL, pp. 99–113.

Hart, A., Brown, C.D., Lewis, K.A. and Tzilivakis, J. (2003) p-EMA (II): Evaluating ecological risks of pesticides for a farm-level risk assessment system. *Agronomie* 23, 85–96.

Hattis, D. and Anderson, E. (1999) What should be the implications of uncertainty, variability, and inherent 'biases' in 'conservatism' for risk management decision making? *Risk Analysis* 19, 95–107.

Hattis, D. and Burmaster, D.E. (1994) Assessment of variability and uncertainty distributions for practical risk analyses. *Risk Analysis* 14, 713–730.

Hattis, D. and Minkowitz, W.S. (1996) Risk evaluation: Criteria arising from legal traditions and experience with quantitative risk assessment in the United States. *Environmental Toxicology and Pharmacology* 2, 103–109.

Hertwich, E.G., McKone, T.E. and Pease, W.S. (1999) Parameter uncertainty and variability in evaluative fate and exposure models. *Risk Analysis* 19, 1193–1204.

Higley, L.G. and Wintersteen, W.K. (1992) A new approach to environmental risk assessment of pesticides as a basis for incorporating environmental costs into economic injury levels. *American Entomologist* 38, 34–39.

Hornsby, A.G. (1992) Site-specific pesticide recommendations: The final step in environmental impact prevention. *Weed Technology* 6(3), 736–742.

Jänsch, S., Frampton, G.K., Römbke, J., van den Brink, P.J. and Scott-Fordsmand, J.J. (2006) Effects of pesticides on soil invertebrates in model ecosystem and field studies: A review and comparison with laboratory toxicity data. *Environmental Toxicology and Chemistry* 25, 2490–2501.

Kovach, J., Petzoldt, C., Degni, J. and Tette, J. (1992) A method to measure the environmental impact of pesticides. *New York's Food and Life Sciences Bulletin* 139, 1–8.

Levitan, L. (1997) An overview of pesticide impact assessment systems (a.k.a. 'pesticide risk indicators') based on indexing or ranking pesticides by environmental impact. Background paper presented at the Organisation of Economic Co-operation and Development (OECD) workshop on pesticide risk indicators, 21–23 April 1997, Copenhagen, Denmark, 89pp.

Levitan, L. (2000) 'How to' and 'why': Assessing the enviro-social impacts of pesticides. *Crop Protection* 19, 629–636.

Levitan, L., Merwin, I. and Kovach, J. (1995) Assessing the relative environmental impacts of agricultural pesticides: The quest for a holistic method. *Agricuture, Ecosystems and Environment* 55, 153–168.

Lewis, K.A., Newbold, M.J., Hall, A.M. and Broom, C.E. (1997) Eco-rating system for optimising pesticide use at farm level – Part 1: Theory and development. *Journal of Agricultural Engineering Research* 68, 271–279.

Lewis, K.A., Newbold, M.J. and Broom, C.E. (1997b) Eco-rating systems for optimizing pesticide use at farm level. Part 2: Evaluation, examples and piloting. *Journal of Agricultural Engineering Research* 68, 281–289.

Luo, Y. and Yang, X. (2007) A multimedia environmental model of chemical distribution: Fate, transport, and uncertainty analysis. *Chemosphere* 66, 1396–1407.

Luo, Y. and Zhang, M. (2009) Multimedia transport and risk assessment of organophosphate pesticides and a case study in the northern San Joaquin Valley of California. *Chemosphere* 75, 969–978.

Luo, Y., Gao, Q. and Yang, X. (2007a) Dynamic modeling of chemical fate and transport in multimedia environments at watershed scale – I: theoretical considerations and model implementation. *Journal of Environmental Management* 83, 44–55.

Luo, Y., Gao, Q. and Yang, X. (2007b) Dynamic modeling of chemical fate and transport in multimedia environments at watershed scale – II: trichloroethylene test case. *Journal of Environmental Management* 83, 56–65.

McKone, T.E., Castorina, R., Harnly, M.E., Kuwabara, Y., Eskenazi, B. and Bradman, A. (2007) Merging models and biomonitoring data to characterize sources and pathways of human exposure to organophosphorus pesticides in the Salinas Valley of California. *Environmental Science and Technology* 41, 3233–3240.

Mineau, P. (2002) Estimating the probability of bird mortality from pesticide sprays on the basis of the field study record. *Environmental Toxicology and Chemistry* 21, 1497–1506.

Mineau, P. (2005) A review and analysis of study endpoints relevant to the assessment of 'long term' pesticide toxicity in avian and mammalian wildlife. *Ecotoxicology* 14(8), 775–799.

Mitchell, R.R., Summer, C.L., Blonde, S.A., Bush, D.M., Hurlburt, K., Snyder, E.M. and Giesy, J.P. (2002) SCRAM: A scoring and ranking system for persistent, bioaccumulative, and toxic substances for the North American Great Lakes – resulting chemical scores and ranking. *Human and Ecological Risk Assessment* 8, 537–557.

Monte Carlo (2001) A technique for characterizing the uncertainty and variability in risk estimates by repeatedly sampling the probability distributions of the risk equation inputs and using these inputs to calculate a range of risk values. US EPA, Washington, DC.

Morra, P., Bagli, S. and Spadoni, G. (2006) The analysis of human health risk with a detailed procedure operating in a GIS environment. *Environment International* 32, 444–454.

National Agri-Environmental Standards Initiative (NAESI) (2005) National Agri-Environmental Standards Initiative. Environment Canada Gatineau. Development of comparative environmental risk assessment tools for pesticides in support of standard development at Environment Canada. Technical Series. Report No. 1–17. Environment Canada, Gatineau, QC, Canada.

National Research Council (1993) *Pesticides in the Diets of Infants and Children*. National Academy Press. Washington, DC. Available at: https://www.ncbi.nlm.nih.gov/books/NBK236275 (accessed 1 October 2018).

Norwegian Agricultural Inspection Services (2000) Pesticide risk indicators for health and environment – Norway. Norwegian Agricultural Inspection Services.Oslo, Norway. Available at: www.oecd.org/chemicalsafety/pesticides-biocides/1934217.pdf (accessed 1 October 2018).

Office of Pesticide Programs (OPP) (2002) Guidance on cumulative risk assessment of pesticide chemicals that have a common mechanism of toxicity. 14 January 2002, US Environmental Protection Agency, Washington, DC.

Organisation for Economic Co-operation and Development (OECD) (1999–2000) Survey of National Pesticide Risk Indicators. Available at: www.oecd.org/netherlands/1934202.pdf (accessed 1 October 2018).

Organisation for Economic Co-operation and Development (OECD) (2004) DRAFT. Summary report of the OECD project on pesticide terrestrial indicators. OECD, Paris.

Pease, W.S., Liebman, J., Landy, D. and Albright, D. (1996) Pesticide use in California: Strategies for reducing environmental health impacts. An Environmental Health Policy Program Report, Center for Occupational and Environmental Health, School of Public Health, California Policy Seminar, University of California, Berkeley, CA. Risk assessment model in appendix 1, pp. 63–93.

Penrose, L.J., Thwaite, W.G. and Bower, C.C. (1994) Rating index as a basis for decision making on pesticide use reduction and for accreditation of fruit produced under integrated pest management. *Crop Protection* 13, 146–152.

Penrose, L.J., Thwaite, W.G. and Maguire, M.J. (1995) Final report: A rating index as a basis for decision making on pesticide use reduction and IPM accreditation. NSW Agriculture, Horticultural Research and Development Corporation, Orange, Australia, June 1995a, p. 28.

Peterson, R.K.D. (2006) Comparing ecological risks of pesticides: The utility of a risk quotient ranking approach across refinements of exposure. *Pest Management Science* 62, 46–56. DOI:10.1002/ps.1126.

Peterson, R.K.D. and Schleier III, J.J. (2014) A probabilistic analysis reveals fundamental limitations with the environmental impact quotient and similar systems for rating pesticide risks. *PeerJ* 2, e364, DOI:10.7717/peerj.364.

Pussemier, L. (1999) A system for predicting the environmental impact of pesticides in Belgium. In: Del Re, A.A.M., Brown, C.D., Capri, E., Errera, G., Evans, S.P. and Trevisan, M. (eds) *Human and Environmental Exposure to Xenobiotics*. La Goliardica Pavese, Pavia, Italy, pp. 851–859.

Reus, J.A.W.A. (1996) An environmental yardstick for pesticides. Poster presented at the Third National IPM Symposium, February 1996, Washington DC.

Reus, J.A.W.A. and Leendertse, P.C. (2000) The environmental yardstick for pesticides: A practical indicator used in the Netherlands. *Crop Protection* 19, 637–641.

Reus, J.A.W.A. and Pak, G.A. (1993) An environmental yardstick for pesticides. *Mededelingen Faculteit Landbouwwetenschappen Rijksuniversiteit Gent* 58/2a, 249–255.

Reus, J., Leendertse, P., Bockstaller, C., Fomsgaard, I., Gutsche, V., Lewis, K., Nilsson, C., Pussemier, L., Trevistan, M., van der Werf, H., Alfarroba, F., Blümel, S., Isart, J., McGrath, D. and Seppälä, T. (1999) Comparing environmental risk indicators for pesticides. *Results of the European CAPER Project*. Centre for Agriculture and Environment (CML), The Netherlands, 184pp.

Roussel, O., Cavelier, A. and van der Werf, H.M.G. (2000) Adaptation and use of a fuzzy expert system to assess the environmental effect of pesticide applied to field crops. *Agricultural Ecology and Environment* 80, 143–158.

RTI (2005) Methodology for predicting cattle biotransfer factors. Prepared for US Environmental Protection Agency Office of Solid Waste. EPA Contract No. 68-W-03-042. Research Triangle Institute, Research Triangle Park, North Carolina.

Sánchez-Bayo, F., Baskaran, S. and Kennedy, I.R. (2002) Ecological relative risk (EcoRR): Another approach for risk assessment to pesticides in agriculture. *Agriculture, Ecosystems and Environment* 91 (1–3), 37–57.

Schäffer, A., Brink, P.J. van den, Heimbach, F., Hoy, S., Jong, F. de, Römbke, J., Sousa, J.P. and Roß-Nickoll, M. (2008) Semi-field methods are a useful tool for the environmental risk assessment of pesticides in soil. *Environmental Science and Pollution Research* 15(3), 176–177.

Schaeffer, A., Brink, P.J. van den, Heimbach, F., Hoy, S.P., Jong, F.M.W. de, Römbke, J., Roß-Nickoll, M. and Sousa, J.P. (2011) *Semi-field Methods for the Environmental Risk Assessment of Pesticides in Soil*. CRC Press, Abingdon, UK.

Sørensen, P.B., Morgensen, B.B., Gyldenkærne, S. and Rasmussen, A.G. (1998) Pesticide leaching assessment method for ranking both single substances and scenarios of multiple substance use. *Chemosphere* 36, 2251–2276.

Strang, D. (2006) Annapolis river watershed pesticide risk ranking. Clean Annapolis River Project, Annapolis Royal, Nova Scotia, Canada.

Swanson, M.B., Davis, G.A., Kincaid, L.E., Schultz, T.W., Bartmess, J.E., Jones, S.L. and George, E.L. (1997) A screening method for ranking and scoring chemicals by potential human health and environmental impacts. *Environmental Toxicology and Chemistry* 16, 372–383.

Teed, R.S. (2004) DRAFT. APPLES, a pesticide priority list: evaluation scheme for the CCME Water Quality Task group. Draft report prepared for the National Guidelines and Standards Office of Environment Canada and the Canadian Council of Ministers of the Environment and the National Guidelines and Standards Office, Gatineau, Canada.

Trevisan, M., Errera, G., Capri, E., Padovani, L. and Del Re, A.A.M. (1999) Environmental potential risk indicator for pesticides. In: Reus, J., Leendertse, P., Bockstaller, C., Fomsgaard, I., Gutsche, V., Lewis, K., Nilsson, C., Pussemier, L., Trevisan, M., van der Werf, H., Alfarroba, F., Blümel S., Isart J., McGrath D. and Seppälä T. (eds) *Comparing Environmental Risk Indicators for Pesticides: Results of the European CAPER Project*. Centre for Agriculture and the Environment, Utrecht, The Netherlands, pp. 141–147.

United States Environmental Protection Agency (US EPA) (1997) Guiding principles for Monte Carlo analysis, EPA/630/R-97/001. Risk Assessment Forum, US EPA, Washington, DC. Available at: https://www.epa.gov/sites/production/files/2014-11/documents/montecar.pdf (accessed 1 October 2018).

United States Environmental Protection Agency (US EPA) (1999a) Draft document. The Office of Pesticide Programs' Policy on determination of the appropriate FQPA safety factor(s) for use in the tolerance-setting process. 64 FR 48617 (10 May 1999). Office of Pesticide Programs, Office of Prevention, Pesticides, and Toxic Substances, Washington, DC. Available at: www.epa.gov/fedrgstr/EPA-PEST/1999/September/Day-07/o-p23197.htm (accessed 1 October 2018).

United States Environmental Protection Agency (US EPA) (1999b) Guidance for identifying pesticide chemicals and other substances that have a common mechanism of toxicity. 29 January 1999. 64 FR 5796. Office of Pesticide Programs,

Office of Prevention, Pesticides, and Toxic Substances. US EPA, Washington, DC.

United States Environmental Protection Agency (US EPA) (2000) The use of data on cholinesterase inhibition for risk assessments of organophosphorous and carbamate pesticides. US Environmental Protection Agency, Office of Pesticide Programs, Washington, DC.

United States Environmental Protection Agency (US EPA) (2002) Determination of the appropriate FQPA safety factor(s) in tolerance assessment. Office of Pesticide Programs, US Environmental Protection Agency. Washington, DC. Available at: https://www.epa.gov/sites/production/files/2015-07/documents/determ.pdf (accessed 1 October 2018).

United States Environmental Protection Agency (US EPA) (2005) Human health risk assessment protocol for hazardous waste combustion facilities. EPA530-R-05-006. US Environmental Protection Agency, Office of Solid Waste and Emergency Response, Washington, DC.

United States Environmental Protection Agency (US EPA) (2006) Organophosphate pesticides: Revised cumulative risk assessment. US Environmental Protection Agency, Washington, DC.

Van den Brink, P.J. and Ter Braak, C.J.F. (1998) Multivariate analysis of stress in experimental ecosystems by principal response curves and similarity analysis. *Aquatic Ecology* 32, 163–178.

Van der Werf, H.M.G. (1996) Assessing the impact of pesticides on the environment. *Agriculture, Ecosystems and Environment* 60, 81–96.

Van der Werf, H.M.G. and Zimmer, C. (1998) An indicator of pesticide environmental impact based on a fuzzy expert system. *Chemosphere* 36, 2225–2249.

Vercruysse, F. and Steurbaut, W. (2002) POCER, the pesticide occupational and environmental risk indicator. *Crop Protection* 21, 307–315.

Villa, S., Vighi, M., Finizio, A. and Sereni, G.B. (2000) Risk assessment for honeybees from pesticide-exposed pollen. *Ecotoxicology* 9, 287–297.

Wijnands, F.G. (1997) Integrated crop protection and environment exposure to pesticides: Methods to reduce the impact of pesticides in arable farming. *European Journal of Agronomy* 7, 251–260.

World Wildlife Fund (WWF) (2003) Recommendations for prioritizing risk reduction strategies based on toxicity loading by commodity, by region, and by pesticide: A discussion draft. WWF, Gland, Switzerland.

8 Co-Relation of Pesticide Risk and Body Characteristics

8.1 Introduction

We are all exposed to pesticides at some level, even if we take action to try to avoid them. However, because of their specific situations and/or characteristics, some people are particularly highly exposed to pesticides or particularly sensitive to their adverse effects. Farmers (and their families) who spray pesticides (pesticide applicators) can have a higher level of exposure to pesticides than the general population. These farmers and even greenhouse workers suffer the highest levels of exposure. Often, even when pesticide applicators are using the correct safety precautions, they are still subject to elevated levels of exposure that can be detected in their body tissues (Damalas and Koutroubas, 2016). This may be due to pesticide sprays drifting through the air from fields near farmers' homes, in addition to farm workers bringing their contaminated clothes and shoes back into the home after work. This is particularly of concern for infants and children because they may be more vulnerable to the toxic effects of some pesticides than adults (Arcury *et al.*, 2007).

8.2 Gender, Health and Work

8.2.1 Occupational health-related biological, psychological and social differences

The reproductive system of women may be affected by exposure to pesticides; for instance, prolonged standing in agricultural fields treated with pesticides during pregnancy can affect the birthweight of children and may produce some abnormalities. Reproductive problems such as miscarriages, low birth weight and malformations can arise from exposure to pesticides, solvents and organic pollutants, heavy workload, postural factors and shift work.

Due to sex differences in a number of important health-related variables such as average blood volume and oxygen consumption, the same physical load may exert greater strain on the average woman than on the average man, since women's average lifting strength is only 50% of men's (WHO, 2004). As far as the metabolism of toxins is concerned, it has been hypothesized that the average woman is at greater risk of harm from fat-soluble chemicals because of a higher proportion of fat tissue, thinner skin and slower metabolism. However, the percentage of fat varies in both sexes according to age, physical fitness and training. A positive association between occupational exposure to complex pesticide mixtures and the presence of chromosomal aberrations (CA), sister-chromatid exchanges (SCE) and micronuclei (MN) has been detected in the majority of studies examining these relationships. Most studies on cytogenetic biomarkers in pesticide-exposed workers have indicated the presence of some dose-dependent effects, worsening with increasing duration or intensity of exposure. Chromosomal damage induced by pesticides appears to have been transient in acute or discontinuous exposure, but cumulative in continuous exposure to complex agrochemical mixtures (Bolognesi, 2003).

Psychologically it has been suggested that men usually have higher self-esteem and confidence, while women are more emotionally expressive. Male–female differences in education, socialization and upbringing may lead to differences in the way workers manage their illnesses, their perception of risk and the propensity to take sick leave or to seek treatment. The sex-specific patterns of occupational health problems may, therefore, be due to differences in exposure to risk factors, in combination with biological, psychological and social differences. It is therefore important to examine occupational health research, implementation, policies, programmes and projects through a gender lens.

8.2.2 Health implications

An estimated 2.2 million men and women die from work-related injuries and diseases each year according to the International Labour Organization (ILO); however, these figures probably underestimate the true levels. The World Health Organization (WHO) estimates that in Latin America, for example, only 1–4% of all occupational diseases are reported. It is ironic that women's occupational health problems are underdiagnosed and women's claims for compensation for some health problems are preferentially refused; therefore, women's work-related injuries and diseases are even less accurately reported, since women's work is generally regarded as safe. Figure 8.1 shows Swedish data on compensation by sex. Women's work in many countries is invisible to health care personnel.

In summary, many of women's work-related accidents and diseases are not recorded as occupational, not compensated by work insurance systems and not included in thinking about occupational health. Cordier *et al.* (2015) examined the levels of thyroid stimulating hormone (TSH), free tri-iodothyronine (FT3) and free thyroxine (FT4) in child blood at 3 months ($n = 111$).

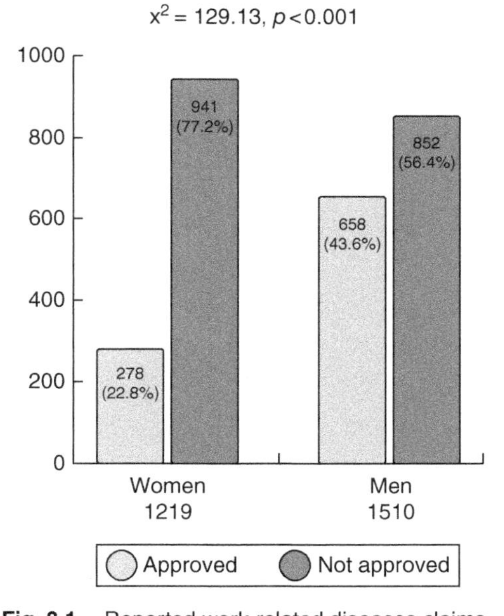

Fig. 8.1. Reported work-related diseases claims in Sweden, 1994–1997. (From WHO, 2004.)

Cord chlordecone was found to be associated with an increase in TSH in boys, whereas postnatal exposure was associated with a decrease in FT3 overall, and in FT4 among girls. Higher TSH level at 3 months was positively associated with the Ages and Stages Questionnaire (ASQ) score for fine motor development at 18 months in boys, but TSH did not modify the association between prenatal chlordecone exposure and poorer ASQ fine motor score.

Specific problems for women

Despite the positive effects of women's employment on their health, many jobs, especially those available to women in low-income countries or to poor, less-educated women in high-income countries, expose women to harmful working environments (WHO, 2004). In developed countries, women are more often exposed to some physical risks caused by highly repetitive movements, awkward postures and biological agents in jobs where there is higher exposure. Women are exposed more often to some psychosocial risk factors, such as psychological and sexual harassment and monotonous work. Because of their low status, women exert less control over their work environment, a condition associated with cardiovascular, mental and musculoskeletal ill health. The combination of paid and unpaid working conditions particularly affects women's health. Some women are known to shoulder extremely heavy physical workloads, such as carrying heavy loads and walking long distances, both at home and at work in low-income countries. This may lead to certain disorders and reproductive problems. Washing and cooking expose women to water-related diseases such as schistosomiasis, malaria and worms. Moreover, cooking on open stoves risks burns and exposure to smoke containing toxic pollutants in India, for example. There is increased risk of tuberculosis, particularly in Indian rural areas from the use of biomass fuels.

In many low-income countries, cash crop production of fruits, vegetables and flowers involves exposure to toxic chemicals. The adverse health effects of pesticide exposure include poisoning, cancer, skin diseases, miscarriages, premature births and malformed babies, as has been shown among floricultural workers in Colombia. Pesticides and chemicals are also widely used in higher-income countries, where agricultural workers are often excluded from occupational health and safety legislation.

Although the WHO strongly supports breastfeeding, it recognizes the potential health risks posed by the presence of environmental toxicants in breast milk, due to chemical exposure. Contamination of human milk is widespread due to persistent pesticides or chemical solvents. These chemicals tend to degrade slowly in the environment, to bioaccumulate in the food chain and to have long half-lives in humans. Many of these environmental pollutants have oestrogen-like activities and, thus they are called environmental oestrogen disruptors or xenoestrogens. Certain adverse health and reproductive outcomes are attributed to these chemicals in laboratory animals and in wildlife as well as in humans. Massart *et al.* (2005) highlighted those environmental chemicals that may affect child health through breastfeeding. Anything that limits breastfeeding is of concern in poor populations, since breast milk is vital for infant health. Chemicals can also interfere with the fertility-suppressing effects of breastfeeding and increase a woman's chance of conceiving before she is ready.

Breast milk is a non-invasive specimen that can be used to assess maternal and infant exposure to polychlorinated biphenyls (PCBs) and organochlorine pesticides (OCPs). Yalçın *et al.* (2015) studied the effect of OCPs on maternal–infant characteristics and evaluated the relationship between OCPs and maternal psychopathologies (Brief Symptom Inventory (BSI), Postpartum Bonding Questionnaire (PBQ), Mother-to-Infant Bonding Scale (MIBS) and Edinburgh Postnatal Depression Scale (EPDS)) and infant growth. Dichlorodiphenyltrichloroethane (DDT), beta-hexachlorocyclohexane (HCH), aldrin and heptachlor were detected in 89.3, 70.7, 58.7

and 34.7% of the samples, respectively. Mothers with low monthly family income had detectable DDTs less frequently. The frequencies of detectable heptachlor epoxide were significantly higher in mothers with gestational nausea. Anaemic mothers had more frequently detectable alpha-HCH. Z scores of head circumference were inversely correlated with beta-HCH and DDT levels. The heptachlor epoxide levels were positively correlated with PBQ, MIBS and indexes of BSI. No relationship was detected between EPDS and OCPs.

Gyalpo *et al.* (2012) investigated the effects of breastfeeding duration, parity and mother's age on DDT concentrations in mother and infant. Model results show that primiparous mothers have greater DDT concentrations than multiparous mothers, which causes higher DDT exposure of first-born children. DDT in the body mainly originates from diet. An increase in concentrations of OCP was observed in breast milk samples collected from populations of mothers in Manhiça, Mozambique, during 2002 and 2006 (Manaca *et al.*, 2011). However, the observed increases were not uniform and in some individuals high values (5100 ng/g lipid) were found. Significant differences were found between the concentrations of DDT and related compounds in breast milk according to parity, with higher concentrations in primiparae than multiparae women. These differences overcome the age effect in DDT accumulation between the two groups and indicate that women transfer a significant proportion of their body burden of DDT and its metabolites to their infants. The median concentrations of total PCB, DDT, ß-HCH and hexachlorobenzene (HCB) showed a clear rise with increasing age of mothers, whereas an increasing number of breastfed infants per mother led to a decrease. Zietz *et al.* (2008) observed a known declining trend of important xenobiotic substances in human milk of German mothers.

Higher body mass index (BMI) was associated with higher amounts of HCB and lower amounts of higher chlorinated PCBs, according to a study conducted by Mikeš *et al.* (2012). The concentrations of OCP in milk and placenta were found to be linearly correlated in similar patterns for Danish and Finnish samples (Shen *et al.*, 2007). Milk samples had higher levels of these pollutants than the placenta. However, the apparently uncorrelated compounds, such as α-HCH, pentachlorobenzene, pentachloroanisole and methoxychlor, generally accumulated more in the placenta, which might suggest a tissue-specific metabolic activity. Damgaard *et al.* (2006) observed an association between congenital cryptorchidism and some persistent pesticides in breast milk as a proxy for maternal exposure, suggesting that testicular descent in the foetus may be adversely affected. The pesticide levels in breast milk were noted to be significantly higher in boys with cryptorchidism. Solomon and Weiss (2002) observed diet as a major factor that influenced breast milk levels of persistent organic pollutants, with patterns in fish consumption playing a particularly significant role. The authors concluded improved global breast milk monitoring programmes would allow for more consistent data on trends over time, detection of new xenobiotics in breast milk and identification of disproportionately exposed populations. Snedeker (2001) explored the association between DDT, dichlorodiphenyldichloroethylene (DDE) and breast cancer risk. The factors influencing the interpretation included the effect of analytic methods, dietary factors, menopausal status, use of different types of control populations, lactation history, oestrogen receptor status, ethnic/racial subgroups, breast tumour characteristics and polymorphisms. The levels of analytes of OCP increased significantly depending on mother's age, as investigated by Nakagawa *et al.* (1999).

As expected by their differential lipid content, the milk samples had higher levels of OCP than the placenta. In agreement, when the distribution of the OCP levels in the maternal, placental and foetal compartments was analysed in a Spanish population, the concentration of endosulfan I and II in female adipose tissue was similar to that in breast milk, but higher than that of the placenta or cord blood.

Specific problems for men

The spermatogenesis system in men is also very sensitive to exposure to pesticides. Men have many more occupational accidents than women, and more fatal injuries in all jurisdictions where data are available. In developed countries, they are reported to have more exposure than women to chemicals.

Since reproduction has been viewed as a woman's domain, male reproductive health related to occupational exposures has been neglected. However, many chemicals, ionizing radiation, toxic contamination, high temperatures and sedentary work have been identified as hazardous to the male reproductive system.

8.2.3 Pesticide-induced health issues in men, women and infants

The detection rates of 27 OCP compounds analysed in breast milk samples collected from 142 pregnant mothers during lactation, in 2011–2012 in Shanghai, China, was in the range of 65.5 to 100% (Lu *et al.*, 2015). In particular, metabolites of DDT such as 2-chloro-1,1-bis(4-chlorophenyl)ethylene (DDMU), 2,2-bis(4-chlorophenyl)ethanol (DDOH), bis(4-chlorophenyl)ketone (DBP) and 4,4′-dichlorodiphenylmethane (DDM) were detected in most milk samples. DDTs, HCB and HCH were the dominant OCPs with mean levels of 316, 49.8 and 41.5 ng/g lipid content, respectively, whereas levels of methoxychlor, $\sum$drins, $\sum$heptachlor, $\sum$chlordane and $\sum$endosulfan were fairly low (0.87–5.6 ng/g lipid content). $\sum$OCPs in this study were much lower than those in human breast milk samples collected in 2002 and 2007. Consumption of higher amounts of fish was associated with higher milk levels of OCPs. Specific OCP patterns in breast milk samples from migrant mothers in Shanghai reflected features of OCP production, use and exposure in their home provinces. During a study by Lee *et al.* (2013) of 206 breast milk samples collected from 87 participants during lactation at <7, 15, 30 or 90 days postpartum in four cities in Korea, some OCP compounds were correlated with maternal age, BMI, parity and delivery mode. Certain types of dietary habits such as seafood and noodle consumption were significantly associated with $\sum$PCB and $\sum$OCP.

The majority of the 30 articles published between 1980 and 2013 on OCP residue in breast milk indicated the presence of two or more OCPs in the collected samples of breast milk. Pirsaheb *et al.* (2015) observed a statistically significant positive correlation between mother's age, her multiparity and concentration of OCPs in breast milk. The concentrations of a range of persistent organic pollutants (POPs) (i.e. polychlorinated biphenyls (PCBs), polybrominated diphenyl ethers (PBDEs) and OCPs) were measured using matched breast milk and infant faecal samples obtained from 10 mother–child pairs in a pilot study conducted by Chen *et al.* (2015). All infants were breastfed, with eight of them also receiving solid food at the time of faecal sampling. The faecal concentrations were observed to be strongly associated with milk concentrations. Different sources (external or internal) of exposure appeared to directly influence faecal concentrations of different chemicals based on different inter-individual variability in the faeces-to-milk concentration ratio Rfm. Overall, the matrix of faeces as an external measure of internal exposure in infants looks promising for some chemicals and is worth assessing further in larger datasets. Among the breast milk samples collected from 39 first-time mothers aged 20–30 years during 2007–2010 in New Zealand and analysed for POP including dioxin-like compounds and OCP, the estimated daily intake (EDI) of DDT (principally in the form of its metabolite p,p′-DDE) was the highest (1.6 mcg/kg/day), and above the tolerable daily intake (TDI) of 0.5 mcg/kg/day. 't Mannetje *et al.* (2014) observed the age of the mother was positively associated with higher EDIs for the infant, particularly for total toxic equivalency (TEQ) and total DDT. Infant daily intakes of chlorinated POPs through breast milk estimated for New Zealand were low or average by international comparison, and five times lower than 25 years ago.

Residues of OCP, organophosphate and synthetic pyrethroids were detected in 127 breast milk samples from Punjab, in a monitoring study by Sharma *et al.* (2014). Cyfluthrin was the leading pesticide detected in breast milk, contributing 31.28% to the total residue load. It was observed that the residue levels were decreasing with increase in parity and age of the mother, and cyfluthrin had the highest mean concentration of 90.63 ng/g in the first parity and 21.11 ng/g in the youngest age group. Residue levels were higher in the urban population than the rural population, although the difference between them was statistically non-significant ($P > 0.05$). The older age groups had higher concentrations of OCP (except for the comparison between the 21–23/24–28 years age groups) in breast milk in Guerrero, Mexico (Chávez-Almazán *et al.*, 2014).

Ostrea *et al.* (2014) monitored the outcome of long-term exposure to propoxur and pyrethroids in young children by analysing children's hair in the Philippines. At ages 2 and 4 years, the prevalence of propoxur exposure increased from 12.4% to 24.1% but dramatically decreased to 1.7% at 6 years. For bioallethrin, the prevalence of exposure steadily increased from 2 years to 4 years and to 6 years. Exposure to transfluthrin significantly increased from 4 years to 6 years. Between 4 and 6 years, there was a higher median concentration of propoxur at 4 compared to 6 years, and for transfluthrin and bioallethrin, at 6 compared to 4 years. Changes in the prevalence and concentration of exposure to propoxur and pyrethroids in children at 2, 4 and 6 years of age are related to the progress in ambulation of young children and to changes in the formulation of home spray pesticides. Qin, *et al.* (2011) observed positive correlations between concentrations of OCPs, PCBs, PBDEs and polycyclic aromatic hydrocarbons (PAHs), with respect to seafood diet habit, BMI and age. Concentrations of HCHs and DDTs in blood plasma of healthy Hong Kong residents were greater than those of other countries, and it was found that smoking, consumption of a seafood diet, BMI, and age could influence concentrations in human blood.

Wohlfahrt *et al.* (2011) observed that 247 children were born to women working in greenhouses in early pregnancy, and 168 were categorized as prenatally exposed to pesticides. The body fat percentage of the children was calculated at 3 months and at 6–11 years of age. Compared to unexposed children, birth weight and weight for gestational age were lower in the highly exposed children and medium exposed children. Exposed children had significantly larger increase in BMI Z score from birth to school age and highly exposed children had 15.8% larger skin folds and higher body fat percentages. If prenatally exposed to both pesticides and maternal smoking, the sum of four skin folds was 46.9% and body fat percentage 29.1% higher.

A study by Gaspari *et al.* (2011) observed a unique case where a 4-month-old girl presented with sexual development, including breast enlargement, menstruation, uterine length of 69 mm at ultrasonography and dramatically high oestrogen bioactivity, but no growth acceleration, pubic hair, pelvis masses or adrenal tumours. Gas chromatography with an electron capture detector and mass spectrometry detected pesticides (p,p-DDD, p,p-DDT, lindane and endosulfan sulfate) in plasma from the infant, the mother and the 38-year-old father, who reported a dramatic decrease in libido, and in soil samples from their farm. The precocious sexual development was probably caused by the oestrogen activity owing to environmental contamination by tonnes of pesticides stored on the family farm. In a pilot study by Berman *et al.* (2011) 12 women (63%) were reported as using some form of household pest control during their pregnancy, and five (26%) reported using household pest control during the past month. Women with a graduate degree had significantly higher geometric mean concentrations of total urinary dimethyl (DM) metabolite concentrations compared to other women. The carbamate bendiocarb was detected in 89% of the plasma samples, while the organophosphate insecticide chlorpyrifos was detected in 42% of the samples. It is unclear why total DM metabolite concentrations were much higher in this population compared to other populations of pregnant women in the United States and The Netherlands.

Barr *et al.* (2010a) evaluated *in utero* exposures to pesticides by measuring maternal and cord serum biomarkers in a New Jersey cohort of pregnant women and the birth outcomes of their neonates. Authors evaluated chlorpyrifos, diazinon, carbofuran, chlorothalonil, dacthal, metolachlor, trifluralin and diethyl-m-toluamide (DEET) in both maternal and umbilical cord sera. Of these compounds, chlorpyrifos, carbofuran, chlorothalonil, trifluralin, metolachlor and DEET were the pesticides most frequently detected in the serum samples. High (≥75th percentile) metolachlor concentrations in cord blood were related to birth weight. An increase in abdominal circumference with increasing cord dichloran concentrations was also observed. These observations suggest that *in utero* exposures to certain pesticides may alter birth outcomes. A survey by Barr *et al.* (2010b) assessed human exposure via biomonitoring to pyrethroid insecticides in a representative sample of the general US population ≥6 years of age. Researchers measured five urinary metabolites of pyrethroid insecticides in 5046 samples collected as a part of the 1999–2002 National Health and Nutrition Examination Survey (NHANES). Results detected 3-phenoxybenzoic acid (3PBA), a metabolite common to many pyrethroid insecticides, in more than 70% of the samples. Non-Hispanic blacks had significantly higher 3PBA concentrations than did non-Hispanic whites and Mexican Americans in the 2001–2002 survey period and in the combined 4-year survey periods but not in the 1999–2000 survey period. Children had significantly higher concentrations of 3PBA than did adolescents in both NHANES periods and than adults in NHANES 1999–2000. Cis- and trans-(2,2-dichlorovinyl)-2,2-dimethylcyclopropane-1-carboxylic acid were highly correlated with each other and with 3PBA, suggesting that urinary 3PBA was derived primarily from exposure to permethrin, cypermethrin or their degradates. The study concludes that pyrethroid insecticide exposure in the US population is widespread, and the presence of its metabolites in the urine of US residents indicates that children may have higher exposures than adolescents and adults.

A study by Payne-Sturges *et al.* (2009) suggests that approximately 40% of children in the United States may have had insufficient margins of exposure (MOEs) for neurological impacts from cumulative exposures to organophosphate pesticides (MOE less than 1000). Koppen *et al.* (2009) measured Cd, Pb, PCB congeners, p,p′-DDE – a key metabolite of DDT – and HCB in cord blood or plasma. The individual PCB congeners could be detected in 40–81% of the newborns. HCB and dioxin-like compounds were above the detection limit in more than 75% of the samples. Although age and smoking habits of the mothers did not influence the cord blood Pb and Cd levels, mothers had 2.6% lower PCB levels in cord blood for each unit increase in pre-pregnancy BMI. The season of delivery, breastfeeding previous children or consumption of local dairy products were minor determinants. Up to 20% of the variability in organochlorine concentrations was explained by residence area. It was concluded that the place of birth (Flanders) is an important determinant of the load of pollutants measured at the start of life. Lu *et al.* (2008) assessed young urban/suburban children's longitudinal exposure to organophosphate pesticides in the Children's Pesticide Exposure Study (CPES). It was demonstrated that dietary intake of organophosphate pesticides represents the major source of exposure in young children.

An association between pesticide exposure and childhood cancer was observed by Infante-Rivard and Weichenthal (2007). Parental occupational exposure to pesticides during pregnancy and carry-home residues also contribute to children's cumulative burden. A number of epidemiological studies have consistently reported associations between pesticide exposures and increased risk of childhood leukaemia, brain cancer, neuroblastoma, non-Hodgkin's lymphoma, Wilms' tumour and Ewing's sarcoma. The results were supported by a longitudinal study conducted by Lu *et al.* (2006), which found an association between parents' self-reported pyrethroid use in the residential environment and elevated pyrethroid metabolite levels found in their children's urine. Children were also exposed to pyrethroids through their conventional diets, although the magnitude was smaller than for residential exposure. Children's ages appear to be significantly associated with pyrethroid exposure, which is likely attributed to the use of pyrethroids around the premises or in the facilities where older children engaged in outdoor activities. The National Research Committee (National Research Council, 1993) found both quantitative and occasionally qualitative differences in toxicity of pesticides between children and adults; that quantitative differences in toxicity between children and adults are usually less than a factor of approximately 10-fold; and that infants and children differ both qualitatively and quantitatively from adults in their exposure levels to pesticide residues in foods.

Kokkinaki *et al.* (2014) evaluated the exposure of rural residents (control group) and an occupational exposed population group of sprayers to organophosphorus pesticides by measuring their non-specific dialkylphosphate metabolites (DAPs) in hair and in urine samples. The observations revealed significantly higher DAP levels in the urine of sprayers than in the urine of the control group and this is justified since sampling occurred during spraying periods. DAP levels in hair samples of the sprayers were also significantly higher than in the hair samples from the control group, confirming long-term exposure to organophosphates. DAP levels in urine and hair samples of subjects were significantly correlated. Woodruff *et al.* (2011) analysed biomonitoring data from NHANES for pregnant women exposed to PCBs, OCPs, perfluorinated compounds (PFCs), phenols, PBDEs, phthalates, PAHs and perchlorate. The percentage of pregnant women with detectable levels of an individual chemical ranged from 0 to 100%. Generally, levels in pregnant women were similar to or lower than levels in non-pregnant women; adjustment for covariates tended to increase levels in pregnant women compared to non-pregnant women. A marginally significant trend of increasing concentration was observed with decreasing age among applicator children, and younger children within these families had significantly higher concentrations when compared to their older siblings. Loewenherz *et al.* (1997) reported dimethylthiophosphate (DMTP) as the dominant metabolite, and levels significantly higher in applicator children than in reference children, with median concentrations of 0.021 and 0.005 μg/ml, respectively; the maximum concentrations were 0.44 and 0.10 μg/ml, respectively.

8.3 Obesity

A connection between widespread exposure to organophosphate pesticides and the global epidemics of type 2 diabetes and obesity has been reported by Peris-Sampedro *et al.* (2015). The report highlighted gene–environment interactions: mice expressing the human apolipoprotein E3 (apoE3) isoform were more prone to developing obesity than those expressing apoE2 or apoE4 upon dietary challenge with chlorpyrifos, the most used organophosphate worldwide. Chlorpyrifos exposure generally increased food ingestion, glucose and total cholesterol concentrations, and tended to elevate acyl ghrelin levels. Nonetheless, excess weight gain and increased leptin levels were inherent to apoE3 mice. Moreover, the propensity towards a diabetic profile was markedly higher in these animals than in C57BL/6N, as they showed a higher homeostatic model assessment for insulin resistance index and higher insulin levels. A study by Buser *et al.* (2014) found a statistically significant positive association between both 2,5-dichlorophenol and 2,4-dichlorophenol with BMI Z score, waist circumference (WC) and obesity in children and adolescents. Abdominal obesity is defined by WC. After stratification by age, the significant associations remained only in adolescents (ages 12–19 years). In another study by Warner *et al.* (2014), *in utero* exposure to DDT and its metabolite, DDE was hypothesized to increase the risk of obesity at 9 years of age in boys but not in girls. The study was conducted to examine the associations between maternal serum concentrations of DDT and DDE during pregnancy and BMI, obesity, WC and percentage of body fat in 9-year-old children ($n = 261$) in the Center for the Health Assessment of Mothers and Children of Salinas (CHAMACOS) study, a longitudinal birth cohort study in the Salinas Valley, California (2000–2010).

Slotkin (2011) observed a contribution of organophosphate exposure to the increased incidence of obesity and diabetes. The study was conducted on neonatal rats with chlorpyrifos, diazinon or parathion doses. The animals developed metabolic dysfunction resembling prediabetes. When the organophosphate-exposed animals consumed a high-fat diet in adulthood, metabolic defects were exacerbated and animals gained excess weight compared to unexposed rats on the same diet. At the same time, the high-fat diet ameliorated many of the central synaptic defects caused by organophosphate exposure, pointing to the possibility of non-pharmacological therapeutic interventions to offset neurodevelopmental abnormalities, as well as towards fostering dietary choices favouring high fat intake. Lee *et al.* (2011) reported that POP exposure may contribute to excess adiposity and other features of dysmetabolism. BMI, triglycerides, HDL-cholesterol, LDL-cholesterol and homeostasis model assessment value for insulin resistance (HOMA-IR) were studied in relation to POP exposure, including 8 OCPs, 22 PCBs and 1 polybrominated biphenyl (PBB). Among OCPs, p,p′-DDE most consistently predicted higher BMI, triglycerides and HOMA-IR and lower HDL-cholesterol at year 20 after adjusting for baseline values. Simultaneous exposure to various POPs in the general population may contribute to the development of obesity, dyslipidemia and insulin resistance, common precursors of type 2 diabetes and cardiovascular diseases.

Peaches containing added residues of chlorpyrifos-methyl, fenitrothion, procymidone and vinclozolin were used for simulated industrial processing in the manufacture of baby food puree. Residues were determined in raw material, in intermediate products at crucial steps of the processing procedure and in the final products. The results of the study were interpreted with respect to enforcement of the stringent maximum residue limit (MRL) of 0.01 mg/kg established by the European Commission (EC) for any pesticide in baby food. Peeling was identified as the most effective procedure in reducing residues. Thermal treatment (concentration and sterilization) substantially reduced organophosphate (chlorpyrifos-methyl, fenitrothion) residues, whereas procymidone and vinclozolin residue levels were increased in peach puree (Balinova *et al.*, 2006). Biological monitoring has been widely used to assess exposures, susceptibility and effects of chlorpyrifos and malathion. In order to biomonitor exposure, chlorpyrifos and malathion have been measured in blood, but most typically their urinary metabolites have been measured. Although a lot of biologic monitoring data have been generated and published on these chemicals, their interpretation is not straightforward (Barr *et al.*, 2006). For example, exposure to environmental degradates of chlorpyrifos and malathion may potentially increase urinary metabolite levels, thus leading to overestimation of exposure. Also, the temporal nature of the exposures makes the evaluation of both exposure and effects difficult.

8.4 Symptoms and Harmful Effects of Some Pesticide Families

8.4.1 Topical effects

Pesticide poisoning can have either topical or systemic symptoms. Topical effects could be developed at the site of pesticide contact, resulting from either the pesticide's irritant properties or an allergic response by the victim. Dermatitis, the most commonly reported topical effect due to pesticide exposure, is associated with symptoms ranging from reddening of the skin to rashes and/or blisters. Individuals react differently to pesticide exposure, depending on their body structure and immunity. Some individuals tend to cough, wheeze or sneeze when exposed to pesticide sprays; others react to the strong odour and irritating effects of the carriers in pesticide formulations. Mostly the eyes, mucous membranes of the nose and even the sensitive linings of the mouth and back of the throat feel raw and scratchy. Development of severe acute respiratory symptoms, often resembling an asthmatic condition, is more likely to be a true hypersensitivity or allergic reaction (Lorenz, 2009).

8.4.2 Systemic effects

Systemic effects often occur away from the original point of contact because of the pesticide being absorbed into and distributed throughout the body. These effects often include nausea, vomiting, fatigue, headache and intestinal disorders. The individuals may experience changes in heart rate, difficulty breathing, convulsions and coma, leading to death. These often occur immediately after exposure, but they could be delayed for up to 24 h. The individual with these symptoms must notify medical personnel about their exposure to pesticides while seeking medical attention. The victim should carry details of the pesticide container; that is, the product name, active ingredient(s) and percentage, and the US Environmental Protection Agency registration number. If the Material Safety Data Sheet is available, it should be taken to the medical personnel, as it contains additional information.

8.4.3 Harmful effects of pesticides

Herbicides have a low acute toxicity to humans because plants and humans have different physiology. However, some herbicides can be dermal irritants as they can be strong acids, amines, esters and phenols. Their inhalation may cause coughing and a burning sensation in the nasal passages and chest, and sometimes dizziness. The acute toxicity of fungicides to humans is generally considered low, compared with insecticides, but they can be irritating to the skin and eyes. Inhalation of spray mist or dust from these pesticides may cause throat irritation, sneezing and coughing. Chronic exposure to lower concentrations of fungicides can cause adverse health effects. Insecticides cause the greatest number of pesticide poisonings globally. The most serious pesticide poisonings usually result from acute exposure to organophosphate and carbamate insecticides, the cholinesterase enzyme inhibitors, causing a disruption of the nervous system. All life forms with cholinesterase in their nervous system, such as insects, fish, birds, humans and other mammals, can be poisoned by these chemicals. The signs and symptoms of acute exposure to commonly used pesticides are summarized in Tables 8.1, 8.2 and 8.3 (Lorenz, 2009).

8.5 Conclusion

Several indicators have been documented and used to assess the potential risk of pesticides to human health and the environment. Because of the uncertainty regarding their possible adverse effects on human health and the environment, there is a need for the development of alternative indicators with increased accuracy and reliability in pesticide risk assessment. Alternatives to synthetic chemicals, with novel modes of action and improved safety profiles may be developed for pest management in agriculture. Alternative cropping systems relying less on pesticides may be implemented. Proper pesticide application technology with well-maintained spraying equipment along with precautions taken during all stages of pesticide handling may also reduce exposure to pesticides. Handling pesticides strictly according to the regulations and also considering public concerns about pesticide residues in food and drinking water could contribute to a reduction in the adverse effects of pesticides on human health and the environment. The selection and use of proper personal protective equipment (PPE) based on the hazards present at a site could

Table 8.1. Signs and symptoms of acute exposure to several insecticide active ingredients.

Active ingredient	Brand name	Signs and symptoms
Acephate (organophosphate)	Orthene	Headache, excessive salivation and tearing, muscle twitching, nausea, diarrhoea, respiratory depression, seizures, loss of consciousness, pinpoint pupils
Aldicarb (N-methyl carbamate)	Temik	Malaise, muscle weakness, dizziness, sweating, headache, salivation, nausea, vomiting, abdominal pain, diarrhoea, nervous system depression, pulmonary oedema in serious cases
Carbaryl (N-methyl carbamate)	Sevin	Malaise, muscle weakness, dizziness, sweating, headache, salivation, nausea, vomiting, abdominal pain, diarrhoea, nervous system depression, pulmonary oedema in serious cases
Chlorpyrifos (organophosphate)	Dursban	Headache, excessive salivation and tearing, muscle twitching, nausea, diarrhoea, respiratory depression, seizures, loss of consciousness, pinpoint pupils
Endosulfan (organochlorine)	Thiodan	Itching, burning, tingling of skin, headache, dizziness, nausea, vomiting, lack of coordination, tremor, mental confusion, seizures, respiratory depression, coma
Malathion (organophosphate)	Cythion	Headache, excessive salivation and tearing, muscle twitching, nausea, diarrhoea, respiratory depression, seizures, loss of consciousness, pinpoint pupils
Methyl parathion (organophosphate)	Penncap-M	Headache, excessive salivation and tearing, muscle twitching, nausea, diarrhoea, respiratory depression, seizures, loss of consciousness, pinpoint pupils
Phosmet (organophosphate)	Imidan	Headache, excessive salivation and tearing, muscle twitching, nausea, diarrhoea, respiratory depression, seizures, loss of consciousness, pinpoint pupils
Pyrethrins (natural origin)		Irritating to skin and upper respiratory tract, contact dermatitis and allergic reactions – asthma
Pyrethroids (synthetic pyrethrin)	Cypermethrin, permethrin	Abnormal facial sensation, dizziness, salivation, headache, fatigue, vomiting, diarrhoea, irritability to sounds or touch, seizures, numbness

help in reducing hazards from the exposure to pesticides. For those substances or hazards where there is no published exposure level, published reports and material safety data sheets should be used for evaluation.

Legislation should guarantee all workers, regardless of sex, age, employment status, type of contract, location of work, size of enterprise, type of job and industry, prevention of occupational health risks, minimum labour wages, prevention of discrimination and compensation in the case of illness or injury. All occupational health effects should be covered, including psychological disability and chronic pain. International and national health policies should be implemented for workers in both the public and domestic domains, instead of waiting for demonstrations by unions. Unions should ensure that women participate at all levels in health and safety activities and should create structures to facilitate their participation in union life. Women workers should receive equal wages to men, and have an active role in developing policies to promote healthier workplaces. At an international level, close collaboration is needed between WHO and ILO and other international agencies concerned with gender equality and with work-related health. Women and men should be hired at all levels of the occupational health prevention and compensation systems.

Table 8.2. Signs and symptoms of acute exposure to several fungicide active ingredients.

Active ingredient	Brand name	Signs and symptoms
Azoxystrobin	Abound, Quadris	Irritating to skin, eyes, respiratory tract
Captan	Captol, Orthocide	Irritating to skin, eyes, respiratory tract
Chlorothalonil	Bravo, Daconil	Irritation to skin, mucous membranes of the eye, respiratory tract, allergic contact dermatitis
Copper compounds	Bordeaux mixture, copper sulfate	Irritating to skin, eyes, respiratory tract, salts are corrosive to mucous membranes and cornea, metallic taste, nausea, vomiting, intestinal pain
Mancozeb	Dithane M-45, Manzate 200	Irritating to skin, eyes, respiratory tract
Maneb	Dithane M-22, Manzate	Irritating to skin, eyes, respiratory tract, skin disease in occupationally exposed individuals
Pentachloronitrobenzene (PCNB)	PCNB, Terraclor	Allergic reactions
Sulfur	Cosan, Thiolux	Irritating to skin, eyes, respiratory tract, breath odour of rotten eggs, diarrhoea, irritant dermatitis in occupationally exposed individuals
Thiram	Polyram-Ultra, Spotrete-F	Irritating to skin, eyes, respiratory mucous membranes
Ziram	Cuman, Vancide	Irritating to skin, eyes, respiratory tract, prolonged inhalation causes neural and visual disturbances

Table 8.3. Signs and symptoms of acute exposure to several herbicide active ingredients.

Active ingredient	Brand name	Signs and symptoms
2,4-dichlorophenoxyacetic acid	2,4-D, Barrage	Irritating to skin, mucous membranes, vomiting, headache, diarrhoea, confusion, bizarre or aggressive behaviour, muscle weakness in occupationally exposed individuals
Acetochlor	Harness, Surpass	Irritating to skin, eyes, respiratory tract
Atrazine	Aatrex, Atranex	Irritating to skin, eyes, respiratory tract, abdominal pain, diarrhoea, vomiting, eye irritation, irritation of mucous membranes, skin reactions
Dicamba	Banvel, Metambane	Irritating to skin, respiratory tract, loss of appetite (anorexia), vomiting, muscle weakness, slowed heart rate, shortness of breath, central nervous system effects
Glyphosate	Rodeo, Roundup	Irritating to skin, eyes, respiratory tract
Mecoprop	Kilporp, MCPP	Irritating to skin, mucous membranes, vomiting, headache, diarrhoea, confusion, bizarre or aggressive behaviour, muscle weakness in occupationally exposed individuals
Metolachlor	Bicep, Dual	Irritating to skin, eyes
Paraquat	Gramoxone	Burning in mouth, throat, chest, upper abdomen, diarrhoea, giddiness, headache, fever, lethargy, dry, cracked hands, ulceration of skin
Pendimethalin	Prowl, Stomp	Irritating to skin, eyes, respiratory tract
Propanil	Propanex, Stampede	Irritating to skin, eyes, respiratory tract

References

Arcury, T.A., Grzywacz, J.G., Barr, D.B., Tapia, J., Chen, H. and Quandt, S.A. (2007) Pesticide urinary metabolite levels of children in eastern North Carolina farmworker households. *Environmental Health Perspectives* 115(8), 1254–1260.

Balinova, A.M., Mladenova, R.I. and Shtereva, D.D. (2006) Effects of processing on pesticide residues in peaches intended for baby food. *Food Additives and Contaminants* 23(9), 895–901.

Barr, D.B. and Angerer, J. (2006) Potential uses of biomonitoring data: A case study using the organophosphorus pesticides chlorpyrifos and malathion. *Environmental Health Perspectives* 114(11), 1763–1769.

Barr, D., Ananth, C.V., Yan, X., Lashley, S., Smulian, J.C., Ledoux, T.A., Hore, P. and Robson, M.G. (2010a) Pesticide concentrations in maternal and umbilical cord sera and their relation to birth outcomes in a population of pregnant women and newborns in New Jersey. *Science of the Total Environment* 408(4), 790–795.

Barr, D.B., Olsson, A.O., Lee-Yang, W., Simeon, U., Samuel, E.B., Ralph, D.W. Jr, Melina, S.M., Bryan, L.W. and Larry, L.N. (2010b) Urinary concentrations of metabolites of pyrethroid insecticides in the general U.S. population: National health and nutrition examination survey 1999–2002. *Environmental Health Perspectives* 118(6), 742–748.

Berman, T., Hochner-Celnikier, D., Barr, D.B., Needham, L.L., Amitai, Y., Wormser, U. and Richter, E. (2011) Pesticide exposure among pregnant women in Jerusalem, Israel: Results of a pilot study. *Environment International* 37(1), 198–203.

Bolognesi, C. (2003) Genotoxicity of pesticides: A review of human biomonitoring studies. *Mutation Research/Reviews in Mutation Research* 543(3), 251–272.

Buser, M.C., Murray, H.E. and Scinicariello, F. (2014) Association of urinary phenols with increased body weight measures and obesity in children and adolescents. *Journal of Pediatrics* 165(4), 744–749.

Chávez-Almazán, L.A., Diaz-Ortiz, J., Alarcón-Romero, M., Dávila-Vazquez, G., Saldarriaga-Noreña, H. and Waliszewski, S.M. (2014) Organochlorine pesticide levels in breast milk in Guerrero, Mexico. *Bulletin of Environmental Contamination and Toxicology* 93(3), 294–298.

Chen, Y., Wang, X., Li, Y., Toms, L.M., Gallen, M. *et al.* (2015) Persistent organic pollutants in matched breast milk and infant faeces samples. *Chemosphere* 118, 309–314.

Cordier, S., Bouquet, E., Warembourg, C., Massart, C. *et al.* (2015) Perinatal exposure to chlordecone, thyroid hormone status and neurodevelopment in infants: The Timoun cohort study in Guadeloupe (French West Indies). *Environmental Research* 138, 271–278.

Damalas, C.A. and Koutroubas, S.D. (2016) Farmers' exposure to pesticides: toxicity types and ways of prevention. In: Bellinger, D. (ed.) *Toxics*, vol. 4(1), p. 1. National Center for Biotechnology Information, US National Library of Medicine, Bethesda, Maryland. DOI: 10.3390/toxics4010001

Damgaard, I., Skakkebaek, N.E., Toppari, J., Virtanen, H.E., Shen, H., Schramm, K.W., Petersen, J.H., Jensen, T.K. and Main, K.M.; Nordic Cryptorchidism Study Group (2006) Persistent pesticides in human breast milk and cryptorchidism. *Environmental Health Perspectives* 114(7), 1133–1138.

Gaspari, L., Paris, F.O., Jeandel, C. and Sultan, C. (2011) Peripheral precocious puberty in a 4-month-old girl: Role of pesticides? *Gynecological Endocrinology* 27(9), 721–724.

Gyalpo, T., Fritsche, L., Bouwman, H., Bornman, R. *et al.* (2012) Estimation of human body concentrations of DDT from indoor residual spraying for malaria control. *Environmental Pollution* 169, 235–241.

Infante-Rivard, C. and Weichenthal, S. (2007) Pesticides and childhood cancer: An update of Zahm and Ward's 1998 review. *Journal of Toxicology and Environmental Health. Part B. Critical Reviews* 10(1–2), 81–99.

Kokkinaki, A., Kokkinakis, M., Kavvalakis, M.P., Tzatzarakis, M.N., Alegakis, A.K., Maravgakis, G., Babatsikou, F., Fragkiadakis, G.A. and Tsatsakis, A.M. (2014) Biomonitoring of dialkylphosphate metabolites (DAPs) in urine and hair samples of sprayers and rural residents of Crete, Greece. *Environmental Research* 134, 181–187.

Koppen, G., Den Hond, E., Nelen, V., Van De Mieroop, E., Bruckers, L., Bilau, M., Keune, H., Van Larebeke, N., Covaci, A., Van De Weghe, H., Schroijen, C., Desager, K., Stalpaert, M., Baeyens, W. and Schoeters, G. (2009) Organochlorine and heavy metals in newborns: Results from the Flemish Environment and Health Survey (FLEHS 2002–2006). *Environment International* 35(7), 1015–1022.

Lee, D.H., Steffes, M., Sjödin, A., Jones, R.S., Needham, L.L. and Jacobs, D.R. Jr (2011) Low dose organochlorine pesticides and polychlorinated biphenyls predict obesity, dyslipidemia, and insulin resistance among people free of diabetes. *PLoS ONE* 6, e15977.

Lee, S., Kim, S., Lee, H.K., Lee, I.S., Park, J., Kim, H.J., Lee, J.J., Choi, G., Choi, S., Kim, S., Kim, S.Y., Choi, K., Kim, S. and Moon, H.B. (2013) Contamination of polychlorinated biphenyls and organochlorine pesticides in breast milk in Korea: Time-course variation, influencing factors, and exposure assessment. *Chemosphere* 93(8), 1578–1585.

Loewenherz, C., Fenske, R.A., Simcox, N.J., Bellamy, G. and Kalman, D. (1997) Biological monitoring of organophosphorus pesticide exposure among children of agricultural workers in central Washington State. *Environmental Health Perspectives* 105(12), 1344–1353.

Lorenz, E.S. (2009) Potential health effects of pesticides. Pesticide Safety Fact Sheets. Pesticide Education Program in Penn State's College of Agricultural Sciences, the Pennsylvania State University.

Lu, C., Barr, D.B., Pearson M., Bartell, S. and Bravo, R. (2006) A longitudinal approach to assessing urban and suburban children's exposure to pyrethroid pesticides. *Environmental Health Perspectives* 114(9), 1419–1423.

Lu, C., Barr, D.B., Pearson, M.A. and Waller, L.A. (2008) Dietary intake and its contribution to longitudinal organophosphorus pesticide exposure in urban/suburban children. *Environmental Health Perspectives* 216(4), 537–542.

Lu, D., Wang, D., Ni, R., Lin, Y., Feng, C., Xu, Q., Jia, X., Wang, G. and Zhou, Z. (2015) Organochlorine pesticides and their metabolites in human breast milk from Shanghai, China. *Environmental Science and Pollution Research International* 22(12), 9293–9306.

Manaca, M.N., Grimalt, J.O., Sunyer, J., Mandomando, I. *et al.* (2011) Concentration of DDT compounds in breast milk from African women (Manhiça, Mozambique) at the early stages of domestic indoor spraying with this insecticide. *Chemosphere* 85(3), 307–314.

Massart, F., Harrell, J.C., Federico, G. and Saggese, G. (2005) Human breast milk and xenoestrogen exposure: A possible impact on human health. *Journal of Perinatology* 25, 282–288.

Mikeš, O., Cupr, P., Kohut, L., Krsková, A. and Cerná, M. (2012) Fifteen years of monitoring of POPs in the breast milk, Czech Republic, 1994–2009: trends and factors. *Environmental Science and Pollution Research International* 19(6), 1936–1943.

Nakagawa, R., Hirakawa, H., Iida, T., Matsueda, T. and Nagayama, J. (1999) Maternal body burden of organochlorine pesticides and dioxins. *Journal of AOAC International* 82(3), 716–724.

National Research Council (1993) *Pesticides in the Diets of Infants and Children.* National Academy Press, Washington, DC.

Ostrea, E.M. Jr, Villanueva-Uy, E., Bielawski, D., Birn, S. and Janisse, J.J. (2014) Trends in long term exposure to propoxur and pyrethroids in young children in the Philippines. *Environmental Research* 131, 13–16.

Payne-Sturges, D., Cohen, J., Castorina, R., Axelrad, D.A. and Woodruff, T.J. (2009) Evaluating cumulative organophosphorus pesticide body burden of children: A national case study. *Environmental Science and Technology* 43(20), 7924–7930.

Peris-Sampedro, F., Cabré, M., Basaure, P., Reverte, I., Domingo, J.L. and Teresa Colomina, M. (2015) Adulthood dietary exposure to a common pesticide leads to an obese-like phenotype and a diabetic profile in apoE3 mice. *Environmental Research* 142, 169–176.

Pirsaheb, M., Limoee, M., Namdari, F. and Khamutian, R. (2015) Organochlorine pesticides residue in breast milk: A systematic review. *Medical Journal of the Islamic Republic of Iran* 29, 228.

Qin, Y.Y., Leung, C.K., Lin, C.K., Leung, A.O., Wang, H.S., Giesy, J.P. and Wong, M.H. (2011) Halogenated POPs and PAHs in blood plasma of Hong Kong residents. *Environmental Science and Technology* 45(4), 1630–1637.

Sharma, A., Gill, J.P., Bedi, J.S. and Pooni, P.A. (2014) Monitoring of pesticide residues in human breast milk from Punjab, India and its correlation with health associated parameters. *Bulletin of Environmental Contamination and Toxicology* 93(4), 465–471.

Shen, H., Main, K.M., Virtanen, H.E., Damggard, I.N., Haavisto, A.M., Kaleva, M., Boisen, K.A., Schmidt, I.M., Chellakooty, M., Skakkebaek, N.E., Toppari, J. and Schramm, K.W. (2007) From mother to child: Investigation of prenatal and postnatal exposure to persistent bioaccumulating toxicants using breast milk and placenta biomonitoring. *Chemosphere* 67(9), S256–S262.

Slotkin, T.A. (2011) Does early-life exposure to organophosphate insecticides lead to prediabetes and obesity. *Reproductive Toxicology* 31, 297–301.

Snedeker, S.M. (2001) Pesticides and breast cancer risk: A review of DDT, DDE, and dieldrin. *Environmental Health Perspectives*, 109(Suppl. 1), 35–47.

Solomon, G.M. and Weiss, P.M. (2002) Chemical contaminants in breast milk: Time trends and regional variability. *Environmental Health Perspectives* 110(6), A339–47.

't Mannetje, A., Coakley, J., Bridgen, P., Smith, A.H., Read, D., Pearce, N. and Douwes, J. (2014) Estimated infant intake of persistent organic pollutants through breast milk in New Zealand. *New Zealand Medical Journal* 127(1401), 56–68.

Warner, M., Wesselink, A., Harley, K.G., Bradman, A., Kogut, K. and Eskenazi, B. (2014) Prenatal exposure to dichlorodiphenyltrichloroethane and obesity at 9 years of age in the CHAMACOS study cohort. *American Journal of Epidemiology* 179(11), 1312–1322.

World Health Organization (WHO) (2004) Gender, health and work. Department of Gender, Women and Health (GWH). Available at: www.who.int/gender-equity-rights/knowledge/gender-health-and-work.pdf?ua=1 (accessed 30 July 2018).

Wohlfahrt-Veje, C., Main, K., Schmidt, I., Boas, M., Jensen, T. and Grandjean, P. (2011) Lower birth weight and increased body fat at school age in children prenatally exposed to modern pesticides: A prospective study. *Environmental Health* 10, 79.

Woodruff, T.J., Zota, A.R. and Schwartz, J.M. (2011) Environmental chemicals in pregnant women in the US: NHANES 2003–2004. *Environmental Health Perspectives* 119(6), 878–885.

Yalçın, S.S., Örün, E., Yalçın, S. and Aykut. O. (2015) Organochlorine pesticide residues in breast milk and maternal psychopathologies and infant growth from suburban area of Ankara, Turkey. *International Journal of Environmental Health Research* 25(4), 364–372.

Zietz, B.P., Hoopmann, M., Funcke, M., Huppmann, R., Suchenwirth, R. and Gierden, E. (2008) Long-term biomonitoring of polychlorinated biphenyls and organochlorine pesticides in human milk from mothers living in northern Germany. *International Journal of Hygiene and Environmental Health* 211(5–6), 624–638.

9 Co-Relation of Toxicity and Classification of Pesticides

9.1 Introduction

The toxicity of a pesticide is its ability to cause injury or illness. The toxicity of a particular pesticide is determined by subjecting test animals to varying dosages of the active ingredient (a.i.) and each of its formulated products. There can be acute and chronic toxicity. Acute toxicity of a pesticide is the chemical's ability to cause injury to a person or animal from a single exposure, generally of short duration. Acute toxicity is determined by examining the dermal toxicity, inhalation toxicity and oral toxicity of test animals. In addition, eye and skin irritation are also examined. Acute toxicity is usually expressed as LD_{50} (lethal dose 50) or LC_{50} (lethal concentration 50); that is, the amount or concentration of a toxicant – the a.i. – required to kill 50% of the animals in a test population. Additionally, the LD_{50} and LC_{50} values are also recorded in milligrams of pesticide per kilogram of body weight (mg/kg) of the test animal or in parts per million (ppm). The lower the LD_{50} or LC_{50} of a pesticide product, the greater its toxicity to humans and animals. The long-term exposure of test animals to the active ingredients determines chronic toxicity of a pesticide. Any harmful effects that occur from small doses repeated over a period of time are termed chronic effects. Some of the suspected chronic effects from exposure to certain pesticides include birth defects, production of tumours, blood disorders and neurotoxic effects (nerve disorders). The chronic toxicity of a pesticide is more difficult to determine through laboratory analysis than acute toxicity (Hock and Lorenz, 2006).

The adverse impacts of used pesticides are unavoidable for at least two reasons: (i) over 90% of applied pesticides reach a different environmental destination other than the target pests (Pimentel *et al.*, 1993); and (ii) all pesticides are toxicologically and/or structurally designed to affect/kill some form(s) of life (Olympic Horticultural Products, 2018) and are then expected to affect human and non-human health in many different ways.

9.2 Restricted Interval of Time Versus Pesticide Classification

To avoid exposure of pesticide residues to workers or other persons not involved in the application of a pesticide, restricted entry intervals (REI) are established to protect them against poisoning if they enter a treated area too soon after application without proper protective equipment. The REI is the time immediately after a pesticide application when entry into the treated area is restricted. The EPA Worker Protection Standard now requires minimum 12-h reentry times for all Category III (CAUTION) pesticides, 24-h minimum reentry times for all Category II (WARNING) pesticides and 48-h minimum reentry times for all Category I (DANGER) pesticides. In New Jersey, the New Jersey Department of Environmental Protection Pesticide Control Program has designated 48-h reentry times for some pesticides, which the United States Environmental Protection Agency (US EPA) has assigned 12- or 24-h reentry times. The REI information is listed on the pesticide labelling; however, California regulations may require longer REIs than listed on the labelling for certain pesticides when used on certain crops. In addition, some labelling provides for 'activity-based REIs' that allow certain, described activities sooner than other activities.

9.2.1 Factors affecting REI

Different pesticides will have different reentry intervals. Some pesticides have one REI, such as 12 h, for all crops and uses. Other pesticides have different REIs depending on the crop, method of application or the post-application activity to be performed. In addition, each pesticide may have one or more different reentry intervals. When two or more pesticides are applied at the same time and have different REIs, the longer REI must be followed. These intervals are established by considering the following (CCOHS, 2018):

1. The toxicity of the active ingredient (actual poison) of the pesticide.
2. If it is a mixture, what remains in the residue after application.
3. How the pesticide is formulated (including adjuvants) – for example, organic solvents in pesticide formulations may increase the skin toxicity of some insecticides.
4. If the pesticides are converted into more toxic compounds under certain environmental conditions.
5. The rate and method of application including whether it is applied outdoors or in a confined space (such as greenhouses or barns).
6. Characteristics of the plant being sprayed (type of crop, height of crop at maturity, density of leaves, how the plant will be handled and how close the pesticide is applied to harvest). Since different rates are used for each crop, the REI can vary between crops. For example, the REI for raspberries could be 7 days, while the REI for apples could be 14 days.
7. Weather conditions may change how the pesticide disperses and dissipates (such as variations in temperature, sunshine, moistness and wind). For example, organophosphates and carbamates in hot and dry climates take longer to break down and longer reentry intervals (1–2 weeks) are sometimes necessary to prevent acute poisoning of field workers.
8. The type of work being done after the pesticide is applied. Human contact with treated plants can vary. For example, a REI could be 14 days for thinning, but only 48 h for irrigating.

9.3 Classification of Pesticides Based on Toxicity

The pesticides have been classified in different categories according to their toxicity (LD/LC_{50} values) and have been assigned certain symbols on their label accordingly (Table 9.1). A number of acaricides, bactericides, fungicides, herbicides, insecticides, nematicides, insect growth regulators and plant growth regulators have been categorized by Hock and Lorenz (2006) as per the toxicity of their active ingredients, based on their LD/LC_{50} values.

9.3.1 Classification of acaricides

The following list presents the acaricides that are commonly used in the general category (except fenbutatin oxide) with oral and dermal LD_{50} values (mg/kg), along with

Table 9.1. Toxicity categories of the active ingredients. (From US EPA, 2018, adapted from 40 CFR Part 156.)

Hazard indicators	US EPA toxicity categories and pesticide label signal words			
	I	II	III	IV
Oral LD_{50}	<50 mg/kg	50–500 mg/kg	500–5000 mg/kg	>5000 mg/kg
Dermal LD_{50}	<200 mg/kg	200–2000 mg/kg	2000–5000 mg/kg	>5000 mg/kg
Inhalation LC_{50}	≤0.2 mg/l	0.2–2 mg/l	2–20 mg/l	>20 mg/l
Eye effects	Corrosive (irreversible destruction of ocular tissue) or corneal opacity not reversible within 7 days	Corneal opacity reversible within 7 days; irritation persisting for 7 days	No corneal opacity; irritation reversible within 7 days	Minimal effects clearing in less than 24 h
Skin effects	Corrosive (tissue destruction into the dermis and/or scarring)	Severe irritation at 72 h (severe erythema or oedema)	Moderate irritation at 72 h (moderate erythema)	Mild or slight irritation (no irritation or erythema)
Dermal sensitization	Positive – product is a sensitizer or is positive for sensitization		Negative – product is not a sensitizer or is negative for sensitization	
Signal word	 DANGER POISON	WARNING	CAUTION	CAUTION

The Hazardous Substances Labeling Act of 2003 legally defines a poison as any substance that causes death in 50% of laboratory animals exposed to doses of less than 50 mg/kg of body weight.

REI (h), respectively (Hock and Lorenz, 2006):

- dicofol – 570–595 (oral), >2000 (dermal), 12 (REI);
- disulfton – 2–12, 3.6–15.9, 48;
- endosulfan – 160, 359, 24;
- etoxazole – >5000, >2000, 12;
- fenbutatin oxide (restricted use) – 2631, >2000, 48;
- formetanate hydrochloride – 21, >10,200, 48, 72
- hexythiazox – >5000, >5000, 12;
- pyridaben – 820–1350, >2000, 12;
- sulfur – >5000, >5000, 12, 24, 48;
- tetradifon – >10,000, >10,000, 12.

9.3.2 Classification of bactericides

The following list presents commonly used bactericides in the general category with oral and dermal LD_{50} values (mg/kg), along with REI (h), respectively (Hock and Lorenz, 2006):

- acibenzolar-S-methyl – >5000, >2000, 12;
- dimanin A – 290, dermal LD_{50} value, REI not available;
- hydrogen dioxide – 330, 1410, 0;
- oxytetracycline – >5000, >2000, 12;
- *Pseudomonas fluorescens* A506 – oral and dermal LD_{50} values not available, 4;
- streptomycin, Agri-Mycin-17 – 9000, dermal LD_{50} value not available, 12.

9.3.3 Classification of nematicides

The insecticide carbofuran is used as a nematicide in many countries like India. However, most of the commonly used nematicides are in the restricted category. The commonly used nematicides in the general (G) and restricted (R) categories with oral and dermal LD_{50} values (mg/kg), along with REI (h), respectively (Hock and Lorenz, 2006) are:

- chloropicrin (R) – 250, dermal value not available, 72;
- DCP, dichloropropene (R, G) – 300, 333, 72;
- ethoprop (R) – 61.5, 2.4, 48;

- fenamiphos (R) – tech 3, 200, 48;
- metam-sodium (G) – 1891, >3074, 48;
- methyl bromide (R) – potential to cause adverse reproductive effects, dermal value not available, 48
- oxamyl (R) – 37, 2960, 48

9.3.4 Classification of plant growth regulators

The majority of the plant growth regulators are classified in the general category (except maleic hydrazide), with varying REI values. The commonly used plant growth regulators in the general (G) and restricted (R) categories with oral and dermal LD_{50} values (mg/kg), along with REI (h), respectively (Hock and Lorenz, 2006) are:

- BAP – 3980 (6-benzylaminopurine or benzyl adenine), dermal value not available, 4;
- chlormequat chloride – 883, >4000, 12;
- daminozide – >5000, >5000, 24;
- dikegulac sodium – 31,000, >1000, no REI;
- ethephon – 4229, dermal value not available, 48;
- flurprimidol – 709, dermal value, REI not available;
- gibberellic acid – 1000–25,000, dermal value not available, 4;
- lactic acid – 3543, >2000, 48;
- maleic hydrazide (RG) – >5000, >5000, 12;
- mepiquat chloride – 464, dermal value not available, 12;
- maphthaleneacetamide – 1690, 2000, 48;
- 1-naphthaleneacetic acid – 2520, dermal value not available, 48;
- paclobutrazol – 5346, >1000, 12;
- plant extract 620 – >20,000, dermal value, REI not available;
- prohexadione calcium – >5000, >2000, 12;
- trinexapac ethyl – >5050, 2020, 0;
- uniconazole-P – 2020, >2000, 12.

9.3.5 Classification of fungicides

The majority of the fungicides are classified in the general category, with varying REI values of 12–48 h (exception of 4 h for azoxystrobin, *Coniothyrium minitans* and fenhexamid; 96 h for captan). Among the major fungicides used in the general category are acibenzolar-s-methyl, azoxystrobin, *Bacillus subtilis*, boscalid, captan, carboxin, chlorine, chloroneb, chlorothalonil, *Coniothyrium minitans*, copper hydroxide, cymoxanil, dicloran, difenoconazole, dimethomorph, dodine, etridiazole, famoxodone, fenbuconazole, fenhexamid, fluazinam, fludioxonil, flutolanil, fosetyl Al, iprodione, kresoxim-methyl, mancozeb, maneb, metalaxyl, metalaxyl-M, metam potassium, metiram, myclobutanil, pentachloronitrobenzene (PCNB), propamocarb hydrochloride, propiconazole, pyraclostrobin, streptomycetes, sulfur, tebuconazole, thiabendazole, thiophanate-methyl, thiram, triadimefon, trifloxystrobin, triflumizole, vinclozolin and zoxamide (Hock and Lorenz, 2006). The pesticide fact sheets published by PennState (Hock and Lorenz, 2006) document most of the fungicides under the general category with oral and dermal LD_{50} values in the range of 2000–5000 mg/kg. A few exceptions are reported: copper hydroxide (oral LD_{50} 1000 mg/kg), cymoxanil (oral LD_{50} 1100 mg/kg), difenoconazole (oral LD_{50} 1453 mg/kg), dodine and etridiazole (oral LD_{50} 1000 and 1077 mg/kg, respectively), tech metalaxyl (oral LD_{50} 669 mg/kg), myclobutanil (oral LD_{50} 1600 mg/kg), propiconazole (oral LD_{50} 1517 mg/kg), pyraclostrobin (oral LD_{50} more than 500 mg/kg), tech thiram (oral LD_{50} 1000 mg/kg) and triadimefon (oral LD_{50} 812 mg/kg).

Only a few fungicides are used in the restricted category: chloropicrin, with oral LD_{50} 250 mg/kg and REI 72 h; dichloropropene, with oral and dermal LD_{50} values 127 and 423 mg/kg, respectively, and REI 120 h; etridiazole and fenbuconazole are used in both the general and restricted categories; methyl bromide with acute vapour toxicity of 200 ppm and REI 48 h; and triphenyltin hydroxide is reported with oral and dermal LD_{50} values of 156–345 and 1600 mg/kg, respectively, and REI 48 h (Hock and Lorenz, 2006).

9.3.6 Classification of herbicides

Out of a list of 76 herbicides only nine are in the restricted category. The herbicides in the general category include acifluorfen, ametryn, asulam, bensulide, bentazon, bromoxynil, butylate, carfentrazone-ethyl, chlorimuron ethyl, chlorpropham, clethodim, clomazone, clopyralid, cycloate, dalapon, dimethyl tetrachloroterephthalate (DCPA), desmedipham, dicamba, dimethenamid, diquat, diquat dibromide, diuron, S-ethyl-N,N-dipropylthiocarbamate (EPTC), ethalfluralin, fenoxaprop-ethyl, fluazifop-P-butyl, flumetsulam, flumiclorac-pentyl, fomesafen, foramsulfuron, glufosinate-ammonium, glyphosate, halosulfuron-methyl, hexazinone, imazamox, imazaquin, imazethapyr, lactofen, linuron, 2-methyl-4-chlorophenoxyacetic acid (MCPA), mesotrione, S-metolachlor, metsulfuron-methyl, napropamide, naptalam, nicosulfuron, solicam, oryzalin, oxyfluorfen, pebulate, pendimethalin, phenmedipham, primisulfuron-methyl, propachlor, propanil, prosulfuron, quizalofop-P-ethel, rimsulfuron, sethoxydim, simazine, sulfentrazone, terbacil, thifensulfuron-methyl, triasulfuron, triclopyr, trifluralin and 4-(2,4-dichlorophenoxy) butyric acid (2,4-DB or 4-butyric acid). The REI values for most of the herbicides have been reported as 12–48 h. However, butylate has an REI value of 2 h; and metsulfuron-methyl, nicosulfuron, rimsulfuron, thifensulfuron-methyl and triasulfuron have REI values of 4 h (Hock and Lorenz, 2006).

The Pennsylvania State University publication on pesticides (Hock and Lorenz, 2006) has documented general category herbicides with oral LD_{50} values of more than 2000 mg/kg, and dermal LD_{50} values in the range of 2000–20,000 mg/kg. A few exceptions are reported: bensulide (oral LD_{50} tech 271–1470 mg/kg), bromoxynil (oral LD_{50} tech 260 mg/kg), clomazone (oral LD_{50} 1369 mg/kg), dimethenamid (oral LD_{50} 849 mg/kg), diquat (oral and dermal LD_{50} 215–235 mg/kg and 400 mg/kg, respectively), diquat dibromide (oral and dermal LD_{50} 600 mg/kg and 260 mg/kg, respectively), glufosinate-ammonium (oral LD_{50} 1620 mg/kg), halosulfuron-methyl (oral LD_{50} 1287 mg/kg), MCPA (oral LD_{50} 900–1160 mg/kg), pendimethalin (oral LD_{50} 1250 mg/kg), propachlor (oral LD_{50} 500–1700 mg/kg) and quizalofop-P-ethyl (oral LD_{50} 1210 mg/kg).

Herbicides used that are in the restricted category are acetochlor, alachlor, atrazine, endothall, metribuzin, paraquat, picloram, pronamide and 2,4-D. All the herbicides under restricted use have 12 h of REI, except endothall with 48 h; paraquat and 2,4-D with two REIs, 12 h and 48 h, 12 h and 24 h, respectively. Alachlor is reported to have an oral LD_{50} value of 930–1350 mg/kg; atrazine 1869 mg/kg; metribuzin 1100–2300 mg/kg; paraquat 150 mg/kg; 2,4-D 375 mg/kg; and endothall oral and dermal LD_{50} of 233 mg/kg and 481 mg/kg, respectively.

9.3.7 Classification of insecticides

The Pesticide Safety Fact Sheets produced by the Pennsylvania State University (Hock and Lorenz, 2006) document 64 insecticides under general and restricted use. Most of the insecticides are reported to have 12–48 h REI values. However, biological insecticides of *Bacillus* species have 4 h REI, the reason being its toxicity – it is harmless to humans, animals and useful insects. Other exceptions with lower REI values are spinosad, tebufenozide and methoxyfenozide (4 h), fipronil (2 h) and tefluthrin (0 h).

The general category insecticides include acephate, acetamiprid, azadirachtin, four species of *Bacillus thuringiensis*, bifenazate, carbaryl, cryolite, dicofol, fipronil, imidacloprid, indoxacarb, insecticidal soap, malathion, metaldehyde, methoxychlor, methoxyfenozide, piperonyl butoxide (PBO), pymetrozine, pyrethrins, pyrethrum, rotenone, spinosad, sulfur, tebufenozide, thiamethoxam and thiodicarb. More than 50% of insecticides are under restricted use. Most of the insecticides belonging to the organophosphates, organochlorines and synthetic pyrethroids are categorized under restricted use. Among this category, the insecticides are abamectin, aldicarb, azinphos-methyl, bifenthrin, carbofuran, chlorethoxyfos, chlorpyrifos, cyfluthrin, cyhalothrin-lambda, cypermethrin, deltamethrin, diazinon, diflubenzuron, dimethoate, disulfoton, emamectin, endosulfan, esfenvalerate, ethoprop, fenamiphos, fenproparthrin, lindane, hexachlorocyclohexane (HCH), methamidophos, methomyl, methyl parathion, oxamyl, oxydemeton-methyl, permethrin, phorate, phosmet, tefluthrin, terbufos, tetramethrin and zeta-cypermethrin.

The oral LD_{50} values for restricted insecticides are quite low, in some cases extremely low: aldicarb has a value of 5 mg/kg, carbofuran 8 mg/kg, chlorethoxyfos 1.8–4.8 mg/kg, azinphos-methyl 5–20 mg/kg, bifenthrin 262 mg/kg, chlorpyrifos 92–276 mg/kg, cyhalothrin-lambda 79 mg/kg, thiodicarb and fenproparthrin 66 mg/kg, terbufos tech 4.5 mg/kg, phosmet tech 147–316 mg/kg, phorate tech 2–4 mg/kg, oxydemeton-methyl tech 50 mg/kg, oxamyl 37 mg/kg, methyl parathion 6 mg/kg, methomyl 17 mg/kg, methamidophos tech 20 mg/kg, metaldehyde 630 mg/kg, lindane 88–125 mg/kg, indoxacarb 268 mg/kg, imidacloprid tech 450 mg/kg, fenamiphos 10 mg/kg, ethoprop 61 mg/kg, esfenvalerate 458 mg/kg, endosulfan tech 160 mg/kg, disulfoton tech 4 mg/kg and dimethoate tech 235 mg/kg. The dermal LD_{50} for a few insecticides are quite low. For instance, azinphos-methyl 220 mg/kg, chlorethoxyfos 12.5–18.5 mg/kg, cyhalothrin-lambda 632 mg/kg, dimethoate >400 mg/kg, disulfoton 10 mg/kg, endosulfan 359 mg/kg, ethoprop 2 mg/kg, methamidophos 130 mg/kg, methyl parathion 50 mg/kg, oxydemeton-methyl 150 mg/kg, phorate 20–30 mg/kg and terbufos 1.1 mg/kg.

Most of the insecticides used in the general category have oral LD_{50} values ranging from 500–7500 mg/kg; however, a few insecticides have quite low values, such as acephate tech 980 mg/kg, acetamiprid 1064 mg/kg, dicofol 570 mg/kg, imidacloprid tech 450 mg/kg, indoxacarb 268 mg/kg, metaldehyde 630 mg/kg and thiodicarb 66 mg/kg. Similarly, dermal LD_{50} values reported for insecticides in this category are >2000 mg/kg. An exception is carbaryl with a value of 850 mg/kg.

9.4 Hazard-based Classification of Active Ingredient Using Environmental Impact Quotient Tool

Ibrahim (2016) suggests that all pesticide labels should include an index or indicator of their overall health and environmental impacts. Unfortunately, the labelling information and colour coding currently recommended by the World Health Organization (WHO, 2009) and the Food and Agriculture Organization (FAO, 2014), are based on the mammalian *acute* toxicity. Despite its importance, the acute toxicity figure only reflects a minor component that may not be the most critical one in determining the overall hazard of pesticides to human health and the environment. In addition, acute toxicity is only important when accidental exposure of undiluted formulations happens. However, once the pesticide is diluted and sprayed, the risk of acute toxicity to humans becomes minimal, while the risk of chronic and environmental toxicity builds up and may become unavoidable. Most of the end points of chronic and environmental toxicities require repetitive exposure or multiple doses of the pesticide(s) over a certain period of time. Because it is a summative value of the adverse impacts of pesticides on seven major components of human health and the environment (Table 9.2), Environmental Impact Quotient (EIQ) is recommended here to be the basis for the hazard classification, colour coding and labelling system of pesticides. The EIQ concept was developed by Kovach *et al.* (1992) at Ohio University to rate pesticides based on various environmental parameters. The EIQ values of active ingredients have been evaluated based on their toxicity to the consumer, picker, terrestrial, aquatic and ecological components of the environment. EIQ is a simple, metric indicator that can be included in the label of the pesticide either alone, as a measure of potential hazard, or compounded with the field use rating (FUR), as a measure of actual risk. These two measures can help growers and integrated pest management (IPM) practitioners make environmentally friendly or sound choices/decisions and ultimately know which pesticide(s) or pest management strategy/programme is likely to have lower health and environmental impacts. Table 9.2 shows EIQ values of 210 kg a.i./ha for the most hazardous pesticide and 6.7 kg a.i. for the least hazardous pesticide. These two values were used to draw a full scale from which an EIQ-based hazard classification and colour coding system was established (Fig. 9.1).

Because the range of this new hazard classification system (from 6.7–210) is narrower than that of the WHO system (from >5000 to <5), it can be used without any further adjustment to the percentage of active ingredients in individual pesticide formulations. It is worth noting that the new system is highly desirable from a practical point of view, logically acceptable from the conceptual point of view, yet subjectively debatable, especially with respect to the EIQ distance between the hazard categories. In fact, this debate is also valid with the WHO classification (WHO, 2009).

Ibrahim (2016) proposed the guidelines on 'Extended Pesticide Producer and User

Table 9.2. The main EIQ components with their upper and lower limits. (From Kovach *et al.*, 1992.)

EIQ component	Calculated impact	Minimum scores	Maximum score	Maximum/minimum ratio
Applicator	C × DT × 5	5	125	25
Picker	C × DT × P	1	125	125
Consumer	(C × ((S + P)/2) × SY) + (L)	2	80	40
Fish	F × R	1	25	25
Birds	(D × ((S+P)/2) × 3)	3	75	25
Honeybee	Z × P × 3	3	75	25
Natural enemies	B × P × 5	5	125	25
Total score	N/R	20	630	31.5
EIQ = Total score/3	N/R	6.7	210	31.3

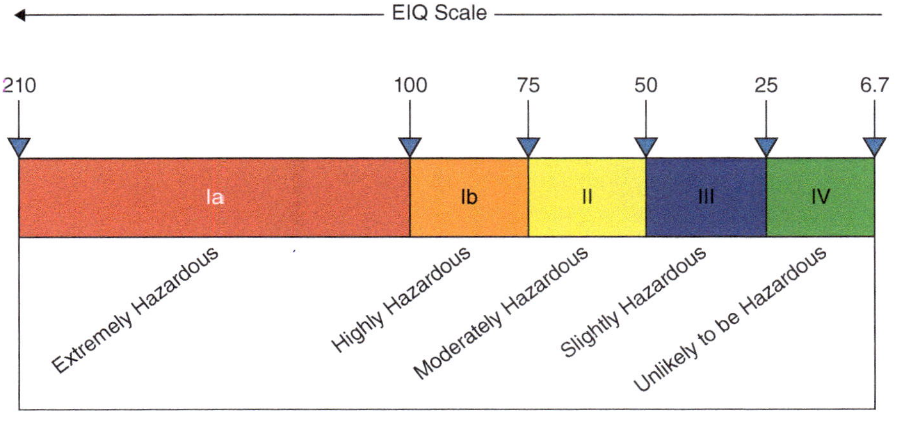

Fig. 9.1. An EIQ-based hazard classification and colour-coding system proposed for eco-labelling of pesticides. Note that this system is based on the Environmental Impact Quotient (EIQ) as a measure of the toxicological impacts of pesticides on human health and the environment. (From Ibrahim, 2016.)

Responsibility' (EPPUR), which aim at a fair and justified distribution of responsibility between the producer and user of pesticides for the negative impacts and associated costs of these pesticides throughout their life cycles. There are at least three categories of pesticide negative impacts: (i) the human health impact; (ii) the environmental impact; and (iii) the end-of-life impact. The first two impacts should be priced based on toxicological measures (health and environmental damages), while the third one may be resolved by physical measures, the purest of which is a take-back measure. To implement an EPPUR policy, three 'toxi-economic' requirements should be fulfilled. The first is to assess the negative impacts of individual pesticides. This requirement can be fulfilled using the EIQ. The second is to monetize the negative external costs. This requirement can be fulfilled by using the author's new Pesticide Negative Externality Assessment (PNEA) system. The third is to fully or proportionally internalize the external cost through a taxation instrument that takes into consideration the external cost of each pesticide per kg a.i. and its total amount produced or used during a certain period of time. Obviously, this requirement cannot be fulfilled without government intervention. With regard to internalization, Pretty (2005) argues against the 'one-size-fits-all' idea, whether in the form of a constant baseline cost or in the form of certain percentages corresponding to the WHO classification of acute hazard (Agne *et al.*, 2000). As a complement to the PNEA system, Pretty (2005) established a new hazard classification and colour-coding system that seems preferable to that of the WHO (2009) and FAO (2014), at least from the conceptual point of view. Instead of just relying on the acute toxicity, the new system is based on acute, chronic and environmental toxicities.

9.5 Conclusion

Pesticides cause acute or chronic toxicity depending on the level of exposure and its duration. Acute toxicity is determined by examining the dermal, inhalation and oral toxicity of test animals, in addition to eye and skin irritation. It is usually expressed as LD_{50} or LC_{50}, which is the amount or concentration of a toxicant, the a.i., required to kill 50% of the test population. Any harmful effects from exposure to small doses repeated over a period of time are termed 'chronic effects'. The chronic effects include birth defects, production of tumours, blood disorders and neurotoxic effects. The chronic toxicity of a pesticide is more difficult to determine through laboratory analysis than acute toxicity. The acute toxicity is responsible for accidental exposure to undiluted formulations of the pesticides, after which dilution and spray is minimized. However, the risk of chronic and environmental toxicity builds up and becomes unavoidable after spray. The EIQ concept has reported it as providing summative values of adverse impacts of pesticides on different components of environment and human health. Under the EPPUR concept, the producer should solely bear the upstream responsibility (toxicological, physical, informative, financial, legal, etc.) for the negative impacts of pesticides from the time of their production till their end of life. However, the downstream (post-consumption phase) responsibility should be distributed between the producer and the consumer/user. For toxicological impacts on human health and environment, another novel system, PNEA, simply and accurately calculates the external cost of individual pesticides using the EIQ concept.

References

Agne, S., Fleischer, G. and Waibel, H. (2000) Economics of pesticide policy reform in developing countries – selected results of a multi-country project. *Deutscher Tropentag*. 2000 in Hohenheim. Session: Technologies, Institutions and Policies for Natural Resource Conservation. Available at: https://ftp.gwdg.de/pub/tropentag/proceedings/2000/Full%20Papers/Section%20I/WG%20c/Agne%20S.pdf (accessed 1 October 2018).

Canadian Centre for Occupational Health and Safety (CCOHS) (2018) Pesticides – re-entry Interval. Government of Canada. Available at: www.ccohs.ca/oshanswers/chemicals/pesticides/reentry.html (accessed 30 July 2018).

Food and Agriculture Organization (FAO) (2014) The International Code of Conduct on Pesticide Management. Food and Agriculture Organization of the United Nations. Available at: www.fao.org/agriculture/crops/thematic-sitemap/theme/pests/code/en (accessed 31 July 2018).

Hock, W.K. and Lorenz, E.S. (2006) Toxicity of pesticides. Pesticide Safety Fact Sheets. Pesticide Education Program in Penn State's College of Agricultural Sciences, The Pennsylvania State University.

Ibrahim, Y.A. (2016) Health and environmental impacts of pesticides: A responsibility principle and two novel systems for hazard classification and external cost determination. *Journal of Toxicology and Health*. DOI:10.7243/2056-3779-3-1.

Kovach, J., Petzoldt, C., Degnil, J. and Tette, J. (1992) A method to measure the environmental impact of pesticides. *New York's Food and Life Sciences Bulletin* 139, 1–8.

Olympic Horticultural Products (OHP) (2018) Chemical class chart: greenhouse and nursery production. Available at: www.ohp.com/Literature/pdf/CCC_XIII.pdf (accessed 31 July 2018).

Pimentel, D., Mclaughlin, L., Zepp, A., Lakitan, B., Kraus, T., Kleinman, P., Vancini, F., Roach, W.J., Graap, E., Keeton, W.S. and Selig, G. (1993) Environmental and economic effects of reducing pesticide use in agriculture. *Agriculture Ecosystems and Environment* 46, 273–288.

Pretty, J. (2005) *The Pesticide Detox: Towards a More Sustainable Agriculture*. Earthscan, Abingdon, UK.

Thomson, L. (2012) Pesticide impacts on beneficial species. Factsheet, May 2012. Grape and Wine Research and Development Corporation, Adelaide, Australia. Available at: www.mvwi.com.au/items/511/2012-05-FS-Pesticide-Impacts2.pdf (accessed 30 July 2018).

United States Environmental Protection Agency (US EPA) (2018) Chapter 7: Precautionary statements – March, 2018. In: Label Review Manual. Office of Pesticide Programs. Available at: https://www.epa.gov/sites/production/files/2018-04/documents/lrm-complete-mar-2018.pdf (accessed 30 July 2018).

World Health Organization (WHO) (2009) The WHO Recommended Classification of Pesticides by Hazard and Guidelines to Classification. World Health Organization. Available at: www.who.int/ipcs/publications/pesticides_hazard/en (accessed 31 July 2018).

III Remediation to Reduce Pesticide Risk

10 Pesticide Risk – Case Studies

10.1 Introduction

Pesticides have numerous beneficial effects, such as crop protection, preservation of food and materials, and prevention of vector-borne diseases. Pesticides have been used in the prevention of malaria, which used to kill up to 1 million children per year, and for preventing other vector-borne diseases such as dengue fever, leishmaniasis and Japanese encephalitis. Pesticides are designed to kill, reduce or repel insects, weeds, rodents, fungi or other organisms that can threaten public health and the economy. Their mode of action is to target systems or enzymes in the pests that may be identical or very similar to systems or enzymes in human beings and, therefore, they pose risks to human health and the environment. According to the World Health Organization (WHO), 3 million cases of pesticide poisoning occur every year, resulting in more than 250,000 deaths (Yang and Deng, 2007). The active ingredients in pesticides are formulated to allow better application, coverage, handling and safety. The formulations are generally only assessed for acute toxicity to humans (by oral and dermal routes, by inhalation, skin irritation, eye irritation and skin sensitization), as well as for organisms in the environment that may come into direct contact with the product. Very often pesticide formulations are designed to be more toxic than the active ingredient itself, through the use of synergists and surfactants that increase the bioavailability/toxicity of the formulation (Kienzler *et al.*, 2014).

10.2 Pesticide Exposure

There is growing concern about children's exposure to pesticides as they are especially susceptible to their effects. A large variety of chemicals and mixtures are used as pesticides in rural areas, and many pesticides are used at the same time in the same place (agricultural regions). The physical structure and strength of children are different from those of adults, and they may have higher exposures and greater vulnerability at both high and low levels of exposure (National Resource Council, 1993). Pesticides are ubiquitous in the environment and most of them are synthetic. Multiple exposures may occur from the preconception period throughout the child's growth into adolescence and adulthood. Children in urban areas are not spared from the hazards of these pesticides, as these may be heavily used indoors, for domestic pest control.

A survey by the US Environmental Protection Agency (US EPA) (CDC, 1999–2002; Lemus and Abdelghani, 2000; Bass *et al.*, 2001; Fields, 2001; Forrester *et al.*, 2004) regarding pesticides used in and around the home revealed some significant findings:

- Almost half (47%) of all households with children under the age of five had at least one pesticide stored in an unlocked cabinet, less than 4 ft off the ground (i.e. within the reach of children).
- Approximately 75% of households without children under the age of five also stored pesticides in an unlocked cabinet, less than 4 ft off the ground. This number is especially significant because 13% of all pesticide poisoning incidents occur in homes other than the child's home.

Insecticides are mostly used in developing countries and fungicides/herbicides in developed countries. The percentage of active ingredients in the formulation generally varies from 2% to 80%. The toxic effects depend on the physical state of the product as well as on the characteristics of the solvent or other substances contained in the formulation. Volatile pesticides and fine powders are more easily inhaled than dense products and coarse granulated materials. Concentrated solutions are much more dangerous than diluted ones; solid baits may be colourful, attractive and sweet, and may be easily ingested by toddlers (Tomlin, 1997; IPCS, 2000–2002).

In some cases, if the concentration of the active compound is less than 2% (as is generally the case for pesticides for household use), the toxicity in cases of human exposure may be due to the solvent (e.g. kerosene or paraffin) and not to the active ingredient.

10.2.1 Pesticide application as source of exposure

- Professional application of pesticides both indoors and outdoors is on the increase and is used commonly for the control of rodents, cockroaches, ants, termites, earwigs and other pests. Signs and symptoms of pesticide-related illnesses have been described after indoor and outdoor spraying.
- Domestic use of insecticides is also a source of exposure. Insecticides formulated as sprays, strips and baits are widely available.
- Certain topical pharmaceuticals for direct application on children's skin or scalps contain insecticides to control lice or scabies. Their use carries a risk of acute (chronic if repeated) overexposure. High doses or applications lasting a long time have caused acute poisoning. There are many pharmaceuticals described in the *International Pharmacopoeia* that contain organophosphorus (malathion) and organochlorine (lindane) pesticides still being used to treat head lice. Such products may contain up to 2% of pesticide.
- Dogs and cats are often treated with insecticides to control fleas or ticks. Veterinary products can be as concentrated as agricultural ones. Pet dips for treatment of flea infestations usually contain organophosphate and pyrethroid pesticides. Children are often involved in pet care.

10.2.2 Pesticide residues as source of exposure

- Insecticides may persist in house dust, in soil tracked in from outdoors, in carpets, toys, food and furniture.
- High levels of insecticides have been measured for weeks after professional application.
- Residues of organophosphorus insecticides sprayed in indoor environments have been reported to occur on floors, carpets, children's toys, furniture, bed covers and in dust.
- Poor hygiene habits or houses that are difficult to clean increase the risk of exposure.

- Playgrounds, playing fields, lawns and gardens may be routinely sprayed in order to keep insects away.
- Pesticides are found in recreational waters (lakes, rivers and in pools (algaecides)).
- Persistent wood preservatives such as arsenic/copper/chromium mixtures have been used on play structures.
- The persistent organic pollutants include nine pesticides.

10.2.3 Exposure to crop beneficials/ pollinators

Beneficial organisms include various insects, mites, nematodes, fungi, bacteria and other microorganisms that feed on or parasitize pest species. Bees and other pollinators are also beneficial organisms, including wild bees, honeybees, certain wasps and other insects, which are important pollinating agents of crops. Some pesticides are harmful to these pollinators and beneficials. Fishel (2017) observed that pesticide exposure can result in direct losses of the insect populations and indirect losses of crop yield because of the lack of adequate pollination. The value of honey alone produced in Florida during 2012 was more than $23,000,000, according to the Florida Department of Agriculture and Consumer Services. It has been reported that if there is some toxicity concern with the pollinators, pesticide labels should have the statement 'Environmental Hazards' on them.

Thomson (2012) reported relative impacts of pesticides used in the Australian viticulture industry on beneficial insects. Insect and mite pest management in vineyards can be achieved through applications of chemicals and through the action of the natural enemies of pests, including numerous invertebrate predators and parasitoids. Based on the factsheets generated, the growers can decide on a spray programme. Control methods using pesticides and beneficial organisms are not compatible for simultaneous use, although pest control using beneficial organisms has the potential to reduce costs, and support the wine industry's focus on sustainability, as compared with pesticides. Moreover, using beneficial organisms in viticulture will bring about both environmental and economic advantages. Thomson (2012) evaluated the impact of individual chemical constituents of pesticide applications on a few representative non-target species (beneficial organisms) through standardized sets of laboratory tests. The mortality of individuals is scored at given concentrations of active constituents to provide information on a lethal dosage, or percentage mortality at field concentrations. The impact of some 'softer' pesticides was also evaluated. The toxicity ratings for individual chemicals are derived from laboratory and semi-field-based tests in accordance with the guidelines of the International Organization for Biological and Integrated Control (IOBC) Working Group on Pesticides and Beneficial Organisms. Toxicity ratings indicate the reduction in the ability of the beneficial species tested to provide pest control, and range from 1 to 4 (Table 10.1).

1. The chemical is 'harmless' and kills fewer than 25% of the beneficial species of interest.
2. The chemical is 'slightly harmful' and kills between 25 and 50% of the beneficial species of interest.
3. The chemical is 'moderately harmful' and kills between 50 and 75% of the beneficial species of interest.
4. The chemical is 'harmful' and kills more than 75% of the beneficial species of interest.

Róbert Chlebo (2006) reported organochlorines, organophosphate pesticides and carbamates as the most common insecticides examined in European honeys. Blasco *et al.* (2003) examined residues of 42 different pesticides in 50 honey samples from Spain and Portugal. Most of the pesticides found in honey were organochlorines. Among them, gamma-hexachlorocyclohexane (HCH, also known as lindane) was found in 50% of the samples and was the most frequently detected substance, followed by hexachlorobenzene (HCB) in 32% of the samples and other isomers of HCH (value 0.03–4.31 mg/

Table 10.1. Toxicity effects on specific beneficial insects of insecticides and miticides registered for use in Australian vineyards. Reduction in population (mortality) following application is represented by shading (see legend below). (Data are sourced from scientific literature and databases including International Organization for Biological and Integrated Control (IOBC), as compiled on IMPACT for Viticulture, http://cesar.org.au/index.php?option=com_collateral_manage, reproduced from Thomson, 2012.)

			Predators							
Active ingredient	Grouping	Parasitoids	Beetles	Ladybirds	Predatory bugs	Predatory flies	Predatory mites	Lacewings	Spiders	Overall rating
Bacillus thuringiensis (Bt)	11C	0	–	0	0	0	0	0	–	0
Sulfur <3000 g ac/ha[a]	M2	1	–	2	–	–	1	0	1	1
Sulfur 3000–4000 g ac/ha	M2	1	–		0	1	1	0	–	2
Sulfur 4000–6000 g ac/ha	M2	2	–	–	0	1	2	0	–	3
Buprofezin	16	0	–	1	0	1	0	–	–	1
Chlorantraniliprole	28	0	0	–	–	–	0	–	–	1?
Methoxyfenozide	18	1	–	1	1	–	0	–	–	1
Emamectin	6A	–	–	–	–	–	–	0	–	1?
Petroleum/paraffinic oil	Unspecified	–	0	0	–	3	0	2	0	2?
Clothianidin	4A	3	3	–	–	–	3	–	–	3
Carbaryl and methomyl	1A	3	–	3	3	3	3	3	–	3
Fipronil	2C	3	–	–	3	3	3	3	–	3
Dicofol	UN	2	–	0	–	3	3	1	–	3
Indoxacarb	22A	3	–	–	3	0	2	–	–	3
Organophosphates[b]	1B	3	–	2	3	3	3	3	–	3
Pyrethroids[c]	3A	3	–	3	3	3	3	3	–	3
Spinosad	5A	3	–		3	3	3	3	–	3

[a]ac/ha, active constituents per hectare.

[b]Organophosphates (1B) include azinphos-methyl chlorpyrifos, diazinon, dimethoate, fenamiphos, fenthion, fenitrothion, malathion, parathion-methyl, trichlorfon.

[c]Pyrethroids (3A) include alpha-cypermethrin, bifenthrin, esfenvalerate and permethrins + piperonyl butoxide.

No shading or dash represents no literature found at time of writing; ? indicates that the rating was based on limited data. 0 = harmless. Values represent reduction in ability to provide pest control, as modified from IOBC toxicity ratings.

Shading represents the mortality of beneficial insects.

<25% 25–50%; 50–75% >75%

kg), but most of them were below 0.5 mg/kg. Bogdanov (2006) observed very small residues of imidacloprid in honey, perhaps due to a filtering effect of bees; however, the effect of imidacloprid on bee health is highly controversial. Schur and Wallner (2000) reported that bees reduce high pesticide nectar concentrations initially, so that the final concentration in honey is much lower, by a factor of about 1000. Generally, there are no maximum residue limits (MRLs) for pesticides in honey. In the European Union (EU) an action level of 0.01 mg/kg is often considered for pesticides with no fixed MRL. To avoid residues, pesticides should be used outside the bloom period or at least not during the foraging time of bees, and also their hives should be placed >3 km from agricultural plants treated with pesticides.

10.2.4 Exposure to aquatic organisms

A study conducted by Kumar and Chapman (1998) observed that inhibition of acetylcholinesterase (AChE) and the body residues of profenofos in the rainbow fish were concentration-dependent during 96-h lethal and sublethal exposures. Profenofos exposures resulted in increased AChE inhibition, which could be related to greater accumulation of profenofos residues in the fish. Sublethal exposure of rainbow fish for longer duration resulted in lower body burdens in rainbow fish, and it was speculated to be associated with biotransformation of profenofos. Sublethal exposures to profenofos affected food intake, growth rates, feeding rates and food conversion efficiency in rainbow fish, which could be related to significant AChE inhibition. Significant behavioural changes such as spontaneous swimming activity and photokinesis during sublethal exposure of fish to profenofos could be related to significant AChE inhibition due to the accumulation of profenofos residues in the bodies of the fish.

The increased use of pyrethroids is reported due to their low mammalian and avian toxicity, better selectivity to target species than organophosphates and lower persistence than organochlorine insecticides. The environmental fate and bioavailability of pyrethroids is governed by their strong adsorption to organic matter, soils and clay. Palmquist *et al.* (2012) reported non-target aquatic invertebrate species, such as fish and amphibians, to be more sensitive to pyrethroid insecticides than targeted terrestrial invertebrates. The route of entry may be through drift and runoff from the sprayed outdoor areas. The pyrethroid exposure may lead to short-term adverse population and community level effects, leading to rapid recovery from these. In terrestrial systems, spray drift generally results in non-homogeneous distribution of pyrethroids in edge-of-field and off-crop areas. Hence, impacts on non-target invertebrates in these areas are spatially variable, and recovery can occur via immigration from non-impacted populations. The organophosphate applicators were observed to be more exposed to oxidative stress, based on increased malondialdehyde (MDA) production and reduced activities of AChE, catalase (CAT) and myeloperoxidase (MPO) in pesticide applicators, than other farmers, requiring increased levels of antioxidant supplements compared with non-organophosphate farmers. Through inverse correlation, Yaqub *et al.* (2015) observed significant negative correlation between AChE and duration of exposure in organophosphate applicators: the longer the duration of exposure to organophosphate, the lower the AChE activity and the more the risk of adverse effects of organophosphates.

10.2.5 Genotoxicity risk of pesticide exposure

Bolognesi *et al.* (2011) determined the genetic risk associated with pesticide exposure in lymphocytes or buccal cells as a target to determine the induction of genotoxic damage. It was observed that pesticides or their mixtures can reduce or potentiate their risk to produce genetic effects including cancer and other chronic pathologies in humans, mainly in work places, such as the chemical industry or in crop fields. Thus, people working in the pesticide manufacturing industries, spraying pesticides, floriculture or agricultural works in general are the subject of specific sections. Studies on workers from pesticide industries showed a statistically significant increase of micronucleus (MN) frequency in peripheral blood lymphocytes. The MN frequency in 41 workers exposed to chlorinated compounds, including HCB in Sao Paulo (Brazil), was significantly higher than in controls, showing also a correlation with working time and with serum concentration of HCB (da Silva Augusto *et al.*, 1997). Another study carried out in Croatia in workers exposed to 2,4-dichlorophenoxyacetic acid (2,4-D), atrazine, alachlor, cyanazine and malathion, during the process of production, showed significant increases in the MN frequency after 8 months of high exposure (Garaj-Vrhovac and Zeljezic, 2001; 2002). In a recent study carried out in Pakistan with workers from an industry producing pesticides belonging to the organophosphate and pyrethroid classes, significant increases in the MN frequency were observed in workers, showing a linear correlation with length of exposure (Bhalli *et al.*, 2006). In a study conducted in California with malathion-exposed sprayers involved in the Mediterranean Fruit Fly Eradication Program, a slight but significant increase in the MN frequency was observed in workers exposed to malathion for >50 h during the last 8 months or with levels of malathion diacid >100 p.p.b. (Windham *et al.*, 1998).

10.3 Regulation of Pesticide Exposure

Under the Federal Insecticide, Fungicide, and Rodenticide Act (FIFRA), the US EPA has the authority to regulate the use of pesticides to prevent unreasonable adverse human health effects associated with pesticide exposure. Accordingly, the US EPA requires pesticide registrants to perform studies evaluating the potential for pesticide handler exposure. Geer *et al.* (2004) demonstrated the value of the pesticide registrant database for the purpose of evaluating pesticide worker exposure in the workplace, by examining methods of assessment, routes and determinants of exposure and dose to the pesticide chlorpyrifos, to close the unexplained variability gap between measures of exposure and biological monitoring. It highlights the significance of the dermal exposure pathway and identifies the need for methods and research to close the gap between external and internal exposure measures. The observed doses were highly variable and differed by job class ($P < 0.05$). A simple linear regression of urinary 3,5,6-trichloro-2-pyridinol (3,5,6-TCP) against exposure-derived absorbed total dose (EDADtot) indicated a positive association explaining 29% of the variability in the 3,5,6-TCP derived estimate of dose. A multiple linear regression model including the variables exposure-derived absorbed dermal dose (EDADderm), exposure-derived absorbed inhalation dose (EDADinh) and application type explained 46% of the variability (R2 = 0.46) in the urinary dose estimate. EDADderm was marginally significant ($P = 0.066$) while EDADinh was not ($P = 0.57$). The EDADderm regression coefficient (0.0007) exceeded the coefficient for EDADinh (0.00002) by a factor of 35.

Woollen (1993) observed a non-linear relationship between external exposure estimates and biomonitoring due to factors such as differences in personal hygiene habits (i.e. washing) and individual differences in metabolism. The rate of exposure of a pesticide and its duration influences the rate of metabolism in the worker. The absorption process is thought to be saturable, not conforming to a linear dose–response model (Zendzian, 2000). Others suggest that the variability seen in passive dosimetry estimates might be attributable to lack of standardization of methods and exclusion of the ingestion pathway. Fenske and Elkner (1990) identified similar positive associations of the 3,5,6-TCP urinary metabolite and external exposure measures (R2 = 0.86) in a study of residential pesticide applicators. Hines and Deddens (2001) found positively and highly correlated weekly means of ln(TCP) and ln(chlorpyrifos air concentrations) ($R2 = 0.73$, $P < 0.0001$) in 41 termiticide applicators in North Carolina. In an environmental in-home exposure assessment, Buckley *et al.* (1997) found 3,5,6-TCP levels in urine moderately correlated with indoor air ($R2 = 0.55$; $P < 0.01$) and dust ($R2 = 0.46$; $P < 0.01$) concentrations of chlorpyrifos during two seasons of sampling. The fact that dermal exposure represents a substantial portion of pesticide workers' exposure and yet there are no standard methods of assessment highlights a significant gap in the protection of worker health and safety.

Geer *et al.* (2004) highlighted a significant gap in the protection of worker health and safety, emphasizing the fact that there is no standard method of assessment of dermal exposure of pesticide workers. Although this gap is addressed through biological-based exposure indices (BEI) for some chemicals, BEI is not available for chlorpyrifos exposure other than non-specific cholinesterase inhibitor (ACGIH, 2001). Although all studies used a common set of methodological guidelines designed to minimize measurement differences across studies (US EPA, 1998), each study was conducted independently and it is possible that there were subtle methodological differences in sampling and/or analysis that resulted in an unknown bias. Overall, the US EPA's Pesticide Registrant Database provides useful information for evaluating pesticide worker exposure and worker exposure assessment methods. These data help demonstrate the significance of the dermal route of exposure in pesticide workers. This analysis also served to demonstrate the public health utility of the US EPA's Pesticide Registrant Database.

Four of five studies carried out with sprayers applying complex mixtures of pesticide in cultivation of different crops during sanitization and indirect exposure by aerial spraying gave positive results. Significant increases of MN associated with the duration of exposure were observed in a study carried out with sprayers from central Italy (Pasquini *et al.*, 1996). A study conducted in vineyard workers from Serbia, applying mainly insecticides and fungicides, showed higher MN frequency compared with controls 1 month after the start of the spraying period, with a further increase at the end of the spraying season (Joksic *et al.*, 1997). Positive effects were also reported in a group of sanitation workers from Belo Horizonte (Brazil), using different pesticides including organophosphates and pyrethroid insecticides, as well as hydroxycoumarinic rodenticides. No time–exposure association was found (Kehdy *et al.*, 2007).

In Mexico, a study with 30 subjects working as floriculturists in greenhouses showed an increase in MN frequency in buccal cells (Remor *et al.*, 2009). A further study in Mexico (Sinaloa State) reported a clear increase in MN frequency in agricultural workers using mainly organophosphates and carbamates without any correlation with age, gender or exposure length to pesticides (Sailaja *et al.*, 2006). A study carried out in Hyderabad (India) in a chemical industry producing organophosphates, carbamates and pyrethroids showed significant increases in the MN frequency in subjects who had worked in the industry for >10 years (Ergene *et al.*, 2007). Slight but significant increases in the frequency of MN were also reported in the Göksu Delta region (Turkey), a wetland area with intensive agriculture, where rice, cotton and groundnuts are grown all year (Martínez-Valenzuela *et al.*, 2009). Significant increases in the frequency of MN were observed in workers involved in soybean culture in the state of Rio Grande do Sul (Brazil); however, these increases were not related to the use of protective measures or the time of exposure (Bortoli *et al.*, 2009).

In a case study of soil fumigation, Sande *et al.* (2011) estimated hazards from methyl bromide and other alternative fumigants to workers, consumers, beneficial arthropods, birds, fish and bees, using the Environmental Impact Quotient (EIQ). Iodomethane was observed to have the lowest relative risk with the lowest EIQ index value and field use rating. Among environmental categories, workers and beneficial arthropods experienced the highest relative risks from all fumigants, and fish and consumers the least risks. The workers were engaged in the mixing and application of fumigants, so were more exposed. Use of chloropicrin in the fumigant formulation, as a warning agent, was shown to greatly increase its relative risk. The impetus behind the methyl bromide phase-out is its ozone depleting potential.

A case report of a 74-year-old male was presented by Baur (2016) during a workshop on 'EU pesticide risk assessment system: The case of glyphosate' (Baur, 2016). He was diagnosed with abdominal non-Hodgkin's lymphoma in 2009. He was involved in application of various herbicides and other pesticides in cultivation of grain on a large scale, without using any preventive measures to avoid exposure to pesticides. The IARC (2015) mentioned a positive association between glyphosate and human cancer in some case studies. Two case studies have mentioned instances of kidney tumours and hemangiosarcoma while using pesticides (Workshop, 2016).

10.4 Conclusion

Pesticides have become a necessity in crop protection, preservation of food and materials, and prevention of vector-borne diseases. Their mode of action is to target systems or enzymes in pests that may be identical or very similar to systems or enzymes in human beings and, therefore, they pose risks to human health and the environment. Three million cases of pesticide poisoning occur every year, resulting in more than 250,000 deaths, according to a WHO report (WHO, 2003). Children have more exposure and greater vulnerability at both high and low levels of exposure of pesticides. The major source of exposure is the application of pesticides, both indoors and outdoors, and pesticide residues in food or the food chain, water, soil and other environments. Beneficial organisms and pollinators also get exposed to these pesticides in the environment. A non-linear relationship is observed between external exposure estimates and biomonitoring due to factors such as differences in personal hygiene habits and individual differences in metabolism. The rate of exposure of a pesticide and its duration influence the rate of metabolism in the worker. Pesticides or their mixtures can also reduce or potentiate their risk to produce genetic effects including cancer and other chronic pathologies in humans.

References

American Conference of Governmental Industrial Hygienists (ACGIH) (2001) TLVs and BEIs threshold limit values for chemical substances and physical agents. ACGIH, Cincinnati, Ohio.

Bass, J.K., Ortega, L., Rosales, C., Petersen, N.J. and Philen R.M. (2001) What has been used at home: A household pesticide survey. *Revista Panamericana de Salud Publica* 9(3), 138–144.

Baur, X. (2016) Options based on clinical daily practice: overview. Presentation during a workshop: EU's pesticide risk assessment. The case of glyphosate. EFSA Meeting, 24 May 2016. PE 587.309.

Bhalli, J.A., Khan, Q.M., Haq, M.A., Khalid, A.M. and Nasim, A. (2006) Cytogenetic analysis of Pakistani individuals occupationally exposed to pesticides in a pesticide production industry. *Mutagenesis* 21, 143–148.

Blasco, C., Fernandes, M., Pena, A., Lino, C., Silveira, M.I., Font, G. and Pico, Y. (2003) Assessment of pesticide residues in honey samples from Portugal and Spain. *Journal of Agricultural and Food Chemistry* 51, 8132–8138.

Bogdanov, S. (2006) Contaminants of bee products. *Apidologie* 37, 1–18.

Bolognesi, C., Creus, A., Ostrosky-Wegman, P. and Marcos, R. (2011) Micronuclei and pesticide exposure – review. *Mutagenesis* 26(1), 19–26.

Bortoli, G.M., Azevedo, M.B. and Silva, L.B. (2009) Cytogenetic biomonitoring of Brazilian workers exposed to pesticides: Micronucleus analysis in buccal epithelial cells of soybean growers. *Mutation Research* 675, 1–4.

Buckley, T.J., Liddle, J., Ashley, D.C., Paschal, V.W., Needham, L.L. and Ackland, G. (1997) Environmental and biomarker measurements in nine homes in the lower Rio Grande valley: Multimedia results for pesticides, metals, PAHs, and VOCs. *Environment International* 23, 705–732.

Centers for Disease Control and Prevention (CDC) (1999–2002) Surveillance for acute insecticide related illness associated with mosquito control efforts in nine states. Available at: www.cdc.gov/mmwr/preview/mmwrhtml/mm5227a1.htm, www.epa.gov/pesticides/factsheets/childsaf.htm (accessed 21 August 2018).

Ergene, S., Celik, A., Cavas, T. and Kaya, F. (2007) Genotoxic biomonitoring study of population residing in pesticide contaminated regions in Goksu delta: Micronucleus, chromosomal aberrations and sister chromatid exchanges. *Environment International* 33, 877–885.

Fenske, R.A. and Elkner, K.P. (1990) Multi-route exposure assessment and biological monitoring of urban pesticide applicators during structural control treatments with chlorpyrifos. *Toxicology and Industrial Health* 6, 349–371.

Fields, S. (2001) Caution – children at play: How dangerous is CCA? *Environmental Health Perspectives* 109, A262. Available at: http://ehp.niehs.nih.gov/docs/2001/109-6/focus.html (accessed 21 August 2018).

Fishel, F.M. (2017) Pesticide effects on nontarget organisms. This document is PI-85, one of a series of the Agronomy Department, UF/IFAS. Original publication date November 2005. Revised June 2010, April 2011, March 2014 and March 2017. Available at: http://edis.ifas.ufl.edu (accessed 21 August 2018).

Forrester M.B., Sievert, J.S. and Stanley, S.K. (2004) Epidemiology of lindane exposures for pediculosis reported to poison centers in Texas, 1998–2002. *Journal of Toxicology – Clinical Toxicology* 42(1), 55–60.

Garaj-Vrhovac, V. and Zeljezic, D. (2001) Cytogenetic monitoring of Croatian population occupationally exposed to a complex mixture of pesticides. *Toxicology* 165, 153–162.

Garaj-Vrhovac, V. and Zeljezic, D. (2002) Assessment of genome damage in a population of Croatian workers employed in pesticide production by chromosomal aberration analysis, micronucleus assay and comet assay. *Journal of Applied Toxicology* 22, 249–255.

Geer, L.A., Cardello, N., Dellarco, M.J., Leighton, T.J., Zendzian, R.P., Roberts, J.D. and Buckley, T.J. (2004) Comparative analysis of passive dosimetry and biomonitoring for assessing chlorpyrifos exposure in pesticide workers. *Annals of Occupational Hygiene* 48(8), 683–695.

Hines, C.J. and Deddens, J.A. (2001) Determinants of chlorpyrifos exposures and urinary 3,5,6-trichloro-2-pyridinol levels among termiticide applicators. *Annals of Occupational Hygiene* 45, 309–321.

International Agency for Research on Cancer (IARC) (2015) IARC monographs on the evaluation of carcinogenenic risks to humans, vol. 112. EP, Glyphosate workshop, 24 May 2016.

International Programme on Chemical Safety (IPCS) (2000–2002) The WHO recommended classification of pesticides by hazard, guidelines to classification. Available at: www.inchem.org/documents/pds/pdsother/class.pdf (accessed 21 August 2018).

Joksic, G., Vidakovic, A. and Spasojevic-Tisma, V. (1997) Cytogenetic monitoring of pesticide sprayers. *Environmental Research* 75, 113–118.

Kehdy, F.S.G., Cerqueira, E.M.M., Bonjardim, M.B., Camelo, R.M. and Castro, M.C.L. (2007) Study of the cytogenetic effects of occupational exposure to pesticides on sanitation workers in Belo Horizonte, Brazil. *Genetics and Molecular Research* 6, 581–593.

Kienzler, A., Berggren, E., Bessems, J., Bopp, S., van der Linden, S. and Worth, A. (2014) Assessment of mixtures – review of regulatory requirements and guidance. *JRC Science and Policy Reports.* European Commission. Available at: https://eurl-ecvam.jrc.ec.europa.eu/news/assessment-mixures-report (accessed 31 July 2018).

Kumar, A. and Chapman, J.C. (1998) Profenofos toxicity to the eastern rainbow fish (*Melanotaenia duboulayi*). *Environmental Toxicology and Chemistry* 17(9), 1799–1806.

Lemus R and Abdelghani A.(2000) Chlorpyrifos: An unwelcome pesticide in our homes. *Reviews on Environmental Health* 15, 421.

Martínez-Valenzuela, C., Gómez-Arroyo, S., Villalobos-Pietrini, R., Waliszewski, S., Calderón-Segura, M.E., Félix-Gastélum, R. and Alvarez-Torres, A. (2009) Genotoxic biomonitoring of agricultural workers exposed to pesticides in the north of Sinaloa State, Mexico. *Environment International* 35, 1155–1159.

Minister of the Environment (2010) Science and technology into action to benefit Canadians. A new approach to pesticide management: finding safer alternatives in response to effects on birds. Available at: http://publications.gc.ca/collections/collection_2011/ec/En4-135-3-2010-eng.pdf, website: www.ec.gc.ca/scitech (accessed 2 October 2018).

National Resource Council (1993) *Pesticides in the diets of infants and children.* National Academy Press, Washington, DC.

Palmquist, K., Salatas, J. and Fairbrother, A. (2012) Pyrethroid insecticides: Use, environmental fate, and ecotoxicology. In: Perveen, F. (ed.) *Insecticides – Advances in Integrated Pest Management.* InTechOpen, London. Available at: https://www.intechopen.com/books/insecticides-advances-in-integrated-pest-management/pyrethroid-insecticides-use-environmental-fate-and-ecotoxicology (accessed 31 July 2018).

Pasquini, R., Scassellati-Sforzolini, G., Angeli, G., Fatigoni, C., Monarca, S., Beneventi, L.,

DiGiulio, A.M. and Bauleo, F.A. (1996) Cytogenetic biomonitoring of pesticide-exposed farmers in central Italy. *Journal of Environmental Pathology, Toxicology and Oncology* 15, 29–39.

Remor, A.P., Totti, A.C., Moreira, D.A., Dutra, G.P., Heuser, V.D. and Boeira, J.M. (2009) Occupational exposure of farm workers to pesticides: Biochemical parameters and evaluation of genotoxicity. *Environment International* 35, 273–278.

Róbert Chlebo (2006) Bees and pesticides. PAN Germany and CEPTA: Pesticide residues in food – regulation, monitoring, policy. Seminar 23–24 June 2006, Modra-Harmónia, Slovakia.

Sailaja, N., Chandrasekhar, M., Rekhadevi, P.V., Mahboob, M., Rahman, M.F., Vuyyuri, S.B., Danadevi, K., Hussain, S.A. and Grover, P. (2006) Genotoxic evaluation of workers employed in pesticide production. *Mutation Research* 609, 74–80.

Sande, D., Mullen, J., Wetzstein, M. and Houston, J. (2011) Environmental impacts from pesticide use: A case study of soil fumigation in Florida tomato production. *International Journal of Environmental Research and Public Health* 8, 4649–4661.

Schur, A. and Wallner, K. (2000) Bewertung des individuellen Wirkstoffeintrages von Sammelbienen nach Pestizidapplikation in blühenden Kulturen. *Apidologie* 31, 615–616.

da Silva Augusto, L.G, Lieber, S.R., Ruiz, M.A. and de Souza, C.A. (1997) Micronucleus monitoring to assess human occupational exposure to organochlorides. *Environmental and Molecular Mutagenesis* 29, 46–52.

Thomson, L. (2012) Pesticide impacts on beneficial species. Factsheet, May 2012. Grape and Wine Research and Development Corporation, Adelaide, Australia. Available at: www.mvwi.com.au/items/511/2012-05-FS-Pesticide-Impacts2.pdf (accessed 30 July 2018).

Tomlin, C. (1997) *The Pesticide Manual* (11th edn). British Crop Protection Council, Farnham, UK.

United States Environmental Protection Agency (US EPA) (1998) Harmonized OPPTS test guidelines series 875. Occupational and residential exposure test guidelines, group B post application exposure monitoring test guidelines. US EPA Office of Prevention, Pesticides, and Toxic Substances, Washington, DC.

World Health Organization (WHO) (2003) The physical school environment, information series on school health, Document 2. World Health Organization, Geneva.

Windham, G.C., Titenko-Holland, N., Osorio, A.M., Gettner, S., Reinisch, F., Haas, R. and Smith, M. (1998) Genetic monitoring of malathion-exposed agricultural workers. *American Journal of Industrial Medicine* 33, 164–174.

Woollen, B.H. (1993) Biological monitoring for pesticide absorption. *Annals of Occupational Hygiene* 37, 525–526.

Workshop (2016) EU's pesticide risk assessment system: The case of glyphosate. Directorate General for internal policies. EFSA Meeting. 24 May 2016. PE 587.309.

Yang, C.C. and Deng, J.F. (2007) Intermediate syndrome following organophosphate insecticide poisoning. *Journal of the Chinese Medical Association* 70, 467–472.

Yaqub, A.S., Rahamon, K.S., Kosoko, M.A. and Arinola, G.O. (2015) Oxidative stress indices in Nigerian pesticide applicators and farmers occupationally exposed to organophosphate pesticides. *International Journal of Applied Basic Medical Research* 4(1), S37–S40.

Zendzian, R.P. (2000) Dermal absorption of pesticides in the rat. *American Industrial Hygiene Association Journal* 61, 473–483.

11 Alternatives for Reducing Pesticide Risk

11.1 Introduction

Millions of tonnes of pesticides are applied annually in modern agriculture to increase production by managing target organisms including insects, fungi, bacteria, viruses and weeds (Liu and Xiong, 2001). However, less than 5% of these products are estimated to reach the target organisms. One of the most important problems related to pesticide use is their possible persistence in the environment, leading to their possible entry into the food chain affecting the ecosystem and, potentially, all human beings (Liu and Xiong, 2001). These toxic compounds have been implicated in various disorders and diseases discussed in previous chapters (Okon, 1985). Pretty (2005), for example, reported that Europeans and North Americans are now concerned about chronic effects of pesticides due to persistent low-level exposure. Farmers and pesticide applicators are particularly prone to adverse effects due to their direct exposure to pesticides. The pesticides can pollute soil and water resources due to drift in air, and can contaminate large areas. The most alarming exposure to pesticides, however, is through residues in food. Exposure is often to multiple mixtures of chemicals, the toxic effects of which are unknown, particularly over longer time scales (Reffstrup *et al.*, 2010). In some cases, these substances can interact such that mixtures may have unpredictable and higher toxicities than the individual components themselves (Leu, 2014). Most research on pesticides is done on the active ingredients. So-called inert ingredients in pesticide formulations can enhance the effect of the active ingredient but can also cause substantial health effects (Mesnage *et al.*, 2014). In addition, metabolites of active and inert ingredients can be of even higher toxicity than the original substances (Leu, 2014; Eyhorn *et al.*, 2015).

Debates around pesticides are usually highly polarized, with opponents demanding an immediate ban on pesticides because they are unsafe, and proponents arguing that we cannot maintain food production without pesticides. As a consequence of this deadlock, progress to address pesticide issues is slow and fragmented. However, the risk of these pesticides can be avoided by adopting certain strategies.

11.2 Strategies for Reduction of Pesticides and Associated Risks

Ideally, agricultural systems should be designed in a way to ensure that pests, diseases and weeds are managed before crossing their economic threshold level (ETL) so as not to cause significant damage to the crop. Suitable agronomic practices, the use of resistant varieties and integrated pest management (IPM) are key preventive measures. Biocontrol and the use of botanical pesticides can complement these efforts. Safe and need-based application of synthetic pesticides should be part of IPM. The following points provide an overview of applied approaches, which in practice, are often combined.

11.2.1 Agronomic practices

Suitable agronomic practices are essential to achieve healthy crops and to prevent build-up of pest, disease and weed pressure (FAO, 2011).

11.2.2 Resistant varieties

Crop varieties differ in their susceptibility to pests and diseases and in their ability to compete with weeds. The selection of appropriate crop varieties is therefore fundamental to a preventive pest management system. Growing resistant varieties along with rotations of non-susceptible crops can substantially limit pest build-up within a field (Pretty, 2005).

11.2.3 Biocontrol and botanical pesticides

Biocontrol through bioagents makes use of pathogens (bacteria, fungi, viruses), insect predators or parasitoids, pheromones and insect traps to keep pest populations low (IAASTD, 2009). The total eradication of a pest, which results from the use of synthetic pesticides, would reduce the food supply of the pest's natural enemies, undermining a key element in system resilience. The aim, therefore, should be to manage insect pest populations to the point where natural predation operates in a balanced way and crop losses to pests are kept to an acceptable minimum (FAO, 2011).

11.2.4 Integrated pest management

The Food and Agriculture Organization (FAO) defines IPM as the careful consideration of all available pest control techniques and subsequent integration of appropriate measures that discourage the development of pest populations and keep pesticides and other interventions to levels that are economically justified and reduce or minimize risks to human health and the environment (FAO, 2015).

11.2.5 Agroecology

Agroecology is a discipline that defines, classifies and studies agricultural systems from an ecological and socio-economic perspective, and applies ecological concepts and principles to the design and management of sustainable agroecosystems (Miguel, 1995). It is an integrative way of farming that focuses on working with and understanding the interactions between plants, animals, humans and the environment. In agroecology, pest control seeks to reinforce interactions of pests and natural enemies with the aim of maintaining a natural balance in the ecosystem (FAO, 2014).

11.2.6 Organic farming

The production system that sustains the health of soils, ecosystems and people is organic farming. It relies on ecological processes, biodiversity and cycles adapted to local conditions, rather than the use of inputs with adverse effects (Definition of Organic Agriculture, 2008). The organic standards strictly prohibit any use of synthetic pesticides. Crop protection in organic agriculture builds on good agronomic practices such as crop rotation and intercropping, the use of organic manures, resistant varieties and biocontrol to prevent pests, diseases and weeds causing significant damage. Organic farming makes use of techniques similar to IPM and agroecology, with the only difference being that synthetic chemicals cannot be used as a last resort.

11.2.7 Use of less hazardous pesticides

There are various systems to classify pesticides according to their toxicity for humans and the environment. Avoiding the use of highly hazardous pesticides and replacing them with less hazardous ones is the most obvious way to reduce the negative side-effects of pesticides (FAO and WHO, 2014). This approach needs to be combined with safe handling of pesticides so that their impact on people and the environment is minimized. The use of protective gear and the observation of waiting periods before harvest are the most important measures in this regard (Pretty, 2005).

11.3 International Policies and Instruments to Reduce Pesticide Use and Risks

Eyhorn *et al.* (2015) quoted that international codes, treaties, conventions, commissions and advisory bodies play an important role in plant protection and pesticide management. Through the ratification of international conventions, governments accept obligations to incorporate them into national policies. The following international policies and instruments are most relevant with regard to reducing the risk associated with pesticide use.

11.3.1 The Rotterdam Convention

The Rotterdam Convention covers international trade in hazardous chemicals (most of them being pesticides) with the aim of protecting human health and the environment. If all parties agree that a specific pesticide constitutes a severe health or environmental hazard, it can be listed for prior informed consent (PIC) procedures. These require countries that export these chemicals to provide importing authorities with data on known hazards. As any party can veto the listing of a pesticide, the process is rather slow. Currently, the convention lists 33 pesticides.

11.3.2 The Stockholm Convention

The Stockholm Convention aims to eliminate or restrict the production and use of persistent organic pollutants (POPs), some of which are pesticides. Based on a specified review process, pesticides that fulfil the criteria for POPs can be listed for elimination or restriction.

11.3.3 The International Code of Conduct on Pesticide Management

The International Code of Conduct on Pesticide Management is a voluntary framework that has been endorsed by the FAO members, and is supported by key pesticide industry associations and civil society organizations (FAO and WHO, 2014). It became a role model for the development of pesticide legislation, and the major pesticide companies have agreed to abide by the Code of Conduct.

11.3.4 The Strategic Approach to International Chemicals Management (SAICM)

The Strategic Approach to International Chemicals Management (SAICM) is a voluntary policy framework and strategy facilitated by the United Nations Environmental Programme (UNEP) to promote chemical safety around the world. It brings together stakeholders and sectors that include agriculture, environment, health, industry, labour, economics, science and academia to catalyse achievement of the goal by 2020 'that chemicals are used and produced in ways that minimize adverse effects on human health and the environment'.

11.3.5 The Joint Meeting on Pesticide Residues (JMPR)

The Joint Meeting on Pesticide Residues (JMPR) is an expert ad hoc body administered jointly by FAO and the World Health Organization (WHO) with the purpose of harmonizing the requirements and risk assessments for pesticide residues. It recommends maximum residue levels (MRLs) in food and feed commodities and provides guidance on pesticide product quality parameters for regulatory and trade purposes.

11.4 Reduced Risk (RR) Pesticides

The risk of synthetic pesticides can also be avoided by adopting reduced risk (RR) pesticides. The concept of US Environmental Protection Agency's (US EPA) Conventional Reduced Risk Pesticide Program (RRPP) was initiated in the Food Quality Protection Act (FQPA) of 1996. The FQPA has introduced the US EPA's determination of residue tolerances by limiting the amount of a given pesticide that can be used for any particular purpose/situation (the limits vary according to the type of pesticide) (Olexa and Broome, 2011). Its purpose is to expedite the review and registration process of conventional pesticides that pose less risk to human health and the environment than existing conventional alternatives. Riskier conventional alternatives are those pesticides the US EPA deems as having neurotoxic, carcinogenic, reproductive and developmental toxicity, or groundwater contamination effects. It serves as a means to ensure that RR pesticides enter the channels of trade and are available to growers as soon as possible. RR decisions are made at the use level. The RR pesticides may provide the following benefits as compared with existing conventional pesticides:

- low impact on human health;
- lower toxicity to non-target organisms (e.g. birds, fish, plants);
- low potential for groundwater contamination;
- low use rates;
- low pest resistance potential; and
- compatibility with IPM practices.

Carbamates and organophosphates are a group of closely related neurotoxic pesticides targeting the cholinesterase system. These cholinesterase inhibitors can cause acute or delayed effects, ranging from mild to severe poisoning. Each person has a certain baseline level of cholinesterase enzyme that is considered normal for that individual. The continuous inhibition of cholinesterase can result in continual overexcitation of nerve-to-nerve and nerve-to-muscle communication (Fishel, 2012). Of the cholinesterase-inhibiting insecticides, 73% of these compounds most widely used in the United States fall into the highest toxicity class of US EPA and none is in the safest class. In contrast, 64% of the RR insecticides fall into the highest safety class, and the rest are in the next safest group III. The main concern is their acute toxicity. These are among US EPA's first priority group of pesticides for review under the FQPA. US EPA made alternatives to organophosphate pesticides the first priority for review and regulatory decision making. The conventional RRPP screens organophosphate alternatives for this initiative. Table 11.1 (US EPA, 2016) provides a list of RR and organophosphate alternative pesticides currently registered for use in the United States. Some active ingredients listed in Table 11.1 are not registered for use in Florida. US EPA determines if a candidate is a potentially significant organophosphate alternative by an approach that includes, but is not limited to, consideration of the following factors:

- The affected organophosphates collectively have a significant market share for the specified use pattern.
- Currently registered alternatives, if any exist, have constraints that prevented their widespread adoption as alternatives to the affected organophosphates, such as inferior efficacy or pest resistance issues.
- The proposed RR alternative appears to overcome many of the constraints of the alternatives.

Table 11.1. Reduced risk (RR)/organophosphate alternative pesticides registered in the United States. (From US EPA, 2016.)

Year	Pesticide[a]	Pesticide type	Site	Reduced risk (RR)/ organophosphate (OP) alternative[b]
1994	*Hexaflumuron*	Insecticide	Belowground bait station (termites)	RR
	Methyl anthranilate	Repellent	Cherry, blueberry, grape, forestry	RR
1995	*Flumiclorac-pentyl*	Herbicide	Corn, soybean	RR
	Tebufenozide	Insecticide	Walnut	RR
	Hymexazol	Fungicide	Sugarbeet (seed treatment)	RR
1996	*Fludioxonil*	Fungicide	Maize	RR
	Imazapic	Herbicide	Groundnut	RR
	Mefenoxam	Fungicide	All metalaxyl uses	RR
1997	*Azoxystrobin*	Fungicide	Non-residential turf	RR
	Spinosad	Insecticide	Cotton	RR
	Alpha-metolachlor	Herbicide	All metolachlor uses	RR
	Imazamox	Herbicide	Soybean	RR
	Hexaflumuron	Insecticide	Aboveground bait station (termites)	RR
	Azoxystrobin	Fungicide	Grape, banana, peach, tomato, pecan, groundnut	RR
1998	Fludioxonil	Fungicide	Potato and seed treatments (many crops)	RR
	Diflubenzuron	Insecticide	Belowground bait station (termites)	RR
	Cyprodinil	Fungicide	Stone fruit	RR
	Spinosad	Insecticide	Almond, apple, citrus, brassica leafy vegetables, fruiting vegetables and leafy vegetables	RR
	Pyriproxyfen	Insecticide	Cotton	RR
	Tebufenozide	Insecticide	Pecan	RR
	Carfentrazone-ethyl	Herbicide	Wheat, corn	RR
1999	Azoxystrobin	Fungicide	Turf (residential), almond, cucurbit vegetables, rice, wheat, canola, potato, stone fruit	RR
	Diflufenzopyr	Herbicide	Maize	RR
	Tebufenozide	Insecticide	Leafy, brassica, and fruiting vegetables, cranberry, forestry, ornamentals, berry crop group, mint, pome fruit, cotton, sugarcane, turnip, canola	RR/OP
	Pyriproxyfen	Insecticide	Pome fruit, walnut	RR/OP
	Glyphosate	Herbicide	Glyphosate-tolerant maize, canola, sugarbeet	RR
	S-dimethenamid	Herbicide	Maize, soybean, groundnut	RR
	Spinosad	Insecticide	Maize, cucurbit and legume vegetables, stone fruit, cereal grains	RR/OP
	Fenhexamid	Fungicide	Grape, strawberry, ornamentals	RR

continued

Table 11.1. Continued.

Year	Pesticide[a]	Pesticide type	Site	Reduced risk (RR)/ organophosphate (OP) alternative[b]
	Bifenazate	Insecticide	Ornamentals	RR/OP
	Trifloxystrobin	Fungicide	Pome fruit, grape, cucurbit vegetables, groundnut, turf, banana, ornamentals	RR
	Fipronil	Insecticide	Outside home use (termites)	OP
	Pymetrozine	Insecticide	Tuberous and corm vegetables, ornamentals, tobacco	RR/OP
2000	Pyriproxyfen	Insecticide	Citrus, fruiting vegetables, tree nuts	RR/OP
	Tebufenozide	Insecticide	Ornamentals (residential), tree nuts	RR/OP
	Ecolyst	Herbicide/ insecticide/ plant growth regulator	Orange	RR
	Spinosad	Insecticide	Non-grass animal feed crop group, grain amaranth, coriander, grass, buckwheat, rye, pistachio, oat, barley, millet, apple, popcorn, ti leaves, watercress, tropical fruit, teosinte, turnip greens	RR/OP
	Fenhexamid	Fungicide	Almond, stone fruit	RR
	Prohexadione calcium	Herbicide/ plant growth regulator	Apple	RR
	Methoxyfenozide	Insecticide	Cotton, pome fruit	RR/OP
	Trifloxystrobin	Fungicide	Almond, fruiting vegetables, hops, potato, sugarbeet, wheat, ornamentals	RR
	Carfentrazone-ethyl	Herbicide	Cereal grains	RR
	Buprofezin	Insecticide	Cucurbit vegetables, head lettuce	RR/OP
	Fenpyroximate	Insecticide	Ornamentals (greenhouse)	RR/OP
	Indoxacarb	Insecticide	Cotton, fruiting and brassica leafy vegetables, lettuce, maize, pome fruit	RR/OP
	Flucarbazone-sodium	Herbicide	Wheat	RR
	Glyphosate	Herbicide	Many: refer to www.epa.gov/opprd001/workplan/completionsportrait.pdf	RR
	Azoxystrobin	Fungicide	Barley, onion, citrus, maize (field, sweetcorn, popcorn), cotton, leafy, root, and tuberous vegetables, soybean	RR

2001	Fipronil	Insecticide	Home lawn, golf course, commercial and recreational turf and sod farms (fire ant), potting medium mixtures (fire ant)	OP
	Thiamethoxam	Insecticide	Barley, canola, cotton, sorghum, wheat (all seed treatment), cotton, pome fruit, cucurbit, fruiting, tuberous and corm vegetables (all foliar)	OP
	Fludioxonil	Fungicide	Strawberry, bulb vegetables, turf	RR
	Pyriproxyfen	Insecticide	Food handling establishments	RR
			Pistachio	RR/OP
	Imidacloprid	Insecticide	Leaf petioles, citrus	OP
	Zoxamide	Fungicide	Grape, cucurbit vegetables, tomato	RR
	Prohexadionc calcium	Plant growth regulator	Grass (grown for seed)	RR
	Pyriproxyfen	Insecticide	Pistachio	RR/OP
	Mesotrione	Herbicide	Maize (field)	RR
	Cyprodinil	Fungicide	Onion (dry, bulb, and green), strawberry	RR
	Buprofezin	Insecticide	Almond, citrus, cotton, grape, tomato	RR/OP
	Carfentrazone-ethyl	Herbicide	Cotton (defoliant use)	OP
			Turf	RR
	Fluazinam	Fungicide	Groundnut, potato	RR
	Zeta-cypermethrin	Insecticide	Alfalfa, maize (field, popcorn, sweetcorn), head and stem brassica vegetables, leafy brassica greens, leafy vegetables, onion (green), sugarbeet, sugarcane, rice	OP
	Azoxystrobin	Fungicide	Leafy brassica greens, blueberry, aubergine, grass (grown for seed), jackfruit, juneberry, lingonberry, loquat, mint (spearmint, peppermint), okra, pawpaw, pepper, persimmon, salal, strawberry, tamarind, tropical fruit, turnip (greens), watercress, wax jambu, white sapote	RR
	Novaluron	Insecticide	Ornamentals (indoors, non-food)	RR
	Spinosad	Insecticide	Artichoke (globe), asparagus, bushberry, cranberry, foliage of legume vegetables, garden beet (root), juneberry, leaves of root and tuber vegetables, lingonberry, okra, pistachio, pome fruit, salal, strawberry, sugarbeet (root), tree nuts	RR/OP
2002	Chlorfenapyr	Insecticide	Post-construction control of termites	OP
	Imazamox	Herbicide	Alfalfa, canola, legume vegetables, wheat	RR
	Pymetrozine	Insecticide	Cotton, leafy, head and stem brassica, and leafy brassica vegetables, hops	RR/OP
			Pecans	OP
	Bifenazate	Insecticide	Cotton, grapes, hops, nectarine, peach, plum, pome fruit, strawberry	RR/OP

continued

Table 11.1. Continued.

Year	Pesticide[a]	Pesticide type	Site	Reduced risk (RR)/ organophosphate (OP) alternative[b]
	Acetamiprid	Insecticide	Cotton, pome fruit, citrus, grapes, brassica leafy, leafy (excl. brassica) and fruiting vegetables, ornamentals	RR/OP
	Trifloxystrobin	Fungicide	Citrus, maize (field, popcorn), pecan, rice, stone fruit	RR
	Cyhalofop-butyl	Herbicide	Rice	RR
	Indoxacarb	Insecticide	Alfalfa, groundnut, potato, soybean	RR/OP
	Fludioxonil	Fungicide	Caneberry, pistachio, stone fruit, watercress	RR
	Pyriproxyfen	Insecticide	Stone fruit, blueberry, lychee, guava	RR/OP
	Imazethapyr	Herbicide	Rice	RR
	Diflufenzopyr	Herbicide	Maize (popcorn, sweetcorn), grass (forage, hay)	RR
	Macalayea extract	Fungicide	Greenhouse ornamentals	RR
	Azoxystrobin	Fungicide	Legume vegetables	RR
	Methoxyfenozide	Insecticide	Fruiting, leafy and brassica leafy vegetables, grapes, maize (field, sweetcorn), stone fruit, tree nuts	RR/OP
	Fenamidone	Fungicide	Lettuce	RR
	Lambda-cyhalothrin	Insecticide	Legume and fruiting vegetables, sugarcane	RR (sugarcane)/ OP (all)
	Spinosad	Insecticide	Berry group, fig, grape, herbs, groundnut, root and tuber vegetables	RR/OP
2003	Lambda-cyhalothrin	Insecticide	Termite barrier	RR
	Pyriproxyfen	Insecticide	Brassica leafy and cucurbit vegetables, olive	RR/OP
	Cyprodinil	Fungicide	Bushberry, caneberry, pistachio, watercress, brassica leafy vegetables, carrot, herbs, lychee fruits	RR
	EH-2001 Rodenticide	Rodenticide	Richardson/Wyoming ground squirrel	RR
	Mesotrione	Herbicide	Maize (popcorn)	RR
	Noviflumuron	Insecticide	Aboveground bait station	RR/OP
	Pyriproxyfen	Insecticide	Avocado fruits, fig, okra, sugar-apple fruits	RR/OP
	Clothianidin	Insecticide	Canola, maize (seed treatments)	OP
	Methoxyfenozide	Insecticide	Cranberry, cucurbits, okra, peas (black-eyed, southern), turnip (greens)	RR/OP
	Azoxystrobin	Fungicide	Artichoke (globe), asparagus, head and stem brassica subgroup, herbs	RR
	Emamectin benzoate	Insecticide	Cotton, fruiting vegetables, tobacco	OP
	Buprofezin	Insecticide	Bean (succulent), lychee fruits, pistachio	RR/OP

	Boscalid	Fungicide	Berries, bulb, fruiting, legume (root except sugarbeet, garden beet, radish, turnip), tuberous and corm vegetables, grape, lettuce (head, leaf), groundnut, stone fruit, strawberry, tree nuts, turf	RR
	Thiamethoxam	Insecticide	Ornamentals, succulent beans (seed), stone fruit, sunflower (seed)	OP
	Trifloxystrobin	Fungicide	Root vegetables leaf petioles (except sugarbeet) subgroup, except radish	RR
	Flonicamid	Insecticide	Ornamentals (greenhouse)	OP
	Acequinocyl	Insecticide	Ornamentals (greenhouse)	RR
	Bifenazate	Insecticide	Cucurbits, fruiting vegetables, mint, pistachio, tomato (greenhouse), tree nuts	RR
	Fenhexamid	Fungicide	Cucumber (greenhouse), fruiting vegetables (except non-bell pepper), kiwi fruit, leafy green subgroup (except spinach), stone fruit	RR
	Etoxazole	Insecticide	Cotton, pome fruit, strawberry	RR
	Quinoxyfen	Fungicide	Grape, hops, cherry	RR
	Glufosinate-ammonium	Herbicide	Rice	RR
2004	Fluroxypyr	Herbicide	Maize (field, sweetcorn)	RR
	Mesosulfuron-methyl	Herbicide	Wheat	RR
	Gamma-cyhalothrin	Insecticide	Alfalfa, brassica head and stem subgroup, canola, maize (field, sweetcorn), cotton, fruiting and legume (edible-podded) subgroup vegetables, garlic, lettuce (head, leaf), tree nuts, onion (dry bulb), pea and bean dry shelled (except soybean) subgroup, pea and bean succulent shelled subgroup, groundnut, pome fruit, rice, sorghum, soybean, stone fruit, sugarcane, sunflower, wheat	OP
	Novaluron	Insecticide	Cotton, pome fruit	OP
	Fenpyroximate	Insecticide	Cotton, grape, pome fruit	RR
	Acequinocyl	Insecticide	Strawberry, almond, citrus, pome fruit, field ornamentals	RR
	Lufenuron	Insecticide	Termite bait station	RR
	Indoxacarb	Insecticide	Fire ant bait	RR/OP
	Pyrimethanil	Fungicide	Almond, grape, onion (dry bulb, green), pome and stone fruit, strawberry, tomato, tuberous and corm vegetables	RR
	Dinotefuran	Insecticide	Leafy vegetables	RR/OP
	Penoxsulam	Herbicide	Rice	RR
	Tebufenozide	Insecticide	Citrus, grape, tuberous and corm vegetables	RR
	Fenamidone	Fungicide	Cucurbit vegetables, onion (dry bulb, green), potato, tomato	RR
	Cyazofamid	Fungicide	Cucurbit vegetables, potato, tomato	RR
	Bispyribac-sodium	Herbicide	Turf	RR
	Deltamethrin	Insecticide	Maize (field), cucurbit, fruiting, root and tuber vegetables, onion (dry, bulb, green), sorghum, tree nuts	OP

continued

Table 11.1. Continued.

Year	Pesticide[a]	Pesticide type	Site	Reduced risk (RR)/ organophosphate (OP) alternative[b]
2005	Fenamidone	Fungicide	Ornamentals	RR
	Diflubenzuron	Insecticide	Horse oral larvicide feed-through treatment	RR
	Dinotefuran	Insecticide	Public health use, cotton, brassica head and stem subgroup, cucurbit and fruiting vegetables, grape, potato	RR/OP
	Clothianidin	Insecticide	Turf, ornamentals, pome fruit, tobacco	OP
	Thiamethoxam	Insecticide	Mint	OP
	Clofentezine	Insecticide	Grape	RR
	Mesotrione	Herbicide	Maize (sweetcorn)	RR
	Buprofezin	Insecticide	Avocado, guava, peach, pome fruit, sugar-apple	RR/OP
	Acetamiprid	Insecticide	Potato	RR/OP
	Spiromesifen	Insecticide	Brassica leafy, fruiting, tuberous, and corm vegetables, maize (field), cotton, cucurbits, leafy greens, ornamentals, strawberry	RR
	Pymetrozine	Insecticide	Asparagus	OP
	Etoxazole	Insecticide	Grape, tree nuts	RR/OP
	Pinoxaden	Herbicide	Barley, wheat	RR
	Aminopyralid	Herbicide	Range and pasture lands, rights of way, roadsides, industrial vegetation management	RR
	Flonicamid	Insecticide	Cotton, cucurbit and fruiting vegetables, pome and stone fruit, potato, nursery and landscape ornamentals	OP
2006	Boscalid	Fungicide	Celery, spinach	RR
	Flumiclorac-pentyl	Herbicide	Cotton defoliant use	RR/OP
	Spinosad	Insecticide	Alfalfa, fruit fly bait, mint, onion (green)	RR
	Fenhexamid	Fungicide	Ginseng, pear, cilantro, pepper (non-bell), pomegranate	RR
	Flonicamid	Insecticide	Head and stem brassica	OP
	Trifloxystrobin	Fungicide	Barley, oats	RR
	Azoxystrobin	Fungicide	Herbs, spices, safflower, sunflower	RR
	Methoxyfenozide	Insecticide	Soybean	RR/OP
	Fenpyroximate	Insecticide	Citrus, hops, mint, pistachio, tree nuts	RR
	Quinoxyfen	Fungicide	Lettuce (head, leaf), melons, pepper (bell, non-bell), strawberry	RR
	Bifenazate	Insecticide	Stone fruit, edible-podded pea, tuberous and corm vegetables	RR/OP

2007	Fluthiacet-methyl	Herbicide	Cotton	OP
	Spiromesifen	Insecticide	Tomato (greenhouse)	RR
	Flazasulfuron	Fungicide	Turf	RR
	Penoxsulam	Herbicide	Turf, aquatic use	RR
	Indoxacarb	Insecticide	Grape	RR
	Spinosad	Insecticide	Mosquito larvicide use	RR
	Spinetoram	Insecticide	Many: refer to www.epa.gov/opprd001/workplan/completionsportrait.pdf	RR
2008	Mandipropamid	Fungicide	Brassica leafy, bulb, cucurbit, fruiting, tuberous and corm, and leafy vegetables, grape	RR
	Mesotrione	Herbicide	Berry group, cranberry, flax, turf (sod farm, golf courses)	RR
	Chlorantraniliprole	Insecticide	Cotton, grape, pome and stone fruit, potato, turf, ornamentals, brassica leafy, cucurbit, fruiting and leafy vegetables	RR
	Spirotetramat	Insecticide	Almond, citrus, grape, hops, onion (bulb), brassica head and stem, brassica leafy greens, cucurbits, fruiting, leafy, and tuberous and corm vegetables	RR
	Etofenprox	Insecticide	Mosquito adulticide use	RR
2009	Etofenprox	Insecticide	Rice	RR
	Mesotrione	Herbicide	Turf (commercial, residential)	RR
	Spiromesifen	Insecticide	Maize (popcorn, sweetcorn), low-growing berry group	RR
	Penoxsulam	Herbicide	Grape, tree nuts	RR
	Chlorantraniliprole	Insecticide	Tree nuts, pistachio	RR
	Cyazofamid	Fungicide	Fruiting vegetables (regional tolerance), okra	RR
	Saflufenacil	Herbicide	Cereal grains, citrus, cotton, foliage of legume vegetables, forage, fodder, and straw of cereal grains, grape, legume vegetables, pome and stone fruit, sunflower, tree nuts	RR
2010	Dinotefuran	Insecticide	Brassica leafy greens, turnip (greens)	RR
	Chlorantraniliprole	Insecticide	Artichoke, asparagus, caneberry, cacao, citrus, coffee, maize (field, sweetcorn, popcorn), fig, forage, fodder and straw of cereal grains, grass forage, fodder and hay, herbs and spices, hops, legume vegetables (e.g. soybean), mint, non-grass animal feeds, oilseed crops, okra, olive, groundnut, persimmon, pomegranate, prickly pear cactus, rice, small vine-climbing fruits, strawberry, sugarcane, tea, tobacco, tropical fruits, tuberous and corm vegetables, termiticide use	RR
	Tolfenpyrad	Insecticide	Ornamentals (greenhouse)	RR
	Cyazofamid	Fungicide	Brassica leafy vegetables, hops, spinach, turnip (greens)	RR
	Spiromesifen	Insecticide	Pea (dry)	RR
	Pyriproxyfen	Insecticide	Artichoke (globe), asparagus, foliage of legume, vegetables, leafy vegetables, leaves of root and tuber vegetables, small vine-climbing fruits (e.g. grape), watercress	RR
	Endothall	Herbicide	Irrigation canal use	RR

continued

Table 11.1. Continued.

Year	Pesticide[a]	Pesticide type	Site	Reduced risk (RR)/ organophosphate (OP) alternative[b]
2011	Spirotetramat	Insecticide	Cotton, legume vegetables, tropical fruits	RR
	Chlorantraniliprole	Insecticide	Bushberry, cucurbit vegetables, fruiting vegetables, okra, onion, bulb, root and tuber vegetables, sugarbeet, tea	RR
2012	Ametoctradin	Fungicide	Brassica leafy vegetables, bulb vegetables, cucurbits, fruiting vegetables, grape, hops, leafy vegetables, tuberous and corm vegetables, turf (golf courses), ornamentals, residential/ commercial landscapes	RR
	Acetamiprid	Insecticide	Asparagus, brassica leafy greens, citrus, fruiting vegetables, pome fruit, turnip (greens)	RR
	Azoxystrobin	Fungicide	Berry and small fruits (13-07A, B, F, G), bulb vegetables, citrus, dragon fruit, fruiting vegetables, oilseeds, tuberous and corm vegetables, wasabi	RR
	Novaluron	Insecticide	Termite bait station use	RR
	Penthiopyrad	Insecticide	Alfalfa, bulb vegetables, brassica leafy vegetables, canola, cereal grains, maize (field, popcorn, sweetcorn), cotton, cucurbits, fruiting vegetables, leafy vegetables, legumes, low-growing berry, groundnut, pome fruit, root and tuber vegetables, soybean, stone fruit, sunflower, tree nuts, turf, ornamentals	RR
2013	Flonicamid	Insecticide	Cucumber (GH), low-growing berry, rapeseed (20A)	RR
	Pyriproxyfen	Insecticide	Herbs	RR
	Acetamiprid	Insecticide	Maize (sweetcorn)	RR
	Dinotefuran	Insecticide	Rice, food handling establishments	RR
2014	Clothianidin	Insecticide	Indoor cockroach gel bait	RR
	Cyantraniliprole	Insecticide	Brassica leafy vegetables, bulb vegetables, bushberries, citrus, cotton, cucurbit vegetables, fruiting vegetables, leafy vegetables, oil seeds, groundnut, pome fruit, stone fruit, tree nuts, tuberous and corm vegetables, fly bait use, turf, ornamentals (greenhouse and nursery), and indoor and outdoor commercial and residential use sites to control public health pests	RR
	Etofenprox	Insecticide	Mosquito adulticide use (removing crop restriction)	RR

[a]New active ingredient reduced risk/organophosphate alternative actions are indicated by italics.
[b]Organophosphate alternative status was not considered by the Reduced Risk Pesticide Program for conventional pesticides until 1999.

Based on a 2009 report and publicly available data from the California Department of Pesticide Regulation and the CropLife Foundation, organophosphate use in the United States has shown an overall decline by approximately 50%. During the same period, carbamate use declined 70%. US EPA (2016) stated a direct benefit of the reduction of these organophosphates (Fishel, 2011) and carbamates to the environmental load (application of chemicals (lbs/acre) in the environment). The RR pesticides are generally used at significantly lower application rates than the conventional compounds they are replacing, which has the effect of decreasing the amount of chemical applied to the environment (Viray and Hollingworth, 2009). The trend from 1994 to 2006 has shown a 45% combined decrease in the environmental load for the carbamate and organophosphate insecticides.

In Canada, Agriculture and Agri-food Canada's (AAFC) Pest Management Centre (PMC) and Health Canada's Pest Management Regulatory Agency (PMRA) have jointly initiated the Pesticide Risk Reduction Program (PRRP) to reduce the risks to human health and the environment from pesticides used in the agriculture and agri-food industry (www.agr.gc.ca/eng/?id=1288277891464). The PRRP aims to provide RR tools to Canadian growers, including biological controls and integrated approaches for both major and minor crops, and to encourage the adoption of alternative pest management approaches, practices and technologies. The PRRP also provides funding and regulatory support for the implementation of projects identified through strategy work and other consultations. In order to improve access to and promote adoption of RR tools and practices by Canadian growers, the PRRP:

- Coordinates strategy discussions and brings together key players to identify pest management issues and solutions.
- Supports research, data generation and demonstration projects.
- Encourages development and adoption of RR alternatives.
- Provides regulatory assistance to industry collaborators.
- Develops and disseminates information on RR pest management options.

The reduced-use or non-use of chemical pesticides in favour of other methods sounds as if it may be detrimental to the crop yield and/or the quality, but this is not correct. In Texas, for example, the Texas Pest Management Association (TPMA) says that agricultural producers have been working with an IPM programme since 1972.

11.5 Ecological Ways to Remove Pesticides

IPM is a well-known strategy including biological controls, cultural methods, pest monitoring and other applicable practices. However, selection of a pesticide has to be based on effectiveness, toxicity to non-target species, cost and site characteristics, as well as its solubility and persistence. Half-lives and partition coefficients of a pesticide are particularly important when the application site of a pesticide is near surface waters or is underlain with permeable subsoil and a shallow aquifer. Among various techniques for mitigation and reduction of environmental pollution by synthetic pesticides, phytoremediation, bioaugmentation, biostimulation and biodegradation are the main ones.

Phytoremediation is the direct use of living green plants for *in situ*, in place, removal, degradation or containment of contaminants in soils, sludges, sediments, surface water and groundwater. Phytoremediation is, therefore, a low-cost, solar energy-driven clean-up technique.

11.5.1 Bioaugmentation

Bioaugmentation is the practice of adding cultured microorganisms into the subsurface for the purpose of biodegrading specific soil and groundwater contaminants.

Bioaugmentation is defined as a technique for improving the degradative capacity of contaminated soil, sediment, sludge or any area with specific competent strains or consortia of microorganisms with specific organic compound-degrading capabilities to enhance *in situ* or *ex situ* bioremediation applications. The efficiency of bioaugmentation is determined by many abiotic and biotic factors. It is one of the *in situ* bioremediation strategies, which improves the biodegradative capacities of contaminated sites by the introduction of single strains or consortia of microorganisms with desired catalytic capabilities (Mrozik and Piotrowska-Seget, 2010). Frequently, the terms bioremediation and bioaugmentation are used interchangeably. In bioremediation, selected microorganisms are used to accomplish a biological clean-up of a specified contaminated area, such as soil or water, while bioaugmentation is the application of selected microorganisms to enhance the microbial populations of an operating waste treatment facility to improve water quality or reduce operating costs. In other words, bioremediation deals with a definite project or area, while bioaugmentation involves working to improve a continuous process.

It is thought that the bioaugmentation approach should be applied when biostimulation and bioattenuation have failed (Vogel, 1996; Iwamoto and Nasu 2001; El Fantroussi and Agathos 2005). Biostimulation is the addition of nutrients to a polluted site in order to encourage the growth of naturally occurring chemical-degrading microorganisms. Bioattenuation is a method that relies on natural processes to dissipate contaminants through biological transformation.

Ellis *et al.* (2000) reported a successful anaerobic bioaugmentation carried out on a trichloroethene (TCE)-contaminated aquifer using a microbial enrichment culture for dechlorinating TCE to ethane. Soils contaminated with the herbicide atrazine are successfully remediated by using a bioaugmentation strategy. *Pseudomonas* sp. strain ADP combined with citrate or succinate biostimulation markedly increased *Pseudomonas* ADP cell survival and atrazine mineralization (Silva *et al.*, 2004). Clausen *et al.* (2002) reported the potential of using *Pseudomonas* ADP at both aerobic and denitrifying conditions for bioaugmentation and the use of competitive polymerase chain reaction (cPCR) as a tool for monitoring the bacteria independent of culturability. Rousseaux *et al.* (2003) observed inoculation of *Chelatobacter heintzii* did not accelerate the rate of atrazine mineralization in the case of soils adapted to atrazine mineralization, which was essentially performed by the indigenous microflora. However, with soils that did not mineralize atrazine, the introduction of 10^4 colony-forming units (CFU)/g soil resulted in a 3-fold increase in atrazine mineralization capacity. Runes *et al.* (2001) studied the potential to establish pesticide biodegradation in constructed wetland sediments. They showed that bioaugmentation could be a simple method to promote pesticide degradation in nursery runoff channelled through constructed wetlands, if persistence of degraders in the absence of pesticides is not a serious constraint. A study conducted by Strong *et al.* (2000) indicated that genetically engineered bacteria overexpressing catabolic genes significantly increased degradation in a soil that was heavily contaminated with atrazine.

Al-Qurainy and Abdel-Megeed (2009) studied phytoremediation of the most widely used synthetic organophosphate pesticides, unacceptable levels of environmental residues of which have been found in many countries worldwide. Many of the organophosphate pesticides, generally nonpersistent, are mobile in the environment, resulting in diffuse contamination, leading to an accidental release or disposal that has caused heavy pollution in certain areas. High concentrations of the organophosphate pesticides (e.g. malathion and dimethoate) are degraded in soil to acceptable levels only after many years (Zeinat *et al.*, 2008). These pesticides are acetylcholinesterase (AChE) inhibitors, mostly liposoluble and pose a hazard to humans. The loss of AChE results in acetylcholine accumulation, which interferes with muscular responses and produces serious symptoms in vital organs, eventually leading to death (Diurak and Kazanicif, 2001).

On the other hand, these pesticides may induce oxidative stress and generate reactive oxygen species (ROS) in insect systems (Zhao *et al.*, 1989). It has been demonstrated that ROS may cause DNA damage, which could lead to single-strand breaks and mutation. Zhao and colleagues' results showed that pesticide treatments induce DNA mutations in different sites of the plants, compared with the untreated control. These data suggest that direct binding of dimethoate and malathion to DNA is unlikely to be a mechanism through which any genotoxic effects are produced. Their study also suggested that these pesticides have the capacity to alter genetic material, particularly chromosomes, in the tested plants. This investigation showed that isolated bacteria from soil could assist plants to increase their ability to degrade dimethoate and malathion. In the analysis of malathion and dimethoate degradation in plants, soil and growth media, there was a reduction of total pesticide residues in the cultivated soil compared with the non-cultivated one. This could be because of the natural activity of the microorganisms in the rhizospheric region (Cho *et al.*, 2000). Plants have evolved interactions and association with microorganisms that can cause accelerated breakdown or transformation of certain pollutants in the plant root zones to products that no longer pose environmental hazards. Brimecombe *et al.* (2001) reported that understanding pesticide metabolism in plants and microorganisms is necessary to develop pesticides for safe and efficient use, as well as for developing pesticide bioremediation strategies for contaminated soil and water. Pesticide biotransformation may occur via multistep processes known as metabolism or co-metabolism (Sharaf *et al.*, 2006).

Ecological farming

Ecological agriculture is the only effective and workable solution to protect human health and the environment. Ecological farming is already being successfully practised in many areas of the world, and world experts agree that agriculture must become more sustainable. The recent increase in organic farming practices in Europe demonstrates that farming without pesticides is entirely feasible, scalable, economically profitable and environmentally safe. Land under organic cultivation increased from 5.7 million hectares (ha) in 2002 to 9.6 million ha in 2011, and includes arable crop and orchard as well as animal sectors (EU, 2013). Food produced by ecological farming is safe for our health. No synthetic chemical pesticides are used and sustainable practices boost plant and animal (functional) biodiversity in agricultural landscapes. With the implementation of ecological farming it is possible to create a toxin-free future and a safer environment for children. Implementation of ecological farming across the world will provide the ability for communities to feed themselves and ensures a future of healthy farming and healthy food for all the people. Greenpeace's seven principles of ecological farming (Reyes, 2015) are: (i) food sovereignty; (ii) rewarding rural livelihoods; (iii) smarter food production and yields; (iv) biodiversity; (v) sustainable soil health; (vi) ecological pest protection; and (vii) resilient food systems.

11.5.2 Bioremediation

The Gila Indian Reservation in Arizona is the site of an abandoned airfield that was at one time used for aerial application activities (US EPA, 1994). Frazer (2000) reported to US EPA that the project involved the treatment of about 80,000 yd^3 (61,164 m^3) of soil using *in situ* land treatment in order to remove parathion and toxaphene (US EPA, 1994). Following chemical oxidation treatment and prior to biological treatment, toxaphene was present at 1470 p.p.m. and parathion was present at 110 p.p.m. (Federal Remediation Technologies Roundtable, 1999). The soil was amended with water and nutrients, in an *in situ* manner, to enhance the rate of biodegradation (US EPA, 1994). Parathion degradation proceeded rapidly, but toxaphene was not degraded below the desired clean-up levels (Richard Martyn, US EPA Region 9, San Francisco, California, 2000, personal communication). In order to promote degradation, a variety of amendments were added, including alfalfa, manure, sludge, sulfuric acid and hydrogen peroxide (US EPA, 1994). Remediation was also attempted with and without water (R. Martyn, San Francisco, 2000, personal communication). Finally, the site was capped, anaerobic conditions were established and the pH was adjusted, which promoted the biodegradation of toxaphene to acceptable levels (R. Martyn, San Francisco, 2000, personal communication).

Table 11.2 presents a number of sites where bioremediation or phytoremediation have been selected as the remedy for groundwater contamination by dichlorodiphenyltrichloroethane (DDT), lindane and other pesticides. The table indicates the particular treatment method along with the contaminated media. The remediation projects listed in Table 11.2 are either demonstrations or Superfund sites.

11.5.3 Phytoremediation of pesticides

Lots of research is being conducted on the interaction between microorganisms and plants in the rhizosphere and the potential to use this for the remediation of pesticide-contaminated media. According to preliminary studies, enhanced degradation of atrazine, metolachlor and trifluralin have been observed in contaminated soils where plants of the *Kochia* sp. have been planted (Coats and Anderson, 1997). Increased degradation occurs in the rhizosphere of this herbicide-tolerant plant, suggesting that rhizosphere interactions between the plant and microorganisms have led to the increased degradation of the pesticides present (Coats and Anderson, 1997). Arthur and Coats (1998) observed promising phytoremediation of pesticide-contaminated soils and groundwater by conducting studies using the *Kochia* sp. (Arthur and Coats, 1998). In laboratory studies, the quick-growing and deep-rooted poplar tree has been shown to be successful in the remediation of groundwater (Schnoor and Licht, 1993). Its rapid growth requires high volumes of water, which are pulled from the saturated zone. Contaminated groundwater is absorbed by the plant and the pollutants are subsequently transformed into organic molecules for plant growth. This technique

Table 11.2. Bioremediation and phytoremediation implementation at several pesticide-contaminated sites. (From Frazer, 2000.)

Site name		Treatment method	Commercial technology	Contaminated media
Aberdeen pesticide dumps[a]	S	Phytoremediation	N/A	Groundwater and soil
Baird and McGuire	S	Pump and treat	N/A	Groundwater, soil and sediment
Bower's field[a]	D	Bioreactor	SABRE	Soil
Ciba-Geigy Corp.	S	Pump and treat, *in situ* bioremediation	N/A	Groundwater, soil, surface water, sludge and sediment
Charleston, SC	D	Composting[b]	DARAMEND	Soil
Gila Indian Reservation	S	Composting[b]	N/A	Soil
Helena Chemical Co.	S	Composting	XENOREM	Soil
Navajo dip vats[a]	S	Bioreactor	N/A	Soil
Novartis[a]	D	Composting	DARAMEND	Soil
SMC Bucks, AL	S	In design process		Groundwater, soil, sediment, sludges and wetlands
SMC Tampa, FL[a]	D	Composting	XENOREM	Groundwater, soil, surface water and sediment

[a]Indicates case study is available in Appendix 1.
(S) indicates Superfund project, (D) indicates technology demonstration.
[b]Indicates *in situ* treatment process.

has already proven successful for the remediation of atrazine-contaminated soil and groundwater (Schnoor and Licht, 1993; Environmental News Network, 1998). Hybrid poplars are being used in the remediation of a farm chemical site with high levels of nitrate, atrazine and arochlor in the groundwater (Frazer, 2000; Phytoremediation of Organics Action Team, 2000).

Approximately 1.5 acres (0.6 ha) are being treated and results to date show reduction of both the nitrates and herbicides in the groundwater (Frazer, 2000).

However, phytoremediation is not the answer to all contamination problems. Plants can generally only remediate soil or sediment in the top 1 m of the soil because of their root lengths. Phytoremediation is generally limited to a depth of 3 m for groundwater remediation. As with bioremediation, phytoremediation is a research-intensive technology, and it can require long periods of time to effectively remediate a site. Plants that take up the contaminants from the soil and transport them to their stems or leaves without biodegrading them to non-toxic compounds could potentially harm herbivores (Arthur and Coats, 1998). Phytoextraction and phytostabilization do not actually result in the destruction of the contaminant. Instead they accumulate or immobilize the contaminant. Although these are beneficial outcomes, destruction of the contaminant is generally preferred. Phytoremediation is still relatively unproven and its capabilities are still being discovered.

The SABRETM (Simplot Anaerobic Biological Remediation) process was developed by the University of Idaho and the J.R. Simplot Company. Under anaerobic conditions, this technology is capable of biodegrading nitroaromatic compounds, such as TNT (2,4,6-trinitrotoulene) and dinoseb. Under aerobic conditions, dinoseb undergoes oxidation and is transformed to its amino and acetamido forms, which are also toxic (Roberts *et al.*, 1993; R. Satterfield, Technology Transfer Office, University of Idaho, 2000, personal communication). These compounds can then form polymers, which can be used by microorganisms to form other toxic compounds (R. Satterfield, Idaho, 2000, personal communication). The SABRE process uses microorganisms in an anaerobic environment to enzymatically attack nitroaromatics in the soil and biodegrade them into acetate and organic acids. The J.R. Simplot Company conducted a demonstration of their SABRE process at the Bowers field site during the summer of 1993 through US EPA's Superfund Innovative Technology Evaluation (SITE) programme (US EPA, 1995). A second demonstration of this technology was also conducted at Weldon Spring Ordnance Works on TNT, also supported by the SITE programme.

Phytoremediation is an attractive alternative to conventional treatments of soil, due to advantages such as low cost, large application areas and the possibility of *in situ* treatment. Sandra *et al.* (2015) assessed phytoremediation process for evaluating *Ricinus communis* L., tropical plant species, for the degradation of 15 POPs, including hexachlorocyclohexane (HCH), DDT, heptachlor, aldrin and others in a 66-day period. The roots of the studied species indicated a reduction in the concentration of pesticides. Results obtained during this study indicated that the higher the hydrophobicity of the organic compound and its molecular interaction with soil or root matrix, the greater its tendency to concentrate in root tissues and the research showed the following trend: HCHs < diclofop-methyl < chlorpyrifos < methoxychlor < heptachlor epoxide < endrin < o,p′-DDE < heptachlor < dieldrin < aldrin < o,p′-DDT < p,p′-DDT by increasing the order of log K_{ow} values. The experimental results confirmed the importance of vegetation in removing pollutants, obtaining remediation from 25% to 70%, and demonstrated that *Ricinus communis* L. can be used for the phytoremediation of such compounds. Briggs *et al.* (1982) observed plants that could take up moderately hydrophobic compounds with K_{ow} values of 0.5–3.0 rather easily, while more lipophilic compounds may partition into root tissue but are more difficult for the plants to transport. Such plant uptake may prove valuable for controlling the contaminant plume by decreasing leaching of the compound.

Phytoremediation of pesticides using degrading enzymes in transgenic plants

Using phytoremediation technology appears to be a promising approach to remediate a pesticide-polluted environment. The most important factor is a suitable plant species that can be used to take up the contaminant. Tripathy *et al.* (2014) listed a few species with the potential to accelerate phytoremediation along with their gene source (Table 11.3). Phytoremediation using 'green plants' has potential benefits in restoring the balance in stressed surroundings. It is a promising, low-cost tool, non-intrusive and aesthetically pleasing, using the outstanding capability of green plants to metabolize diverse essentials and compounds from the atmosphere. Phytoextraction and phytostabilization, the processes that take place during phytoremediation, are most reliable to ameliorate pesticide pollutions and promote the healthy growth of plants. Tripathy *et al.* (2014) reported this cost-effective and innovative technology employing plants to clear wide-ranging organic and inorganic wastes. This study was conducted to give an overview of the role of this green technology in eliminating further contamination caused by the application of pesticides, sewage sludge, fertilizers and manure. The evidence from several reports leads to the conclusion that phytoremediation is a cheap, effective method, taking advantage of a plant's ability to remove pollutants from the environment or to render them less harmful.

Plants generally require catabolic pathways to execute degradation/mineralization of xenobiotic compounds. Therefore, the transfer of genes involved in catabolism of xenobiotics from microbes or other eukaryotes to plants will further augment their potential for remediation (Eapen *et al.*, 2007). An extracellular fungal enzyme, the laccase of *Coriolus versicolor*, in tobacco transgenic plants, was able to remove pentachlorophenol by creating laccase, which was secreted into the rhizosphere. Similarly, a bacterial organophosphorus hydrolase (OPH) gene has been added to tobacco plants to allow the decontamination of organophosphate pesticide. The transgenic plants degraded more than 99% of methylparathion after 14 days of growth, and illustrated a considerably higher tolerance to alachlor, suggesting their potential role in phytoremediation of herbicide-contaminated agricultural fields. A multiplicity of plant enzymes, including peroxygenases, cytochrome P450, peroxidase, glutathione-S-transferases, carboxylesterases, N- and O-glucosyltransferases and N- and O-malonyltransferases, are involved in the phytotransformation of xenobiotics in plant cells (Karavangeli *et al.*, 2005). Glutathione-S-transferases have been observed as a multigene family of detoxifying enzymes for herbicides. The case is similar with arylamine N-acetyltransferases (NAT) as xenobiotic-metabolizing enzymes that catalyse the transmission of an acetyl group from acetyl CoA (AcCoA) to the

Table 11.3. Enzymes introduced in plant cells to accelerate phytoremediation. (From Tripathy *et al.*, 2014.)

Pesticide	Plant species	Source of gene	Enzyme
Alachlor	Tobacco	Maize (*Zea mays*)	Glutathion-S-transferase (GST) I
Atrazine	Tobacco	*Homo sapiens sapiens*	P450$^{\#}$ (CYP1A1, CYP1A2)
Metamitron	Tobacco	*Homo sapiens sapiens*	P450$^{\#}$ (CYP1A1, CYP1A2)
Acetochlor, metolachlor, triazine	Arabidopsis seedling	*Arabidopsis*	GSTs, AtGSTF2, AtGSTU1 and AtGSTU24
Organophosphates	Tobacco	*Pseudomonas pseudoalcaligenes*	Hydrolase
Herbicides	Rice	Pig (*Sus scrofa*)	CYP2B22 and CYP2C49
Herbicides	Rice	*Homo sapiens sapiens*	P450$^{\#}$ (CYP1A1, CYP2B6 and CYP2C19)

nitrogen or oxygen atom of arylamine-based xenobiotics and their hydroxylated metabolites. They have been reported in soil bacteria, such as *Pseudomonas* sp. de la Pena Mattozzi *et al.* (2006) constructed a strain of *Pseudomonas putida* that can efficiently degrade paraoxon and use it as a carbon, energy and phosphorus source.

In a separate study, Inui *et al.* (1999) urbanized transgenic potato plants expressing human CYP1A1/yeast NADPH–cytochrome P450 reductase (YR)-fused enzyme generated from an *Agrobacterium* transformation system. They reported that P450-dependent monooxygenase activity of the transgenic potato plants was 3.5 to 4.2 times higher in 7-ethoxycoumarin 0-deethylation *in vitro* and 5.3 to 6.4 times in chlortoluron metabolism *in vivo* than those of the control plants. The manifestation of deisopropylated deethylated metabolite, a non-phytotoxic metabolite of atrazine, was more distinct in transgenic plants than the control. The transgenic plants demonstrated tolerance towards both atrazine and pyriminobac-methyl, while the control died when treated with both herbicides. These plants showed tolerance to the herbicides acetochlor, atrazine, chlortoluron, methabenzthiazuron, metolachlor, norflurazon and pyributicarb, and established better metabolism of chlortoluron, atrazine, pyributicarb, and methoxychlor than non-transgenic control plants. Inui *et al.* (1999) reported in their studies that three P450 species expressed in T1977 coordinately functioned and actively metabolized the herbicides and the insecticide, and, thus, they could be used for the development of transgenic plants for phytoremediation of pesticide-contaminated soil environments.

The phytoremediation capabilities of plants could also be extended to pesticides by generating transgenic plants expressing antibodies specific to a particular group of pesticides. Transgenic plants have been used to produce immunoglobulin G (IgG) antibody, multimeric secretory antibody and a wide range of functional antibody fragments used as a model system of hydroponic tobacco plant cultures expressing the murine monoclonal antibody Guy's 13 to demonstrate the feasibility of using antibody-expressing plants for phytoremediation purposes. Guy's 13, a neutralizing antibody, recognizes a 185-kDa cell surface protein (SA I/II) of *Streptococcus* mutants. In the first plant type, functional antibody was rhizosecreted and shown to bind with antigen in the surrounding medium to form an immune complex (Drake *et al.*, 2002). In the second, a transmembrane sequence retained monoclonal antibodies in the plants on the plasma membrane.

Antigen added to the nutrient medium around the roots of mIgG plants was transported within 24 h to the topmost leaves of the plant, where it was sequestered as an immune complex by binding to antibody on the cell membrane (Drake *et al.*, 2002). The concentration of immune complex in the leaf tissue remained constant over a 72-h period after removal of the antigen from the nutrient medium. Moreover, Drake *et al.* (2003) demonstrated the secretion of a functional, full-length monoclonal antibody complex from transgenic *Nicotiana tabacum* roots. Initially, seeds were germinated on nitrocellulose membranes and antibody secretion detected from the developing roots. Plants were then established in hydroponic culture and secretion into the growth medium measured over 25 days. Western blotting indicated that full-length antibody was present in the medium along with other fragments. The secreted antibody was shown to be functional by binding to antigen in enzyme-linked immunosorbent assay (ELISA) studies. In contrast, no antibody could be detected from transgenic *Nicotiana* in which the same antibody was expressed as a membrane protein in the plasmalemma. These results strongly indicate that such rhizosecretion of antibodies in transgenic plants may prove a novel approach for removal or neutralization of environmental pollutants such as pesticides (Hussain *et al.*, 2009).

Phytoremediation of pesticides in soil

Degradation of xenobiotics may occur in the rhizosphere of plants, via release of degrading enzymes from plants, or via microbial degradation. Yoshitomi and Shann (2001) reported increased mineralization of pyrene with the addition of maize root exudates. Microorganisms, including those residing in plant rhizospheres, have the enzymatic ability to degrade environmental contaminants. Bacteria isolated from pesticide-contaminated soils have the ability to degrade atrazine (Mandelbaum *et al*., 1995; Radosevich *et al*., 1996; Struthers *et al*., 1998; Zhao *et al*., 2003). Several studies have shown the plant rhizosphere provides an ideal residence for such microbial communities and suggest the use of plants for pesticide degradation (Arthur and Coats, 1998; McGrath *et al*., 2001). Other mechanisms of phytoremediation include uptake by the plant, either through the transpiration process or through a more active process, and subsequent sequestration and/or metabolism of the compounds by enzymes within the plant tissues.

Soils at agrochemical dealer sites are often contaminated with pesticide residues from decades of accidental and incidental spillage. Henderson *et al*. (2006) reported that a mixed stand of big bluestem, switch grass and yellow indiangrass developed a rhizosphere with microflora that can readily detoxify pesticide residues. Specific atrazine-degrading bacteria or the free enzyme atrazine chlorohydrolase can also enhance the rate of biotransformation of atrazine in soil. Metolachlor degradation can be accelerated significantly by the rhizosphere effect in the prairie grass native to the Midwestern US. Because of their extensive root zone and general hardiness, prairie grasses have been used on roadsides and in vegetative filter strips, and have been suggested as a phytoremediation technique (Anderson *et al*., 1993; Comis, 1996; Arthur and Coats, 1998).

These grasses are suitable for phytoremediation because of their tolerance to most herbicides and climatic extremes, such as heat, cold, drought and flooding. Belden and Coats (2004) and Beldon *et al*. (2004) have shown decreased mobility and bioavailability of several herbicides in prairie grassed soil columns compared with unvegetated control soil columns. Several grasses used in filter strips have also been evaluated for their pesticide degradation capabilities. The authors observed earthworm 8-day bioaccumulation assays and root growth assays to assess the bioavailability of the pesticide residues. Mass balance experiments have utilized radiolabelled atrazine and metolachlor to ascertain the complete metabolism and binding profile of those two pesticides in phytoremediation studies

11.5.4 Biostimulation

Scow and Hicks (2005) defined the term 'biostimulation' as the addition of electron acceptors, electron donors or nutrients to stimulate naturally occurring microbial populations. Couto *et al*. (2010) perceived biostimulation as the introduction of adequate amounts of water, nutrients and oxygen into the soil in order to enhance the activity of indigenous microbial degraders, or to promote co-metabolism (Lorenzo, 2008). Kanissery and Sims (2011) considered biostimulation paired under the 'enhanced bioremediation' techniques along with 'bioaugmentation', which is merely the introduction of specific microorganisms (indigenous or non-indigenous) aimed at enhancing the biodegradation of a target compound or serving as donors of catabolic genes. The concept of biostimulation is to boost the intrinsic degradation potential of a polluted matrix (perhaps polluted with pesticides) through the accumulation of amendments, nutrients or other limiting factors, and it has been used for a wide variety of xenobiotics (Kadian *et al*., 2008). The diverse natural microbial populations have the potential for contaminant remediation at polluted sites; however, factors such as lack of electron acceptors or donors, low nitrogen or phosphorus availability, or a lack of stimulation of the metabolic pathways responsible for degradation can inhibit or delay the remediation. Cosgrove *et al*. (2010) reported accumulation of exogenous nutrients enhancing the degradation of toxic materials in such cases. Chen *et al*. (2012) explored biostimulation mechanisms with an electron donor and a shuttle for accelerating pentachlorophenol (PCP) transformation in iron-rich soils. The results indicated that indigenous microbial communities are important for PCP transformation in soils. Biostimulation of indigenous microbial communities by the addition of lactate and anthraquinone-2,6-disulfonate (AQDS) led to enhanced rates of PCP dechlorination by the dechlorinating- and iron-reducing bacteria in soils. Electrochemical studies using cyclic voltammograms and microbial current measurements confirmed the high reduction potential and the large amount of electrons generated under biostimulation conditions, which were responsible for the higher rates of PCP transformation. After biostimulation treatments consisting of the addition of lactate and/or AQDS during PCP dechlorination processes, microbial community analysis by the terminal restriction fragment length polymorphism (T-RFLP) method showed an abundance of terminal restricted fragments (T-RFs), an indicator of bacterial abundance, which represents the dechlorinating- and iron-reducing bacteria, suggesting their critical roles in PCP dechlorination in soils.

Kanissery and Sims (2011) reported threatening of groundwater by residues from historical use of pesticides in Denmark. Subsoil pollution with pesticides as well as their degradation products was of great public concern in the country, as groundwater is the main source of drinking water. *In situ* aerobic bioremediation in the anaerobic aquifers may be an alternative approach to the treatment of contaminated groundwater at waterworks. The major threat to drinking water supplies is through groundwater contamination by pesticides, making it a widespread environmental problem. Low pesticide concentrations (nanogram–microgram per litre) in large volumes of water are regarded as one of the major threats to groundwater quality originating from agriculture, roads and railways. Levi (2013) observed that these large volumes of water in combination with the low concentrations of pesticides caused difficulties in preventing contamination of drinking water supplies, making it a challenge to develop remediation solutions. He reported that even if only one well is contaminated, this water mixes with uncontaminated groundwater from the other wells and necessitates the treatment of excessive volumes of water at the waterworks. It was observed that biostimulation by oxygen addition even at relatively low concentrations is a promising remediation technology for groundwater contaminated by pesticides at abstraction well fields. The results of a study by Levi (2013) contributed to the development of treatment strategies to protect drinking water wells and suggested an alternative solution to the closure of abstraction wells due to pesticide contamination. Zuzana Frková (2014), during her PhD studies, focused on the microbial degradation and mineralization of a range of herbicides in contaminated soils using biostimulation and bioaugmentation approaches – stimulating indigenous microbial communities, and introducing efficient exogenous microorganisms to partially degrade or mineralize the pesticides. The increased use of pesticide-degrading microorganisms for the remediation of polluted soil and groundwater appeared to be a powerful, cost-efficient and sustainable alternative to conventional remediation techniques. Tortella *et al.* (2010) investigated the degradation of the insecticide chlorpyrifos (160 a.i/mg/kg), using a biomix of a biobed system biostimulated with inorganic fertilizer (NPK). The chlorpyrifos was dissipated efficiently (>75%) after 40 days of incubation and no additional dissipation was obtained with increasing concentrations of NPK after 20 days of incubation. The results demonstrated that biomix prepared with Andisol and biostimulated with NPK nutrient can be recommended in biobeds as a viable alternative to chlorpyrifos dissipation avoiding the possibility of soil and water contamination.

Bioremediation is the best method for decontamination of environments that have been contaminated by various activities including pesticide-formulating industries. Chemical and physical methods are not only costly but also not very effective in open-field systems. Qureshi *et al.* (2009) demonstrated the study of *in situ* bioremediation of organochlorine-contaminated soil by combined biostimulation and bioaugmentation strategies, followed by evaluation using a molecular method. Both the biostimulation and the bioaugmentation treatments showed an initial lag phase of 80 days towards colony-forming units. A decline of up to 85–90% was observed in the concentrations of residual pesticides in the soil, indicating their utilization with time. The bioaugmentation along with biostimulation was observed to be effective in the open-field systems. Rani and Dhania (2014) reviewed different processes in the removal of pollutants (Tables 11.4 and 11.5), along with bioremediation and biodegradation techniques of pesticides from contaminated soil and water.

11.5.5 Biodegradation

Pesticide degradation is the process by which a pesticide is transformed into a benign substance that is environmentally compatible with the site to which it was applied. The concepts of biodegradation and biotransformation overlap extensively, and are synonymous in appearance. Biodegradation involves the biological reactions that modify the chemical structure of the compound, implying a decrease in toxicity (Velázquez-Fernández *et al.*, 2012), while biotransformation reduces the pollutant concentration by either modification or translocation. Thus, biotransformation could end up decreasing or increasing the undesirable effects. Parkinson (2001) reported that the biotransformation concept has been developed for biological detoxification systems and is a key concept in bioremediation strategies because they both are intended to eliminate undesirable effects of pollutants to organisms.

Different approaches of biodegradation are:

1. Fungal.
2. Bacterial.
3. Enzymatic.

The biochemical reactions involved in biodegradation are:

- Detoxification: conversion to non-toxic compound.
- Degradation: breaking down/transformation of a complex to simple molecules leading to complete mineralization.

Table 11.4. Removal of pollutants from soil and water. (From Rani and Dhania, 2014.)

Process	Function	Pollutant	Medium	Plants
Phytoextraction	Remove metals pollutants that accumulate in plants. Remove organics from soil by concentrating them in plant parts	Cd, Pb, Zn, As, petroleum, hydrocarbons and radionuclides	Soil and groundwater	*Viola baoshanensis*, *Sedum alfredii*, *Rumex crispus*
Phytotransformation	Plant uptake and degradation of organic compounds, plants and associated microorganisms degrade organic pollutants	Xenobiotic substances	Soil	Cannas
Rhizofiltration	Roots absorb Zn, Pb, Cd, and As. Groundwater adsorbs pollutants, mainly metals, from water and aqueous waste streams	Zn, Pb, Cd, As	Groundwater	*Brassica juncea*
Phytostabilization (immobilization)	Use of plants to reduce the bioavailability of pollutants in the environment	Cu, Cd, Cr, Ni, Pb, Zn	Soil	*Anthyllis vulneraria*, *Festuca arvernensis*

- Conjugation: an organism makes the substrate more complex or combines the pesticides with the cell metabolites.

Environmental physical parameters such as temperature, pH, water potential, oxygen and substrate availability influence the biodegradation efficiency. Sometimes, there is a need for other substrates to degrade the pollutants, like organochlorine compounds (Alexander, 1999). This phenomenon is called co-metabolism. Contrarily, the presence of other carbon sources decreases organophosphate biodegradation (Hayatsu *et al.*, 2000). For adequate pesticide degradation, more than one microorganism is required, and each microorganism contributes to biodegradation reactions on pesticides, but no reports of mineralization by a single strain have been cited. Reported microbiodegraders belong to basidiomycetes or bacterial classes: gamma-proteobacteria (v.gr.: *Pseudomonas*, *Aerobacter*, *Acinetobacter*, *Moraxella*, *Plesiomonas*), beta-proteobacteria (v.gr.: *Burkholderia*, *Neisseria*), alpha-proteobacteria (v.gr.: *Sphingomonas*), actinobacteria (*Micrococcus*) and flavobacteria (*Flavobacterium*). Besides bacteria, some fungi, plants and algae could also biodegrade pesticides.

The possible degradation of pesticides may involve reactions like dechlorination, cleavage, oxidation and reduction by different enzymes. The shorter the duplication time of the organism, the more adequate the organism is and the easier to obtain biodegraders. In bacterial degradation, both aerobic and anaerobic degradation should happen if a mineralization is expected to occur (Langerhoff *et al.*, 2001). Anaerobic metabolism is reported as more adequate for dechlorination (Barragán-Huerta *et al.*, 2007; Baczynski *et al.*, 2010), while aerobic metabolism is better for cleavage in aromatic or aliphatic cyclic metabolites. The higher persistence of organochlorine in aerobic conditions (Singh *et al.*, 1999) might be due to the absence of enzymes or oxidative damage following organochlorine metabolism. In biodegradation, the removal of heteroatoms (like halogens) or heteroatom-containing groups is catalysed by monooxygenases, dioxygenases or peroxidases (Wackett, 1995; Singh *et al.*, 1999), which in aerobic conditions could generate large quantities of free radicals. The US EPA

Table 11.5. Bioremediation of pollutants using different organisms. (From Rani and Dhania, 2014.)

Process	Source	Pollutant	Microbes/plants	References
Phytoremediation	Soil	Pb, Cd	*Vetiveria zizanioides* and *Eichornia crassipes*, plus applications of zeolite	Rani and Dhania, 2014
Biodegradation	Garden, beach and mud	Saw	*Aspergillus* sp., *Trichoderma* sp.	Sivaramanan, 2014
Degradation	Cellulosic materials such as corncobs, sawdust and sugarcane	Blue dye 2B	*Bacillus* sp.	Bhoosreddy, 2014
Phytoremediation, adsorption	Sewage irrigated soil wastewater	Heavy metals, lead, iron, manganese, copper and zinc, etc.	Flagellate sp. of *Dunaliella* algae	Qari and Hassan, 2014
Adsorption	Soil	Cu, Mn, Zn, Pb, Cr and Pd	Three herbaceous plants (*Plantago major* L., *Taraxacum officinale* L. and *Urtica dioica* L.) and one leguminous (*Trifolium pratense* L.)	Maliziad *et al.*, 2010

(2006) outlined that anaerobic conditions are better for biodegradation of organochlorine pesticides, while aerobic are better for biodegrading hydrocarbon metabolites from pesticides. However, some examples of organochlorine pesticide bioremediation could be accomplished *in situ* (Langerhoff *et al.*, 2001; Qureshi *et al.*, 2009).

The enzymes involved in biodegradation are promiscuous and present in organisms even before the exertion of the evolutionary pressure that could have induced genetic recombination or mutation leading to enzymes with better biodegradation ability. Copley (2009) has excellently reviewed the evolution of metabolic pathways and those factors affecting the efficiency of pollutant biodegradation. Knowing the metabolism of those biodegrader species or strains improves the selection of bioremediation strategy for each site either by biostimulating the indigenous biodegraders (biostimulation) or adding exogenous species (microbial population) to the site (bioaugmentation). Biodegradation enzymes play a key role in bioremediation processes and knowledge of them could help in designing or choosing the most adequate strategy. Phase I of biotransformation consists of those enzymes catalysing reactions that modify pollutant functional groups. In phase II, those enzymes catalysing transfer reactions of whole groups or biomolecules to pollutants are classified. Phase III includes translocation processes rendering pollutants or their metabolites non-bioavailable. The oxidoreductases, hydrolases, transferases and translocases are four types of biochemical-based biotransformation enzymes responsible for bioremediation. The common oxidoreductases like monooxygenases (i.e. cytochrome P450), dioxygenases, peroxidases and oxidases are responsible for bioremediation purposes, while hydrolases (i.e. A-esterases) are involved in biodegradation pathways. Various types of transferases are: methyl-transferases, acetyltransferases, glutathione-s-transferases and others. The translocases are reported to translocate herbicides or glutathione-conjugates to vacuoles.

Microbes in soil responsible for microbial degradation of pesticides

The soil is the ultimate 'sink' of the pesticides applied in agriculture and public health care. Soil is considered as the storehouse of multitudes of microbes, in quantity and quality; it receives chemicals in various forms and acts as a scavenger of harmful substances. The efficiency and competence to handle chemicals vary with the soil and its physical, chemical and biological characteristics. Depending upon structure and properties of the constituents in the pesticide and availability of moisture in soil, the persistence of pesticides in soil varies from a week to several years. For instance, the highly toxic phosphates do not persist for more than 3 months, while chlorinated hydrocarbon insecticides (e.g. DDT, aldrin, chlordane, etc.) are known to persist for at least 4–5 years and sometimes more than 15 years (http://agriinfo.in/?page=topic&superid=5&topicid=174). In addition to the physical and chemical forces that degrade pesticides to some extent, microorganisms play a major role in the degradation of pesticides. The most common type of degradation in soil occurs through the activity of microorganisms, especially the fungi and bacteria; however, under microbial degradation of pesticides, microorganisms simply supply a medium and an energy source for rather simple chemical reactions to take place. They, in return, obtain food, essential elements or energy to carry on their life functions. These soil microorganisms have the ability to act upon pesticides converting them into simpler non-toxic compounds. However, not all pesticides are biodegraded by soil microbes; such chemicals showing complete resistance to biodegradation are called 'recalcitrant'. The conversion of pesticides to non-toxic compounds is termed as detoxification, while breaking down/ transformation of a complex substrate into simpler products leading finally to mineralization is degradation. Sometimes, an additional product is formed by those organisms catalysing the reaction on addition of an amino acid, or organic acid to the substrate; for example, in the microbial metabolism of sodium dimethyl dithiocarbamate, the organism combines the fungicide with an amino acid molecule normally present in the cell, thereby inactivating the pesticides/ chemical.

In contrast, a non-toxic substrate may be converted into a toxic molecule; for example, the herbicide, 4-(2,4-dichlorophenoxy) butyric acid (2,4-DB or 4-butyric acid) and the insecticide phorate are transformed and activated microbiologically in soil to give metabolites that are toxic to weeds and insects. Some pesticides are metabolized to yield products inhibitory to entirely dissimilar groups of organisms; for example, the fungicide pentachloronitrobenzene (PCNB) is converted in soil to chlorinated benzoic acids that kill plants.

However, the biodegradation of pesticides is greatly influenced by soil factors (Chowdhary *et al.*, 2008), such as structure and concentration of pesticide, soil type, its moisture, temperature, pH, salinity and organic matter, in addition to microbial population and pesticide solubility. Optimum temperature, moisture and organic matter in soil provide a congenial environment for the breakdown or retention of any pesticide added in the soil. Most of the organic pesticides degrade within a short period (3–6 months) under tropical conditions. Metabolic activities of bacteria, fungi and actinomycetes have a significant role in the degradation of pesticides. According to Gale's (1952) principle of microbial infallibility, for every naturally occurring organic compound there is a microbe/enzyme system capable of its degradation.

Herbicide degradation in soil has already been elaborated upon in the section on 'Biostimulation' (involving dilution of contaminated soil with uncontaminated soil). Such dilution of the contaminated soil would stimulate the biodegradation, owing to the increase in activity of soil dehydrogenase (Felsot and Dzantor, 1995). Later various organic amendments and bioprocessed materials have been investigated through the course of time as potential agents for the stimulation of native microflora for the enhanced degradation of herbicides (Table 11.6).

In one of the earliest reports of successful biostimulation of herbicide degradation, Hance (1973) demonstrated the consequence of inorganic nutrient addition on the breakdown of atrazine in the soil. The addition of inorganic salts like ammonium nitrate, potassium nitrate and ammonium phosphate significantly decreased the half-life of herbicides in soil. Thereafter, the concept of nutrient supplementation for enhanced degradation of contaminants was brought into the limelight by various researchers, and the prospects of microbial biostimulation through the manipulation of organic and inorganic nutrient status in the soil have since been investigated (Table 11.7). Inorganic nitrogen starvation has been reported to be effective in promoting degradation of atrazine and other heterocyclic compounds (Bichat *et al.*, 1999; Sims, 2006). This can potentially be accomplished by supplying excess carbon to make nitrogen limiting.

The pathogens in degradation of pesticides

Due to high persistence and illegal applications of synthetic pesticides, their residues contaminate the environments that serve as the last destination for agricultural drainage. However, certain microorganisms can survive in this contaminated place and have the potential to degrade pesticides through different mechanisms (Aly *et al.*, 2017). Moreover, degradation of a pesticide by either external or internal enzymes stops at any step if an appropriate enzyme is not present. Absence of an appropriate enzyme is one of the common reasons for persistence of any pesticide. The presence of pesticides in soil may reduce the biodegrading microbial population due to their toxicity. However, a specific microorganism can be added or introduced to soil to enhance the activity of the existing population (Singh, 2008). The use of bacteria for the degradation and detoxification of numerous toxic chemicals such as pesticides is an effective tool to decontaminate the polluted sites. Mervat (2009) reported environmentally friendly means of *in situ* detoxification by isolating indigenous bacteria capable of metabolizing pesticides. The degradation of

Table 11.6. Organic amendments and bioprocessed materials for the degradation of herbicides. (From Kanissery and Sims, 2011.)

Amendment(s)	Target herbicide(s)	Major finding	Reference
Animal manure and sewage sludge	Atrazine and alachlor	Herbicide degradation has been stimulated by the amendments	Guo *et al.*, 1991
Activated sludge	Atrazine and simazine	Achieved a biodegradation of 25.2 and 32.8%, respectively, in 9 h	Leoni *et al.*, 1992
Sewage sludge and cornmeal	Alachlor and trifluralin	Significant dissipation of herbicide achieved in amended treatments	Dzantor *et al.*, 1993
Maize straw	Methabenzthiazuron	Significantly enhanced the degradation and mineralization of the herbicide	Printz *et al.*, 1995
Dairy manure	Atrazine	Identified as a successful technique to enhance the remediation of herbicide-contaminated soil	Gan *et al.*, 1996
Dairy manure	Atrazine	Lag period prior to herbicide degradation was significantly shortened in manure-treated plot	Topp *et al.*, 1996
Cornmeal, rye grass and poultry litter	Cyanazine and fluometuron	All the amendments enhanced the herbicide degradation and facilitated mineralization	Wagner and Zablotowicz, 1997
Plant residues, ground seed or commercial meal	Alachlor, metolachlor, atrazine and trifluralin	Acknowledged the use of organic amendments as an inexpensive option for the disposal of herbicide waste	Felsot and Dzantor, 1997
Cellulose, straw and compost	Atrazine	Organic matter addition stimulated herbicide dealkylation in non-adapted soils	Abdelhafid *et al.*, 2000b
Compost, maize stalks, maize fermentation by-product, peat, manure and sawdust	Atrazine, trifluralin and metolachlor	5% addition of amendments resulted in significant increase in bacterial population and dehydrogenase activity	Moorman *et al.*, 2001
Raw olive cake	Chlorsulfuron, prosulfuron and bensulfuron	Stimulated microbial dehydrogenase activity	Delgado-Moreno and Pena, 2007
Biogas slurry, spent mushroom compost and farmyard manure	Atrazine	Atrazine dissipation was observed to be highest (34%) with biogas slurry	Kadian *et al.*, 2008
Rice straw, farmyard manure, sawdust and charcoal	Atrazine	Farmyard manure was found to be the most effective for the degradation of herbicide (89.5% degradation in 60 days)	Mukherjee, 2009

pesticides by microorganisms depends not only on the presence of enzymes, but also on a wide range of environmental parameters, such as temperature, pH, water potential, nutrients and the amount of pesticide or metabolite in soil (Singh, 2008).

Fungi and bacteria in degradation of pesticides

The problems caused by the common pesticides endosulfan, chlorpyrifos and cypermethrin have been overcome by an enrichment culture technique used to isolate bacterial strains from garden soil tolerating high concentrations of the selected pesticides (Table 11.8).

The persistence of monocrotophos, an extensively used organophosphate pesticide, in the environment and its toxic effect on biota necessitate its removal. Baghel and Pandey (2013) observed bioremediation activity of monocrotophos in the environment using two fungal and one bacterial

Table 11.7. Nutrient supplements for stimulating the biodegradation of herbicides in soil. (From Kanissery and Sims, 2011.)

Nutrient(s)	Target herbicide(s)	Reference
Ammonium nitrate, potassium nitrate and ammonium phosphate	Atrazine	Hance, 1973
Phosphorus	Dichlobenil and atrazine	Qiu *et al.*, 2009
Minimal nutrient medium, casamino acid, glucose and phosphate	2,4-Dichlorophenoxyacetic acid (2,4-D) and mecoprop	Lipthay *et al.*, 2007
Nitrate	(R)-mecoprop	Harrison *et al.*, 2003
Nitrate and phosphorus	Isoproturon	Perrin-Ganier *et al.*, 2001
Tryptic soy broth	Diclofop	Wolfaardt *et al.*, 1994
Mannitol	Atrazine	Assaf and Turco, 1994
Glucose	Atrazine	McCormick and Hiltbold, 1966; Wagner and Chahal, 1996; Abdelhafid *et al.*, 2000a

isolate. Eighty-two bacterial isolates with potential carbofuran degradation activity were obtained from soils cultivated with the potato (*Solanum tuberosum*) in Colombia with different records of pesticide application. The bacteria were selected for their ability to grow in a medium containing 200 mg/l carbofuran as the sole source of carbon and/or nitrogen (Castellanos *et al.*, 2013). Biodegradation of chlorophenols by the bacterial consortium was observed in 50 days in the surface and subsurface environments, and treatment of seeds with a microbial consortium resulted in 100% germination, indicating the efficiency of the treatment (Logakanthi *et al.*, 2013). Mahiuddin *et al.* (2014) used enrichment culture technique and minimal salt growth medium in which diazinon served as a sole carbon source, where *Pseudomonas peli*, *Burkholderia caryophylli* and *Brevundimonas diminuta* were able to completely degrade 20 mg/l diazinon within 12 days of incubation.

Moreno-Medina *et al.* (2013) observed the removal of organophosphate pesticides, methyl parathion and coumaphos in mineral salt medium supplemented with glucose, by degradation using immobilized bacterial consortium, from agricultural soils. Botero *et al.* (2012) degraded methyl parathion using an active consortium isolated from organic solid waste.

There is a dire need to develop strategies for the disposal of pesticides that are eco-friendly and economically viable and use bioremediation methods for cleaning soil and waters. One of the most accepted treatments is application of biological systems, leading to degradation and detoxification of pesticides, that is, organophosphates, without harming the environment or human health (Singh, 2009; Ortiz-Hernández *et al.*, 2011). Interest in biological methods is increasing due to their simplicity, high efficiency and cost effectiveness compared to other methods (Chandran and Das, 2011). Biodegradation of pesticides provides a cheap and efficient solution for their final disposal or for the treatment of agricultural soils, contaminated water or polluted ecosystems. To date, a number of different microorganisms have been identified, along with the enzymes involved in biodegradation. Among biological treatments, cell immobilization has been a successful approach ensuring catalytic activity is maintained for longer periods (Manohar *et al.*, 2001; Cheng *et al.*, 2003; Yáñez-Ocampo *et al.*, 2009). The most widely studied method for the practical application of immobilized cell techniques is entrapping cells in polymer gels, such as alginate and carrageenan. However, this method has been limited by problems with gel stability and by the mass transfer limitations of gel beads. Entrapment in natural polymeric gels has become the preferred technique for the immobilization of cells, due to the toxicity problems associated with the synthesis of polymeric materials (Lusta *et al.*, 1990).

Table 11.8. The fungi and bacteria involved in degradation of pesticides.

Microorganisms	Source	Pesticides degraded	Reference
Fungi	*Aspergillus niger*, *Ganoderma austral*, *Trichosporon*, *Verticillium dahaliae*	Malathion, endosulfan	Hussaini *et al.*, 2013
Bacteria	*Flavobacterium* sp. ATCC 27551 from a soil sample from the Philippines	Organophosphate	Sethunathan and Yoshida, 1973
	Pseudomonas aeruginosa, *P. putida*, *Stenotrophomonas maltophilia*, *Flavimonas oryzihabitans* and *Morganella morganii*. *Pseudomonas aeruginosa* and *F. oryzihabitans* isolated from coffee beans	Dichlorodiphenyltrichloroethane (DDT) and endosulfan	Barragán-Huerta *et al.*, 2007
	The genera *Neisseria*, *Veillonella*, *Micrococcus* and *Staphylococcus* from soil were recorded in a bioremediation technique used to degrade pesticide		
	A bacterial consortium consisting of six pure strains, *Stenotrophomonas malthophilia*, *Proteus vulgaris*, *Vibrio metschinkouii*, *Serratia ficaria*, *Serratia* spp. and *Yersinia enterocolitica* isolated from agricultural soil	Tetrachlorvinphos and chlorpyriphos	Ortiz-Hernández and Sánchez-Salinas, 2009 Kumar, 2011
	The bacterial strain *Pseudomonas* sp. 2C and *Alcaligenes* sp. XDC22, from the soil, led to complete mineralization (Sarat and Barathi, 2013)	Endosulfan	Sarat and Barathi, 2013
	A bacterial consortium of *Proteus vulgaris*, *Vibrio* sp., *Serratia* sp. and *Acinetobacte*r sp. isolated from agricultural soil	Dichlorvos	Agarry *et al.*, 2013
	Five bacterial strains of *Bacillus amyloliquefaciens*, *Staphylococcus sciuri*, *Pseudomonas earuginosa*, *Bacillus pseudomycoides* and *Bacillus licheniformis* from agricultural soil	Malathion	Thabit and El-Naggar, 2013
	Sphingomonas paucimobilis, *Pseudomonas aeruginosa* and *Yersinia pseudotuberculosis*	Hydrolytic degradation of carbofuran	Castellanos *et al.*, 2013
	Pseudomonas isolated from *Solanum melongena* (aubergine) cultivated soil	Bioremediation of carbosulfan	Sharif and Mollick, 2013
	Pseudomonas peli, *Burkholderia caryophylli* and *Brevundimonas diminuta*	Diazinon	Mahiuddin *et al.*, 2014
Microflora	*Pennisetum pedicellatum* and *Cenchrus setigerus*	Chlorpyrifos, cypermethrin and fenvalerate	

Rhizoremediation (microflora) is the most evolved process of bioremediation, and involves remediation of contaminants by the interactions of microflora associated with plant roots in the mycorrhizosphere. *Pennisetum pedicellatum* and *Cenchrus setigerus* have been selected for rhizosphere remediation of chlorpyrifos, cypermethrin and fenvalerate. A consortium survived in a pesticide environment, which was identified as *Stenotrophomonas maltophila* and *Sphingobacterium thalpophilum* (Fulekar, 2014).

11.6 Conclusion

Of the millions of tonnes of pesticides applied on crops for pest management and increased production, less than 5% reach the target organisms. Research carried out on pesticides has focused on the active ingredients, ignoring the inert ingredients in pesticide formulations that enhance the effect of the active ingredients, and can also cause substantial health effects. The ideal strategy in agriculture is to manage pest populations before they cross their ETL, using IPM techniques. IPM techniques include resistant varieties, cultural practices, biocontrol and biological methods. Need-based application of less hazardous pesticides is the chemical component of IPM. US EPA initiated RRPP to expedite the review and registration process of conventional pesticides that pose less risk to human health and the environment than existing conventional alternatives. The RR pesticides are generally used at significantly lower application rates than the conventional compounds they are replacing, which has the effect of decreasing the amount of chemical applied to the environment. Almost 50% of the organophosphates have been phased out in the United States and carbamate use has declined 70%. The ecological ways to remove pesticides include phytoremediation, bioaugmentation, biostimulation and biodegradation. Phytoremediation uses green plants for degradation or removal of pesticides from soils, sludges, sediments, surface water and groundwater, and is a low-cost technology. In bioaugmentation cultured microorganisms are added to the subsurface for the purpose of biodegrading specific soil and groundwater contaminants. Bioremediation promotes degradation by adding amendments like alfalfa, manure or sludge. The plants generally require catabolic pathways to carry out degradation/mineralization of xenobiotic compounds. Therefore, the transfer of genes involved in the catabolism of xenobiotics from microbes or other eukaryotes to plants will further augment their potential for remediation. Similarly, the degradation of xenobiotics may occur in the rhizosphere of plants, via release of degrading enzymes from plants, or via microbial degradation. In bioaugmentation, the degradative capacity of contaminated soil, sediment, sludge or any area, is improved with specific competent strains or consortia of microorganisms, with specific organic compound-degrading capabilities to enhance *in situ* or *ex situ* bioremediation applications. It is thought that the bioaugmentation approach should be used when biostimulation and bioattenuation have failed. Electron acceptors, electron donors or nutrients are added to stimulate naturally occurring microbial populations in the biostimulation technique. However, bioremediation has been reported as the best method for decontamination of environments contaminated by pesticide-formulating industries. The degradation may be achieved using fungi, bacteria or enzymes to transform a pesticide into a benign substance that is environmentally compatible with the site to which it was applied. Biodegradation efficiency is influenced by temperature, pH, water potential, oxygen and substrate availability. Degradation of pesticides may involve reactions like dechlorination, cleavage, oxidation and reduction by different enzymes. Soil is considered as the storehouse of multitudes of microbes, in quantity and quality; it receives chemicals in various forms and acts as a scavenger of harmful substances. The physical, chemical and biological characteristics of soil influence its efficiency and competence to handle chemicals. Depending upon the structure and properties of the constituents in the pesticide and availability of moisture in the soil, the persistence of pesticides in soil varies from a week to several years.

References

Abdelhafid, R., Houotmethly, S. and Barriuso, E. (2000a) How increasing availabilities of carbon and nitrogen affect atrazine behaviour in soils. *Biology and Fertility of Soils* 30(4), 333–340.

Abdelhafid, R., Houot, S. and Barriuso, E. (2000b) Dependence of atrazine degradation on C and N availability in adapted and non-adapted soils. *Soil Biology and Biochemistry* 32(3), 389–401.

Agarry, S.E., Olu-Arotiowa, O.A., Aremu, M.O. and Jimoda, L.A. (2013) Biodegradation of dichlorovos (organophosphate pesticide) in soil by bacterial isolates. *Journal of Natural Sciences Research* 3(8), 12–16.

Al-Qurainy, F. and Abdel-Megeed, A. (2009) Phytoremediation and detoxification of two organophosphorous pesticides residues in Riyadh area. *World Applied Sciences Journal* 6(7), 987–998.

Alexander, M. (1999) *Biodegradation and bioremediation* (2nd edn). Academic Press, New York, p. 453.

Aly, M.M., Al-Aidaroos, B.A. and Alfassi, F.A. (2017) Pesticides characters, importance and microbial degradation. *IOSR Journal of Pharmacy and Biological Sciences* 12(2), 20–28.

Anderson, T.A., Guthrie, E.A. and Walton, B.T. (1993) Bioremediation in the rhizosphere. *Environmental Science and Technology* 27, 2630–2636.

Arthur, E.L. and Coats, J.R. (1998) Phytoremediation. In: Kearney, P.C. and Roberts, T. (eds) *Pesticide Remediation in Soil and Water*. John Wiley & Sons, Washington DC, pp. 251–281.

Assaf, N.A. and Turco, R.F. (1994) Influence of carbon and nitrogen application on the mineralization of atrazine and its metabolites in soil. *Pesticide Science* 41(1), 41–47.

Baczynski, T.P., Pleissner, D. and Grotenhuis, T. (2010) Anaerobic biodegradation of organochlorine pesticides in contaminated soil – significance of temperature and availability. *Chemosphere* 78(1), 22–28.

Baghel, P.S. and Pandey, B. (2013) Isolation of microorganism for mioremediation of monocrotophos pesticide. *International Journal of Current Microbiology and Applied Sciences* 2(11), 202–205.

Barragán-Huerta, B.E., Costa-Pérez, C., Peralta-Cruz, J., Barrera-Cortés, J., Esparza-García, F. and Rodríguez-Vázquez, R. (2007) Biodegradation of organochlorine pesticides by bacteria grown in microniches of the porous structure of green bean coffee. *International Biodeterioration and Biodegradation* 59, 239–244.

Belden, J.B. and Coats, J.R. (2004) Effect of grasses on herbicide fate in the soil column: Infiltration of runoff, movement, and degradation. *Environmental Toxicology and Chemistry* 23, 2251–2258.

Belden, J.B., Phillips, T.A. and Coats, J.R. (2004) Effect of prairie grass on the dissipation, movement, and bioavailability of selected herbicides in prepared soil columns. *Environmental Toxicology and Chemistry* 23, 125–132.

Bhoosreddy, G.L. (2014) Decolorization and biodegradation of direct blue 2B by mix consortia of Bacillus. *IOSR Journal of Pharmacy and Biological Sciences* 9(2), 34–40.

Bichat, F., Sims, G.K. and Mulvaney, R.L. (1999) Microbial utilization of heterocyclic nitrogen from atrazine. *Soil Science Society of America Journal* 63(1), 100–110.

Botero, L.R., Nagles, N., Barriuso, E. and Peñuela, G.A. (2012) Liquid culture for isolating microorganisms with potential to degrade methyl parathion from agricultural soils and organic waste. *Revista Ingenierías Universidad de Medellín* 11(21).

Briggs, G.G., Bromilow, R.H. and Evans, A.A. (1982) Relationships between lipophilicity and root uptake and translocation of non-ionised chemicals by barley. *Pesticide Science* 13495–13504.

Brimecombe, M.J., DeLeij, F.A. and Lynch, J.M. (2001) The effects of root exudates on rhizosphere microbial populations. In: Pinta, R., Varanini, Z. and Nannipieri, P. (eds) *The rhizosphere: Biochemistry and Organic Substances at the Soil-Plant Interface*. Marcel Dekker Inc., New York, pp. 95–137.

Castellanos, R.J., Nieves, J.S., Uribe Vélez, D., Moreno Chacón, L. and Melgarejo Muñoz, L.M. (2013) Characterization of carbofuran degrading bacteria obtained from potato cultivated soils with different pesticide application records. *Revista Facultad Nacional de Agronomía Medellín* 66(1), 6899–6908.

Chandran, P. and Das, N. (2011) Degradation of diesel oil by immobilized *Candida tropicalis* and biofilm formed on gravels. *Biodegradation* 22, 1181–1189.

Chen, M., Shih, K., Hu, M., Li, F., Liu, C., Wu, W. and Tong, H. (2012) Biostimulation of indigenous microbial communities for anaerobic transformation of pentachlorophenol in paddy soils of southern China. *Journal of Agricultural and Food Chemistry* 60(12), 2967–2975.

Cheng, K., Wu, J., Yang, W. and Hwang, S. (2003) Evaluation of effective diffusion coefficient and intrinsic kinetic parameters on azo dye biodegradation using PVA-immobilized cell beads. *Biotechnology and Bioengineering* 83, 821–832.

Cho, T.H., Wild, J.B. and Donnelly, K.C. (2000) Utility of organophosphorus hydrolase for the remediation of mutagenicity of methyl parathion. *Environmental Toxicology and Chemistry* 19, 2022–2028.

Chowdhary, A., Pradhan, S. Saha, M. and Sanyal, N. (2008) Impact of pesticides on soil microbiological parameters and possible bioremediation strategies. *Indian Journal of Microbiology* 48, 114–127.

Clausen, G.B., Larsen, L., Johnsen, K., de Lipthay, J.R. and Aamand, J. (2002) Quantification of the atrazine-degrading *Pseudomonas* sp. strain ADP in aquifer sediment by quantitative competitive polymerase chain reaction. *FEMS Microbiology Ecology* 41(3), 221–229.

Coats, J.R. and Anderson, T.A. (1997) *The Use of Vegetation to Enhance Bioremediation of Surface Soils Contaminated with Pesticide Wastes.* US EPA, Office of Research and Development, Washington, DC.

Comis, D. (1996) Green remediation: Using plants to clean the soil. *Journal of Soil and Water Conservation* 51, 184–187.

Copley, S.D. (2009) Evolution of efficient pathways for degradation of anthropogenic chemicals. *Nature Chemical Biology* 5(8), 559–566.

Cosgrove, L., McGeechan, P.L., Handley, P.S. and Robson, G.D. (2010) Effect of biostimulation and bioaugmentation on degradation of polyurethane buried in soil. *Applied and Environmental Microbiology* 76(3), 810–819.

Couto, M.N.P.F.S., Monteiro, E. and Vasconcelos, M.T.S.D. (2010) Mesocosm trials of bioremediation of contaminated soil of a petroleum refinery: Comparison of natural attenuation, biostimulation and bioaugmentation. *Environmental Science and Pollution Research* 17(7), 1339–1346.

Definition of Organic Agriculture (2008) IFOAM. Vignola: IFOAM, 2008. Organic World Congress 2008.

Delgado-Moreno, L. and Pena, A. (2007) Organic amendments from olive cake as a strategy to modify the degradation of sulfonylurea herbicides in soil. *Journal of Agricultural and Food Chemistry* 55(15), 6213–6218.

Diurak, L. and Kazanicif, M. (2001) Effect of some organophosphorus insecticides on soil microorganisms. *Turkey Journal of Biology* 25, 51–58.

Drake, P.M.W., Chargelegue, D.M., Vine, N.D., van Dolleweerd, C.J., Obregon, P. and Ma, J.K. (2002) Transgenic plants expressing antibodies: a model for phytoremediation. *FASEB Journal* 16, 1855–1860.

Drake, P.M.W., Chargelegue, D.M., Vine, N.D., van Dolleweerd, C.J., Obregon. P. and Ma, J.K. (2003) Rhizosecretion of a monoclonal antibody protein complex from transgenic tobacco roots. *Plant Molecular Biology* 52, 233–241.

Dzantor, E.K., Felsot, A.S. and Beck, M.J. (1993) Bioremediating herbicide-contaminated soils. *Applied Biochemistry and Biotechnology* 39–40(1), 621–630.

Eapen, S., Singh, S. and D'Souza, S.F. (2007) Advances in development of transgenic plants for remediation of xenobiotic pollutants. *Biotechnology Advance* 25, 442–451.

El Fantroussi, S. and Agathos, S.N. (2005) Is bioaugmentation a feasible strategy for pollutant remediation? *Current Opinion in Microbiology* 8, 268–275.

Ellis, D.E., Lutz, E.J., Odom, J.M., Buchanan, R.J., Bartlett, C.L., Lee, M.D., Harkness, M.R. and Deweerd, K.A. (2000) Bioaugmentation for accelerated in-situ anaerobic bioremediation. *Environmental Science and Technology* 34(11), 2254–2260.

Environmental News Network (1998) Poplar trees may remove contaminants. Available at: www.enn.com/enn-newsarchive/1998/09/093098/poplar30.asp (accessed 29 June 2000).

European Union (EU) (2013) Facts and figures on organic agriculture in the European Union. European Union, DG Agriculture and Rural Development, Unit Economic Analysis of EU Agriculture. Available at: http://ec.europa.eu/agriculture/markets-and-prices/more-reports/pdf/organic-2013_en.pdf (accessed 28 August 2018).

Eyhorn, F., Roner, T. and Specking, H. (2015) Reducing pesticide use and risks – what action is needed? Briefing Paper, supported by Mercator Foundation Switzerland, the Swiss Federal Office for Agriculture (BLW) and HELVETAS Swiss Intercooperation.

Food and Agriculture Organization (FAO) (2011) *Save and Grow. A Policy Maker's Guide to the Sustainable Intensification of Smallholder Crop Production.* FAO, Rome.

Food and Agriculture Organization (FAO) (2014) *International Symposium on Agroecology for Food Security and Nutrition.* FAO, Rome.

Food and Agriculture Organization (FAO) and World Health Organization (WHO) (2014) *International Code of Conduct on Pesticide Management.* FAO and WHO, Rome.

Food and Agriculture Organization (FAO) (2015) Pest and pesticide management. AGP – integrated pest management. [Online] 21 August

2015. Careful consideration of all available pest control techniques. Available at: www.fao.org/agriculture/crops/core-themes/theme/pests/ipm (accessed 2 October 2018).

Federal Remediation Technologies Roundtable (1999) EPA Removal Action Gila Indian Reservation, AZ. Remediation technologies screening matrix and reference guide, version 3.0. Available at: www.frtr.gov/matrix2/section3/table3_4_nfr.html (accessed 10 August 2000).

Felsot, A.S. and Dzantor, E.K. (1995) Effect of alachlor concentration and an organic amendment on soil dehydrogenase activity and pesticide degradation rate. *Environmental Toxicology and Chemistry* 14(1), 23–28.

Felsot, A.S. and Dzantor, E.K. (1997) Potential of biostimulation to enhance dissipation of aged herbicide residues in land-farmed waste. In: Kruger, E.L., Anderson, T.A. and Coats, J.R. (eds) *Phytoremediation of Soil and Water Contaminants*. American Chemical Society, Washington, DC, pp. 77–91.

Fishel, F.M. (2011) Pesticide toxicity profile: Organophosphate pesticides. PI-50. University of Florida Institute of Food and Agricultural Sciences, Gainesville. Available at: http://edis.ifas.ufl.edu/pi087 (accessed 31 July 2018).

Fishel, F.M. (2012) Pesticides and cholinesterase. PI-221. University of Florida Institute of Food and Agricultural Sciences, Gainesville, Florida. Available at: http://edis.ifas.ufl.edu/p.221 (accessed 31 July 2018).

Frazer, C. (2000) The bioremediation and phytoremediation of pesticide contaminated sites. National Network of Environmental Studies (NNEMS) Fellow under US EPA. US Environmental Protection Agency, Office of Solid Waste and Emergency Response, Technology Innovation Office, Washington, DC.

Frková, Z. (2014) Biostimulation and bioaugmentation – potential tools for cleaning up pesticide-polluted sites? PhD defence, Monday 14 July 2014. Bio-degradation of herbicides by axenic bacterial cultures and soil microbial communities.

Fulekar, M.H. (2014) Rhizosphere bioremediation of pesticides by microbial consortium and potential microorganism. *International Journal of Current Microbiology and Applied Sciences* 3(7), 235–248.

Gale, E.F. (1952) *The Chemical Activities of Bacteria*. Academic Press, New York, p. 234.

Gan, J., Becker, R.L., Koskinen, W.C. and Buhler, D.D. (1996) Degradation of atrazine in two soils as a function of concentration. *Journal of Environmental Quality* 25(5), 1064–1072.

Guo, L., Bicki, T.J., Felsot, A.S. and Hinesly, T.D. (1991) Phytotoxicity of atrazine and alachlor in soil amended with sludge, manure and activated carbon. *Journal of Environmental Science and Health* Part B. 26(5–6), 513–527.

Hance, R.J. (1973) The effect of nutrients on the decomposition of the herbicides atrazine and linuron incubated with soil. *Pesticide Science* 4(6), 817–822.

Harrison, I., Williams, G.M. and Carlick, C.A. (2003) Enantioselective biodegradation of mecoprop in aerobic and anaerobic microcosms. *Chemosphere* 53(5), 539–549.

Hayatsu, M., Hirano, M. and Tokuda, S. (2000) Involvement of two plasmids in fenitrothion degradation by *Burkholderia* sp. strain NF100. *Applied and Environmental Microbiology* 66(4), 1737–1740.

Henderson, K.L.D., Belden, J.B., Zhao, S. and Coats, J.R. (2006) Phytoremediation of pesticide wastes in soil. *Zeitschrift fur Naturforschung* C 61(3–4), 213–221.

Hussain, S., Siddique, T., Arshad, M. and Saleem, M. (2009) Bioremediation and phytoremediation of pesticides: Recent advances. *Critical Reviews in Environmental Science and Technology* 39, 843–907.

Hussaini, S.Z., Shaker, M. and Iqbal, M. (2013) Isolation of fungal isolates for degradation of selected pesticides. *Bulletin of Environment, Pharmacology and Life Science* 2(4), 50–53.

International Assessment of Agricultural Knowledge, Science and Technology for Development (IAASTD) (2009) *International Assessment of Agricultural Knowledge, Science and Technology for Development: Global Report.* Island Press, Washington, DC.

Inui, H., Ueyama, Y., Shiota, N., Ohkawa, Y. and Ohkawa, H. (1999) Herbicide metabolism and crosstolerance in transgenic potato plants expressing human CYP1A1. *Pesticide Biochemistry and Physiology* 64, 33–46.

Iwamoto, T. and Nasu, M. (2001) Current bioremediation practice and perspective. *Journal of Bioscience and Bioengineering* 92, 1–8.

Kadian, N., Gupta, A., Satya, S., Mehta, R.K. and Malik, A. (2008) Biodegradation of herbicide (atrazine) in contaminated soil using various bioprocessed materials. *Bioresource Technology* 99(11), 4642–4647.

Kanissery, R.G. and Sims, G.K. (2011) Biostimulation for the enhanced degradation of herbicides in soil. *Applied and Environmental Soil Science*. Article ID 843450, p. 10. DOI:http://dx.doi.org/10.1155/2011/843450.

Karavangeli, M., Labrou, N.E., Clonis, Y.D. and Tsaftaris, A. (2005) Development of transgenic tobacco plants overexpressing maize glutathione-S-transferase I for chloroacetanilide herbicides phytoremediation. *Bimolecular Engineering* 22, 121–128.

Kumar, S. (2011) Bioremediation of chlorpyrifos by bacteria isolated from the cultivated soils. *International Journal of Pharma and Bio Sciences* 2(3), 359–366.

Langerhoff, A., Charles, P., Alphenaar, A., Zwiep, G. and Rijnaarts, H. (2001) Intrinsic and stimulated in situ biodegradation of hexachlorocyclohexane (HCH). 6th International HCl and Pesticides Forum. Poland 2001. pp. 132–185.

Leoni, V., Cremisini, C., Giovinazzo, R., Puccetti, G. and Vitali, M. (1992) Activated sludge biodegradation test as a screening method to evaluate persistence of pesticides in soil. *Science of the Total Environment* 123–124, 279–289.

Leu, A. (2014) *The Myths of Safe Pesticides*. Acres USA, Austin, Texas.

Levi, S. (2013) Biostimulation and enhancement of pesticide degradation around water abstraction fields. PhD thesis. Department of Environmental Engineering, Technical University of Denmark.

de Lipthay, J.R., Sørensen, S.R. and Aamand, J. (2007) Effect of herbicide concentration and organic and inorganic nutrient amendment on the mineralization of mecoprop, 2,4-D and 2,4,5-T in soil and aquifer samples. *Environmental Pollution* 148(1), 83–93.

Liu, Y.Y. and Xiong, Y. (2001) Purification and characterization of a dimethoate-degrading enzyme of *Aspergillus niger* ZHY256 isolated from sewage. *Applied Environmental Microbiology* 67, 3746–3749.

Logakanthi, S., Arulazhagan, P. and Vasudevan, N. (2013) Degradation of pentachlorophenol by a bacterial consortia and the effect of cured soil on *Phaseolus mungo* L. *International Journal of Current Microbiology and Applied Sciences* 6(2), 11–23.

Lorenzo, V de (2008) Systems biology approaches to bioremediation. *Current Opinion in Biotechnology* 19(6), 579–589.

Lusta, K.A., Starostina, N.G. and Fikhte, B.A. (1990) Immobilization of microorganisms: Cytophysiological aspects. In: De Bont, J.A.M., Visser, J., Mattiasson, B. and Tramper, J. (eds) *Proceedings of an International Symposium: Physiology of Immobilized Cells*. Elsevier, Amsterdam, pp. 557–562.

Mahiuddin, M., Fakhruddin, A.N.M., Al-Mahin, A., Chowdhury, M.A.Z., Rahman, M.A. and Alam, M.K. (2014) Degradation of the organophosphorus insecticide diazinon by soil bacterial isolate. *The International Journal of Biotechnology* 3(1), 12–23.

Maliziad, D., Giuliano, A., Ortaggi, G. and Masotti, A. (2010) Environmental monitoring of heavy metal contamination in soil with common plants used as potentially reliable analytical instruments. In: *Application of Multivariate Analysis and Chemometry to Cultural Heritage and Environment* (3rd edn). Taormina, Sicily, pp. 26–29.

Mandelbaum, R.T., Allan, D.L. and Wackett, L.P. (1995) Isolation and characterization of a *Pseudomonas* sp. that mineralizes the s-triazine herbicide atrazine. *Applied and Environmental Microbiology* 61, 1451–1457.

Manohar, S., Kim, C.K. and Karegoudar, T.B. (2001) Enhanced degradation of naphthalene by immobilization of *Pseudomonas* sp. strain NGK1 in polyurethane foam. *Applied Microbiology and Biotechnology* 55, 311–316.

McCormick, L.L. and Hiltbold, A.E. (1966) Microbiological decomposition of atrazine and diuron in soil. *Weeds* 14(1), 77–82.

McGrath, S.P., Zhao, F.J. and Lombi, E. (2001) Plant and rhizosphere processes involved in phytoremediation of metal-contaminated soils. *Plant Soil* 232, 207–214.

Mervat, S.M. (2009) Degradation of methomyl by the novelbacterial strain *Stenotrophomonas maltophilia* M1. *Electronic Journal of Biotechnology* 12(4), 1–6.

Mesnage, R., Defarge, N., de Vendômois, J.S. and Séralini, G.E. (2014) Major pesticides are more toxic to human cells than their declared active principles. *BioMed Research International* Article ID 179691. DOI:10.1155/2014/179691.

Miguel, A. (1995) *Agroecology: The Science of Sustainable Agriculture*. Westview Press, Boulder, Colorado.

Moorman, T.B., Cowan, J.K., Arthur, E.L. and Coats, J.R. (2001) Organic amendments to enhance herbicide biodegradation in contaminated soils. *Biology and Fertility of Soils* 33(6), 541–545.

Moreno-Medina, D.A., Sanchez- Salinas, E. and Ortiz-Hernández, M.L. (2013) Removal of methyl parathion and coumaphos pesticides by a bacterial consortium immobilized in luffa cylindrical. *Revista Internacional de Contaminación Ambiental* 30, 51–63.

Mrozik, A. and Piotrowska-Seget, Z. (2010) Bioaugmentation as a strategy for cleaning up of soils contaminated with aromatic compounds. *Microbiological Research* 165, 363–375.

Mukherjee, I. (2009) Effect of organic amendments on degradation of atrazine. *Bulletin of Environmental Contamination and Toxicology* 83(6), 832–835.

Okon, Y. (1985) *Azospirillum* as a potential inoculant for agriculture. *Trends in Biotechnology* 3, 223–228.

Olexa, M.T. and Broome, Z. (2011) *Handbook of Florida Water Regulation: Food Quality Protection Act.* FE589. University of Florida Institute of Food and Agricultural Sciences, Gainesville, Florida. The document is one of a series of the Food and Resource Economics Department, UF/IFAS Extension. Original publication date October 1998. Revised June 2017. Available at: http://edis.ifas.ufl.edu/fe589 (accessed 2 October 2018).

Ortiz-Hernández, M.L. and Sánchez-Salinas, E. (2009) Biodegradation of the organophosphate pesticide tetrachlorvinphos by bacteria isolated from agriculture soils in Mexico. *Revista internacional de contaminación ambiental* 26(1), 27–38.

Ortiz-Hernández, M.L., Sánchez-Salinas, E., Olvera-Velona, A. and Folch-Mallol, J.L. (2011) Pesticides in the environment: Impacts and its biodegradation as a strategy for residues treatment. In: Stoytcheva, M. (ed.) *Pesticides-Formulations, Effects, Fate.* InTech, Rijeka, Croatia, pp. 551–574.

Parkinson, A. (2001) Biotransformation of xenobiotics. In: Curtis, K., (ed.) *Casarett and Doull's Toxicology: The Basic Science of Poisons* (6th edn). McGraw-Hill, New York, pp. 133–224.

de la Pena Mattozzi, M., Tehara, S.K., Hong, T. and Keasling, J.D. (2006) Mineralization of paraoxon and its use as a sole C and P source by a rationally designed catabolic pathway in *Pseudomonas putida. Applied and Environmental Microbiology* 72(10), 6699–6706.

Perrin-Ganier, C., Schiavon, F., Morel, J.L. and Schiavon, M. (2001) Effect of sludge-amendment or nutrient addition on the biodegradation of the herbicide isoproturon in soil. *Chemosphere* 44(4), 887–892.

Phytoremediation of Organics Action Team (2000) Phytoremediation Site Profiles: Cantrall. US EPA, Washington, DC.

Pretty, J.S. (2005) *The Pesticide Detox – Towards a More Sustainable Agriculture.* Earthscan, London.

Printz, H., Burauel, P. and Fuhr, F. (1995) Effect of organic amendment on degradation and formation of bound residues of methabenzthiazuron in soil under constant climatic conditions. *Journal of Environmental Science and Health, Part B* 30(4), 435–456.

Qari, H.A. and Hassan, I.A. (2014) Removal of pollutants from waste water using *Dunaliella* algae. *Biomedical and Pharmacological Journal* 7(1), 147–151.

Qiu, Y., Pang, H., Zhou, Z., Zhang, P., Feng, Y. and Sheng, G.D. (2009) Competitive biodegradation of dichlobenil and atrazine coexisting in soil amended with a char and citrate. *Environmental Pollution* 157(11), 2964–2969.

Qureshi, A., Mohan, M., Kanade, G.S., Kapley, A. and Purohit, H.J. (2009) In situ bioremediation of organochlorine-pesticide-contaminated microcosm soil and evaluation by gene probe. *Pest Management Science* 65(7), 798–804.

Radosevich, M., Traina, S.J. and Tuovinen, O.H. (1996) Biodegradation of atrazine in surface soils and subsurface sediments collected from an agricultural research farm. *Biodegradation* 7, 137–149.

Rani, K. and Dhania, G. (2014) Bioremediation and biodegradation of pesticide from contaminated soil and water – a novel approach. *International Journal of Current Microbiology and Applied Sciences* 3(10), 23–33.

Reffstrup, T.K., Larsen, J.L. and Meyer, O. (2010) Risk assessment of mixtures of pesticides: Current approaches and future strategies. *Regulatory Toxicology and Pharmacology* 56, 174–192.

Reyes Tirado (2015) Ecological farming: The seven principles of a food system that has people at its heart. Baker, M (ed.). Greenpeace Published in May 2015 by Greenpeace International Ottho Heldringstraat, Amsterdam, The Netherlands.

Roberts, D.J., Kaake, R.H., Funk, S.B., Crawford, D.L. and Crawford, R.L. (1993) Anaerobic remediation of dinoseb from contaminated soil: An on-site demonstration. *Applied Biochemistry and Biotechnology* 39/40, 781–789.

Rousseaux, S., Hartmann, A., Lagacherie, B., Piutti, S., Andreux, F. and Soulas, G. (2003) Inoculation of an atrazine-degrading strain, *Chelatobacter heintzii* Cit1, in four different soils: Effects of different inoculum densities. *Chemosphere* 51(7), 569–576.

Runes, H.B., Jenkins, J.J and Bottomley, P.J. (2001) Atrazine degradation by bioaugmented sediment from constructed wetlands. *Applied Microbiology and Biotechnology* 57(3), 427–432.

Sandra, R.R., Mário, S.G., João, R.F., Marli, G., Homero, M.G., Renata, R. and Vinícius de Almeida, M. (2015) Evaluation of *Ricinus communis* L. for the phytoremediation of polluted soil with organochlorine pesticides. *BioMed Research International* Article ID 549863. DOI:http://dx.doi.org/10.1155/2015/549863

Sarat, N.S. and Barathi, S. (2013) Enrichment and isolation of endosulfan degrading microorganisms in cashew plantations of Kasargod District, Kerala. *International Journal of ChemTech Research* 5(1), 6–14.

Schnoor, J.L. and Licht, L.A(1993) Deep-rooted poplar trees as an innovative treatment technology for pesticide and toxic organics removal from soil and groundwater. 22 February 1990 through 1 February 1993. Final report, US EPA. Organic Chemical Contamination of Soil/Water, Land and Waste Management. US EPA, Washington, DC.

Scow, K.M. and Hicks, K.A. (2005) Natural attenuation and enhanced bioremediation of organic contaminants in groundwater. *Current Opinion in Biotechnology* 16(3), 246–253.

Sethunathan, N. and Yoshida, T. (1973) *Flavobacterium* sp. that degrades diazinon and parathion. *Canadian Journal of Microbiology* 19, 873–875.

Sharaf, A.N., Soliman, M.H., Ibrahim, M.A., Moustafa, S.A. and Abd El-Hadi, A.A. (2006) Molecular identification and cloning of organophosphate degradation gene in some bacterial isolates. *Arab Journal of Biotechnology* 10, 259–274.

Sharif, D.I. and Mollick, M. (2013) Selective isolation of a gram negative carbamate pesticide degrading bacterium from brinjal cultivated soil. *American Journal of Agricultural and Biological Sciences* 8(4), 249–256.

Silva, E., Fialho, A.M., Sa-Correia, I., Burns, R.G. and Shaw, L.J. (2004) Combined bioaugmentation and biostimulation to clean up soil contaminated with high concentrations of atrazine. *Environmental Science and Technology* 38(2), 632–637.

Sims, G.K. (2006) Nitrogen starvation promotes biodegradation of N-heterocyclic compounds in soil. *Soil Biology and Biochemistry* 38(8), 2478–2480.

Singh, D.K. (2008) Biodegradation and bioremediation of pesticide in soil: Concept, method and recent developments. *Indian Journal of Microbiology* 48, 35–40.

Singh, B.K. (2009) Organophosphorus-degrading bacteria: Ecology and industrial applications. *Nature Reviews Microbiology* 7(2), 156–164.

Singh, B.K., Kuhad, R.C., Singh, A., Lal, R. and Tripathi, K.K. (1999) Biochemical and molecular basis of pesticide degradation by microorganisms. *Critical Reviews in Biotechnology* 19(3), 197–225.

Sivaramanan, S. (2014) Biodegradation of saw in plant fertilizer. *Research Journal of Agriculture and Forestry Sciences* 2(2), 13–19.

Strong, L.C., McTavish, H., Sadowsky, M.J. and Wackett, L.P. (2000) Field-scale remediation of atrazine-contaminated soil using recombinant Escherichia coli expressing atrazine chlorohydrolase. *Environmental Microbiology* 2(1), 91–98.

Struthers, J.K., Jayachandran, K. and Moorman, T.B. (1998) Biodegradation of atrazine by *Agrobacterium radiobacter* J14a and use of this strain in bioremediation of contaminated soil. *Applied and Environmental Microbiology* 64, 3368–3375.

Thabit, T.M.A. and El-Naggar, M.A.H. (2013) Malathion degradation by soil isolated bacteria and detection of degradation products by GC-MS. *International Journal of Environmental Sciences* 3(5), 1467–1476.

Topp, E., Tessier, L. and Gregorich, E.G. (1996) Dairy manure incorporation stimulates rapid atrazine mineralization in an agricultural soil. *Canadian Journal of Soil Science* 76(3), 403–409.

Tortella, G.R., Rubilar, O., Cea, M., Wulff, C., Martínez, O. and Diez, M.C. (2010) Biostimulation of agricultural biobeds with NPK fertilizer on chlorpyrifos degradation to avoid soil and water contamination. *Journal of Soil Science and Plant Nutrition* 10(4), 464–475.

Tripathy, S., Bhagyasree, P. and Ranajit, K.K. (2014) Phytoremediation: proficient to prevent pesticide pollution. *IJISET – International Journal of Innovative Science, Engineering and Technology* 1(10), 282–287.

United States Environmental Protection Agency (US EPA) (1994) Bioremediation in the Field Search System v2.1: Gila Indian Reservation.

US EPA (1995) SITE Superfund Innovative Technology Evaluation. SITE Technology Capsule: J.R. Simplot Ex Situ Bioremediation Technology: Dinoseb. EPA/540/r-94/508a.

US EPA (2006) In situ treatment technologies for contaminated soil. US EPA, 2006 Contract No.: EPA 542/F-06/013.

US EPA (2016) The EPA Conventional Reduced Risk Pesticide Program. Available at: https://www.epa.gov/pesticide-registration/conventional-reduced-risk-pesticide-program (accessed 31 July 2018).

Velázquez-Fernández, J.B., Martínez-Rizo, A.B., Ramírez-Sandoval, M. and Domínguez-Ojeda, D. (2012) Biodegradation and bioremediation of organic pesticides. In: *Pesticides-Recent Trends in Pesticide Residue Assay*, pp. 253–272.

Viray, F.A. and Hollingworth, R. (2009) The use and benefits of reduced risk pesticides since the passage of the food quality protection act. *The IR-4 Project Newsletter*, 40(4). New Jersey Agricultural Experiment Station. Available at: http://ir4.rutgers.edu/Newsletter/vol40no4.pdf (accessed 1 March 2016).

Vogel, T.M. (1996) Bioaugmentation as a soil bioremediation approach. *Current Opinion in Biotechnology* 7, 311–316.

Wackett, L.P. (1995) Recruitment of co-metabolic enzymes for environmental detoxification of organohalides. *Environmental Health Perspectives* 103 (Suppl. 5), 45–48.

Wagner, G.H. and Chahal, K.S. (1996) Decomposition of carbon-14 labelled atrazine in soil samples from Sanborn field. *Soil Science Society of America Journal* 30(6), 752–754.

Wagner, S.C. and Zablotowicz, R.M. (1997) Effect of organic amendments on the bioremediation of cyanazine and fluometuron in soil. *Journal of Environmental Science and Health, Part B* 32(1), 37–54.

Wolfaardt, G.M., Lawrence, J.R., Robarts, R.D. and Caldwell, D.E. (1994) The role of interactions, sessile growth, and nutrient amendments on the degradative efficiency of a microbial consortium. *Canadian Journal of Microbiology* 40(5), 331–340.

Yáñez-Ocampo, G., Sánchez-Salinas, E., Jiménez-Tobon, G.A., Penninckx, M. and Ortiz-Hernández, M.L. (2009) Removal of two organophosphate pesticides by a bacterial consortium immobilized in alginate or tezontle. *Journal of Hazardous Materials* 168, 1554–1561.

Yoshitomi, K.J. and Shann, J.R. (2001) Corn (*Zea mays* L.) root exudates and their impact on 14C-pyrene mineralization. *Soil Biology and Biochemistry* 33, 1769–1776.

Zeinat, K.M., Fetyan, N.A., Ibrahim, M. and El-Nagdy, S. (2008) Biodegradation and detoxification of malathion by *Bacillus thuringiensis* MOS-5. *Australian Journal of Basic and Applied Sciences* 2, 724–732.

Zhao, S., Arthur, E.L. and Coats, J.R. (2003) Influence of microbial inoculation (*Pseudomonas* sp. Strain ADP), the enzyme atrazine chlorohydrolase, and vegetation on the degradation of atrazine and metolachlor in soil. *Journal of Agricultural and Food Chemistry* 51, 3043–3048.

Zhao, X., Urano, K. and Ogasawara, S. (1989) Characteristics of clay minerals of montmorillonites in the wet state. *Nippon Kagaku Kaishi* 8, 209–215.

12 Awareness for Reducing Pesticide Risk

12.1 Introduction

According to the Food and Agriculture Organization (FAO) Programme Officer for the Rotterdam Convention (RC) (Barden, 2012), Elisabetta Tagliati:

> The livelihoods of rural families are dependent on their crops and their harvests. They rely on these to feed their children, themselves and to sell at market. Often, this is their sole form of survival. So, when a farmer identifies a pest threatening their only source of food or money, their immediate reaction is that a 'ready-to-use' solution like a pesticide 'is exactly' what they need.

Insecticides are designed to destroy insects and this means they are also likely to be toxic to humans. The toxic nature of many pesticides can be harmful to the environment and to human health. Herbicides are widely used, and over time, low doses of exposure can increase the risk of Parkinson's disease, cancers, diabetes, gluten intolerance, infertility and reproduction disorders said Tagliati. The ultimate goal of using pesticides should be to manage pests in a way that allows sustainable crop production without negative impacts on the environment and human health. These externalities are particularly grave in some developing countries because of the widespread use of highly hazardous pesticides, the lack of awareness of risks and the lack of protective equipment. But even in countries with strict registration processes and the application of these pesticides as prescribed, multiple pesticide residues are found in food and water bodies. The effects of pesticide residues are difficult to accurately assess due to the large number of active ingredients and commercially available pesticide formulations, the even larger number of metabolites that are sometimes more toxic than the parent compound and the synergistic effects of multiple pesticides. It becomes complicated to understand the long-term effects of multiple residues of pesticides (Eyhorn *et al.*, 2015). It would be desirable for pesticides to be used only when essential and with the least side effects possible. However, often the cost of the negative impacts of pesticides (i.e. the impact on health and the environment and the costs of avoiding them) is higher than the value generated by using them (i.e. yield increase and net cost reduction).

Every year, around 5000 chemicals are added to the existing list of about 500,000 chemicals used in industrial processes. Most of them are extremely beneficial. They are necessary for industrial processes to produce fertilizers and plant protection products, which are required to sustain a certain level of agricultural production. According to Gerold Wyrwal (FAO Agricultural Officer for the RC), about 200 million farmers apply these substances around the world (Barden, 2012). There is also growing concern about the health effects caused by exposure to mixtures of pesticides – known as the 'cocktail effect'. Exposure to mixtures of pesticides can be far more damaging than exposure to just one.

Often farmers are not aware of the correct ways to handle pesticides, from their purchase and sale, through to transporting them, in addition to the precautions to take to store them safely. The risks to those spraying fields without adequate equipment are high, and ensuring instructions can be understood by those coming into contact with pesticides is essential. The labels on pesticide containers are often a barrier to safe use because many of those utilizing the chemicals are unable to read or understand the languages in which guidelines are produced (Barden, 2012). Many of these farmers are women and these women often become the victims of disturbing experiences. The RC has also noted that children commonly play in fields where pesticides are present and that women frequently wash contaminated clothes with their bare hands.

12.2 Educating Farmers by Raising Their Awareness

Ideally, agricultural systems should be designed in such a way that pests, diseases and weeds do not build up to a level at which they cause significant damage to crops. Suitable agronomic practices, the use of resistant varieties and integrated pest management (IPM) are key preventive measures. Biocontrol and the use of natural substances can complement these efforts. The safe and need-based application of minimal toxic synthetic pesticides should be used as a last resort. Farmers and farm workers may be aware of the risks of pesticides, but not always of the alternatives to using them. They should be made more aware of the health risks associated with their injudicious use. Poor farmers in Africa, based on the false notion that pesticides act as a kind of 'insurance' against pests, treat their crops routinely with pesticides, regardless of the existence of any serious pest threat. However, the weight of scientific evidence shows the vast majority of these pesticides do little or nothing to increase production and most often reduce farm profits and cause major negative side effects for human and environmental health.

Yang *et al.* (2014) investigated the levels of knowledge and awareness of the dangers of pesticide use for the environment and human health in farmers and retailers in two rural regions (Qianyang County (S1) and Chencang County (S2)) of the Wei River catchment in China. The results showed that farmers learned about the usage and dangers of pesticides mainly by oral communication. Protective measures were inadequate: farmers in both regions did not use any protective measures during spraying. Washing hands was the most common mode of personal hygiene, relative to wearing masks, showering and changing clothes, but no significant differences were observed between the selected regions. Most pesticide wastes were dumped directly onto the land or into water, suggesting that educational measures should be taken to address the potential risks from the residues in the wastes. Over 85% of farmers (S1 and S2) claimed to use illegal pesticides, but the reasons for their use varied. Retailers were well-informed and highly conscious of their responsibility for the safe use of pesticides, especially in S2. A canonical correspondence analysis indicated that educational level and age differed between the two regions and contributed greatly to the risks from pesticide use. The levels of knowledge of pesticide safety were insufficient in older and less educated interviewees. There was a need for educational programmes targeted to age groups, proper disposal of pesticide waste and sufficient supervision from authorities to improve the levels of knowledge about and awareness of the dangers of pesticides to human health and environmental pollution in the Wei River catchment, China. Recena *et al.* (2006)

and Ibitayo (2006) reported that a low level of education among farmers hampered their ability to follow the hazard warnings provided by the chemical industry and regulatory agencies. Lack of education has been associated with poisonings, suicides, exposure risks and high mortality rates in many rural areas of developing countries (Phillips *et al.*, 2002; Zyoud *et al.*, 2010; Zhou *et al.*, 2011). Salameh (2004) claimed that adequate and reliable sources of information seemed to induce the perception of risk and the adoption of preventive measures, despite the low general level of education of the subjects involved. Oral communication with retailers and among farmers played a dominant role in learning about pesticides and their functions. Although some media channels provide information about the control of pests and diseases in cultivation, media advertising seems an inadequate means to provide information about pesticides. It was observed that information provided by television was difficult to apply because of differences from the real conditions of cultivation. The distrust of pesticide advertisements led to a strong reliance on the information provided by distributors and neighbours and on the personal experience of the farmers. Accordingly, based on their own experiences, farmers would likely overuse or mix pesticides, being unaware of the negative interactions among the various active ingredients. Mixed pesticides not only lose efficacy but may also enhance the hazards to the sprayer and the environment because of their synergistic behaviour (Karunamoorthi *et al.*, 2011).

There is a need for independent robust basic research on the impacts of pesticides particularly with regards to the long-term effects of pesticide formulations and their metabolites and on the synergistic effects of multiple residues on human health and on ecosystems. It is difficult to understand all the effects of pesticides, in particular the long-term effects of multiple residues. Eyhorn *et al.* (2015) emphasized the need to calculate the full external costs of pesticide use in order to assess them in relation to the benefits of pesticide use, which would help in determining what level of pesticide use is acceptable for society. The study, based on observations, also discussed redesigning farming systems based on agroecology, breeding robust varieties, advancing alternative crop protection methods, and funding research and development.

12.3 Alternatives and Safe Use of Pesticides

According to the US Environmental Protection Agency (US EPA), all pesticides are toxic to some degree, and the commonplace, widespread use of pesticides is both a major environmental problem and a public health issue (Lu *et al.*, 2001). The US EPA also believes that most pesticides – despite having a US EPA registration – have not been adequately tested to determine their effects on people or the environment (Taylor, 2000). In many developing countries, farmers are still applying highly hazardous, phased out pesticides and are very far from 'safe use' of pesticides. They are ignorant of the risks associated with injudicious use of these pesticides, and they should be equipped with realistic and feasible measures to reduce these risks. They need to undergo compulsory training about the risk of pesticides, their avoidance and on safety measures. The same kind of training can be imparted to home gardeners.

To facilitate crop health management, farmers can adopt crop rotation for locally available crops following methods enhancing soil fertility and biodiversity and therefore facing fewer pest, disease and weed problems. Agroecology could be integrated into vocational education and training curricula. In France, agroecology is being mainstreamed in agriculture. Resilient farming systems and better management practices can be designed by identifying the best practice from different systems. Conventional farmers could be encouraged to use biocontrol or botanical sprays and precision farming techniques like selective spraying. There is always a preference for food products with low or no pesticides on the market, an essential driver for pesticide reduction. The demand for the same can be translated to the producers. Consumer expectations that food is free of pesticide residues and responsibility for health and the environment are important reasons for processors and retailers to engage in pesticide reduction. One of the best alternatives, IPM, should become part of the recommendations for good agricultural practices and should be integrated into the work of extension or rural advisory services. It is sometimes difficult to convince farmers but demonstrations and on-field trials are convincing ways to facilitate the exchange of knowhow among practitioners. Farmer field schools have proved quite effective in this regard. Information and communication technology (ICT) is another way to improve timely access to forecasts of pest and disease pressure, allowing better and more timely management of crop pests.

Composting can be one of the alternatives to chemical pesticides. Good quality compost, the cornerstone of organic gardening, results in healthy and vigorous plants, which resist pests and diseases. Healthy plants need healthy soil, and the best way to have healthy soil is to use compost. Composting increases the biodiversity of the soil, making it a host to soil-dwelling organisms, which helps plants to resist disease. Biodiversity makes it possible to control pests without chemicals. Lack of chemicals encourages an environment conducive to all species; the pests' natural enemies will keep potential pest populations in check.

Organic practices require tolerance for all creatures and recognition that all have a role to play. Composting supplies beneficial soil organisms to the whole ecosystem, in addition to providing a food source, moisture-holding humic substances, buffering qualities and enough diversity for their survival. That is why using high-quality compost eliminates the need for synthetic fertilizers and pesticides (Pesticide Reduction Resource Guide, 2002).

It is in the best interest of community health for all residents to learn about the hazards of pesticides, and to adopt organic gardening and landscaping techniques as well as an IPM approach to all pest-related problems.

12.4 Storage and Disposing of Pesticides

Methods for storing and disposing of pesticide containers are critical points of intervention to enhance the awareness of safety during the application of pesticides (Matthews, 2008). Storing pesticides in the home can easily contaminate drinking water and food and can threaten human health, especially that of children. Matthews (2008) reported that 62% of 8500 smallholders in 26 countries stored pesticides, and 68% of this group stored the pesticides in locked locations. Similarly, over 50% of the farmers in the survey stored pesticides, but approximately half of this group stored them in their homes. The interesting point noted is that about 40% of the farmers had never stored pesticides, which is a good option for reducing the dangers of pesticides. Some larger countries recycle pesticide containers despite the fact that they have contained dangerous chemicals (Salameh, 2004; Matthews, 2008). Some farmers have often been observed discarding empty plastic bags and bottles in wells or ditches where the pesticide solutions were prepared. This observation was supported by the responses of the interviewed farmers. Pesticide residues in empty containers may be released or washed into the surroundings by rain and/or irrigation. This uncontrolled discharge of pesticide pollutants can easily contaminate agricultural soil and water and also threaten human health and ecosystem quality (Ibitayo, 2006; Hvistendahl, 2013).

12.5 Restricting Use of Hazardous Pesticides and Banned Pesticides

Pesticide dealers and suppliers can be forced to comply with existing laws and codes. Further restrictions can be imposed by prohibiting pesticides included on the list of highly hazardous pesticides of the Pesticide Action Network (PAN). Several pesticides have been banned in the past three decades due to their ecotoxicities and long half-lives, the reasons for which may not be clear to the farmers. In spite of the availability of environmental friendly pesticides, a few banned pesticides are still commercially available on the Asian market for use. There should be regular residue tests in order to ensure compliance with set requirements, and suppliers should be informed of the results of these tests. Yang *et al.* (2014) felt a need for effective legislation and strict supervision to monitor the production and circulation of pesticides. Uri (1998) saw economics as another factor in why farmers ignore the dangers of pesticides, as farmers claimed expenditure was an important factor in the purchase of pesticides. The survey and calculations conducted by Yang *et al.* (2014) indicated that farmers were not likely to reduce pesticide use for economic reasons. Furthermore, pesticide consumption and use involves several stakeholders, such as authorities, producers, retailers and farmers. Local pesticide dealers play an important role in disseminating pesticides and guiding their use for farmers and agricultural workers. Moreover, local pesticide dealers are within their reach when agriculture specialists are not accessible. The dealers also provide guidance on the protective measures available for their application. However, in another study, retailers revealed that farmers were reluctant to adopt these measures, preferring to rely on their own experience and the influence of their neighbours and ancestors (Yassin *et al.*, 2002; Matthews, 2008; Zyoud *et al.*, 2010).

Damalas *et al.* (2008) advised that the perception of risk was an important element in developing effective campaigns for education and communication. After one such safety educational programme, the prevalence of pesticide poisoning decreased from 1.05% to 0.25% (Chen *et al.*, 1998). Farmers and agricultural workers should be acquainted with the dangers of pesticides to the environment and human health, inculcating a consciousness of self-protection and a responsibility to their surroundings (Ibitayo, 2006).

Farmers are the backbone of agriculture in any country. Due to their lack of awareness of the latest technologies in agricultural farming, such as IPM, they may opt for injudicious use of chemical pesticides, posing a risk not only to their health, but to the environment. There is always a need to create awareness among farmers involved in agricultural farming. Proper mechanisms should be followed for dissemination of this knowledge. These mechanisms could be, for example, farmer field schools (FFS) and mass media campaigns (MMC) (Heinrichs *et al.*, 2017). In FFS, the participatory training approach, farmers utilize the experiences and learning principles of the trainers. The approach improves farmers' capacities through a series of regular meetings and discussions with highly trained facilitators (typically extension workers), where farmers conduct their own field experiments, and learn to diagnose and resolve their problems (Van den Berg and Jiggins, 2007). In one FAO programmes, according to Pincus (1999), the FFS participants had reduced insecticide use and increased rice IPM yield over time, results supported by Tripp *et al.* (2005). However, difficulties with financial sustainability and effectively spreading information among non-participants are the major criticisms against FFS. MMC may be used to convey any message to farmers through a participatory workshop that involves researchers, extension and agricultural communication specialists. This may also involve the distribution of leaflets, posters, roadside billboards and radio drama broadcast through local stations and played at coffee shops (Heong *et al.*, 1998). FFS and MMC can ultimately help with reduction of pesticides likely to create risks for the environment and health of farm workers.

In West Africa, farmers were encouraged to use pesticides for crop pest management in the early 1970s, and they have become commonly used throughout the region. Many of the chemicals used are highly toxic and either already banned in northern hemisphere countries, or require protective equipment and conditions for use, outside the reach of most African farmers. The farmers treat their crops routinely with pesticides, regardless of the existence of any serious pest threat, based on the misbelief that pesticides act as a kind of 'insurance' against pests. In fact the vast majority of these pesticides do little or nothing to increase production and most often reduce farm profits and incur major negative side effects for human and environmental health (FAO, 2017). FAO's integrated production and pest management (IPPM) programme has partnered with governments, non-governmental organizations, research institutions and farmer organizations at all levels to raise awareness of the importance of reducing or eliminating the use of toxic pesticides in agriculture. Effective community-based education through FFS opens up the possibility for widespread analysis of pesticides in West African agricultural systems and surface water systems. Training on pest management via season-long FFS raises awareness of the health risks associated with pesticides and encourages farmers to reduce their use in favour of feasible, low-cost and non-toxic or less toxic pesticides. It also touches on how to avoid commercial pressure to use pesticides.

12.6 Use of Risk Indicator Tools

In agriculture, pesticides are typically chosen on the basis of the target pest, effectiveness, cost and pre-harvest interval allowance, and very rarely based on their health and environmental risks. However, the risk of a given product to human health (the health of consumers and pesticide applicators) and the environment is not easy to quantify. In Québec City, an online tool called Québec Pesticide Risk Indicator (QPRI) Express has been used by growers, crop advisers and agribusinesses involved in plant protection since spring 2007 to help them assess pesticide risks. QPRI Express, a diagnostic tool for assessing the risk associated with specific products, is a decision-making tool for optimizing the management of pesticide use. The tool has a health component (QPRI-Health) and an environment component (QPRI-Environment), to calculate the health risk indicator (HRI, which represents acute toxicity and chronic toxicity) and the environmental risk indicator (ERI, which represents physico-chemical and ecotoxicological properties) of the product, respectively. Products with higher theoretical HRIs and ERIs are more toxic than those with lower indicator values. QPRI enables users to select lower-risk products within an IPM context. It can be used to track changes in risks from year to year, including for an entire region. Laniel (2017) conducted a study in Québec on three main aspects: raising stakeholder awareness of the value of QPRI Express, validating QPRI Express using actual farm data and creating risk compilations to analyse the health and environmental risks of pesticides, mainly on five crops (strawberry, maize, dry onion, potato and apple), and 100 businesses were selected on an experimental basis. QPRI Express is considered a useful way to help growers reduce the risks associated with pesticides. However, mostly the advisers used QPRI Express while making recommendations to growers, to make and track risk compilations for their businesses. The tool could be used to make judicious selection of pesticides by enabling lower-risk solutions and control strategies for a given crop.

Another risk indicator tool – environmental impact quotient (EIQ), developed by Kovach *et al.* (1992) – has been used by Muhammetoglu and Uslu (2007) to measure the environmental impacts of pesticide active ingredients used in vegetable and fruit production, in the Kumluca region of Turkey. In this study, some pesticide management scenarios were evaluated to select the least detrimental pesticide by comparing EIQ model results. The model is considered a helpful tool and can be easily applied by pest management practitioners and agricultural specialists. It has been efficiently used to compare different agricultural pest management strategies or programmes. The EIQ model has been observed to reduce pesticide risks in potato production in developing countries. Kromann *et al.* (2011) applied the EIQ model for measuring the potential benefits of integrated pest and disease interventions in potato production in Ecuador and Peru. A substantial variation was observed under different conditions and different crop management approaches. The EIQ values ranged from 40 for an IPM system (including resistant cultivar with less hazardous pesticides) to 1235 for a high-input conventional system (susceptible cultivar plus frequent use of hazardous pesticides). The study demonstrated the utility of the EIQ for assessing health and environmental hazards of potato production in the Andes and potentially other areas in the developing world with diverse conditions. Mazlan and Mumford (2005) compared cabbage farmers and insecticide applications in five zones in the highlands of Malaysia. Bardenes-Perez and Shelton (2006) used the EIQ to compare potential impact of cruciferous vegetables growing in the Kenyan highlands and in three areas in the Kullu Valley in India. Morse *et al.* (2006) assessed the environmental impact of an agricultural shift to genetically modified cotton production in South Africa.

Numerous other tools have been developed to assess environmental and human health hazards or risks associated with pesticide use (Levitan *et al.*, 1995; Pittinger *et al.*, 2003). Levitan *et al.* (1995) classified a number of tools ranging from anecdotal accounts to holistic assessments of impact of agriculture. The recent scoring systems used for pesticides include the Environmental Hazard Index and the Priority Substances List in Canada (Gallivan *et al.*, 2005), the Ecological Relative Risk indicator in Austria (Sánchez-Bayo *et al.*, 2002), the European Risk Ranking method (EURAM), the Chemical Hazard Evaluation for Management Strategies from University of Tennessee and Purdue Research Foundation's Pollution Prevention Progress Measurement Method in the USA (Pittinger *et al.*, 2003).

12.7 Use of Pesticide Risk Reduction Models/Strategies

During a workshop organized by the Organisation for Economic Co-operation and Development (OECD) in Neuchâtel, Switzerland, in 1998 on IPM and Pesticide Risk Reduction (PRR), cost–benefit assessment and other economic models were discussed. Models can be very useful to investigate and compare different optional strategies for risk reduction. The economic consequences supporting PRR strategies can be illustrated and analysed using economic models. The application of these models can contribute to PRR at the farm level, for example, by identifying areas where pesticide use can be reduced with little or no cost to the farmer. The OECD PRR programme encourages the development, adoption and implementation of advanced sustainable pest management strategies and policies, including risk reduction policies/practices and alternative approaches to the use of chemicals, as well as improved risk management and compliance/enforcement activities (in particular via the use of new pesticides with lower-risk properties, of biopesticides and the adoption and implementation of IPM) (PRR, www.oecd.org/chemicalsafety/pesticides-biocides/pesticideriskreduction.htm).

During another OECD workshop (2001) two different approaches were identified: (i) bottom-up strategies with the goal of informing farmers so that better decisions are made by the farmer himself; (ii) top-down strategies, where a pre-set goal is implemented via extension services or educational measures. It is recommended to facilitate international dialogue on strategies and instruments for PRR, transition to integrated and organic agriculture, and development of farmer decision tools, and to facilitate work on defining and valuing externalities and non-market benefits of PRR. Economic assessment models are based on a simulation and simplification of reality that is not always transparent, and may include a variety of assumptions and estimates. It should be clear that economic assessments are just one tool in the pesticide policy toolbox, of particular value in comparing policy options but less suitable for quantifying benefits in absolute terms (PRR, www.oecd.org/chemicalsafety/pesticides-biocides/pesticideriskreduction.htm).

In a case study, Phung *et al.* (2013) applied a comprehensive needs assessment model to evaluate community needs and policy development for and solutions to PRR for farmers in Vietnam. Through discussions with farmers, the needs identified for pesticide regulations included enhancement of pesticide legislation; multisectoral involvement in pesticide regulations; improvement of capacity for sectors involved in pesticide regulations; risk–benefit guidance for pesticide registration; reforms of pesticide regulations relating to the restriction, cancellation, suspension, transport, storage and disposal of pesticides; and the development of occupational hygiene and safety policy and programmes for agricultural activities. The key requests from farmers included technical training for occupational safety and hygiene of pesticide application, and support for safety facilities for pesticide application and protective equipment. Based on the results of comprehensive needs assessment, a range of PRR strategies were developed for legislative improvement, workplace and personal hygiene, information and training, and medical surveillance and pesticide poisoning first aid.

To reduce risks to the environment and human health from pesticide use in agriculture, pesticide risk reduction strategies have been developed under the Pesticide Risk Reduction Program (PRRP), a joint programme of Agriculture and Agri-Food Canada (AAFC) and the Pest Management Regulatory Agency (PMRA) of Health Canada (MUPP, 2017). The programme aims to address grower needs for reduced risk management tools and practices for specific pest issues. Greenhouse floriculture production is an important industry in Canada. The greenhouse provides an environment where biological control can be used as an effective, low-risk management approach. However, it leads to additional challenges such as year-round production and the swift spread of pest problems. The growers need support to implement and maintain a biological control system in the greenhouse. A number of pest management issues were identified through stakeholder consultations; for example, lack of reduced risk pest control products, need for improved availability and accessibility of biological controls, lack of alternative pest management practices and technologies, and insufficient communication and extension support across growing regions. These issues were resolved by improving the access to new lower-risk pest control products, developing new reduced risk pest management tools, technologies and integrated systems, and promoting and facilitating the adoption of reduced risk techniques.

The strategy helped greenhouse floriculture growers reduce reliance on chemical pesticides while remaining an economically competitive industry.

12.8 Role of Decision Makers and Officials

Eyhorn *et al.* (2015) emphasized the need for awareness among decision makers in different fields and sectors. Pesticide reduction is a joint responsibility of all stakeholders and they all need to pull in the same direction. Scientists, researchers and policy makers must be acquainted with the fact-based information on pesticide issues so as to reduce the use of pesticides and their associated risks, if any. Scientists and government officials can research the alternatives to risky pesticides and implement conducive policies. It also helps to strengthen the link between farmers, consumers, the private sector, civil society and governments. Depending on the related issues, training can be arranged for farmers, and ICT-based advisory services can also be strengthened to transmit knowledge and information to farmers.

Based on the scientific evidence for the harm to human health and the environment from current pesticide use, the European Union (EU) introduced a package of new laws to reduce pesticide dependency. However, many governments have been slow to carry out the laws. The Massachusetts Pesticide Awareness Collaborative (MPAC), a collection of citizens and health officers from Massachusetts, is interested in reducing pesticide use in their own communities and in the state. Participants share information and resources to expedite the education of citizens and adoption of municipal pesticide reduction policies.

12.9 Conclusion

The injudicious use of pesticides leaves residues in food, water, soil and the environment. However, it is difficult to accurately assess the impact of residues, due to the large numbers of active ingredients in commercially available pesticide formulations, the even larger number of metabolites that are sometimes more toxic than the parent compound and the synergistic effects of multiple pesticides. It becomes complicated to understand the long-term effects of multiple residues of pesticides. Often farmers are not aware of the risks associated with these pesticides and become victims of their hazards due to ignorance of safety measures in their handing, storage, application, transportation and disposal. Need-based application of pesticides and monitoring of the pest population in the field is most important. There is also a need for independent robust basic research on the impacts of pesticides, particularly with regard to the long-term effect of pesticide formulations and their metabolites, and on synergistic effects of multiple residues on human health and on ecosystems. Farmers need education to raise their awareness about resistant varieties, biocontrol techniques and IPM strategies. An IPM strategy, using integration of all control measures without solely relying on chemical pesticides, produces a safe and economic crop. Crop rotation, one of the IPM components, enhances soil fertility and biodiversity. The demand for food with low or no pesticides can be translated to the producers. Other alternatives include good quality composting, which eliminates the need for synthetic fertilizers and pesticides. Restrictions can be imposed on pesticide dealers and suppliers to force them to comply with existing laws and codes, and by prohibiting highly hazardous pesticides. The newly introduced low-dosage molecules are costly and not affordable for many farmers. Economics has been observed as the major factor in why farmers ignore the dangers of pesticides. Proper mechanisms should be followed for dissemination of knowledge through FFS, where farmers utilize the experiences and learning principles of the trainers, through a series of regular meetings and discussions, where farmers conduct their own field experiments, learn to diagnose and resolve their own problems. A variety of risk indicator tools are available to assess pesticides in a qualitative and quantitative manner. EIQ is one of the tools for impact assessment of pesticides in the environment. To reduce risks to the environment and human health from pesticide use in agriculture, pesticide risk reduction strategies are developed under the PRRP, a joint programme of AAFC and the PMRA of Health Canada. Pesticide reduction is a joint responsibility of all stakeholders and to implement conducive policies, policy makers must be acquainted with fact-based information on pesticides. Scientists report that global reproductive health is being affected and pesticides are at least partly to blame. Moreover, pesticides have been linked to miscarriages, premature births and reduced fertility in both men and women. The evidence indicates that exposure, even to small doses, can be lethal. The pesticide problem calls for renewed and ongoing action.

References

Agriculture and Agri-Food Canada (2017) Government of Canada. Pesticide Risk Reduction Strategy for Greenhouse Floriculture. AAFC. Ottawa, Ontario, Canada.

Badenes-Perez, F.R. and Shelton, A.M. (2006) Pest management and other agricultural practices among farmers growing cruciferous vegetables in the Central and Western Highlands of Kenya and the Western Himalayas of India. *International Journal of Pest Management* 52(4), 303–315.

Barden, S. (2012) Reducing risks from pesticides by empowering rural women. Synergies: Among the Basel, Rotterdam and Stockholm Conventions. FAO Rotterdam Convention. Communications and Advocacy Officer Secretariat. Available at: www.brsmeas.org/Implementation/MediaResources/NewsFeatures/Reducepesticidesrisksbyempoweringruralwomen/tabid/5809/language/en-US/Default.aspx (accessed 1 August 2018).

Chen, S., He, F., Zhang, Z., Gao, Y., Zhou, A., Xie, C., Xiong, L., Chen, D., Wang, S. and Jia,

J. (1998) Evaluation of a safety educational programme for the prevention of pesticide poisoning. *La Medicina del lavoro* 89(Suppl 2), S91–S98.

Damalas, C.A., Telidis, G.K. and Thanos, S.D. (2008) Assessing farmers' practices on disposal of pesticide waste after use. *Science of the Total Environment* 390, 341–345.

Erskine, G., Jensen, G., Stauffer, A. and Smith, D. (2010) Six steps to pesticide reduction. A HEAL (Health and Environment Alliance, HEAL) toolkit for communities and individuals wishing to reduce local pesticide use. Available at: https://www.env-health.org/IMG/pdf/9-_Six_steps_to_pesticide_reduction_-_A_HEAL_toolkit_for_communities_and_individuals_wishing_to_reduce_local_pesticide_use.pdf (accessed 1 August 2018).

Eyhorn, F., Roner, T. and Specking, H. (2015) Reducing pesticide use and risks – what action is needed? Briefing paper. HELVETAS Swiss Intercooperation. Available at: https://www.helvetas.org/Publications-PDFs/Briefing%20paper%20pesticide%20reduction_conclusions.pdf (accessed 1 August 2018).

Food and Agriculture Organization (FAO) (2017) Pesticide risk reduction. Integrated Production and Pest Management Programme in Africa. Available at: www.fao.org/agriculture/ippm/activities/pesticide-risk-reduction/en/ (accessed 1 August 2018).

Gallivan, G.J., Berges, H. and McGee, B. (2005) Evaluation of the changes in pesticide risk. Research Project SR9128, Ontario Ministry of Agriculture and Food, Guelph, Canada.

Heinrichs, E.A., Nwilene, F.E., Stout, M.J., Hadi, B.A.R. and Freitas, T. (2017) *Rice Insect Pests and their Management*. Burleigh Dodds Science Publishing Limited, Cambridge, UK.

Heong, K.L., Escalada, M.M., Huan, N.H. and Mai, V. (1998) Use of communication media in changing rice farmers' pest management in the Mekong Delta, Vietnam. *Crop Protection* 17, 413–425.

Hvistendahl, M. (2013) In rural Asia, locking up poisons to prevent suicides. *Science* 341, 738–739.

Ibitayo, O.O. (2006) Egyptian farmers' attitudes and behaviors regarding agricultural pesticides: Implications for pesticide risk communication. *Risk Analysis* 26, 989–995.

Karunamoorthi, K., Mohammed, A. and Jemal, Z. (2011) Peasant association member's knowledge, attitudes, and practices towards safe use of pesticide management. *American Journal of Industrial Medicine* 54, 965–970.

Kovach, J., Petzoldt, C., Degnil, J. and Tette, J. (1992) A method to measure the environmental impact of pesticides. *New York's Food and Life Sciences Bulletin* 139, 1–8.

Kromann, P., Pradel, W., Cole, D., Taipe, A. and Forbes, G.A. (2011) Use of the environmental impact quotient to estimate health and environmental impacts of pesticide usage in Peruvian and Ecuadorian potato production. *Journal of Environmental Protection* 2, 581–591.

Laniel, C. (2017) Raising awareness to pesticide risk reduction on the farm for the benefit of health and the environment. Project Code PRR07-300. Agriculture and Agri-Food Canada, Available at: www.agr.gc.ca/eng/?id=1299088010821 (accessed 1 August 2018).

Levitan, L., Merwin, I. and Kovach, J. (1995) Assessing the relative environmental impacts of agricultural pesticides: The quest for a holistic method. *Agriculture, Ecosystems and Environment* 55(3), 153–168.

Lu, C., Knutson, D.E., Fisker-Andersen, J. and Richard, A.F. (2001) Biological monitoring survey of organophosphorus pesticide exposure among preschool children in the Seattle metropolitan area. *Children's Health-Article. Environmental Health Perspectives* 109, 299–303.

Matthews, G.A. (2008) Attitudes and behaviours regarding use of crop protection products – a survey of more than 8500 smallholders in 26 countries. *Crop Protection* 27, 834–846.

Mazlan, N. and Mumford, J. (2005) Insecticide use in cabbage pest management in the Cameron Highlands, Malaysia. *Crop Protection* 24(1), 31–39.

Minor Use Pesticides Program (MUPP) (2017) Minor Use Pesticides Program. Agriculture and Agri-Food Canada. Available at: www.agr.gc.ca/eng/?id=1286197216280 (accessed 25 August 2018).

Morse, S., Bennet, R. and Ismael, Y. (2006) Environmental impact of genetically modified cotton in South Africa. *Agriculture, Ecosystems and Environment* 117(4), 277–289.

Muhammetoglu, A. and Uslu, B. (2007) Application of environmental impact quotient model to Kumluca region, Turkey to determine environmental impacts of pesticides. *Water Science and Technology* 56(1), 139–145.

Organisation for Economic Co-operation and Development (OECD) Workshop (2001) Report of the OECD workshop on the economics of pesticide risk reduction in agriculture. OECD, Copenhagen, pp. 28–30.

Pesticide Reduction Resource Guide (2002) *For Citizens and Municipalities of Massachusetts.* May, 2002. Wellesley Natural Resources Commission. Project funding in part by Toxics Use Reduction Institute at the University of Massachusetts, Lowell, the Massachusetts Executive Office of Environmental Affairs and Department of Environmental Protection. Available at: https://www.mass.gov/files/documents/2016/08/ox/rpcguide.pdf, website: www.ci.wellesley.ma.us/nrc/pesticide (accessed 2 October 2018).

Phillips, M.R., Yang, G., Zhang, Y., Wang, L., Ji, H. and Zhou, M. (2002) Risk factors for suicide in China: A national case-control psychological autopsy study. *Lancet* 360, 1728–1736.

Phung, D.T., Connell, D., Miller, G., Rutherford, S. and Chu, C. (2013) Needs assessment for reducing pesticide risk: A case study with farmers in Vietnam. *Agromedicine Journal* 18(4), 293–303.

Pincus, J. (1999) The impact of IPM farmer field schools on farmers' cultivation practices in their own fields. FAO Programme for Community IPM in Asia Report.

Pittinger, C., Brennan, T., Badger, D., Hakkinen, P. and Fehrenbacher, M.C. (2003) Aligning chemical assessment across the hazard–risk continuum. *Risk Analysis* 23(3), 529–535.

Recena, M.C., Caldas, E.D., Pires, D.X. and Pontes, E.R. (2006) Pesticides exposure in Culturama, Brazil – knowledge, attitudes, and practices. *Environmental Research* 102, 230–236.

Salameh, P. (2004) Pesticides in Lebanon: A knowledge, attitude, and practice study. *Environmental Research* 94, 1–6.

Sánchez-Bayo, F., Baskaran, S. and Kennedy, I.R. (2002) Ecological Relative Risk (EcoRR): Another approach for risk assessment of pesticides. *Agriculture, Ecosystems and Environment* 91(1–3), 37–57.

Taylor, D.A. (2000) A new crop of concerns: Congress investigates pesticide safety. *Environmental Health Perspectives* 108(9) , A408–A411.

Tripp, R., Wijeratne, M. and Piyadasa, V.H. (2005) What should we expect from farmer field schools? A Sri Lanka case study. *World Development* 33, 1705–1720.

Uri, N.D. (1998) Government policy and the development and use of biopesticides. *Futures* 30, 409–423.

Van den Berg, H. and Jiggins, J. (2007) Investing in farmers: The impacts of farmer field schools in relation to integrated pest management. *World Development* 35, 663–686.

Yang, X., Wang, F., Meng, L., Zhang, W., Fan, L., Geissen, V. and Ritsema, C.J. (2014) Farmer and retailer knowledge and awareness of the risks from pesticide use: A case study in the Wei River catchment, China. *Science of the Total Environment* 497–498, 172–179.

Yassin, M., Mourad, T.A. and Safi, J. (2002) Knowledge, attitude, practice, and toxicity symptoms associated with pesticide use among farm workers in the Gaza Strip. *Occupational and Environmental Medicine* 59, 387–393.

Zhou, L., Liu, L., Chang, L. and Li, L. (2011) Poisoning deaths in central China (Hubei): A 10-year retrospective study of forensic autopsy cases. *Journal of Forensic Science* 56(Suppl. 1), S234–S237.

Zyoud, S.H., Sawalha, A.F., Sweileh, W.M., Awang, R., Al-Khalil, S.I., Al-Jabi, S.W. and Bsharat, N.M. (2010) Knowledge and practices of pesticide use among farmworkers in the West Bank, Palestine: Safety implications. *Environmental Health and Preventive Medicine* 15(4), 252–261.

13 Regulation of Pesticides and Associated Risks

13.1 Introduction

Although pesticides are developed through very strict regulation processes to function with reasonable certainty and minimal impact on human health and the environment, serious concerns have been raised about health risks resulting from occupational exposure to different pesticides via different routes, including residues in food and drinking water. Moreover, people have also become increasingly concerned about pesticide use and particularly about their impacts on human health and environmental quality (Damalas, 2009). The highly hazardous pesticides in particular, 'cause adverse human health and environmental effects in many countries, particularly in low-income and middle-income countries', and need urgent action to address them, encouraging an emphasis on promoting agroecologically based alternatives (SAICM/ICCM.4/CRP.16) (PAN, 2017). Double standards in the international trade of pesticides from developed countries to developing countries are still prevalent and are present in both the export of hazardous pesticides and the transfer of production facilities (PAN, 2017).

The toxicity of a pesticide, safety during its application, dosage applied, adsorption in soil, weather conditions and its persistence in the environment determine its adverse effects on the environment (water, soil and air contamination from leaching, runoff and spray drift, as well as the detrimental effects on wildlife, fish, plants and other non-target organisms). The pesticide may not be safe under all circumstances, nor can its long-term health and environmental effects in hypothetical situations be predicted with certainty (Damalas and Eleftherohorinos, 2011). However, the risk assessment of chemicals for regulatory purposes does not generally take into account the 'real-life' exposure to these pesticides, mainly relying on the assessment of individual pesticides.

These increased concerns have resulted mainly from reduced trust in agricultural plant protection as well as in the authority's regulations aimed at protecting both the environment and human health. Therefore, evaluation of pesticide safety, scientific data, policy guidelines and professional judgement must be incorporated while assessing a pesticide to determine whether it can be used beneficially within the limits of an acceptable risk. However, the manufacturers are of the opinion that lowering risk implies either decreased output or increased input resulting from the substitution for the pesticide inputs (Morrison Paul *et al.*, 2002). Thus, policies aiming at reducing the risks associated with the use of pesticides will impose costs on the agricultural community, which in turn will affect agricultural commodity prices (Morrison Paul *et al.*, 2002). Substantive costs may be imposed on the agricultural sector by the requirements to reduce environmental risk deriving from pesticide use.

International codes, treaties, conventions, commissions and advisory bodies play an important role in plant protection and pesticide management. Through the ratification of international conventions, governments accept obligations to incorporate them into national policies. The Rotterdam Convention, the Stockholm Convention, the International Code of Conduct on Pesticide Management, the Strategic Approach to International Chemicals Management (SAICM) and the Joint Meeting on Pesticide Residues (JMPR) are the most relevant international bodies for making policies and instruments with regard to reducing the risk associated with pesticide use (Eyhorn *et al.*, 2015).

The European Union (EU) developed a 'Thematic Strategy on Sustainable Use of Pesticides' keeping in view the impacts of pesticide use on human health and the environment (Commission of the European Communities, 2006). Moreover, alternative crop management systems have also been developed to minimize the negative effects of farming on the environment and on human health. A conference report submitted by Stenrød *et al.* (2016) described a need for close collaboration across country borders in Europe to be imposed within the three pesticide authorization zones (designated the north, central and south zones), as well as a need to revise the legal framework for authorization of the use of plant protection products and pesticides within the European Union/European Economic Area (EU/EEA; Regulation EC 1107/2009, Directive 2009/128/EC).

The principles of zonal evaluation and mutual recognition embedded in Regulation EC 1107/2009 concerning marketing of plant protection products are intended to reduce the approval times for pesticides. The 'Guidance Document on Work Sharing in the Northern Zone in the Authorization of Plant Protection Products' addresses the complexities of pesticide modelling in different climatic zones within Europe. Moreover, as per Directive 2009/128/EC, targeted and country-specific pesticide monitoring programmes considering the impact of weather conditions on the risk of leaching and runoff of pesticides are a prerequisite to ensure sustainable use of pesticides. Nordic soil and climate conditions have been reported as challenging due to slow degradation and the risk of mobilization of sorbed pesticides in winter/spring caused by the freezing/thawing of soil (Laitinen *et al.*, 2006; 2009; Siimes *et al.*, 2006; Stenrød *et al.*, 2008; Almvik *et al.*, 2014), leading to loss of pesticides during the winter period. The Nordic-Baltic Pesticide Fate Workshop highlighted the many possibilities within reach, and emphasized the need to improve and harmonize fate and risk assessment of pesticides in the northern zone using information derived from knowledge exchange among researchers.

Under California law (Statutes of 1969, chapter 1169), the Department of Pesticide Regulation (DPR, n.d.) must 'eliminate from use' any pesticide that 'endangers the agricultural and non-agricultural environment, is not beneficial for the purposes, for which it is sold, or is misrepresented'. To do this, the law requires the department to have 'an ordinary program for the continuous evaluation' of registered pesticides. DPR uses various tools to evaluate pesticide products to determine what risks they pose and whether changes to the use or proposed use are necessary. The human health risk assessment is the likelihood of adverse health effects in humans who may be exposed to pesticides now or in the future. Exposures may be in air, water or food, at home or in the workplace. Risk assessments are often the driving force behind new regulations or other use restrictions. The DPR will not register a pesticide if satisfactory controls cannot be put into place to avoid harmful exposures. It can cancel its use if it is already registered. The pesticide risk assessment takes into account the health problems to people caused by exposure to pesticides. It may also address an exposure level, if any, below which risk to humans is negligible.

The persistence of a pesticide plays an important role in risk assessment. It depends on its physical and chemical properties (partition coefficients, degradation rates, deposition rates) and the characteristics of the environment. Climate characteristics also play a role in persistence. Studies in the Arctic have shown that insecticides and herbicides persist three to eight times longer in cold climates than in temperate ones (Health Canada, 1998). In order to reduce the adverse effects of already existing pesticides on human health and the environment, new molecules with novel modes of action with improved safety profiles and safer formulations (e.g. microcapsule suspensions) are required. In addition, the use of appropriate and well-maintained spraying equipment along with taking all precautions that are required at all stages of pesticide handling could minimize human exposure to pesticides and their potential adverse effects on the environment (Damalas and Eleftherohorinos, 2011).

13.2 Registration of a Pesticide

To register a pesticide, the manufacturer has to submit the research data, which are reviewed by the registration authority, followed by their decision either to register the pesticide, based on the merits of the submitted data, or to deny registration. The decisions of the registration authority hinge on a benefit-to-risk analysis of the required data. Therefore, it is essential that all steps in the registration process are transparent, based on sound and published criteria and guidance documents, with full information shared with the applicant on the outcomes of the various steps in the registration procedure (FAO, 2002). Also, the registration authority ensures that each registered pesticide continues to meet the highest standards of safety to protect human health and the environment, as these standards are becoming stricter over the years with regard to our ability to evaluate the potential effects of pesticides. Within this context, older pesticides are being reviewed to ensure that they meet current scientific and regulatory standards. This process, called reregistration, considers the human health and ecological effects of pesticides and results in actions to reduce risks that are of concern.

Indeed, very drastic changes have occurred in the list of legally marketed pesticides over the past years in the EU as a result of the EU legislation on marketed pesticides, which was enacted in 1993 (with Directive 91/414/EEC) and lasted effectively until December 2008. During this period, approximately 704 active substances were banned (Karabelas *et al.*, 2009). Also, the United States Environmental Protection Agency (US EPA) has completed several individual pesticide reregistration and tolerance reassessment decisions, which has improved food safety, human health and environmental protection in the US (US EPA, 2009). The registration process for a pesticide usually requires the manufacturer (registrant) to submit research data on toxicity, bioefficacy and residues. The data have to be generated from multi-locational trials. The conducted scientific tests define the product chemistry, risks to humans and domestic animals, the environmental fate of the pesticide and the impact on non-target organisms (WHO, 2010; FAO, 2002). Data on all relevant aspects of the product during its full life cycle are required to support an application for registration. The data should include the physical and chemical properties of the active ingredient and formulated product, analytical methods, human and environmental toxicity, proposed label and uses, safety data sheets, efficacy for the intended use as well as residues resulting from the use of the pesticide molecule, container management and disposal of the same. Generation of such data for a single compound may take several years and cost millions of dollars.

In addition to the above, the data should also include ecological risk assessments to determine the risks, if any, posed by a pesticide and whether changes to the proposed use(s) of the product are necessary to protect human health, wildlife and the environment. All the data on a pesticide product are scrutinized by the registration authority to evaluate the environmental risks. If there is any likelihood of hazard to wildlife or any phytotoxicity to non-target plants, the authority may require additional testing and extra data, or may not allow its use, or permit the application of the pesticide under restricted use (Damalas and Eleftherohorinos, 2011).

13.3 Management of Harmful and Toxic Chemicals

Harmful and toxic chemicals are divided into 'supervised chemicals' and 'toxic chemicals' according to the extent of their harmfulness (Meng *et al.*, 2009). Toxic chemicals, including banned and severely restricted chemicals, will cause harmful effects to personal health and the environment by means of environmental accumulation, bioaccumulation, biological transformation or chemical reactions, or are seriously harmful to humans. Banned chemicals are those completely prohibited for all purposes due to their harmful effects to humans and the environment. Strictly limited chemicals are those that are prohibited because of their harmful effects to health and the environment, unless their use is authorized for certain special purpose under special conditions. In 1976 the American government promulgated the 'United States Act on the Control of Toxic Substances', which was revised 10 times between 1981 and 1989. In October 1973, Japan promulgated the legal serial No. 117 'Japanese Act for Controlling the Examination and Production of Chemical Substances'. In April 1992, the council of the European Economic Community (EEC) promulgated the Seventh Revised (92/32/EEC) Laws, and Administrative Regulations (67/548/EEC) concerning the common classification, packaging and labelling of dangerous substances. In 1989 the United Nations Environment Programme (UNEP) revised their London Guidelines on Information Exchange on Chemicals in International Trade.

Self-poisoning with agricultural pesticides is a major contributor to the global burden of suicide (WHO, 2014). Pesticide self-poisoning is particularly prevalent in South Asia, Southeast Asia and China. China is the world's most populous country, and previously had relatively high rates (62%) of suicides due to pesticide poisoning (Phillips *et al.*, 2002). An estimate of 11.3% of pesticide suicides, out of all the suicides in the Southeast Asia Region (SEARO), was recorded, contributing to approximately 35,500 suicides annually. The World Health Organization (WHO) estimates 1–5 million cases of pesticide poisonings occur each year, resulting in several thousands of fatalities, including in children. The number of fatal pesticide self-poisonings may be influenced by changes over time in treatment capacity, toxicity, formulations, packaging and storage, as well as absolute sales of pesticides (Mew *et al.*, 2017). India, for example, has experienced marked fluctuations in insecticide use, beginning with a steady decline from more than 51,000 tonnes (t) of active ingredients in 1992 to 3000 t in 2008, followed by an increase to almost 21,000 t by 2010 (FAO, 2015). Globally, the pesticide industry is forecast to grow annually by approximately 6% over the next few years with the highest growth rates occurring in South America and Asia (Reportlinker, 2012; Research and Markets, 2015). In 2000, Brazil's Ministry of Health reported an approximate 300,000 poisonings a year, following 5000 deaths from agricultural pesticides; the cost of treatment and lost work was estimated at US$540 million. The Philippines has estimated that health costs for farmers exposed to pesticides are 61% higher than for those not exposed, while Sri Lanka estimates health costs to farmers from pesticide exposure are equivalent to 10 weeks' income (UNDP, 2012).

In many developing countries, children are exposed to chemicals through agricultural work. In 2002, the International Institute of Tropical Agriculture estimated that over 150,000 children apply pesticides in West African cocoa production. Half of Cambodian farmers allow their children to spray crops, as per the survey report by the United Nations Food and Agriculture Organization (FAO). Improper labelling and storage of chemicals in the home are significant causes of poisoning, including in young children, particularly in developing countries. The Intergovernmental Forum on Chemical Safety (IFCS) in 2003 estimated that each year there are 3 million acute poisonings worldwide, with more than 200,000 fatalities.

Numerous highly hazardous pesticides, such as paraquat, are produced in and exported from countries that do not allow their use. This situation is intensified by the lack of resources for prevention and control of pesticides in developing countries and lack of legislation and inspection by governments. Overall, this factor further contributes to the continued impact of pesticides on children's health and well-being. Many of the pesticides banned in developed countries are still in use in developing countries. Table 13.1 provides information on 21 pesticides highly hazardous to children that are still in use in many Asia-Pacific countries.

Inappropriate use of chemicals can increase the costs to poor farmers: when pesticides are used inappropriately (e.g. where pest resistance exists or is created by killing off predators of pests), they can lower crop output. There are many documented and reported instances of small farmers paying from 30–50% of their total cash outlay for agrochemicals. Pesticide and fertilizer runoff contributes to lowered productivity of freshwater and marine fisheries, which are the main or significant sources of protein in many developing countries. Good Agriculture Practice (GAP) could be the means to decrease the negative influence of farming on the environment and to prevent the impoverishment and irrational use of natural resources (i.e. soil, water, plants, animals and landscape) caused by unnecessary or inappropriate farming practices, including with respect to the use of chemicals. For instance, GAP helps reduce improper use of pesticides that results in acute and chronic poisonings of farmworkers, which imposes extensive health and lost productivity-related costs on our society. GAP is also recommended for agricultural products to avoid any barriers in international markets (UNDP, 2012).

13.4 Application of Pesticides

The presence of individual or mixtures of chemical pesticides in the working environment creates the possibility of occupational exposure for human beings. Agriculture workers who mix, load, transport and apply formulated pesticides are normally considered to receive the greatest exposure because of the nature of their work and are therefore at the highest risk of possible acute intoxication (Fenske and Day, 2005).

The application of a pesticide product at the right time with the right dosage at the right stage of crop growth, targeting the right pest may not be hazardous to the environment. The optimization of pesticide use through adjusted dosage according to pest population density is economical. However, the dosage of active ingredients is not closely related to environmental activity, while environmental friendly and innovative compounds are often more expensive than obsolete, hazardous ones.

The use of appropriate and well-maintained spraying equipment, taking all the precautions required at all stages of pesticide handling, could reduce exposure to pesticides. The overall optimization of pesticide handling, during preparation of formulated pesticide solution, application, storage, transportation and disposal, acting strictly according to regulations and also considering public concerns about pesticide residues in food and drinking water could contribute to reduction of the adverse effects of pesticides on human health and the environment. Special attention is needed to avoid accidental spills of chemicals, leakages or faulty spraying equipment. Workers need to follow the instructions on how to use the pesticides and, particularly, the basic safety guidelines on the use of personal protective equipment (PPE) and fundamental sanitation practices, such as washing hands after pesticide handling or before eating.

Particular attention needs to be paid to the following points while using/handling pesticides:

- Pesticides in liquid form are prone to splashing and occasionally spillage,

Table 13.1. Pesticides highly hazardous to children still in use in the Asia-Pacific region. (From PAN, 2017.)

Pesticide	Type	Primary crops/use	Hazards to children	No. of countries where banned[a]
Atrazine	Herbicide	Maize, soy, sorghum, sugarcane	Birth defects, cancer, endocrine disruption, immunotoxicant	37
Carbaryl	Insecticide	Tomatoes, aubergines, olives, oranges, apples	Birth defects, cancer, endocrine disruption, developmental toxicant, neurotoxicant, immunotoxicant	32
Chlorothalonil	Fungicide	Potatoes, groundnuts, tomatoes	Cancer, endocrine disruption, autoimmune disorders and developmental effects	2
Chlorpyrifos	Insecticide	Cotton, maize, oranges, bananas, apples, vegetables	Acute poisoning, birth defects, cancer, endocrine disruption, neurotoxicant, autoimmune disorders and predisposal to obesity and diabetes	1
Cypermethrin	Insecticide	Onions, garlic, lettuce, broccoli, cereals/grains, oilseeds, fruits	Acute poisoning, cancer, endocrine disruption, behavioural effects and delayed mental development, Parkinson's disease later in life	0
Dichlorodiphenyltrichloroethane (DDT)	Insecticide	Mosquito control	Endocrine disruption, neurotoxicant, predisposal to obesity and diabetes	68
Deltamethrin	Insecticide	Carrots, maize, rice, spinach, wheat	Cancer, endocrine disruption, neurotoxicant, immunotoxicant	0
Diazinon	Insecticide	Chinese kale, tomatoes, spinach, apples, peaches	Acute poisoning, cancer, developmental toxicant, neurotoxicant, endocrine disruption, predisposition to diabetes and Parkinson's disease	29
Dichlorvos	Insecticide	Beans, brassica seedlings, structural and commodity fumigation, poultry houses	Acute poisoning, cancer, neurotoxicant, endocrine disruption, immunotoxicant, predisposition to diabetes and Parkinson's disease	30
Lambda-cyhalothrin	Insecticide	Hay, pistachios, rice, lettuce, soy, wheat	Acute poisoning, cancer, endocrine disruption, neurotoxicant	28[b]
Malathion	Insecticide	Rice, mangoes, aubergines, lettuce	Acute poisoning, birth defects, cancer, endocrine disruption, neurotoxicant, predisposition to attention deficit hyperactivity disorder (ADHD), diabetes and obesity	1
Mancozeb	Fungicide	Potatoes, bananas, lettuce, Asian pears	Acute poisoning, allergic sensitization, birth defects, cancer, developmental toxicant, endocrine disruption	1
Maneb	Fungicide	Potatoes, bananas, lettuce, broccoli	Acute poisoning, behavioural effects, birth defects, cancer, developmental toxicant, endocrine disruption, immunotoxicant, predisposition to Parkinson's disease	1

continued

Table 13.1. Continued.

Pesticide	Type	Primary crops/use	Hazards to children	No. of countries where banned[a]
Methamidophos	Insecticide	Cotton, rice, citrus, maize, grapes, soybeans, tobacco, vegetables, hops, peaches, bananas, pineapples	Acute poisoning, behavioural effects, death, developmental toxicant, neurotoxicant	47
Methyl parathion	Insecticide	Walnuts, potatoes, grapes	Neurotoxicant, endocrine disruption	26
Monocrotophos	Insecticide	Cotton, rice, pulses, groundnuts, tomatoes, aubergines, mangoes, grapes, chillies, cardamom, coconuts, oil palms, coffee, tea, castor, citrus, olives, maize, sorghum, sugarcane, sugarbeet, peas, potatoes, soybeans, cabbages, mustard, onions, peppers, ornamentals, tobacco	Birth defects, cancer, endocrine disruption, neurotoxicant, possible immunotoxicant	57
Paraquat	Herbicide	Cotton, oil palms, bananas, grapes, cereals, pulses, oil seeds, vegetables	Acute poisoning, death, endocrine disruption, immunological effects, neurotoxicant, implicated in diabetes	35
Parathion	Insecticide	Cereals, fruit, nuts, vines, vegetables, ornamentals, cotton, field crops	Acute poisoning, death, birth defects, cancer, neurotoxicant, immunotoxicant, predisposition to diabetes and obesity	26
Permethrin	Insecticide	Pistachios, lettuce, cotton, wheat, maize, alfalfa, vector control	Cancer, endocrine disruption, neurotoxicant, immunological effects	29
Propoxur	Insecticide	Structural, landscape sugarcane, cocoa, grapes, maize, rice, vegetables, cotton, alfalfa, forestry, ornamentals	Acute poisoning, cancer, developmental toxicant, endocrine disruption, immunosuppressant	29
Glyphosate	Herbicide	Rice, soy, maize, cotton, canola, oil palms	Birth defects, cancer, endocrine disruption, immunotoxicant, kidney damage, implicated in Parkinson's disease	1

See http://pan-international.org/resources.

[a]The PAN CL is not complete, as many countries do not publish lists of banned pesticides, and/or do not notify the Secretariat of the Rotterdam Convention, which is the only international body that keeps track of such bans.

[b]Not banned in any country, but is not approved in the EU.

resulting in direct or indirect skin contact through clothing contamination.

- Solids may generate dust while being loaded into the application equipment, resulting in exposure to the face and the eyes and also respiratory hazards.
- Packaging of pesticide products can also affect potential exposure – the size of cans, bottles or other liquid containers may affect the potential for spillage and splashing.
- Adjuvant chemicals (added in pesticide formulations to enhance their biological activity and to facilitate application and reaching target species) may also show toxicity (Surgan *et al.*, 2010) thus contributing to the overall effect of exposure to a commercial pesticide product.
- The volatility of the pesticide may be affected by the air temperature, humidity, the perspiration rate of the human body and the use of PPE by the users (Gomes *et al.*, 1999; Fenske and Day, 2005; Jindal *et al.*, 2007; Gil *et al.*, 2008). Wind velocity increases spray drift and is directly proportional to the distance the pesticide moves. Low relative humidity and high temperature cause more rapid evaporation of spray droplets between the spray nozzle and the target.
- Mixing and spraying of pesticide formulation during windy conditions can increase exposure levels.
- The frequency and duration of pesticide handling both on a seasonal and lifetime basis affect exposure levels (Fenske and Day, 2005).
- Consuming food and water contaminated with pesticides, living close to workplaces that use pesticides and bringing contaminated articles (clothes) home lead to substantial exposure to pesticides (Davis *et al.*, 1992; Jaga and Dharmani, 2003). In addition, delayed exposure can occur through inhalation of residual air concentrations or exposure to residues found on surfaces, clothing, bedding, food, dust, discarded pesticide containers or application equipment (Davis *et al.*, 1992).
- Exposure is likely to occur owing to pesticide spills, improper use or poor storage as a result of use without reading or taking account of the pesticide label. Pesticide mishandling – such as transferring the products from their original packages into household containers – and also lack of compliance with instructions of the label can also lead to exposure (Jaga and Dharmani, 2003).

There is, therefore, an imperative need to consider the adverse effects linked to the use of pesticides in agriculture, and for the use of criteria to select pesticides that are effective, cost efficient and safe for the operator and the environment (Hornsby *et al.*, 1993; Reus *et al.*, 2002; Bockstaller *et al.*, 2009).

The introduction of integrated pest management (IPM) systems would contribute to a significant reduction in the impact of pesticides on human health and the environment without affecting crop productivity or increasing the probability of crop losses (Burger *et al.*, 2008; Mariyono, 2008; Nwilene *et al.*, 2008), leading to a promising and sufficient supply of safe food production within a viable agricultural production system. New low-dosage compounds with novel modes of action and improved safety profiles are being developed, which are less persistent and target-specific (Bhattacharyya *et al.*, 2009). There has been a tremendous change in the assessment of the behaviour of these chemicals in the environment, the residues in crop plants, and of their potential toxicity to humans and the environment (Neumann, 1997; Urech, 1999; Müller, 2002). Moreover, these new agrochemicals in combination with the appropriate measures taken for safer and more effective pesticide application make chemical crop protection one of the most well-established technologies in agriculture. It seems likely that it will continue to play an important role in agribusiness in spite of the rapid emergence of novel biotechnological solutions (Neumann, 1997; Smith *et al.*, 2008).

13.5 Regulation of Pesticides

Pesticide regulation in the US is focused in large part on the carcinogenic risks posed by pesticide residues on food, as determined by laboratory animal tests (Cash *et al.*, 2002). Under the Federal Food, Drug, and Cosmetic Act (FFDCA), US EPA is responsible for setting a pesticide residue tolerance level for food crops and products. For new products, US EPA either sets a tolerance and registers the pesticide under the Federal Insecticide, Fungicide, and Rodenticide Act (FIFRA), or denies the registration request, effectively banning the substance from commercial use. Similarly, US EPA can ban substances currently in use if new toxicological data suggest that the pesticide may pose a significant health risk.

With regard to spray and dust drift label statements for pesticide products, US EPA has drafted a notice to issue new labelling guidance for the manufacturers, producers, formulators and registrants of pesticide products (US EPA, 2007). This will provide improved and more consistent product label statements for controlling pesticide drift in order to be protective of human health and the environment. The notice indicates US EPA's position on drift, a rationale for the label statements and an implementation plan.

13.5.1 Plant protection-based regulation of pesticides

The best method for pesticide regulation is to enforce integrated crop management (ICM) strategies, including guidelines to be used by farmers for safeguarding the environment (Frangenberg, 2000; Baker *et al.*, 2002; Tsakiris *et al.*, 2004; Burger *et al.*, 2008; Nwilene *et al.*, 2008). ICM encourages the use of biological control, cultural or physical measures, varieties resistant against insects and fungi, keeping the pest or weed population below its economic injury level and minimizing pesticide impacts on other components of the agro-ecosystem (Kogan, 1998; Way and Van Emden, 2000). Pesticides are used in IPM programmes as chemical components under ICM (Mariyono, 2008; Nwilene *et al.*, 2008; Chandler *et al.*, 2008). The pesticides selected in IPM programmes should be user and environmentally friendly, and economically viable. They should have high selectivity, optimal residual effect, good plant tolerance and low risk of resistance (Palacios, 2010).

The crops must be protected through a multi-level approach that increases heterogeneity of the landscape so as to provide habitat for pollinators and natural pest control species. This functional biodiversity can be increased through active vegetation management. Diversity of crop types and cultivars, rotation and fallowing increases the fertility of soils and increases resistance to pests. Natural biocontrol agents such as beneficial bacteria, viruses, insects and nematodes have been used successfully in improving crop protection (Forster *et al.* 2013).

13.5.2 Health-based regulation of pesticides

In 1910, the Insecticide Act prohibited the sale of fraudulently labelled pesticides (Brown *et al.*, 2001). Later on, it was decided to ban a pesticide on the basis of it presenting a potential hazard to the public. Since 1972, protecting farmworker and consumer safety has been the primary focus of pesticide regulation, as carried out by the US Food and Drug Administration (FDA) and the US EPA. Under the FFDCA, US EPA is responsible for setting a pesticide residue tolerance level for food crops and products. For new products, US EPA either sets a tolerance and registers the pesticide under FIFRA, or denies the registration request, effectively banning the substance from commercial use. Similarly, US EPA can ban substances currently in use if new toxicological data suggest that the pesticide may pose a significant health risk.

A 'Guide to Pesticide Regulation in California' has been published by the Office of

Environmental Health Hazard Assessment (OEHHA), for 'Assessing Pesticide Risks to Human Health'. In 1984, the Birth Defect Prevention Act (BDPA, chapter 669, SB 950) was passed. The law mandated that registrants of pesticides registered before 1984 bring health effects data on their chemicals up to current specific standards. It also required that the DPR should not register new active ingredients without a full complement of health effects studies. The required studies (primarily done on experimental animals) were chronic toxicity, mutagenicity, neurotoxicity, oncogenicity, reproductive effects and teratology. The BDPA required DPR to use these and other data to determine if a pesticide would cause human health problems. If continued use of a pesticide presents a significant health hazard that cannot be adequately mitigated, DPR is required to cancel the registration of products containing that active ingredient. As of late 2016, there remain only 57 active ingredients for which DPR is making determinations as to whether more studies are needed to close data gaps.

Before registration, DPR conducts a pre-market evaluation of pesticide products based on standards used by US EPA and studies required by California statutes to decide if the product can be used safely. Both pre-market evaluations and full risk assessments are based on a prescribed set of scientific data provided by registrants as well as information from available scientific literature and other sources. The data include exposure monitoring studies conducted by DPR's Worker Health and Safety Branch and air and water monitoring studies conducted by the Environmental Monitoring Branch. The application for product registration must provide all information needed to support the different uses proposed. Only products with a database that includes all required studies are allowed to progress through evaluation. In limited instances, some chronic health effects data may be waived in consultation with OEHHA.

The FDA, US Department of Agriculture (USDA) and DPR all conduct residue analysis of samples of fresh produce, collected randomly. The FDA and USDA also test for residues in cooked and processed foods. Because these samples are analysed closer to the point of consumption, the resulting data can characterize pesticide residues in food to more closely approximate real-world exposures. For more accurate exposure estimates, DPR may rely on field trial data, while USDA conducts nationwide surveys every several years to estimate the kinds and amount of food that people eat. Food consumption is reported for people of different racial and ethnic groups, age groups, genders, geographical regions and seasons. The consumption rate is expressed in terms of body weight and accounts for a potentially higher intake by children, as compared to adults, per pound of body weight. Dietary exposure to a pesticide is based on the estimated food consumption coupled with the estimated pesticide residue levels in the food. These dietary exposure estimates are combined with the toxicity data to assess the risk to various population subgroups, including infants and children, from the exposure to pesticide residues in food. Both chronic and acute dietary exposures are generally considered. Chronic exposure occurs over a long period; therefore, it is calculated using average consumption and residue values. In contrast, acute exposure considers the highest single (acute) exposure. It is calculated using individual consumption data. The resulting information on dietary risk is then included in an overall assessment of the risk posed by the pesticide for all uses.

UNDP (2012) has also proposed the Globally Harmonized System of Classification and Labeling of Chemicals (GHS), which seeks to make broadly available information on health, physical and environmental hazards and toxicity (e.g. safety data sheets for workers) to enhance the protection of human health and the environment during their handling, transport and use, and provide for greater consistency in regulatory requirements (without lowering the level of protection currently afforded by such systems) to facilitate trade in chemicals and reduce the need for testing and evaluation. It may be more difficult to export different products without complying with GHS. A proper use of pesticides in accordance to

GHS may also facilitate the export of agricultural products. The EU has also intended to contribute to the global harmonization of criteria for classification and labelling, at UN level, through implementation of the internationally agreed GHS criteria into EU law (Kienzler *et al.*, 2017). Each substance and/or mixture is classified into hazard classes, depending on the nature of the hazard, and into a hazard category, indicating the severity of the hazard. The detailed human health and environmental hazard classes are depicted in Tables 13.2 and 13.3, respectively, along with their corresponding classification principles.

13.5.3 Relevant legislation and policy

WHO gender health issues

Concerning gender health issues, the occupational health policy and legislation consider the treatment of sex differences, and the methods for handling discrimination, including sexual harassment, as the key areas. These include protection of pregnant and nursing workers exposed to hazardous working conditions as well as provisions concerning access to dangerous work such as night work. Some national legislation has incorporated a 1948 International Labour Organization (ILO) convention prohibiting night work for women (Act No. 11-92/ADP, 1992). While these provisions are now incompatible with European equality law, other countries may still prevent women from performing night work and other work perceived as inappropriate. International legislation governing discrimination in the workplace aims to prevent both sexual harassment and discrimination, including discriminatory hiring practices.

Aside from legislation overtly designed to apply to working conditions of women, it is also important to analyse seemingly gender neutral legislation aimed to prevent or to compensate for occupational injury and disease. Even in countries where equality is guaranteed by law, application of occupational health and safety legislation may have discriminatory effects. Swedish and Canadian studies reveal that women and men are often offered different rehabilitation measures for similar work-related health problems. Men are more often offered training, access to a wider variety of new jobs and are offered more help in the home, while women receive rehabilitation benefits for a shorter time (Messing and Östlin, 2002). In many countries, claims for workers' compensation benefits for psychological problems or musculo-skeletal disorders (more common among women) are sometimes excluded from the purview of the law, so that systemic discrimination may be at work even if the legislation appears to be gender neutral. When prevention priorities are determined by compensation costs, women are less likely to benefit from protective legislation. When women's claims for industrial disease are greeted with scepticism, union attention and organized support groups can be very helpful. Policy analysis should also take into account gender differences in precarious and non-standard employment. Specific health risks are associated with specific types of precarious or contingent work. Home-based work presents very different challenges from those presented by temporary or part-time work. Policy applicable to specific types of contingent work should be scrutinized with regard to gendered effects.

As per OECD workshop (2001) recommendations, governments should analyse and make the costs and benefits of different pest control systems transparent. It is also advised to consider the use of economic instruments as a means to promote risk reduction, and to carefully evaluate the impact of such instruments on different sectors at different scales. The alternative solutions to today's practices must be economically viable. The economic models are no doubt an important element of the decision making process; however, there is also need for motivation, incentives, perceptions, values and knowledge. Farmers could be convinced through demonstrations and field trials to reduce the amount of pesticides. Case studies are valuable to identify motives and incentives. There should be clear communication with farmers for

Table 13.2. Hazard class and category for human health hazard classification under the CLP[a] regulation. (From Kienzler *et al.*, 2017.)

Hazard class	Category				Classification principles
Acute toxicity	Category 1	Category 2	Category 3	Category 4	• Priority is given to the use of data on the mixture, if available • If no data, use of bridging principle from similar mixtures, if available • If no similar mixture, calculation (CA) if data available for all ingredients for estimating the toxicity, considering ingredients present at a concentration of ≥1% (unless there are indications that the compounds are relevant for classification even at lower concentrations, i.e. substance classified as acute toxic, category 1–3, for which a fraction of 0.1% is to be considered for the mixture classification) • If no data available for all ingredients, use other data available (extrapolation, evidence, QSARs, etc.) to estimate conversion values for classification and convey hazards of the known ingredients
Acute oral toxicity (mg/kg/bw)	ATE ≤ 5	5 ≤ ATE ≤ 50	50 ≤ ATE ≤ 300	300 ≤ ATE ≤ 2000	
Acute dermal toxicity (mg/kg/bw)	ATE ≤ 50	50 ≤ ATE ≤ 200	200 ≤ ATE ≤ 1000	1000 ≤ ATE ≤ 2000	
Acute inhalation toxicity: (gases (ppm V), vapours (mg/l), dust and mists (mg/l))	ATE ≤ 100 ATE ≤ 0.5 ATE ≤ 0.05	100 ≤ ATE ≤ 500 0.5 ≤ ATE ≤ 2.0 0.05 ≤ ATE ≤ 0.5	500 ≤ ATE ≤ 2500 2.0 ≤ ATE ≤ 10.0 0.5 ≤ ATE ≤ 1.0	2500 ≤ ATE ≤ 20,000 10.0 ≤ ATE ≤ 20.0 1.0 ≤ ATE ≤ 5.0	
Skin corrosion and irritation	Category 1: corrosive (three subcategories) Corrosive in more than one of three animals: 1A: If response is noted following up to 3 min exposure and up to 1 h observation 1B: If response is noted between 3 min and 1 h of exposure and observation up to 14 days 1C: If response is noted between 1 h and 4 h of exposure and observation up to 14 days		Category 2: irritant (one single category) • At least two of three animals have a mean score between 2.3 and 4 for erythema/eschar or for oedema, or • Inflammation that persists to the end of the observation periods (14 days) in a least two animals, or • If there is a pronounced variability of response among animals, with very definite positive effects in a single animal but less than the criteria above		The mixture will be classified using the criteria for substance • Priority is given to data on the full mixture; moreover, the use of existing alternative tests and of a tiered weight of evidence strategy is encountered in order to avoid unnecessary animal testing • If no data on the full mixture are available, bridging principles are applied, including the use of a summation method for estimating the overall potency of the mixture with cut-off values. The 'Theory of additivity' is mentioned, but no specific guidelines on their application are given
Serious damage to the eye and eye irritation	Category 1: irreversible effect on the eye (three subcategories) If the applicant produces: • In at least one animal, effect on the cornea, iris or conjunctiva that is not expected to reverse or has not fully reversed within 21 days and/or • In at least two of three animals a positive response of corneal opacity ≥3 and/or iritis >1.5 (mean scores following grading at 24, 48 and 72 h)		Category 1: irritating to eyes If the applicant produces: • In at least two of three animals a positive response of corneal opacity ≥1 and/or conjunctival redness ≥2 and/or conjunctival oedema ≥2 (mean scores following grading at 24, 48 and 72 h after application, and which fully reversed within 21 days)		Ingredients are considered relevant if present at a concentration of ≥1% unless there is a presumption that an ingredient present at a concentration of less than 1% can still be relevant for classifying the mixture). If strong acids and bases are present, the pH should be used as a classification criterion.

continued

Table 13.2. Continued.

Hazard class	Category		Classification principles
Respiratory or skin sensitization	Respiratory sensitizer category 1 • If there is evidence in humans that the substance can lead to specific respiratory hypersensitivity/ sensitization by skin contact in a substantial number of persons respectively, and/or • If there is a positive result from an appropriate animal test	Skin sensitizer category 1	• When reliable experimental data or good quality evidence from human experience are available for the mixture, these can be used to classify the mixture by weight of evidence. If not, use bridging principles on similar tested mixtures or individual ingredients • Use of generic concentration limits: should be categorized into this class if at least one of the components is classified into the hazard class 'respiratory or skin sensitization' and is present at >0.2% (gas) or >1% (solid/liquid).
Genotoxicity			
Germ cell mutagenicity	Category 1 Substances to be regarded as if they induce heritable mutations in the germ cells of humans. 1A: based on positive evidence from human epidemiological studies. 1B: based on positive result(s) from *in vivo* heritable germ cell mutagenicity tests in mammals; or from *in vivo* somatic cell mutagenicity tests in mammals and some evidence that the substance has potential to cause mutations to germ cells; or positive results from tests showing mutagenic effects in the germ cells of humans, without demonstration of transmission to progeny	Category 2 Substances that cause concern for humans owing to the possibility that they may induce heritable mutations in the germ cells of humans	• Priority is given to the relative content of components in the mixture that are themselves classified (summation method with cut-off criteria). An ingredient classified as a germ cell mutagen 1A or 1B and present at more than 0.1% triggers the classification of the mixture as germ cell mutagen 1A or 1B; an ingredient classified as germ cell mutagen 2 and present at more than 0.1% triggers the classification of the mixture as germ cell mutagen 2. Data on the mixture itself are only considered on a case-by-case basis when 'demonstrating effects that have not been established from the evaluation based on the individual ingredients' • The bridging principles are also considered, if applicable. CA is not applied for any classification purposes

Carcinogenicity	Known or presumed human carcinogens on the basis of epidemiological and/or animal data. 1A: known to have carcinogenic potential for humans, classification is largely based on human evidence, or 1B: presumed to have carcinogenic potential for humans, classification is largely based on animal evidence	Suspected human carcinogens classified on the basis of evidence obtained from human and/or animal studies, but which is not sufficiently convincing to place the substance in category 1A or 1B	• Priority is given to the relative content of components in the mixture that are themselves classified (summation method with cut-off criteria). An ingredient classified carcinogen category 1A or 1B and present at more than 0.1% triggers the classification of the mixture as carcinogen 1A or 1B; an ingredient classified carcinogen 2 and present at more than 0.1% triggers the classification of the mixture as carcinogen 2. Data on the mixture itself are only considered on a case-by-case basis when 'demonstrating effects that have not been established from the evaluation based on the individual ingredients' The bridging principles are also considered, if applicable. CA is not applied for any classification purposes
Reproductive toxicity	Known or presumed human reproductive toxicant (i.e. clear evidence of an adverse effect on sexual function and fertility or on development). In addition, effects on lactation are allocated to a separate hazard category. 1A: known human reproductive toxicant; classification is largely based on evidence from humans. 1B: presumed human reproductive toxicant; classification is largely based on data from animal studies	Suspected human reproductive toxicant classification based on some evidence from humans or experimental animals, possibly supplemented with other information, of an adverse effect on sexual function and fertility, or on development, and where the evidence is not sufficiently convincing to place the substance in category 1	• Priority is given to the relative content of components in the mixture that are themselves classified (summation method with cut-off criteria). An ingredient classified as reproductive toxicant category 1A, 1B, 2 or 'additional effect on or via lactation' and present at more than 0.3% triggers the classification of the mixture as reproductive toxicant category 1A, 1B, 2 or 'additional effect on or via lactation', respectively. Data on the mixture itself are only considered on a case-by-case basis when 'demonstrating effects that have not been established from the evaluation based on the individual ingredients'. The bridging principles are also considered, if applicable. CA is not applied for any classification purposes

continued

Table 13.2. Continued.

Hazard class	Category			Classification principles
Specific target organ toxicity				
	Category 1	Category 2	Cat 3	
Specific target organ toxicity – single dose	Substances that have produced significant toxicity in humans or that, on the basis of evidence from studies in experimental animals, can be presumed to have the potential to produce significant toxicity in humans following single exposure are classified on the basis of: (i) reliable and good quality evidence from human cases or epidemiological studies; or (ii) observations from appropriate studies in experimental animals in which significant and/or severe toxic effects of relevance to human health were produced at generally low exposure concentrations	Substances that can be presumed to have the potential to be harmful to human health following single exposure are classified on the basis of observations from appropriate studies in experimental animals in which significant toxic effects, of relevance to human health, were produced at generally moderate exposure concentrations	Transient target organ effects – this category only includes narcotic effects and respiratory tract irritation. These are target organ effects for which a substance does not meet the criteria to be classified in categories 1 or 2. These are effects that adversely alter human function for a short duration after exposure and from which humans may recover in a reasonable period without leaving significant alteration of structure or function	• Primary classification criterion is the data on the mixture itself, although specific care is advised by the guideline to ensure that 'the dose, duration, observation or analysis does not render the results inconclusive'. If no reliable data are available on the mixture itself and the bridging principles cannot be used, the mixture is classified for its specific organ toxicity: an ingredient classified as category 1 specific target organ toxicant and present at more than 10% triggers the classification of the mixture as category 1 specific target organ toxicant 1A; if present between 1 to 10%, it triggers the classification category 2. An ingredient classified as category 2 specific targot organ toxicant and present at more than 10% triggers the classification of the mixture as category 2 specific target organ toxicant. When extrapolating toxicity of a mixture that contains specific target organ toxicity – single exposure category 3 ingredient(s), a generic

Specific target organ toxicity – repeated dose	Category 1 Substances that have produced significant toxicity in humans or that, on the basis of evidence from studies in experimental animals, can be presumed to have the potential to produce significant toxicity in humans following repeated exposure. Classification on the basis of: reliable and good quality evidence from human cases or epidemiological studies; or observations from appropriate studies in experimental animals in which significant and/or severe toxic effects, of relevance to human health, were produced at generally low exposure concentrations	Category 2 Substances that can be presumed to have the potential to be harmful to human health following repeated exposure. Classification on the basis of observations from appropriate studies in experimental animals in which significant toxic effects, of relevance to human health, were produced at generally moderate exposure concentrations	concentration limit of 20% is appropriate; however, this concentration limit may be higher or lower depending on the category 3 ingredient(s) and that some effects such as respiratory tract irritation may not occur below a certain concentration, while other effects such as narcotic effects may occur below this 20% value, therefore expert judgement is required. CA is not applied for any classification purposes. Care should be exercised when toxicants affecting more than one organ system are combined that the potentiation or synergistic interactions are considered, because certain substances can cause target organ toxicity at <1% concentration when other ingredients in the mixture are known to potentiate its toxic effect
Aspiration hazards	Category 1: substances known to cause human aspiration toxicity hazards or to be regarded as if they cause human aspiration toxicity hazard. Classification based on: (i) reliable and good quality human evidence; or (ii) if it is a hydrocarbon and has a kinematic viscosity of 20.5 mm^2/s or less, measured at 40°C		The classification of a mixture into the hazard class 'aspiration hazard' follows the same outline: use of data on the whole mixture of interest, bridging principles and cut-off values for individual compounds

[a]CLP, classification, labelling and packaging.

Table 13.3. Hazard class and category for environmental hazard classification under the CLP regulation.

Hazard class	Category		Classification principles		
Acute aquatic toxicity	Category 1 Classification on the basis of acute aquatic toxicity data only (EC_{50} or LC_{50}) • A substance or a mixture (on the bases of a whole-testing approach) is classified as acute category 1 if its EC_{50} is ≤1 mg/l • Summation methods: if the mixture contains ≥25% of compounds classified into acute toxicity category 1, the mixture is also classified into category 1		• Priority is given to the data on the whole mixture (if available) • If no information available, the classification can be based on bridging principle • If this is not possible, the 'summation method' is applied; highly toxic compounds are given increased weight for the summation by the application of M-factors • If the mixture contains not yet classified compounds for which adequate toxicity data are available, CA is used for the assignment of an acute category to that portion of the mixture, which is then used in applying the summation method for the whole mixture If relevant compounds are present in the mixture that are not classified for their acute toxicity and for which no toxicity data are present the classification of the mixture shall be based on the assessable rest of the mixture		
Chronic aquatic toxicity	Category 1 If EC_{50} ≤1 mg/l and the substance is not rapidly biodegradable and/or the experimental determined BCF ≥500 (or log *Kow* ≥4) Summation methods: If Σ(comp. chronic category 1 × M) ≥25% M = multiplying factor, dependent on the LC_{50} of the component	Category 2 If 1 mg/l ≤ EC50 ≤ 10 mg/l and the substance is not rapidly biodegradable and/or the experimental determined BCF ≥500 (or log *Kow* ≥4), unless the chronic toxicity >1 mg/l Summation methods: If Σ[(M × 10 × comp. Chronic Cat 1) + Chronic Cat 2] ≥25%	Category 3 If is 10 mg/l ≤ EC50 ≤ 100 mg/l and the substance is not rapidly biodegradable and/or the experimental determined BCF ≥ 500 (or log *Kow* ≥4), unless the chronic toxicity >1 mg/l Summation methods: If Σ[(100 × M × comp. chronic category 1) + (10 × M × Chronic Cat 2)] ≥ 25%	Category 4 'Safety net' classification for use when the data available do not allow classification under the formal criteria but there are nevertheless some grounds for concern Σ[Chronic Cat 1+ Chronic Cat 2+ Chronic Cat 3+ Chronic Cat 4] ≥ 25%	Classification principles Classification of a substance into the chronic categories combines two types of information, i.e. acute aquatic toxicity data and environmental fate data (degradability and bioaccumulation data) Data on the chronic toxicity data (no observed effect concentration values) of the whole mixture and/or the classification information on the individual compounds are used for the classification of the mixture into chronic aquatic toxicity category 1–4

ATE, acute toxicity estimate (LD/EC_{50}); ppm V; parts per million per volume; CA, concentration addition; Exp, exposure; Obs, observation.

raising awareness to enhance compliance with regulatory measures and advice.

The human hormonal system is affected by the toxic properties of endocrine disruptor pesticides (Mnif *et al.*, 2011). Most risk assessment studies and some epidemiological studies have looked at the exposure and toxicology of a single compound. However, two other considerations must be included: the presence of pesticide byproducts and the cumulative exposure to pesticide multiresidue. Pesticide byproducts have been reported to exhibit greater harmful effects than their parent compounds (Kelce *et al.*, 1994; Blair *et al.*, 2000; Kim *et al.*, 2005; Küster *et al.*, 2007; Salazar-Arredondo *et al.*, 2008). Furthermore, the combined actions of pesticides also need to be addressed in the risk assessment process because mixtures of these substances may cause higher toxic effects than those expected from the single compounds (Larsen *et al.*, 2003). Finally, the impact of synthetic pesticides, due in particular to excessive use (including environmental pollution and implications to human health) have led to modifications in agricultural practices and various national and international regulations limiting their use. The alternative solutions that are safer and non-toxic to the environment and humans are: natural pesticides, such as botanical pesticides, microbial/biological agents and inorganic minerals. These are interesting alternatives; however, further studies are needed on the occurrence, fate and impact of such pesticides on the ecosystem and public health.

The EU, under Regulation No. 283/2013 (EC, 2013a), has asked for sufficient information on active ingredients of plant protection products to permit both 'a risk assessment of consumer exposure' and 'an estimation of the exposure to operators, workers, residents and bystanders' including 'where relevant, the cumulative exposure to more than one active substance'. The need to mention the possible presence of pesticide residues arising 'from sources other than current plant protection uses of active substances (for example use of active substances resulting in common metabolites, use as biocide or veterinary drug)' is also put forward as the necessity to take into account their aggregate exposure and the cumulative exposure to more than one active substance, where relevant (Kienzler *et al.*, 2017). In order to grant authorization to any plant protection products adequate maximum residue limits (MRLs) should be set as per Regulation (EC) No. 1107/2009. The regulation introduces a clear requirement that plant protection products and their residues 'shall not have harmful effect on human health, directly or through drinking water (taking into account substances resulting from water treatment), food, feed or air, taking into account known cumulative and synergistic effects where the scientific methods accepted by the authority to assess such effects are available'. The regulation states that PPP and their residues 'shall not have any unacceptable effect on the environment'. To achieve an adequate consideration of mixture toxicity in the environmental risk assessment of plant protection products for birds and mammals, for which toxicity data with formulated products or mixtures of active substances are not always available, especially for birds, a four-step approach for combined effects of simultaneous exposure to several active substances has been proposed (EFSA, 2009a). This approach does not refer to an assessment of formulation toxicity as such, but of the expected effects from exposure to a mixture of active substances (and possibly also toxic co-formulants) in the environment resulting from the use of the formulation.

Although there is no one acceptable risk level, US EPA is required to set tolerances that are 'safe', defined as 'a reasonable certainty that no harm will result from aggregate exposure to the pesticide chemical residue' (US EPA, 1999). In the absence of carcinogenic effects in laboratory animals, the tolerances are based upon the highest dose at which no toxic effects were observed in animal tests, which is then divided by a safety factor (usually 10 or 100) to determine an acceptable daily dose for humans (Cash *et al.*, 2002).

The registration process for each pesticide set requires the authorized culture, the dose, pre-harvest interval, and the MRL. The legal limit of the residue does not coincide with the toxicological limit and for this reason if the legal limit is exceeded it will not pose a serious risk to human health. The legal limit is determined from toxicological data establishing a lack of risk to human health (NOEL = No Observed Effect Level) commonly corrected by a safety factor of 100. Field residues of pesticides are affected by the environmental conditions (temperature, wind, rain, solar irradiance, etc.). Field residues, if below the corrected acceptable daily intake (ADI), are used to set the legal limit of the pesticide residue (Caboni and Cabras, 2010). Residue limits can vary between countries because of the different climatic conditions, leading to trade difficulties in different countries. When appropriate application methods of agrochemicals are not followed, pesticide residues can be above the MRLs established by the governments of each country, and the time for pesticide dissipation is also high.

Currently there is emphasis on harmonization of the MRLs of pesticides, especially by the EU. In Italy, a total of 137 pesticides are currently registered for use on grapes. Italy is one of the few countries that has set legal limits (MRLs) for 39 pesticides on grape and wine. In other countries, where there is a lack of a legal limit for processed foods, the amount of the raw food corresponding to processed food unit (e.g. 1.5 kg of grapes for 1 l of wine), and the incidence of technological process should be taken into account. Since each active ingredient has its particular behaviour, residue changes during the transformation process should be determined. In the absence of these data, the unique and safe reference is the MRL of the primary food.

Many countries have established regulatory levels for the control of chemical contaminants in food. This can be either an MRL in the case of an agricultural or veterinary chemical, or a maximum level (ML) for a contaminant or natural toxicant in the food supply. For example, the Australia New Zealand Food Standards Code (the Code) outlines the limits that relate to agricultural and veterinary residues or contaminants in food (Szabo *et al.*, 2010). Food containing chemicals at levels greater than the MRL or ML are viewed as breaching the Code and are subject to regulatory enforcement action at the border and within the nation.

Ensuring food safety risks are understood and managed is a joint government and industry responsibility in Australia and New Zealand. The onus on industry is to produce food that is safe and suitable, and on governments to regulate on a risk basis to ensure industry has mechanisms in place to produce safe and suitable food.

13.5.4 MRLs of pesticides

MRLs for pesticide residues are the maximum concentrations of residues to be legally permitted in or on a food. MRLs for pesticides may also be applicable to animal feeds. MRLs are set by the Codex Alimentarius Commission (CAC), acting as the risk manager. Draft MRLs for adoption by the CAC are elaborated by the relevant Codex committees, the Codex Committees on Pesticide Residues (CCPR) and the Residues of Veterinary Drugs in Food (CCRVDF), on the basis of scientific expert advice provided by the risk assessors – that is, JMPR and Joint Expert Committee on Food Additives (JECFA), respectively (Kienzler *et al.*, 2017). The scientific advice developed by JMPR and JECFA aims to provide MRLs for individual crops, plant and animal products, based on the results of scientific studies, so that these levels can be used by the relevant Codex committee to develop the draft MRLs, which may be adopted by the CAC. In 2005, FAO, the National Institute for Public Health and the Environment of the Netherlands (RIVM) and WHO held a workshop entitled 'Updating the Principles and Methods of Risk Assessment: MRLs for Pesticides and Veterinary Drugs' to update the Principles and Methods for the Risk Assessment of Chemicals in Food for the JECFA and the JMPR (FAO/WHO, 2006). The purpose was for recommending MRLs and to harmonize their approaches to the extent useful and possible with emphasis on the general principles applied by both scientific bodies (CAC, 2006).

The CAC was established by WHO and FAO to develop international food standards covering all world diets. It is not a regulatory body, and its standards are voluntary, including Codex MRLs, called CXLs. CXLs are very important for countries without a well-developed MRL system. Some countries use a Default MRL when no MRL has been established for the chemical–commodity combination; for example, the EU uses 0.01 p.p.m., Japan 0.01 p.p.m., Canada 0.1 p.p.m. and Korea 0.01 p.p.m. (in 2017) as Default MRL; www1.agric.gov.ab.ca/$Department/deptdocs.nsf/all/crop15631/$FILE/carol-saunders.pdf)

The MRL challenges for Canada are chlormequat chloride, a plant growth regulator used in wheat crop, registered in 2014; Quinclorac, a herbicide used in canola; crop desiccants in pulses (diquat, glyphosate, glufosinate, saflufenacil, flumixoazin and carfentrazone). No MRL has been established in 2015 or 2016.

The regulation (Regulation 396/2005) for MRLs of pesticides ensures that pesticide residues are not present in food and feed products 'at levels presenting an unacceptable risk to humans and, where relevant, to animals'. It focuses only on human risk assessment and environmental risk assessment is out of scope. The Regulation establishes the maximum quantities of pesticide residues permitted (or MRLs) in products of animal or vegetable origin intended for human or animal consumption. The MRLs included are specific to particular foodstuffs intended for human or animal consumption and, for products and/or pesticides for which no specific MRLs are set, a default value of 0.01 mg/kg applies (Article 18, paragraph 1(b)). Risk assessment is the responsibility of the European Food Safety Authority (EFSA) (Recital 6, Articles 10 and 24), which provides Reasoned Opinions on each intended new MRL, amendment or removal. As a consequence, support measures related to harmonized pesticide MRLs shall include 'studies and other measures necessary for the preparation and development of legislation and of technical guidelines on pesticide residues, aimed, in particular, at developing and using methods of assessing aggregate, cumulative and synergistic effects' (Article 36, § 1 (c)), and Commission decisions concerning the MRLs shall take account of 'the possible presence of pesticide residues arising from sources other than current plant protection uses of active substances, and their known cumulative and synergistic effects, when the methods to assess such effects are available' (Article 14, § 2 (b)).

The established procedures for setting MRLs on the basis of ADI values and food consumption patterns are focused on single substance assessments, although the Regulation explicitly addresses the need for carrying out 'further work to develop a methodology to take into account cumulative and synergistic effects' (Kienzler *et al.*, 2017, p. 17) and states that MRLs should be set in 'view of human exposure to combinations of active substances and their cumulative and possible aggregate and synergistic effects on human health' (Kienzler *et al.*, 2017, p. 41).

EFSA gives independent scientific advice to risk managers based on risk assessments. The European Commission and Member States take risk management decisions on regulatory issues, including approval of active substances and setting of legal limits for pesticide residues in food and feed (MRLs). EFSA's Pesticides Unit is responsible for the EU peer review of risk assessments of active substances used in plant protection products, in close cooperation with EU Member States. The risk assessment of active substances evaluates whether, when used correctly, these substances are likely to have any direct or indirect harmful effects on human or animal health – for example, through drinking water, food or feed – or on groundwater quality. In addition, the environmental risk assessment aims to evaluate the potential impact on non-target organisms (https://www.efsa.europa.eu/en/topics/topic/pesticides). The Pesticides Unit also gives scientific advice to the European Commission on possible risks related to the presence of pesticide residues in food treated with plant protection products and makes proposals regarding the setting of MRLs. In addition, the Unit is responsible for preparing the Annual Report on Pesticide Residues in the EU. Furthermore, the Pesticides Unit provides administrative and scientific support to the Panel on Plant Protection Products and their Residues (PPR) Panel.

EFSA's PPR Panel gives scientific advice on issues that cannot be resolved within the peer review of active substances, MRL applications/MRL reviews or when guidance is needed on more generic issues, commonly in the fields of toxicology, ecotoxicology, fate and behaviour, and the development of risk assessment practice. The PPR Panel and the Pesticides Unit have the task of developing and revising scientific methodologies, including guidance documents, for pesticide risk assessment. In this context, EFSA regularly outsources tasks to external organizations to assist with gathering scientific data and information or developing modelling tools. Stakeholder views on new guidance and methodologies are collected through public consultations. The guidance documents provide advice to applicants and Member States on how to conduct a risk assessment for a particular area in the context of the peer review of active substances or national authorizations of plant protection products.

13.5.5 Pesticide disposal

This programme is for the disposal of unwanted and outdated pesticides currently stored by agricultural producers, homeowners and golf courses. A similar programme was initiated by Virginia Department of Agriculture and Consumer Services Office of Pesticide Services, by giving details of the best methods for their disposal. In many countries obsolete pesticides are a problem. The FAO have developed a set of guidelines for disposal at www.fao.org/WAICENT/FAOINFO/AGRICULT/AGP/AGPP/Pesticid/Disposal/index_en.htm.

The consumption of pesticides increased substantially over time; for example, in the USA, the use of pesticides doubled from 1960 to 1980, but total use has since remained stable or fallen. Most pesticides are used in agriculture, but in 1999 about 74% of households in the USA were reported to use at least one pesticide in the home. Use has risen in developing countries and the fastest-growing markets in Africa, Asia, South and Central America and the eastern Mediterranean. There is high pesticide use on crops grown for export (Dr N. Besbelli, WHO, personal communication, July 2008) (Laborde, 2008). Although developing countries use only 25% of the pesticides produced worldwide, they experience 99% of the deaths. This is because use of pesticides tends to be more intense and unsafe, and regulatory, health and education systems are weaker in developing countries (Jeyaratnam, 1990; Kiely *et al.*, 2004; UNEP 2002; FAO Newsroom, 2002; AAPCEH, 2003).

13.5.6 Regulation in risk reduction for agriculture workers

Solomon *et al.* (2000) proposed an ecological risk assessment method by using probability distribution functions to independently describe exposure concentrations and toxicological responses of aquatic organisms to pesticide compounds. An environmental yardstick had been developed in The Netherlands as a practical indicator for farmers to select pesticides with the least environmental impacts (Reus and Leendertse, 2000). Eight indicators had been compared for the evaluation of pesticide environmental risk in Europe (Reus *et al.*, 2002). Most of these evaluations were conducted on the pesticide level and none was on the grower level.

The EU pesticides legislation has included a new element as a substitution principle for conducting comparative assessments for all uses of plant protection products that contain active substances with certain hazardous properties, the so-called candidates for substitution. Faust *et al.* (2014) reported a more detailed assessment of the workload from comparative risks assessments and proposed that development of an optimal strategy for efficiently dealing with the issue would need to consider these different types of assessment situations.

A new method for pest risk assessment and the identification and evaluation of risk-reducing options is currently in development by the EFSA Plant Health Panel (Gilioli *et al.*, 2017). The draft method has been tested on pests of concern to the EU. The proposed method can focus either on all the steps and substeps of the assessment process or on specific parts, if necessary. It is based on assessing changes in pest population abundance as the major driver of the impact on cultivated plants and on the environment. Responses to the likelihood and magnitude of factors contributing risk can be based on data or expert judgement. Crucially, the approach is quantitative, and it captures uncertainty through the provision by risk assessors of quantile estimates of the probability distributions for the assessed variables and parameters. The assessment is based on comparisons between different scenarios, and the method integrates risk-reducing options where they apply to a scenario. A strategy has been developed to communicate the results of the risk assessment in a clear, comparable and transparent way, with the aim of providing the requestor of the risk assessment with a useful answer to the question(s) posed to the EFSA Plant Health Panel. The method has been applied to four case studies, two fungi, *Ceratocystis platani* and *Cryphonectria parasitica*, the nematode Ditylenchus destructor and the Grapevine flavescence dorée phytoplasma. Selected results from these case studies illustrate the types of output that the method can deliver.

In one of the studies by Phung *et al.* (2013) the solutions to pesticide risk reduction for farmers in Vietnam are identified using a comprehensive needs assessment model applied for evaluating community needs and policy development. A comparative analysis of pesticide regulations was used to identify comparative needs, based on the in-depth interviews conducted with authorities and experts. The needs for pesticide regulations obtained from comparative analysis and experts include enhancement of pesticide legislation; multisectoral involvement in pesticide regulations; improvement of capacity for sectors involved in pesticide regulations; risk–benefit guidance for pesticide registration; reforms of pesticide regulations relating to the restriction, cancellation, suspension, transport, storage and disposal of pesticides; and the development of occupational hygiene and safety policy and programmes for agricultural activities. The expressed needs based on field observations comprise improvement in knowledge and behaviour of farmers about pesticide safety with specific areas, and supports in safety facilities and PPE.

There is a need to conduct technical training for occupational safety and hygiene of pesticide application, and support for safety facilities for pesticide application and protective equipment, for the farmers. This can lead to the development of a range of strategies in legislative improvement, workplace and personal hygiene, information and training, and medical surveillance and pesticide poisoning first aids for pesticide risk reduction for Vietnamese farmers.

Eyhorn *et al.* (2015) proposed some recommendations for raising awareness for pesticide use and revisiting the legislation and policies, for risk reduction of plant protection products. Introduction of more robust or resistant crop varieties strongly depends on whether they are accepted in the market. These measures require proactive awareness raising and information of consumers with suitable marketing activities. Increasing sales of IPM-labelled products will be an important contribution to pesticide reduction. The increasing demand of such products offers an interesting business opportunity. In addition, brands and retailers can convert entire products to compliance with minimum sustainability standards. These standards address pesticide use in the chemical component of IPM, mostly by excluding certain hazardous pesticides and demanding safety measures (training and safety equipment). Gradually strengthening these standards with regard to pesticide reduction and safe use of pesticides, along with the measures indicated earlier, offers scope for continuous improvement in a significant segment of consumption. Some of the knowhow and technologies could also be transferred to other crops and farming systems. The food industry could also play an important role in lobbying for better policy frameworks.

Consumers may be aware of the risks of pesticides, but not always of the consequences of their choices. They should be made more aware of the health risks associated with unsustainably produced food. They may not be aware with regard to what is 'good food' – a product that is safe, healthy, tasty, good for the environment and good for those who produce it, but not necessarily visually perfect. This information needs wider publicity in order to induce change. There is a need for new alliances to achieve this raised awareness, that is, governments, science, the health sector, the food industry, the water sector, environmental groups and media should join forces for this task. Instead of scaring consumers, they should be informed about the alternatives, and motivated to ask and pay for good food. At the same time it is important not to play down consumer fears with regard to pesticide risks.

13.6 Conclusion

Pesticide reduction is a joint responsibility of all stakeholders including decision makers in different fields and sectors. Fact-based information on pesticide issues and on ways to reduce pesticide use and risks needs to be conveyed to scientists, government offices, public health and consumer organizations, the management of relevant companies, investors etc. It is therefore important to make information on the availability of alternatives and on options to reduce pesticides and their risks available to relevant government authorities. Sufficient funds should be made available for development of alternatives to synthetic pesticides. This includes introduction of improved resistant varieties, testing and registration of alternative means such as biocontrol and the use of botanical preparations. By revisiting regulations and policies, governments can create a conducive environment for pesticide reduction, and can encourage the private sector to come forward with innovations in the field of agricultural production. Restrictions may be imposed on the use of certain pesticides, making them unavailable to farmers, and fostering the search for alternatives. It would help in strengthening the link between farmers, consumers, the private sector, civil society and governments. The targets set under national action plans need to be acceptable to stakeholders. The plans should emphasize incentives for alternatives and improved practices for pest management. Voluntary restrictions by the private sector may enable faster results than only trying to impose restrictions.

The current practices for pesticide registration cover acute toxicity of single active ingredients. In Switzerland and the EU, the whole formulation, including the additives, is tested for acute toxicity. Additional risk assessment studies are performed on additives of toxicological concern. However, metabolites of ingredients and synergistic or additive effects with other pesticides or with substances naturally occurring in food items may increase the impact of pesticide exposure. Regulation practices should be based on the latest findings by independent researchers. International codes and conventions and regional collaboration on legislation and regulation processes provide valuable guidelines for continuous improvement. There is a need for transparency on the toxicity data of active ingredients provided by the manufacturer, which may be published online to avoid any conflicts of interest. Like in Denmark, a pesticide tax could be introduced as a worthwhile tool to internalize and minimize externalities of pesticide use. However, to be effective, the tax needs to differentiate between levels of toxicity or hazard. The income generated through the tax may be used to support alternatives and to cover the costs related to monitoring activities. Rewarding the non-use of pesticides with a subsidy or introducing payments for environmental services may be politically easier to introduce.

Governments could initiate agricultural education and training programmes, using modern information and communication technology to strengthen advisory services to farmers. More research is needed to better assess and monitor the impact of pesticides on human health and on ecosystems. Monitoring is also required in order to assess whether the measures taken in order to reduce pesticide use and risks are effective or need to be adapted.

References

Act No. 11-92/ADP (1992) of 22 December 1992 establishing the Labour Code, arts. 80, 83. See also Decree No. 5254 IGTLS/AOF of 19 July 1954 respecting the employment of women and pregnant women, art. 5, and Decree No. 436/ ITLS/HV of 15 July 1953 concerning night work hours, art. 1.

Almvik, M., Bolli, R., Christiansen, A., Eklo, O.M., Holten, R., Senneset, G.V. and Wærnhus, K. (2014) Nedbrytning av soppmidler i norsk klima. [Degradation of fungicides in Norwegian climate]. (In Norwegian). *Bioforsk TEMA* 9(4). Bioforsk, Ås, Norway.

American Academy of Pediatrics Committee on Environmental Health (AAPCEH) (2003) Pesticides. In: Etzel, R.A. (ed.) *Pediatric Environmental health* (2nd edn). American Academy of Pediatrics, Elk Grove Village, Illinois.

Baker, B.P., Benbrook, C.M., Groth, E. III and Lutz Benbrook, K. (2002) Pesticide residues in conventional, integrated pest management (IPM)-grown and organic foods: insights from US data sets. *Food Additives and Contaminants: Part A* 19, 427–446.

Bhattacharyya, A., Barik, S.R. and Ganguly, P. (2009) New pesticide molecules, formulation technology and uses: Present status and future challenges. *Journal of Plant Protection Sciences* 1, 9–15.

Blair, R.M., Fang, H., Branham, W.S., Hass, B.S., Dial, S.L., Moland, C.L., Tong, W., Shi, L., Perkins, R. and Sheehan, D.M. (2000) The estrogen receptor relative binding affinities of 188 natural and xenochemicals: Structural diversity of ligands. *Toxicological Science* 54, 138–153.

Bockstaller, C., Guichard, L., Keichinger, O., Girardin, P., Galan, M.B. and Gaillard, G. (2009) Comparison of methods to assess the sustainability of agricultural systems: a review. *Agronomy for Sustainable Development* 29, 223–235.

Brown, E.C., Holmstead, J., Claassen, A., Powell, T., Hathaway, R.C., Wehrum, W., Weinstein *et al.* (2001) *The Environmental Law Reporter Pesticide Regulation Deskbook*. Environmental Law Institute, Washington, DC.

Burger, J., Mol, F. and Gerowitt, B. (2008) The 'necessary extent' of pesticide use – thoughts about a key term in German pesticide policy. *Crop Protection* 27, 343–351.

Caboni, P. and Cabras, P. (2010) Pesticides' influence on wine fermentation. *Advances in Food and Nutrition Research* 59, 43–62.

Cash, S.B., Sunding, D.L. and Zilberman, D. (2002) Health tradeoffs in pesticide regulation. Paper presented at the Annual Meeting of the American Agricultural Economics Association, Long Beach, California. 28–31 July 2002.

Chandler, D., Davidson, G., Grant, W.P., Greaves, J. and Tatchell, G.M. (2008) Microbial biopesticides for integrated crop management: An assessment of environmental and regulatory sustainability. *Trends in Food Science and Technology* 19, 275–283.

Codex Alimentarius Commission (CAC) (2006) Joint FAO/WHO Food Standards Program Codex Alimentarius Commission, Twenty-ninth Session, International Conference Centre, Geneva (Switzerland), 3–7 July 2006, other matters arising from FAO and WHO prepared by FAO and WHO.

Commission of the European Communities (2006) Proposal for a regulation of the European parliament and of the council concerning the placing of plant protection products on the market. COM, 388 Final; Commission of the European Communities, Brussels, Belgium.

Damalas, A.C. and Eleftherohorinos, I.G. (2011) Pesticide exposure, safety issues, and risk assessment indicators. *International Journal of Environmental Research and Public Health* 8, 1402–1419.

Damalas, C.A. (2009) Understanding benefits and risks of pesticide use. *Scientific Research and Essays* 4, 945–949.

Davis, J.R., Brownson, R.C. and Garcia, R. (1992) Family pesticide use in the home, garden, orchard, and yard. *Archives of Environmental Contamination and Toxicology* 22, 260–266.

Department of Pesticide Regulation (DPR) (n.d.) Assessing pesticide risks to human health. In: A Guide to Pesticide Regulation in California. California Department of Pesticide Regulation, pp. 41–51. Available at: www.cdpr.ca.gov/docs/pressrls/dprguide/chapter5.pdf (accessed 2 October 2018).

European Commission (EC) (2013a) Regulation (EU) No. 283/2013 of 1 March 2013 setting out the data requirements for active substances, in accordance with regulation (EC) no 1107/2009 of the European Parliament and of the Council concerning the placing of plant protection products on the market. *Official Journal of the European Union* L 93, 1–84.

European Food Safety Authority (EFSA) (2009a) Updating the opinion related to the revision of Annexes II and III to Council Directive 91/414/EEC concerning the placing of plant protection products on the market – ecotoxicological studies. Scientific Opinion of the Panel on Plant Protection Products and their Residues (Question No EFSA-Q-2009-00556), Adopted on 18 June 2009. *The EFSA Journal* 1165, 1–24.

Eyhorn, F., Roner, T. and Specking, H. (2015) Reducing pesticide use and risks – what action is needed? Briefing paper. HELVETAS Swiss Intercooperation. Available at: https://www.helvetas.org/Publications-PDFs/Briefing%20paper%20pesticide%20reduction_conclusions.pdf (accessed 1 August 2018).

Faust, M., Vogs, C., Rotter, S., Wöltjen, J., Höllrigl-Rosta, A., Backhaus, T. and Altenburger, R. (2014) Comparative assessment of plant protection products: How many cases will regulatory authorities have to answer? *Environmental Sciences Europe* 26, 11.

Fenske, R.A. and Day, E.W., Jr (2005) Assessment of exposure for pesticide handlers in agricultural, residential and institutional environments. In: Franklin, C.A. and Worgan, J.P. (eds) *Occupational and Residential Exposure Assessment for Pesticides*, John Wiley & Sons, Chichester, UK, pp. 13–43.

Food and Agriculture Organization (FAO) (2002) *Manual on the Submission and Evaluation of Pesticide Residues Data for the Estimation of Maximum Residue Levels in Food and Feed.* Food and Agriculture Organization, Rome.

Food and Agriculture Organization (FAO) (2015) FAOSTAT. Available at: http://faostat3.fao.org/browse/R/RP/E (accessed 5 February 2016).

Food and Agriculture Organization (FAO) Newsroom (2002) New code of conduct on pesticide adopted. Available at: www.fao.org/english/newsroom/news/2002/10525-en.html (accessed 4 November 2002).

Food and Agriculture Organization (FAO)/World Health Organization (WHO) (2006) Updating the principles and methods of risk assessment: MRLs for pesticides and veterinary drugs, Rome. Available at: www.fao.org/fileadmin/templates/agphome/documents/Pests_Pesticides/JMPR/bilthoven_2005.pdf (accessed 2 October 2018).

Forster, D., Adamtey, N., Messmer, M.M., Pfiffner, L., Baker, B., Huber, B. and Niggli, U. (2013) Organic agriculture – driving innovations in crop research. In: Bhuller, G.S. and Bhuller, N.K. (eds) *Agricultural Sustainability: Progress and Prospects in Crop Research.* Elsevier Inc., Oxford, UK.

Frangenberg, A. (2000) Integrated crop management as fundamental basis for sustainable production. *Pflanzenschutz-Nachrichten Bayer* 53, 131–153.

Gil, Y., Sinfort, C., Guillaume, S., Brunet, Y. and Palagos, B. (2008) Influence of micrometeorological factors on pesticide loss to the air during vine spraying: Data analysis with statistical and fuzzy inference models. *Biosystems Engineering* 100, 184–197.

Gilioli, G., Schrader, G., Grégoire, J.C., MacLeod, A., Mosbach-Schulz, O., Rafoss, T., Rossi, V., Urek, G. and van der Werf, W. (2017) The EFSA quantitative approach to pest risk assessment methodological aspects and case studies. *Bulletin OEPP/EPPO Bulletin* 47(2), 213–219.

Gomes, J., Lloyd, O.L. and Revitt, D.M. (1999) The influence of personal protection, environmental hygiene and exposure to pesticides on the health of immigrant farm workers in a desert country. *International Archives of Occupational and Environmental Health* 72, 40–45.

Health Canada (1998) *Health and the Environment: Handbook for Health Professionals.* Ministry of Public Work and Government Services, Ottawa, Canada.

Hornsby, A.G., Buttler, T.M. and Brown, R.B. (1993) Water quality and the environment. Managing pesticides for crop production and water quality protection: Practical grower guides. *Agriculture, Ecosystems and Environment* 46, 187–196.

Jaga, K. and Dharmani, C. (2003) Sources of exposure to and public health implications of organophosphate pesticides. *Pan American Journal of Public Health* 14, 171–185.

Jeyaratnam, J. (1990) Acute pesticide poisoning: A major global health problem. *World Health Statistics Quarterly* 43(3), 139–144.

Jindal, T., Singh, D.K. and Agarwal, H.C. (2007) Effect of UV radiation and temperature on mineralization and volatilization of coumaphos in water. *Journal of Environmental Science and Health B* 42, 367–372.

Karabelas, A.J., Plakas, K.V., Solomou, E.S., Drossou, V. and Sarigiannis, D.A. (2009) Impact of European legislation on marketed pesticides – a view from the standpoint of health impact assessment studies. *Environment International* 35, 1096–1107.

Kelce, W.R., Monosson, E., Gamcsik, M.P., Law, S.C. and Gray, L.E. (1994) Environmental hormone disruptors, evidence that vinclozolin developmental toxicity is mediated by antiandrogenic metabolites. *Toxicology and Applied Pharmacology* 126, 276–285.

Kiely, T., Donaldson, D. and Grube, A. (2004) Pesticides industry sales and usage: 2000 & 2001 market estimates. Biological and Economic Analysis Division, Office of Pesticide Programs, Office of Prevention, Pesticides, and Toxic Substances, U.S. Environmental Protection Agency Washington, DC. May 2004. Available at: https://nepis.epa.gov/Exe/ZyPURL.cgi?Dockey=3000659P.TXT (accessed 2 October 2018).

Kienzler, A., Berggren, E., Bessems, J., Bopp, S., van der Linden, S. and Worth, A. (2017) Assessment of mixtures – review of regulatory requirements and guidance. JRC Science and Policy Reports. European Commission, Luxembourg.

Kim, H.J., Park, Y.I. and Dong, M.S. (2005) Effects of 2,4-D and DCP on the DHT-induced androgenic action in human prostate cancer cells. *Toxicological Science* 88, 52–59.

Kogan, M. (1998) Integrated pest management: Historical perspectives and contemporary development. *Annual Review of Entomology* 43, 243–270.

Küster, E. and Altenburger, R. (2007) Suborganismic and organismic effects of aldicarb and its metabolite aldicarb-sulfoxide to the zebrafish embryo (*Danio rerio*). *Chemosphere* 68, 751–760.

Laborde, A. (2008) Children's Health and the Environment. WHO Training Package for the Health Sector. World Health Organization. Available at: www.who.int/ceh/capacity/Pesticides.pdf (accessed 2 October 2018).

Laitinen, P., Siimes, K., Eronen, L., Rämö, S., Welling, L., Oinonen, S., Mattsoff, L. and Ruohonen-Lehto, M. (2006) Fate of the herbicides glyphosate, glufosinate-ammonium, phenmedipham, ethofumesate and metamitron in two Finnish arable soils. *Pest Management Science* 62, 473–491.

Laitinen, P., Rämö, S., Nikunen, U., Jauhiainen, L., Siimes, K. and Turtola, E. (2009) Glyphosate and phosphorus leaching and residues in boreal sandy soil. *Plant and Soil* 323, 267–283.

Larsen, J.C., Binderup, M.L., Dalgaard, M., Dragsted, L.O., Hossaini, A., Ladefoged, O., Lam, H.R., Madsen, C., Meyer, O., Rasmussen, E.S. *et al.* (2003) combined actions and interactions of chemicals in mixtures. The toxicological effects of exposure to mixtures of industrial and environmental chemicals; Fødevare Rapport 12. Danish Veterinary and Food Administration, Søborg, Denmark. Available at: https://www.foedevarestyrelsen.dk/Publikationer/Alle%20publikationer/2003012.pdf (accessed 1 August 2018).

Mariyono, J. (2008) Direct and indirect impacts of integrated pest management on pesticide use: A case of rice agriculture in Java, Indonesia. *Pest Management Science* 64, 1069–1073.

Meng, Q.Y., Yang, W.X., Xu, X.B. and Sun, L.J. (2009) Environmentally sound management of toxic chemicals – China. In: Honglie, S. and Zhao, S. (eds) *Area Studies – China: Regional Sustainable Development Review – Vol. II.* Available at: https://www.eolss.net/Sample-Chapters/C16/E1-54-22.pdf (accessed 1 August 2018).

Messing, K. and Östlin, P. (2002) Relevant legislation and policy. In: *Gender Equality, Work and Health: A Review of the Evidence.* World Health Organization, Stockholm, Sweden, pp. 20–23. Available at: www.who.int/gender/documents/Genderworkhealth.pdf (accessed 1 August 2018).

Mew, E.J., Padmanathan, P., Konradsen, F., Eddleston, M., Chang, S.S., Phillips, M.R. and Gunnell, D. (2017) The global burden of fatal self-poisoning with pesticides 2006–15: systematic review. *Journal of Affective Disorders* 219, 93–104.

Mnif, W., Hassine, A.I.H., Bouaziz, A., Bartegi, A., Thomas, O. and Roig, B. (2011) Effect of endocrine disruptor pesticides: A review. *International Journal of Environmental Research and Public Health* 8, 2265–2303.

Müller, U. (2002) Chemical crop protection research. Methods and challenges. *Pure and Applied Chemistry* 74, 2241–2246.

Morrison Paul, C.J., Ball, V.E., Felthoven, R.G., Grube, A. and Nehring, R.F. (2002) Effective costs and chemical use in United States agricultural production: Using the environment as a 'free' input. *American Journal of Agricultural Economics* 84, 902–915.

Neumann, R. (1997) Chemical crop protection research and development in Europe. *Developments in Crop Science* 25, 49–55.

Nwilene, F.E., Nwanze, K.F. and Youdeowei, A. (2008) Impact of integrated pest management on food and horticultural crops in Africa. *Entomologia Experimentalis et Applicata* 128, 355–363.

Organisation for Economic Co-operation and Development (OECD) Workshop (2001) Report of the OECD workshop on the economics of pesticide risk reduction in agriculture. Copenhagen, pp. 28–30.

Palacios Xutuc, C.N. (2010) Manual to train trainers on safe and correct use of plant protection products and integrated pest management (IPM), *CropLife Latin America*, Guatemala City, Guatemala.

PAN (2017) Global governance of hazardous pesticides to protect children: Beyond 2020. Pesticide Action Network Asia Pacific, January 2017. Available at: www.indiaenvironmentportal.org.in/files/file/SAICM-PANAP-Children-and-Pesticides.pdf (accessed 1 August 2018).

Phillips, M.R., Yang, G., Zhang, Y., Wang, L., Ji, H. and Zhou, M. (2002) Risk factors for suicide in China: A national case-control psychological autopsy study. *Lancet* 360, 1728–1736.

Phung, D.T., Connell, D., Miller, G., Rutherford, S. and Chu, C. (2013) Needs assessment for reducing pesticide risk: A case study with farmers in Vietnam. *Journal of Agromedicine* 18(4), 293–303.

Reportlinker (2012) Global agrochemical industry 2012–2017: Trend, profit, and forecast analysis. *PR Newswire*. Available at: www.prnewswire.com/news-releases/global-agrochemical-industry-2012-2017-trend-profit-and-forecast-analysis-162559196.html (accessed 1 August 2018).

Research and Markets (2015) Global pesticide market – by regions and vendors – market size, demand forecasts, industry trends and updates, supplier market shares 2014–2020. Available at: https://www.researchandmarkets.com/reports/3338864/global-pesticide-market-by-regions-and-vendors (accessed 1 August 2018).

Reus, J., Leendertse, P., Bockstaller, C., Fomsgaard, I., Gutsche, V., Lewis, K., Nilsson, C., Pussemier, L., Trevisan, M., van der Werf, H.M.G., Alfarroba, F., Blumel, S., Isart, J., McGrath, D. and Seppala, T. (2002) Comparison and evaluation of eight pesticide environmental risk indicators developed in Europe and recommendations for future use. *Agriculture, Ecosystems and Environment* 90(2), 177–187.

Reus, J.A.W.A. and Leendertse, P.C. (2000) The environmental yardstick for pesticides: A practical indicator used in the Netherlands. *Crop Protection* 19(8–10), 637–641.

Salazar-Arredondo, E., Solis-Herediaa, M., Rojas-Garcia, E., Hernandez-Ochoa, I. and Quintanilla-Vega, B. (2008) Sperm chromatin alteration and DNA damage by methyl-parathion; chlorpyrifos and diazinon and their oxon metabolites in human spermatozoa. *Reproductive Toxicology* 25, 455–460.

Siimes, K., Rämö, S., Welling, L., Nikunen, U. and Laitinen, P. (2006) Comparison of the behaviour of three herbicides in a field experiment under bare soil conditions. *Agricultural Water Management* 84, 53–64.

Smith, K., Evans, D.A. and El-Hiti G.A. (2008) Role of modern chemistry in sustainable arable crop protection. *Philosophical Transactions of the Royal Society B* 363, 623–637.

Solomon, K., Giesy, J. and Jones R. (2000) Probabilistic risk assessment of agrochemicals in the environment. *Crop Protection* 19(8–10), 649–655.

Stenrød, M., Perceval, J., Benoit, P., Almvik, M., Bolli, R.I., Eklo, O.M., Sveistrup, T.E. and Kværner, J. (2008) Cold climatic conditions: effects on bioavailability and leaching of the mobile pesticide metribuzin in a silt loam soil in Norway. *Cold Regions Science and Technology* 53, 4–15.

Stenrød, M., Almvik, M., Eklo, O.M., Gimsing, A.L., Holten, R., Künnis-Beres, K., Larsbo, M., Putelis, L., Siimes, K., Turka, I. and Uusi-Kämppä, J. (2016) Pesticide regulatory risk assessment, monitoring, and fate studies in the northern zone: Recommendations from a Nordic-Baltic workshop. *Environmental Science and Pollution Research* 23, 15779–15788.

Surgan, M., Condon, M. and Cox, C. (2010) Pesticide risk indicators: Unidentified inert ingredients compromise their integrity and utility. *Environmental Management* 45, 834–841.

Szabo, E.A., Jansson, E., Miles, D., Hambridge, T., Stanley, G., Baines, J., and Brent, P. (2010) Responding to incidents of low level chemical contamination in food. *Ensuring Global Food Safety*, 411–437. DOI:10.1016/b978-0-12-374845-4.00024-2

Tsakiris, I.N., Danis, T.G., Stratis, I.A., Nikitovic, N., Dialyna, I.A., Alegakis, A.K. and Tsatsakis, A.M. (2004) Monitoring of pesticide residues in fresh peaches produced under conventional and integrated crop management cultivation. *Food Additives and Contaminants, A* 21, 670–677.

United Nations Development Programme (UNDP) (2012) UNDP guide for integrating the sound management of chemicals into development planning. United Nations Development Programme. Available at: www.undp.org/content/dam/aplaws/publication/en/publications/environment-energy/www-ee-library/chemicals-management/integrating-sound-management-of-chemicals-into-mdg-based-development-planning/Integrating%20Sound%20Management%20in%20Chemicals%20Final%20r2.pdf (accessed 1 August 2018).

United Nations Environment Programme (UNEP) (2002) Children in the new millennium:

Environmental impact on health. United Nations Environment Programme, Nairobi; United Nations Children's Fund, New York; World Health Organization, Geneva.

United States Environmental Protection Agency (EPA) (1999) Progress Report: Implementing the Food Quality Protection Act. USEPA archive document. Prevention, Pesticides, And Toxic Substances (7506C), EPA 735-R-99001 August 1999 Available at: https://archive.epa.gov/pesticides/regulating/laws/fqpa/web/pdf/fqpareport.pdf (accessed 17 January 2019)

United States Environmental Protection Agency (EPA) (2007) Pesticide registration (PR) notice 2001-X draft: Spray and dust drift label statements for pesticide products. US EPA, Washington, DC. Available at: https://www.epa.gov/pesticide-registration/prn-2001-x-draft-spray-and-dust-drift-label-statements-pesticide-products (accessed 1 August 2018).

United States Environmental Protection Agency (US EPA) (2009) Registering pesticides. Available at: www.epa.gov/pesticides/regulating/registering/index.htm (accessed 1 April 2011).

Urech, P.A. (1999) The agrochemical industry: Its contribution to crop protection and environmental policy. *Plant Pathology* 48, 689–692.

Way, M.J. and Van Emden, H.F. (2000) Integrated pest management in practice – pathways towards successful application. *Crop Protection* 19, 81–103.

World Health Organization (WHO) (2010) International Code of conduct on the distribution and use of pesticides: Guidelines for the registration of pesticides. WHO, Rome. Available at: www.fao.org/3/a-bc870e.pdf (accessed 1 August 2018).

World Health Organization (WHO) (2014) Preventing suicide: A global imperative. WHO, Geneva. Available at: www.who.int/mental_health/suicide-prevention/world_report_2014/en (accessed 1 August 2018).

Index

Note: Page numbers in **bold** type refer to **figures**
Page numbers in *italic* type refer to *tables*